THE PRIMORDIAL DENSITY PERTURBATION
Cosmology, Inflation and the Origin of Structure

The origin and evolution of the primordial perturbation is the key to understanding structure formation in the earliest stages of the Universe. It carries clues to the types of physical phenomena active in that extreme high-density environment. Through its evolution, generating first the observed cosmic microwave background anisotropies and later the distribution of galaxies and dark matter in the Universe, it probes the properties and dynamics of the present Universe.

This graduate-level textbook gives a thorough account of theoretical cosmology and perturbations in the early Universe, describing their observational consequences and showing how to relate such observations to primordial physical processes, particularly cosmological inflation. With ambitious observational programmes complementing ever-increasing sophistication in theoretical modelling, cosmological studies will remain at the cutting edge of astrophysical studies for the foreseeable future.

DAVID H. LYTH is Professor of Particle Astrophysics in the Physics Department of Lancaster University.

ANDREW R. LIDDLE is Professor of Astrophysics in the Department of Physics and Astronomy at the University of Sussex. They have a long-established research collaboration and have jointly developed some of the key concepts in studies of cosmological perturbations, particularly in relation to the inflationary cosmology. They previously co-authored the Cambridge University Press textbook *Cosmological Inflation and Large-Scale Structure* in 2000.

THE PRIMORDIAL DENSITY PERTURBATION

Cosmology, Inflation and the Origin of Structure

DAVID H. LYTH
University of Lancaster

ANDREW R. LIDDLE
University of Sussex

CAMBRIDGE
UNIVERSITY PRESS

University Printing House, Cambridge CB2 8BS, United Kingdom

Cambridge University Press is part of the University of Cambridge.

It furthers the University's mission by disseminating knowledge in the pursuit of education, learning and research at the highest international levels of excellence.

www.cambridge.org
Information on this title: www.cambridge.org/9780521828499

First published 2009
Reprinted with corrections 2012

A catalogue record for this publication is available from the British Library

ISBN 978-0-521-82849-9 Hardback

To Margaret, John, and Duncan

and to Ed, John, and Rocky

Contents

Frequently used symbols, and their place of definition.

Symbol	*Page*	*Definition*
$a_{\ell m}$	154	Observed multipole of Θ
$\hat{a}_{\mathbf{k}}$ $(\hat{a}^\dagger_{\mathbf{k}})$	255	Annihilation (creation) operator
B	55	Baryon number
B_ℓ	155	A CMB polarization multipole
C_ℓ	155	Spectrum $\langle \|a_{\ell m}\|^2 \rangle$ of CMB anisotropy
ds^2	7	Spacetime interval
E	12	Energy
E_ℓ	182	A CMB polarization multipole
$f_{\rm a}$	283	The axion parameter
f $(f_{\rm a})$	19	Distribution function (of species 'a')
$f_{\rm NL}$	100	Reduced bispectrum of ζ
g $(g_{\rm a})$	19	Number of spin states (of species 'a')
g_{ij}	34	Spatial metric tensor
$g_{\mu\nu}$	27	Metric tensor (generic coordinates)
h_{ij}	38, 199	Gravitational wave amplitude
H	42	Hubble parameter $\dot{a}/a$
H_0	60	Present Hubble parameter
H	216, 217	Hamiltonian
$\hat{H}$	248	Hamiltonian operator
I	78	Unit matrix or unit operator
j^μ	16	Any conserved current
k, $\mathbf{k}$	87	Wavenumber, wave-vector (physical or comoving)
k, $\mathbf{k}$	255	Momentum of a single particle (physical or comoving)
L	55	Lepton number
L	214	Lagrangian
L	87	Comoving box size for Fourier expansion
$\mathcal{L}$	218	Lagrangian density
m	12	Mass of a particle
M	71	Cosmological mass within a comoving sphere
n $(n_{\rm a})$	20	Number density (of species 'a')
n	99	Spectral index of ζ

Symbol	*Page*	*Definition*
N	80	e-folds between any two epochs
N	81	N such that $\zeta = \delta N$
N	313	e-folds of observable inflation
$\mathbf{p}$	12	Momentum
p^μ	12	4-momentum
$P \ (P_\mathrm{a})$	17	Pressure (of species 'a')
P_g	89	Spectrum of a perturbation g
$\mathcal{P}_g$	90	Spectrum of a perturbation g (alternative definition)
Q	55	Electric charge
Q	267	Renormalization scale
r	199	Tensor fraction $\mathcal{P}_h/\mathcal{P}_\zeta$
R	42	Cosmological comoving radius
s	54	Entropy density
S	214	Action
S_i	190	An isocurvature perturbation
t	7, 39	Time, cosmic time
T	24	Temperature (usually of the early Universe or CMB)
T	125, 159	A transfer function
$T^{\mu\nu}$	16	Energy–momentum tensor
u^μ	12	4-velocity
$\mathbf{v} \ (\mathbf{v}_\mathrm{a})$	106, 120	Fluid velocity perturbation (of species 'a')
$V \ (V_\mathrm{a})$	107, 120	Fluid velocity scalar (of species 'a')
V	219	Scalar field potential
W	291	Superpotential
$x \ (x^\mu)$	7	Minkowski spacetime coordinates $(\mu = 0, 1, 2, 3)$
$x \ (x^\mu)$	26	Generic spacetime coordinates $(\mu = 0, 1, 2, 3)$
x^i	34	Generic spatial coordinates $(i = 1, 2, 3)$
$\mathbf{x}$	7	Cartesian coordinates (x^1, x^2, x^3)
$\mathbf{x}$	41	Comoving Cartesian coordinates (x^1, x^2, x^3)
$w \ (w_\mathrm{a})$	122	Ratio P/ρ (for species 'a'), equation of state
z	52	Redshift

Symbol	*Page*	*Definition*
δ ($\delta_{\rm a}$)	76	Density contrast $\delta\rho/\rho$ (of species 'a')
ϵ	316	Slow-roll flatness parameter $\frac{1}{2}M_{\rm Pl}^2(V'/V)^2$
$\epsilon_{\rm H}$	317	Slow-roll parameter $-\dot{H}/H^2$
ζ	78	Primordial curvature perturbation
η	315	Slow-roll flatness parameter $M_{\rm Pl}^2 V''/V$
η	41	Conformal time $d\eta = dt/a$
$\eta_{\rm ls}$	66	Conformal time at photon decoupling
η_0	44	Present conformal time
$\eta_{\mu\nu}$	7	Metric tensor (Minkowski coordinates)
Θ	153	Photon brightness function dT/T
Θ_ℓ	158	Multipole of a Fourier component of Θ
Λ	46	Cosmological constant
μ ($\mu_{\rm a}$)	24	Chemical potential (of species 'a')
Π ($\Pi_{\rm a}$)	120	Anisotropic stress scalar (of species 'a')
Π_{ij}	120	Dimensionless anisotropic stress
ρ ($\rho_{\rm a}$)	16	Energy density (of species 'a')
$\sigma_{\rm T}$	130	Thomson scattering cross-section
Σ_{ij}	17	Anisotropic stress
τ	65	Optical depth
ϕ (ϕ_n)	219	Scalar field (the nth scalar field)
Φ	106	Newtonian peculiar gravitational potential
Φ	121	One of the relativistic gravitational potentials
φ	383	Conformal inflaton field perturbation $a\delta\phi$
Ψ	121	One of the relativistic gravitational potentials
Ω ($\Omega_{\rm a}$)	47, 61	Ratio $\rho/\rho_{\rm crit}$ (for species 'a')

Preface

The beginning of the twenty-first century stands a good chance of being identified in history as the time when humankind first came to grips with the Universe. In a rush of observational progress, the charge led by the 1992 discovery of cosmic microwave background anisotropies by the Cosmic Background Explorer (COBE) satellite, the elements required to build accurate cosmological models were assembled. Complementing this, development of theoretical methods allowed accurate predictions to be made to confront those observations.

The landmark was the 2003 announcement of precision cosmic microwave measurements from the team operating the Wilkinson Microwave Anisotropy Probe (WMAP). Ironically, this success lay in a kind of failure — a failure to uncover anything new and unexpected. In the words of astrophysicist John Bahcall at NASA's announcement press conference, "the biggest surprise is that there are no surprises". Instead, then, the power of the observations became fully focussed on determining the properties of the cosmological model, and for the first time many of its components were determined to a satisfying degree of accuracy: the percent level for quantities such as the geometry of the Universe, its age, and the density of the baryonic material, and the ten percent level for many other aspects.

In 2000, we published a graduate-level textbook, *Cosmological Inflation and Large-Scale Structure*, written during the late 1990s and which described many of the ideas underpinning the modern cosmology. We had been considering a second edition which would bring everything up to date. It became apparent, however, that the subject had already developed too far for a simple update to be possible. The emergence of the standard cosmological model had left by the wayside many concepts which in the late 1990s had still had possible relevance for our Universe. Meanwhile, the community's attention had moved on to the new frontiers, such as cosmic non-gaussianity, where the next discoveries illuminating our Universe may lie. New ideas on the origin of inflationary perturbations had emerged, as had more elegant and streamlined ways of deriving established results.

So, instead, a new book! Some of the subject coverage is the same as the previous one, but the focus is more narrowly fixed on the now-standard cosmology, and as such goes deeper. The emphasis is shifted somewhat in a theoretical direction, and looks forward to the new frontiers of cosmology. After all, setting in place a viable and well-specified cosmological model is only the start, suggesting that we are learning *how* the Universe works. But what we really want to know is *why*!

We have learned much from research collaborators, and from many others acknowledged in our published papers. In addition, we thank Anthony Challinor, Eung Jin Chun, Paolo Creminelli, Wayne Hu, Kazunori Kohri, Eiichiro Komatsu, George Lazarides, Antony Lewis, Andrei Linde, Marta Losada, John McDonald, Ian Moss, Joe Polchinski, Misao Sasaki, Quasar Shafi, Alex Vilenkin, David Wands, Steven Weinberg and Martin White for input relating to matters dealt with in this book that we didn't have the opportunity to acknowledge in published papers.

Of course, we have done our best to ensure that the contents of this book are accurate; however, experience tells us that some errors will have slipped through. Please let us know of any you spot. There will be an up-to-date record of known errors, plus other updates, accessible at the book's World Wide Web Home Page at

http://astronomy.sussex.ac.uk/~andrewl/pdp.html

1
Overview

Although cosmology can trace its beginnings back to Einstein's formulation of his general theory of relativity in 1915, which enabled the first mathematically consistent models of the Universe to be constructed, for most of the following century there was much uncertainty and debate about how to describe our Universe. Over those years the various necessary ingredients were introduced, such as the existence of dark matter, of the hot early phase of the Universe, of cosmological inflation, and eventually dark energy. In the latter part of the last century, cosmologists and their funding agencies came to realize the opportunity to deploy more ambitious observational programmes, both on the ground and on satellites, which began to bear fruit from 1990 onwards. The result is a golden age of cosmology, with the creation and observational verification of the first detailed models of our Universe, and an optimism that that description may survive far into the future. The objective, often described as *precision cosmology*, is to pin down the Universe's properties as best as possible, in many cases at the percent or few percent level. In particular, the landmark publication in 2003 of measurements of the cosmic microwave background by the Wilkinson Microwave Anisotropy probe (WMAP), seems certain to be identified as the moment when the Standard Cosmological Model became firmly established.

The key tool in understanding our Universe is the formation and evolution of structure in the Universe, from its early generation as the primordial density perturbation to its gravitational collapse to form galaxies. As we already argued in the introduction to our book *Cosmological Inflation and Large-Scale Structure*, the complete theory of structure formation, starting with the quantum fluctuations of a free field, continuing with general-relativistic gas dynamics, and ending with the free fall of photons and matter, is perhaps one of the most beautiful and complete in the entire field of physics. It has also demonstrated powerful predictive power, for instance anticipating the oscillatory structure of the cosmic microwave anisotropy spectrum more than twenty years before the anisotropies were measured in any form, and in detail making percent-level predictions that continue to be in accord with what are now percent-level observations.

The purpose of this book is to give a detailed account of the physics of density perturbations in the Universe, focussed around the form and implications of the primordial perturbation. We aim to describe the main astrophysical processes which transform the initial density perturbation into observables, such as the cosmic microwave anisotropies, and to show how these observable consequences can be tracked back to an origin which sheds light on fundamental physical processes in the early Universe.

The book is divided into parts, as follows.

Part I: Relativity gives the basics of general relativity along with the applications needed for cosmology, starting from a basic knowledge of special relativity.

Part II: The Universe after the first second concerns itself with the evolution of perturbations, starting with a primordial density perturbation, whose existence is at this point taken for granted. After a brief overview of the theory of the background (homogeneous) cosmology, density perturbations are defined and characterized, and their evolution studied in both Newtonian and relativistic frameworks. This evolution ultimately leads to the observable consequences of the theory.

Part III: Field theory sets the context for explaining the origin of the primordial density perturbation in terms of fundamental physics. It gives those aspects of field theory that are needed for Part IV, starting from a basic knowledge of quantum mechanics. Among the key ideas developed are scalar field dynamics, internal symmetry, supersymmetry, and the quantization of free fields.

Part IV: Inflation and the early Universe exploits these ideas to explain the leading theory for the origin of perturbations, cosmological inflation. We describe a number of variants on the basic inflationary theme. We conclude by developing the observational consequences of a wide range of inflationary scenarios, setting the challenge to distinguish amongst them using future observations.

The reader will notice that many references are given for the chapters of Part IV while very few are given for earlier chapters. This reflects a profound difference between the material in Part IV and that in Parts I–III. The theories covered in the first three parts have been around for at least several years, and in many cases for far longer. It is true that Nature may have chosen not use some of them. There may be no significant tensor or isocurvature perturbation, no supersymmetry, no axion, and no seesaw mechanism for neutrino masses. But the theories themselves are well established. As a result, most of the additional material consulted by the reader will consist of texts and reviews as opposed to topical research articles. The most appropriate sources of that kind will depend on the reader's background and future intentions, and we mention only a few possibilities.

In contrast, the study of the very early Universe covered in Part IV is at the cutting edge of current research. It is not covered in any text at present, and the coverage of reviews is quite patchy. The situation is also quite complicated, with a large menu of possibilities confronting many different kinds of observation. What we have done in Part IV is to get the reader started on a study of the main possibilities, pointing along the way to reviews and research papers that can be the basis of further study.

Notes on exercises

Most chapters end with a few exercises to allow the reader to practice applying the information given within the chapter. Several of these examples require some simple numerical calculations for their solution; in cosmology these days, it is practically impossible to avoid carrying out some numerical work at some stage. A typical task is the numerical computation of an integral that cannot be done analytically, or the evaluation of some special functions. These can be done via specially written programs, using library packages (e.g., *Numerical Recipes* [1], which is also an invaluable source of general information on scientific computation), or a computer algebra package such as Mathematica or Maple.

Units

In keeping with conventional notation in cosmology, we set the speed of light c equal to one, so that all velocities are measured as fractions of c. Where relevant, we also set the Planck constant $\hbar$ to one, so that there is only one independent mechanical unit. In particular, the phrases 'mass density' and 'energy density' become interchangeable. Often it is convenient to take this unit as energy, and we usually set the Boltzmann constant k_B equal to 1 so that temperature too is measured in energy units. (In normal units $k_B = 8.618 \times 10^{-5}\,\mathrm{eV\,K^{-1}}$.)

Newton's gravitational constant G can be used to define the **reduced Planck mass** $M_{\mathrm{Pl}} = (8\pi G)^{-1/2}$. Thought of as a mass, $M_{\mathrm{Pl}} = 4.342 \times 10^{-6}$ g, which converts into an energy of 2.436×10^{18} GeV. We use the reduced Planck mass throughout, normally omitting the word 'reduced'. It is a factor $\sqrt{8\pi}$ less than the alternative definition of the Planck mass, never used in this book, which gives $m_{\mathrm{Pl}} = 1.22 \times 10^{19}$ GeV. We use M_{Pl} and G interchangeably, depending on the context. Inserting appropriate combinations of $\hbar$ and c, we also can obtain the reduced Planck time $T_{\rm Pl} \equiv \hbar/c^2 M_{\rm Pl} = 2.70 \times 10^{-43}$ s and reduced Planck length $L_{\rm Pl} \equiv \hbar/cM_{\rm Pl} = 8.10 \times 10^{-33}$ cm.

Reference

[1] W. H. Press, S. A. Teukolsky, W. T. Vetterling and B. P. Flannery. *Numerical Recipes: The Art of Scientific Computation*, 3rd edition (Cambridge: Cambridge University Press, 2007).

Part I

Relativity

2
Special relativity

In this chapter we review the basic special relativity formalism, expressed in a way that is readily extended to general relativity in the following chapter. We then study the energy–momentum tensor, both for a generic fluid and for a gas.

2.1 Minkowski coordinates and the relativity principle

A starting point for relativity is provided by the interval ds^2 between neighbouring points of spacetime, known as **events**. The interval may be regarded as a given concept, like the distance between two points in space.

Special relativity assumes the existence of coordinates in which

$$ds^2 = -dt^2 + dx^2 + dy^2 + dz^2 \,. \tag{2.1}$$

(In a different convention, the sign of ds^2 is reversed.) Such coordinates are called **Minkowski coordinates**. They are also referred to as an inertial frame. The spatial coordinates are Cartesian, and the distance $d\ell$ between two points in space with same time coordinates is given by

$$d\ell^2 = ds^2 = dx^2 + dy^2 + dz^2. \tag{2.2}$$

It is convenient to use the index notation $(x^0, x^1, x^2, x^3) \equiv (t, x, y, z)$, and to denote a generic coordinate by x^μ.[1] We also use the notation $x^\mu = (t, \mathbf{x})$. Then we can define a **metric tensor** $\eta_{\mu\nu}$ as the diagonal matrix with elements $(-1, 1, 1, 1)$. Using it, Eq. (2.1) can be written

$$ds^2 = \eta_{\mu\nu}\, dx^\mu dx^\nu. \tag{2.3}$$

We adopt the summation convention: that there is a sum over every pair of identical

[1] We take Greek letters $(\mu, \nu, \dots)$ to run over the values $0, 1, 2, 3$, and italic letters $(i, j, \dots)$ to run over the values $1, 2, 3$.

spacetime indices. A summation of the above form, over the metric tensor, is called a contraction. In this case, μ and ν are contracted.

As in this example, sums over spacetime indices involve one upper index and one lower index, and spacetime coordinates always carry an upper index. In contrast, if an expression involves purely spatial Cartesian coordinates, one can take all indices as lower while still adopting the summation convention. In particular, the distance $d\ell$ between nearby points at a given time is given by

$$d\ell^2 = ds^2 = \delta_{ij} dx_i dx_j, \tag{2.4}$$

where δ_{ij} is the Kronecker delta, equal to 1 for equal indices and to 0 for unequal ones. (We will also denote it by δ^{ij} or δ^i_j according to the context.)

A transformation taking us from one inertial frame to another preserves the form of Eq. (2.3). The time-reversal transformation $t' = -t$ does this, but we fix the sign of t so that it increases going from past to future. A parity transformation, changing right-handedness to left-handedness, also does it, but we choose a right-handed coordinate system. (The parity transformation can be taken as a reversal of all three coordinates, or of just one of them.) With these restrictions, the transformations preserving the form of Eq. (2.1) are a translation of the spacetime origin and/or a Lorentz transformation. This is the Poincaré group of transformations.

A spacetime translation corresponds to new coordinates

$$x'^\mu = x^\mu + X^\mu, \tag{2.5}$$

with X^μ a constant. A Lorentz transformation corresponds to new coordinates

$$x'^\mu = \Lambda^\mu{}_\nu x^\nu, \tag{2.6}$$

with Λ^μ_ν a constant matrix satisfying

$$\Lambda^\alpha{}_\mu \Lambda^\beta{}_\nu \eta_{\alpha\beta} = \eta_{\mu\nu}. \tag{2.7}$$

Note that

$$\Lambda^\mu{}_\nu = \frac{\partial x'^\mu}{\partial x^\nu}. \tag{2.8}$$

It is often convenient to consider an infinitesimal Lorentz transformation,

$$\Lambda^\mu{}_\nu = \delta^\mu_\nu + \omega^\mu{}_\nu. \tag{2.9}$$

Requiring that the form of Eq. (2.3) is preserved, one sees that this is a Lorentz transformation if and only if the infinitesimal quantity $\omega_{\mu\nu} \equiv \eta_{\mu\alpha}\omega^\alpha_\nu$ is antisymmetric.

The most general Lorentz transformation is a rotation and/or a Lorentz boost. A rotation changes just the space coordinates and preserves the form of Eq. (2.4). It has the form

$$x'_i = R_{ij} x_j, \tag{2.10}$$

where the rotation matrix is orthogonal ($R^{-1} = R^{\mathrm{T}}$). For a rotation through angle ϕ about the z-axis,

$$R_{ij} = \begin{pmatrix} \cos\phi & -\sin\phi & 0 \\ \sin\phi & \cos\phi & 0 \\ 0 & 0 & 1 \end{pmatrix}. \tag{2.11}$$

A Lorentz boost along (say) the x-axis mixes t and x components according to

$$x' = \gamma(x - vt), \qquad t' = \gamma(t - vx), \tag{2.12}$$

where $\gamma = (1 - v^2)^{-1/2}$ (recall that we set $c = 1$). The parameter v gives the relative velocity of the old and new frames.

Allowing only boosts with $v \ll 1$ gives the Galilean transformation $t' = t$ and $x' = x - vt$. This is the Newtonian description of spacetime, in which there is a universal time coordinate.

Special relativity ignores gravity and assumes the existence of Minkowski coordinates, which define an inertial frame. According to the **relativity principle** as originally formulated there is no preferred inertial frame, which means that the form of the equations remains the same under every transformation from one set of Minkowski coordinates to another. This was supposed to include time reversal (T) $t \to -t$, and the parity transformation (P) $x^i \to -x^i$ which reverses the handedness of the coordinates. According to quantum field theory, these transformations are related to charge conjugation (C), which interchanges particles with antiparticles in such a way that the combined CPT invariance is guaranteed, and this is verified by observation.

It was found in 1956 that the weak interaction is not invariant under the parity transformation P, though it seemed to be invariant under CP or equivalently T. In 1964 it was found that even these invariances are not exact. Therefore, at a fundamental level, the relativity principle should be applied only after t has been chosen to increase into the future, and (say) a right-handed coordinate system has been specified. This turns out not to be an issue in the usual scenarios of the early Universe perturbations, because in those scenarios there is no mechanism by which the violation of T and P leads to an observable effect. Scenarios have been proposed where that is not the case, so that for example the cosmic microwave background has net left-handed circular polarization, but no such effect has been observed.

2.2 Vectors and tensors with Minkowski coordinates

According to the relativity principle, the laws of physics should take on the same form in every inertial frame. To achieve this, the equations are written in terms of 4-scalars, 4-vectors and 4-tensors. These objects are invariant under spacetime translations, and they transform linearly under the Lorentz transformation.

2.2.1 4-scalars and 4-vectors

A 4-scalar is specified by a single number, and is invariant under the Lorentz transformation.[2] A 4-vector A^μ is specified by four components, transforming like dx^μ:

$$A'^\mu = \frac{\partial x'^\mu}{\partial x^\nu} A^\nu. \tag{2.13}$$

As with ordinary vectors one can use a symbol to denote the vector itself as opposed to its components, such as $\vec{A}$. Also, one can define basis vectors $\vec{e}_\mu$ such that

$$\vec{A} = \sum_\mu A^\mu \vec{e}_\mu. \tag{2.14}$$

In a given inertial frame, each 4-vector is of the form $A^\mu = (A^0, \mathbf{A}) = (A^0, A^i)$ where as usual $\mathbf{A}$ denotes a 3-vector. The inner product of two 4-vectors is defined as $\eta_{\mu\nu} A^\mu B^\nu$. It is a 4-scalar and the 4-vectors are said to be orthogonal if it vanishes. From now on we generally refer to a 4-vector simply as a vector.

For any vector, it is useful to define the lower-component object

$$A_\mu = \eta_{\mu\nu} A^\nu, \tag{2.15}$$

or more explicitly, $A_0 = -A^0$, $A_i = A^i$. It is called a covariant vector while A^μ is called a contravariant vector (more properly one should talk about the covariant or contravariant components of the same vector). The scalar product of two vectors can be written $A_\mu B^\mu$. Going to a new inertial frame,

$$A'_\mu B'^\mu = A_\mu B^\mu. \tag{2.16}$$

The transformation law for A_μ follows from this, remembering that B'^μ is an arbitrary vector. Indeed, let us set every component of that object equal to zero except for one component $B'^\mu = 1$ (with μ either 0, 1, 2 or 3). Inserting this on the left-hand side and putting the transformed quantity $B^\nu = (\partial x^\nu / \partial x'^\mu) B'^\mu$ into the right-hand side, we get

$$A'_\mu = \frac{\partial x^\nu}{\partial x'^\mu} A_\nu. \tag{2.17}$$

[2] If we consider in addition the effect of a parity transformation, we need to distinguish between true scalars (simply called scalars), which are invariant under the parity transformation, and pseudo-scalars which reverse their sign. A similar distinction needs to be made for vectors and tensors. We shall have occasional need of this distinction.

2.2.2 4-tensors

A second-rank 4-tensor is a sixteen-component object $C_{\mu\nu}$, which transforms like a product $A_\mu B_\nu$ of two vectors;

$$C'_{\mu\nu} = \frac{\partial x^\alpha}{\partial x'^\mu}\frac{\partial x^\beta}{\partial x'^\nu} C_{\alpha\beta}. \tag{2.18}$$

Third- and higher-rank 4-tensors are defined in the same way. We can lower any component of a tensor with $\eta_{\mu\nu}$, for instance, $C^\mu{}_\nu = \eta_{\nu\lambda} C^{\mu\lambda}$. A tensor with all lower indices transforms like a product of covariant vectors and a tensor with mixed upper and lower indices transforms like a product of the appropriate mixture of vectors.

It is often useful to regard 4-vectors and 4-scalars as, respectively, first- and zeroth-rank tensors. Depending on the context, the term 'tensor' will either include those, or will mean an object of rank ≥ 2.

As with vectors, one can denote (say) a second-rank tensor by $\vec{C}$. One can also define basis tensors $\vec{e}_\mu \otimes \vec{e}_\nu$, such that

$$\vec{C} = \sum_{\mu,\nu} C^{\mu\nu} \vec{e}_\mu \otimes \vec{e}_\nu. \tag{2.19}$$

If we multiply two tensors and contract any number of indices we get another tensor; for instance $A_{\mu\nu} B^{\mu\alpha}$ is a tensor. An inverse of this statement is true; if the multiplication and contraction of an object with an *arbitrary* tensor gives another tensor, then that object is itself a tensor. We encountered a special case of this 'quotient theorem' in Eq. (2.16). As in that case, it can be proved simply by setting in turn each component of the arbitrary tensor to 1 with the others zero.

From Eq. (2.7), the metric tensor $\eta_{\mu\nu}$ is indeed a tensor, but a very special one. Its components are the same in every inertial frame, and $\eta^\mu{}_\nu = \eta_\nu{}^\mu = \delta^\mu_\nu$, where δ^μ_ν is again the Kronecker delta. Also, $\eta^{\mu\nu}$ has the same components as $\eta_{\mu\nu}$. The Levi–Civita symbol $\epsilon_{\mu\nu\alpha\beta}$ (totally antisymmetric with $\epsilon_{0123} = 1$) also has the same components in every frame. Any other tensor with the same components in every frame has to be constructed by multiplying and/or contracting these two.

If the components of a 4-vector or 4-tensor vanish in one coordinate system, they vanish in all coordinate systems. This means that a 4-vector or 4-tensor is defined uniquely by giving its components in any coordinate system.

In order to satisfy the principle of relativity, laws of physics are written in the form 'tensor = tensor', or in other words in the form 'tensor = 0'. Such an equation is said to be **covariant**; both sides transform in the same way. [3]

[3] At the quantum level one can also allow 'spinor' = 0, where the transformation of spinor components is given by Eq. (15.48).

2.2.3 3-tensors

At a fixed time in a given inertial frame, we can repeat the above analysis for the three space dimensions. Then we deal with 3-scalars, -vectors and -tensors. Under a rotation, a 3-vector transforms like the coordinates:

$$A'_i = R_{ij} A_j. \tag{2.20}$$

A second-rank 3-tensor transforms like a product:

$$A'_{ij} = R_{in} R_{jm} A_{nm}, \tag{2.21}$$

and so on.

2.3 Spacetime lines and geodesics

Starting from a generic inertial frame, we can use a Lorentz boost to bring an interval ds^2 into a transparent form as follows. If ds^2 is negative one can bring it into the form $ds^2 = -dt^2$. Then the interval is said to be timelike and dt is called the proper time interval, which we will denote by $d\tau$. If ds^2 is positive one can make $dt = 0$. Then the interval is said to be spacelike and ds^2 is the (proper) distance-squared. Finally, if ds^2 vanishes it corresponds to equal intervals of time and distance (in every inertial frame) and is said to be lightlike or null.

A line through spacetime is timelike, lightlike, or spacelike if ds^2 has that property for all nearby points. The **lightcone**, illustrated in Figure 2.1, consists of the set of lightlike lines through a given spacetime point. Timelike lines through the point (also called worldlines) go from the past lightcone to the future lightcone and spacelike lines lying outside the cones.

A parametrized line $x^\mu(\lambda)$ has at each point a **tangent vector** $t^\mu \equiv dx^\mu/d\lambda$. The 4-velocity of a worldline is $u^\mu = dx^\mu/d\tau$, where $d\tau$ is the proper time interval. Its components are

$$u^\mu = u^0(1, v^i), \tag{2.22}$$

where $v^i \equiv dx^i/dt$ is the 3-velocity. The time component is $u^0 = dt/d\tau = (1 - v^2)^{-1/2}$, because $\eta_{\mu\nu} u^\mu u^\nu = -1$. The 4-acceleration is $a^\mu \equiv du^\mu/d\tau$. The 4-momentum of a particle with mass $m > 0$ is $p^\mu = m u^\mu$, the mass m being a scalar. The energy and momentum are given by $p^\mu = (E, \mathbf{p})$, with $\mathbf{p} = E\mathbf{v}$. To handle zero mass we can take the limit $m \to 0$ with p^μ constant. Then $v \equiv |\mathbf{v}| = 1$ and $p \equiv |\mathbf{p}| = E$.

Particles with non-zero mass move along timelike lines through spacetime, called worldlines, and those with zero mass move along lightlike lines. Relative to any inertial frame, the former have speed v less than 1 and the latter have speed equal

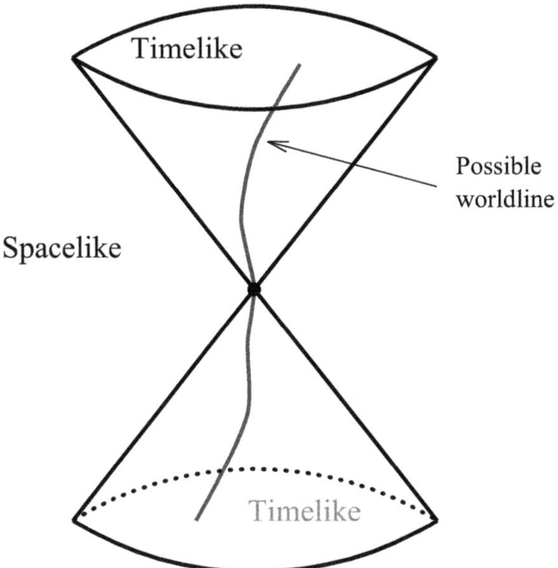

Fig. 2.1. The worldline of a possible observer lies within the lightcone, which divides timelike separations from spacelike separations.

to 1. Causality forbids the existence of particles with speed bigger than 1, which would move along spacelike lines.

For each point on a worldline there is a **local rest frame** in which the velocity is zero. To be precise, there is an infinity of local rest frames, related by rotations. A worldline is said to be orthogonal to a spacelike line intersecting it, if nearby events on the spacelike line are simultaneous in the local rest frame. (One way of defining simultaneity is to bounce a light signal between nearby observers, as in Figure 2.2. The analogous construction for spacelike lines is to bisect an angle into two equal angles.)

A **geodesic** corresponds to a straight line in Minkowski coordinates. In other words, there exists a parameter λ such that $x^\mu(\lambda)$ satisfies $d^2x^\mu/d\lambda^2 = 0$. Such a parameter is called an affine parameter.

For a worldline or a lightlike line we can choose $d\lambda = dt$, and a geodesic has constant 3-velocity:

$$\frac{d^2x^\mu}{dt^2} = 0. \tag{2.23}$$

For a worldline we can choose $d\lambda = d\tau$ and a geodesic has constant 4-velocity:

$$\boxed{a^\mu \equiv \frac{du^\mu}{d\tau} = 0}. \tag{2.24}$$

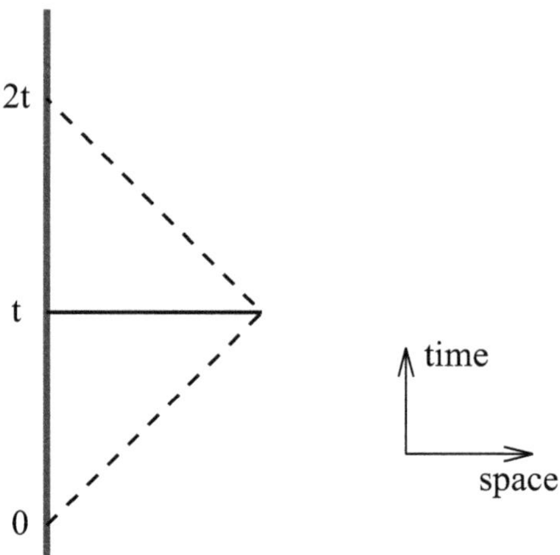

Fig. 2.2. The vertical line represents an observer equipped with a clock, and a device for bouncing a photon off a nearby object. This observer defines the half-time event as simultaneous with the bounce.

According to Newton's second law, free particles have geodesic worldlines. For a spacelike line we can choose $d\lambda = d\ell$, and a geodesic is a straight line in space.

A 4-scalar, -vector or -tensor might be defined in a region of spacetime. Then one can act on it with the 4-gradient $\partial_\mu \equiv \partial/\partial x^\mu$. The 4-gradient is a 4-vector with lower-index components because it transforms according to Eq. (2.17) (the chain rule of partial differentiation). It turns a 4-scalar into a 4-vector, a 4-vector into a second-rank tensor, and so on. One can also define the upper-index quantity $\partial^\mu \equiv \eta^{\mu\nu}\partial_\nu$. From the gradient one can form the d'Alembertian $\Box$ which is a scalar:

$$\Box \equiv \partial^\mu \partial_\mu = -\frac{\partial^2}{\partial t^2} + \nabla^2 = -\frac{\partial^2}{\partial t^2} + \frac{\partial^2}{\partial x^2} + \frac{\partial^2}{\partial y^2} + \frac{\partial^2}{\partial z^2}. \qquad (2.25)$$

2.4 Fluid dynamics

For the rest of this chapter we look at the relativistic description of fluids. A fluid is a medium whose properties can be regarded as smooth functions of spacetime position.

2.4.1 Velocity field

The 4-velocity $u^\mu(\mathbf{x}, t)$ of a fluid element defines the fluid flow. The 4-acceleration can be written

$$a^\mu \equiv u^\nu \partial_\nu u^\mu. \qquad (2.26)$$

To see this one can use the chain rule. More elegantly, one can just notice that it is valid in the local rest frame of each fluid element and hence (being a covariant expression) is valid in any inertial frame for the whole fluid.

The **velocity gradient** is defined by

$$v_{ij}(\mathbf{x},t) \equiv \partial_j v_i. \tag{2.27}$$

It is of the form

$$\boxed{v_{ij} = \theta\delta_{ij} + \omega_{ij} + \sigma_{ij}}, \tag{2.28}$$

where the **vorticity** ω_{ij} is antisymmetric, while the **shear** σ_{ij} is traceless and symmetric. The inverse relations are

$$3\theta = v_{ii} \equiv \partial_i v_i \equiv \nabla \cdot \mathbf{v}, \tag{2.29}$$

$$\omega_{ij} = \frac{1}{2}\left(\partial_j v_i - \partial_i v_j\right), \tag{2.30}$$

$$\sigma_{ij} = \frac{1}{2}\left(\partial_j v_i + \partial_i v_j\right) - \delta_{ij}\theta. \tag{2.31}$$

A worldline which has the fluid velocity is said to be **comoving** (moving with the fluid). A volume is said to be comoving if its boundary is comoving. If $\mathcal{V}$ is the volume of an infinitesimal comoving region, we can integrate Eq. (2.29) using the divergence theorem to find

$$3\theta\mathcal{V} = \int_{\partial\mathcal{V}} \mathbf{v} \cdot d\mathbf{S}. \tag{2.32}$$

The right-hand side is just $d\mathcal{V}/dt$, and so θ measures the average rate of expansion:

$$\boxed{\frac{1}{\mathcal{V}}\frac{d\mathcal{V}}{dt} = 3\theta}. \tag{2.33}$$

The vorticity corresponds to the quantity $\nabla \times \mathbf{v}$. It measures the speed of rotation of a fluid element, whereas the shear measures the anisotropy in its expansion rate.

2.4.2 Conserved currents

A non-relativistic fluid with mass density $\rho(\mathbf{x})$ and velocity $\mathbf{v}(\mathbf{x})$ satisfies the **mass continuity equation**

$$\dot{\rho} + \nabla \cdot \mathbf{j} = 0, \tag{2.34}$$

where $\mathbf{j} = \rho\mathbf{v}$. To uncover the meaning of this equation we can multiply it by a fixed volume element $\mathcal{V}$ and apply the divergence theorem to the second term.

Then we have

$$\frac{dM}{dt} + \int_{\partial \mathcal{V}} \mathbf{j} \cdot \mathbf{dS} = 0, \tag{2.35}$$

where dM is the mass within the volume element and $\partial \mathcal{V}$ its boundary. The mass continuity equation states that M changes only insofar as mass flows across the boundary into or out of the volume element.

Any equation of this form is called a continuity equation, and its physical meaning is always the same; within a given volume the quantity can change only insofar as the corresponding current flows across the boundary of the volume. If there is no flow, the amount of the quantity within the volume is conserved.

In particular, ρ can be the density of electric charge, and $\mathbf{j}$ the electric current. In this case we deal with a relativistic theory, and it makes sense to define a 4-current j^μ, whose components are $(\rho, \mathbf{j})$. Then the charge continuity equation can be written

$$\boxed{\partial_\mu j^\mu = 0}\,. \tag{2.36}$$

The same device works for any relativistic theory, and whenever it does we will call j^μ a conserved current. Note that, despite the use of the word 'conserved', the current j^μ will be time dependent. It is only the 'charge' $\int j^0 d^3x$ that is conserved in the usual sense of that word.

2.5 Energy–momentum tensor

2.5.1 Energy, momentum and stress

The mechanical properties of a relativistic fluid are described using the energy–momentum tensor $T^{\mu\nu}$. It is symmetric, $T^{\mu\nu} = T^{\nu\mu}$. The energy density is T^{00} and the components of the momentum density are T^{0i}, making $T^{0\mu}$ the components of the four-momentum density. The stress is T_{ij}, so that the force exerted on an area element dS_i by the fluid behind it is

$$F_i = T_{ij} dS_j. \tag{2.37}$$

Considering a given point in spacetime it is often convenient to choose the inertial frame so that the fluid velocity vanishes, the local rest frame. In a local rest frame the momentum density vanishes:

$$T^{0i} = 0 \qquad \text{(local rest frame)}. \tag{2.38}$$

In this book we generally use ρ to denote the energy density in a local rest frame:

$$\rho \equiv T^{00} \qquad \text{(local rest frame)}. \tag{2.39}$$

A **perfect fluid** is defined in the context of relativity as one that is isotropic in a local rest frame: $T_{ij} = P\delta_{ij}$, where P is the pressure. (The force on area dS_i is PdS_i, the usual definition of pressure.) In general,

$$T_{ij} = P\delta_{ij} + \Sigma_{ij} \qquad \text{(local rest frame)}, \tag{2.40}$$

where the second term is traceless ($\Sigma_{ii} = 0$) and is called the anisotropic stress. In a frame where the fluid velocity is v_i, Eqs. (2.39) and (2.40) remain valid to first order in v and, to the same order, the momentum density is

$$T^{i0} = (\rho + P)v^i. \tag{2.41}$$

We see that both energy density and pressure contribute to the 'inertial mass density', defined as the momentum density divided by the 3-velocity.

To go to a generic inertial frame it is convenient to lower one of the indices. Then the energy–momentum tensor of a perfect fluid has the form

$$\boxed{T^\mu_\nu = (\rho + P)u^\mu u_\nu + P\delta^\mu_\nu}, \tag{2.42}$$

where u^μ is the fluid 4-velocity. This expression must be correct because it has the required form in a local rest frame. If there is anisotropic stress, Eq. (2.42) becomes

$$T^\mu_\nu = (\rho + P)u^\mu u_\nu + P\delta^\mu_\nu + \Sigma^\mu_\nu, \tag{2.43}$$

where, in a local rest frame, the only non-zero components of Σ^μ_ν are the space–space components.

2.5.2 Energy and momentum conservation

The energy–momentum tensor satisfies

$$\boxed{\partial_\mu T^{\mu\nu} = 0}. \tag{2.44}$$

We can then regard $T^{\mu\nu}$ as a set of conserved currents, one for each ν. These currents correspond to the conservation of energy and momentum:

$$\partial_t T^{00} + \partial_j T^{0j} = 0 \qquad \text{(energy) continuity}, \tag{2.45}$$

$$\partial_t T^{i0} + \partial_j T^{ij} = 0 \qquad \text{Euler}. \tag{2.46}$$

Following some of the cosmology literature, we are calling the momentum continuity equation the Euler equation.[4]

The 4-momentum in a fixed volume $\mathcal{V}$ is $p^\mu = T^{\mu 0}\mathcal{V}$, and its rate of decrease is

$$-\frac{dp^\mu}{dt} = \int_{\partial\mathcal{V}} T^{\mu j}\, dS_j. \tag{2.47}$$

[4] More usually, the term 'Euler equations' is taken to mean the whole set of mechanical continuity equations (mass, momentum and energy) restricted to the non-relativistic domain.

We see that the momentum density T^{0j} corresponds to the flow of *energy*, while the stress T^{ij} corresponds to the flow of the ith component of *momentum*. From Eq. (2.37), the rate of change of the total momentum within the volume is equal to the force exerted on it (Newton's second law). For an isolated system $T^{\mu\nu}$ vanishes on the boundary, so that energy and momentum are conserved.

Now go to a local rest frame. Then Eq. (2.41) is valid, and putting it into Eq. (2.46) we find

$$\boxed{(\rho + P)a_i = -\partial_j T_{ij}} \qquad \text{(Euler)}. \tag{2.48}$$

For a perfect fluid the Euler equation becomes

$$(\rho + P)a_i = -\partial_i P. \tag{2.49}$$

This is the non-relativistic Euler equation, except that the mass density is replaced by $\rho + P$.

In a local rest frame the continuity equation becomes

$$\boxed{\dot\rho = -(\partial_i v_i)(\rho + P) \equiv -3\theta(\rho + P)} \qquad \text{(continuity)}. \tag{2.50}$$

Applying Eq. (2.33) to an infinitesimal comoving volume, the continuity equation becomes

$$\dot\rho = -\frac{\dot{\mathcal{V}}}{\mathcal{V}}(\rho + P). \tag{2.51}$$

This in turn can be written as $dE = -Pd\mathcal{V}$, where $E = \rho\mathcal{V}$ is the energy in the comoving volume. This is the thermodynamic expression, without heat flow. There is no concept of heat flow in our discussion, because we are not making any distinction between different forms of energy and momentum.

2.5.3 Angular momentum conservation

Consider now the tensor

$$M^{\lambda\mu\nu} \equiv x^\mu T^{\lambda\nu} - x^\nu T^{\lambda\mu}. \tag{2.52}$$

It satisfies the continuity equation $\partial_\lambda M^{\lambda\mu\nu} = 0$, which means that the antisymmetric tensor

$$J^{\mu\nu} \equiv \int d^3x M^{0\mu\nu} \tag{2.53}$$

is conserved for an isolated system. We can regard $M^{\lambda\mu\nu}$ as a set of conserved currents, one for each of the six independent choices of the $\mu\nu$.

The space–space components $J^{\mu\nu}$ correspond to the angular momentum:

$$J_i \equiv \frac{1}{2}\epsilon_{ijk}J_{jk} = \epsilon_{ijk}\int d^3x x_j T^0_k. \tag{2.54}$$

The other three components of $J^{\mu\nu}$ correspond to the conservation of $K_i \equiv J^0_i$. It is given by

$$\frac{\mathbf{K}}{E} \equiv \frac{t\mathbf{p}}{E} - \mathbf{X}, \tag{2.55}$$

where $\mathbf{p}$ is the momentum, E is the energy, and

$$\mathbf{X} \equiv \int d^3x\, \mathbf{x}\frac{T^{00}}{E} \tag{2.56}$$

is the 'centre of energy'. The conservation of $\mathbf{K}$ corresponds to the statement that the centre of energy moves with constant velocity $\mathbf{p}/E$. In contrast with the non-relativistic centre of mass, the centre of energy is of only formal interest.

2.6 Gas dynamics

The treatment so far applies to any fluid. Now we take the fluid to be an ideal gas. This means that the interaction of the particles is negligible in the sense that practically all of the particles are free at any instant. The properties of the fluid are to be calculated by taking the spatial average (smoothing) over a volume containing many particles.

A gas may contain one or more components. In ordinary applications the components are different molecules or atoms. In the early Universe they are different species of elementary particle, and we will talk about different species rather than different components.

2.6.1 Distribution function

An ideal gas may contain particles of various species, which we label by letters a, b, ... Its state is specified by giving, for each species, the **distribution function** $f_a(t, \mathbf{r}, \mathbf{p})$. It defines the number density dn_a of particles with momentum in a given range through the relation

$$dn_a = (2\pi)^{-3} g_a f_a(\mathbf{p}) d^3p. \tag{2.57}$$

Here g_a is the number of spin states. The density of states factor $(2\pi)^{-3}$ is inserted so that f_a is the average occupation number of the quantum states, within a cell d^3p centred on $\mathbf{p}$.

For a massive particle species with spin s, there are $2s + 1$ spin states, and the same for the antiparticle. The photon has two spin states (right- and left-handed spin). The ν_e (regarded as massless) has only left-handed spin while the $\bar{\nu}_e$ has only right-handed spin, and similarly for the other two neutrino species.

The energy and momentum of a species with mass $m_{\rm a}$ are related by $E = \sqrt{m_{\rm a}^2 + p^2}$. The quantities Ed^3x and d^3p/E are Lorentz invariant. (Starting in the rest frame of a given particle and boosting along the z-axis, for example, we see that, indeed, $dz \propto 1/E$, corresponding to the Lorentz contraction, and that $dp_z \propto E$.) As a result, d^3xd^3p is Lorentz invariant. That means that $f_{\rm a}$ is invariant, because $f_{\rm a}d^3xd^3p$ is the number of particles with position and momentum in a given range (up to a numerical factor). The number density, which is not Lorentz invariant, is given by

$$\boxed{n_{\rm a}(t, \mathbf{r}) = \frac{g_{\rm a}}{(2\pi)^3} \int f_{\rm a}(t, \mathbf{x}, \mathbf{p})\, d^3p}\,. \tag{2.58}$$

2.6.2 *Electromagnetic current and energy–momentum tensor*

From the distribution function we can calculate the contribution of each species to the electromagnetic current, and to the energy–momentum tensor. We consider just one species and drop the subscript a.

Consider first the contribution to the electromagnetic current, of a species with charge Q, from particles whose momentum is within a cell d^3p centred on momentum $\mathbf{p}$. In a local rest frame, where $\mathbf{p} = 0$, they contribute $\rho(\mathbf{r}) = Qfd^3p$ and $\mathbf{j}(\mathbf{r}) = 0$. In an arbitrary frame, their contribution is therefore

$$j^\mu = \frac{g}{(2\pi)^3} Qfp^\mu \left(\frac{d^3p}{E}\right). \tag{2.59}$$

The last factor is a scalar, making j^μ a 4-vector that reduces to the required form in a local rest frame. Integrating over momentum gives

$$j^\mu = \frac{g}{(2\pi)^3} Q \int fp^\mu \frac{d^3p}{E}. \tag{2.60}$$

To get the energy–momentum tensor, all we have to do is replace the charge Q of each particle by its 4-momentum p^ν. This gives

$$T^{\mu\nu} = \frac{g}{(2\pi)^3} \int fp^\mu p^\nu \frac{d^3p}{E}. \tag{2.61}$$

The 4-momentum density, as it should be, is

$$T^{\mu 0} = \frac{g}{(2\pi)^3} \int fp^\mu d^3p. \tag{2.62}$$

Using Eqs. (2.39), (2.40) and (2.41), which are valid to first order in the fluid velocity, we find that for each species the energy density ρ, momentum density $(\rho + P)\mathbf{v}$, pressure P, and anisotropic stress Σ_{ij} are given in terms of the occupation number by

$$\boxed{\rho = \frac{g}{(2\pi)^3}\int f E\, d^3p}\,, \tag{2.63}$$

$$\boxed{(\rho + P)\mathbf{v} = \frac{g}{(2\pi)^3}\int f\mathbf{p}\, d^3p}\,, \tag{2.64}$$

$$\boxed{P = \frac{g}{3(2\pi)^3}\int f p^2 \frac{d^3p}{E}}\,, \tag{2.65}$$

$$\boxed{\Sigma_{ij} = \frac{g}{(2\pi)^3}\int f\left(p_i p_j - \frac{1}{3}p^2\delta_{ij}\right)\frac{d^3p}{E}}\,. \tag{2.66}$$

Note that for radiation, $p = E$, and we have the familiar result $P = \rho/3$. If f is independent of direction, then Σ_{ij} vanishes and we have a perfect fluid. This is true in particular in thermal equilibrium.

2.7 Boltzmann equation

2.7.1 *Liouville equation*

We now develop equations giving the evolution of the distribution functions. If there are no collisions, each particle has constant momentum $\mathbf{p}$. In that case the distribution function is constant along a given particle trajectory $\mathbf{r}(t)$:

$$\frac{df}{dt} = 0, \tag{2.67}$$

This is the **Liouville equation**, also sometimes called the collisionless Boltzmann equation.

To see why f is constant, consider an initially rectangular region of phase space (i.e. the dynamical configuration space, here defined by the position and momentum coordinates). As time passes, the region moves and its shape is distorted, but the number of particles in it remains constant; thus f will be constant if the volume of the region is constant. To see that this is indeed the case, assume without loss of generality that the particle motion is in the x_1 direction, as illustrated in Figure 2.3, and focus on the initially rectangular pair of faces with sides dx_1 and dp_1. The coordinate p_1 of each corner is constant, but the other coordinate r^1 moves with speed proportional to p_1 ($p_i = m dx_i/dt$). The rectangle therefore becomes a parallelogram with the passage of time, but its area remains constant and so does

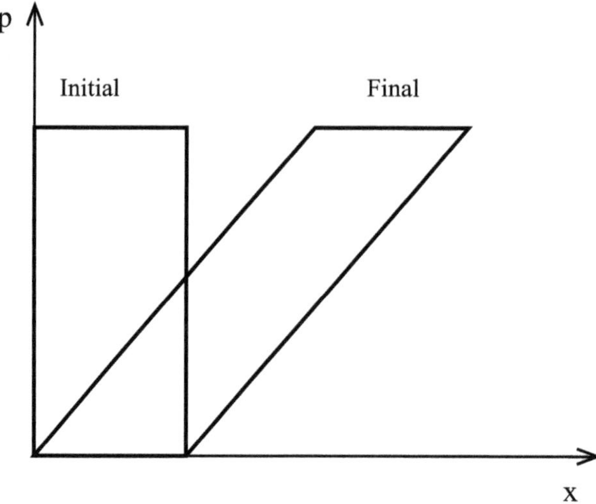

Fig. 2.3. The volume of a phase-space element doesn't change if the particles are moving with constant velocity.

the volume that we are considering. (A more formal way of seeing the constancy of the volume is to calculate the Jacobian of the transformation of variables that corresponds to the elapse of a time interval dt.)

2.7.2 Collision term

If there are collisions, particles will enter and leave the chosen volume of phase space. The same thing will happen if the particles can decay. (We will use the term 'interaction' to indicate a collision or decay.) However, in the ideal gas limit a given particle in the volume travels a long way before it collides or decays. (To be precise, it travels much farther than the mean particle spacing.) As a result, it still makes sense to follow f along the trajectory of a particle that doesn't collide, but now its rate of change no longer vanishes. We call the rate of change the *collision term*, and if we denote it by $C[f]$, the Liouville equation becomes

$$\frac{df}{dt} = C[f]. \tag{2.68}$$

This is the **Boltzmann equation** for f.

The outcome of an interaction depends on the particle spins, but in many situations we can ignore this fact. Then, the collision term for a given particle species can be calculated if we know the distribution functions of all species whose interactions can create or destroy the given species. This leads to a set of coupled Boltzmann equations, which can be solved to give the distribution functions.

Let us write down the collision term appearing in the coupled Boltzmann equations, in fully relativistic form but ignoring the spin dependence of the collision rate. We will consider a particle species a, and suppose for the moment that its only significant interaction is a collision process of the form ab↔cd, with all four species different. We will take a and c be fermions and b and d be bosons. Then the collision term for, say, the species b is

$$\begin{aligned}\frac{d^3p_{\rm b}}{(2\pi)^3}C[f_{\rm b}] &= Dp_{\rm b}\int Dp_{\rm a}Dp_{\rm c}Dp_{\rm d}(2\pi)^4\delta^4(p^\mu_{\rm a}+p^\mu_{\rm b}-p^\mu_{\rm c}-p^\mu_{\rm d})|M|^2 \\ &\quad\times[f_{\rm c}f_{\rm d}(1-f_{\rm a})(1+f_{\rm b})-f_{\rm a}f_{\rm b}(1-f_{\rm c})(1+f_{\rm d})]. \quad (2.69)\end{aligned}$$

The right-hand side of this expression is Lorentz invariant. The Dirac delta function δ^4 ensures 4-momentum conservation. The quantity $Dp_{\rm a}$ is the Lorentz-invariant phase-space element defined by

$$Dp_{\rm a}=\frac{d^3p_{\rm a}}{(2\pi)^3 2E_{\rm a}}, \quad (2.70)$$

with $E_{\rm a}=\sqrt{m^2_{\rm a}+p^2_{\rm a}}$ the energy. We saw earlier that it is indeed Lorentz invariant, and it is equal to the density of momentum states divided by $2E_{\rm a}$.

The crucial factor in this expression is the transition rate $|M|^2$. Working in a finite box, so that the momenta are discrete, $|M|^2$ can be thought of as the probability that particles with given initial momenta create particles with given final momenta. By virtue of time-reversal invariance, the transition rate is the same in both directions.[5] Depending on the circumstances, it may be calculable from the appropriate theory (quantum field theory for elementary particles, atomic physics for atoms, nuclear physics for nuclei) and/or it may be measurable in the laboratory.

Inside the square brackets in Eq. (2.69), the first term represents the process cd → ab, causing a flow of particles *into* the momentum-space element $d^3p_{\rm a}$, and the second term represents the process ab → cd, causing an outward flow of particles. Focusing on the second term, there are factors $f_{\rm a}$ and $f_{\rm b}$ because the rate is proportional to the occupation number of each initial state. There is also a factor $1-f_{\rm c}$ because transition to an occupied fermion state is forbidden (Pauli blocking), and a factor $1+f_{\rm d}$ because transition to an occupied boson state is amplified (stimulated emission, which makes a laser work).

In general, the collision term is a sum of terms, each describing a separate process that may have an arbitrary number of particles in the initial and final states. Provided that neither the final nor the initial state contains identical particles, the collision term is the direct generalization of the one we wrote down, with M the appropriate amplitude and the delta function modified to include all of the

[5] At the end of this chapter we see how to handle a violation of time-reversal invariance.

4-momenta. For each initial species, there is a factor $Dp_n f_n$, and for each final species there is a factor $Dp_n(1 \pm f_n)$ (where n is the species label). If there are identical particles, the expression has a well-known modification corresponding to the symmetry or antisymmetry of the corresponding wavefunction. Our only explicit use of this expression will be for Thomson scattering (the process $\gamma e \to \gamma e$ with the electrons non-relativistic).

The Boltzmann equation for the distribution functions cannot fully handle particle spin, or the mixing of different particle species. Focussing on the spin, we can try to take account of it by letting the particle label 'a' include the spin state. This doesn't work at the quantum level though, because different spin states interfere so that we should be adding the *amplitudes* M and not the rates $|M|^2$. In this situation, we need to replace the occupation number f by the density matrix $\mathcal{D}$, with dimensionality equal to the number of spin states. The expectation value of an observable relating to spin, with matrix A, is equal to the trace of $\mathcal{D}A$. In particular, the occupation number is simply $f = \mathrm{Tr}\,\mathcal{D}$. The density matrix satisfies a Boltzmann equation, with the collision term of each matrix element involving all of the others in general.

The complication described in the previous paragraph can be ignored in many circumstances. In particular, it can be ignored for Thomson scattering which will be our main application of the Boltzmann equation. Also, it can be shown that the complication doesn't affect the thermal equilibrium distribution, to which we now turn.

2.7.3 Thermal equilibrium

If appropriate processes occur with a sufficiently fast transition rate, the Boltzmann equation will drive the distribution functions to a thermal equilibrium value which is independent of time and position. In practice we deal with approximate equilibrium which allows the distribution functions to vary slowly with time and position. We consider here exact equilibrium, and use the Boltzmann equation because the density matrix gives the same result.

Let us suppose first that there is just one process ab$\leftrightarrow$cd, with all four species different. Then Eq. (2.69) holds, and thermal equilibrium requires the square brackets of Eq. (2.69) to vanish. This requires that the distribution functions are of the form

$$\boxed{f_{\mathrm{a}}(p) = \frac{1}{\exp\left(\frac{E_{\mathrm{a}}(p)-\mu_{\mathrm{a}}(T)}{T}\right) \pm 1}}, \tag{2.71}$$

with a minus sign for bosons and a plus sign for fermions, and

$$\mu_{\mathrm{a}} + \mu_{\mathrm{b}} = \mu_{\mathrm{c}} + \mu_{\mathrm{d}}. \tag{2.72}$$

The parameter T is the temperature.[6] The function $\mu_{\rm a}$ is the **chemical potential** of the species. The expression is valid in the frame where the fluid is at rest and not rotating, so that energy is the only mechanical conserved quantity with non-zero value.

If an arbitrary set of processes is in thermal equilibrium, Eq. (2.71) still holds and the sum of the chemical potentials is still conserved by each process:

$$\boxed{\sum \mu_i = \sum \mu_f}, \tag{2.73}$$

where i runs over the particles in the initial state and f runs over the particles in the final state. This constraint relates the number densities of species that can be created or destroyed.

If particle–antiparticle pairs of a species are freely created, the chemical potentials $\mu_{\rm a}$ and $\bar{\mu}_{\rm a}$ must be equal and opposite. By 'freely created' we mean that some processes $X \leftrightarrow Y$ and $X \leftrightarrow Y+$ pair are both in equilibrium, where X and Y each denote any set of particles.

In the early Universe T invariance may be badly violated, but one can usually arrive at the same equilibrium distributions by invoking CPT invariance. That is possible if for each process in equilibrium the CPT related process is also in equilibrium, and if particle–antiparticle pairs are freely created. The first requirement plus CPT invariance means that the condition for the vanishing of the square bracket of Eq. (2.69) can be written

$$\sum(\mu_i + \bar{\mu}_i) = \sum(\mu_f + \bar{\mu}_f), \tag{2.74}$$

and with the second condition we then recover Eq. (2.73).

Exercises

2.1 By specializing Eq. (2.7) to a rotation, show that the rotation matrix is orthogonal.

2.2 Verify that the Lorentz boost satisfies Eq. (2.7). Do the same for a rotation around the z-axis.

2.3 By using Eq. (2.14), show that $\vec{e}_0$ has components $(1, 0, 0, 0)$, and similarly for the other basis vectors. By using Eq. (2.19), show that the basis vector $\vec{e}_0 \otimes \vec{e}_1$ has components $\delta_{\mu 0}\delta_{\nu 1}$.

2.4 Use a Lorentz boost to derive Eq. (2.41).

2.5 Verify that, with the distribution function of the form Eq. (2.71), the collision term given by Eq. (2.69) vanishes if and only if $\mu_{\rm a} + \mu_{\rm b} = \mu_{\rm c} + \mu_{\rm d}$.

[6] Equation (2.71) can be regarded as the definition of temperature for an ideal gas.

3
General relativity

Special relativity assumes the existence of Minkowski coordinates, such that the line element takes the form (2.1). In that case one says that spacetime is flat, otherwise one says that spacetime is curved. Spacetime is flat only insofar as gravity can be ignored.

In this chapter we first see how to write the equations of special relativity using generic coordinates. Then we consider curved spacetime, the equivalence principle and the Einstein field equation. All of these things taken together are called general relativity. We end the chapter by giving a basic description of the particular curved spacetime that corresponds to a homogeneous and isotropically expanding universe.

3.1 Special relativity with generic coordinates: mathematics

To handle curved spacetime we have to learn how to use generic coordinates. It is helpful to do this first in the familiar context of special relativity, where Minkowski coordinates do exist.

Once a coordinate choice x^μ has been made, it defines a **threading** of spacetime into lines (corresponding to fixed x^i) and a **slicing** into hypersurfaces (corresponding to fixed x^0), as shown in Figure 3.1. The threads are chosen to be timelike, so that they are the worldlines of possible observers, and the slices are chosen to be spacelike. The coordinate choice uniquely defines the threading and slicing, but the reverse is not true. Given a slicing and threading there is still freedom in choosing the coordinates which label the slices and threads.

3.1.1 Vectors and tensors

Starting with some coordinates x^μ, we may go to new coordinates x'^μ which are functions of the old ones. Then $dx'^\mu = (\partial x'^\mu / \partial x^\nu)\, dx^\nu$. The components of a

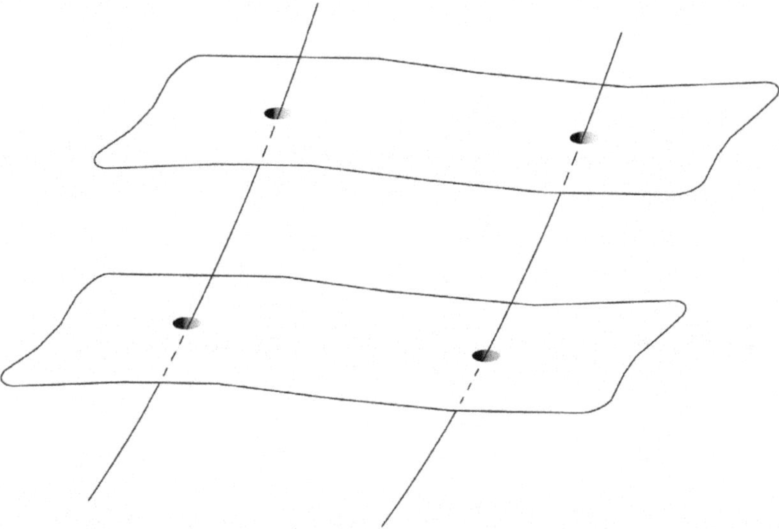

Fig. 3.1. A schematic illustration of slicing and threading of spacetime.

4-vector, *defined at a given spacetime point*, are taken to transform like dx^μ:

$$A'^\mu = \frac{\partial x'^\mu}{\partial x^\nu} A^\nu . \tag{3.1}$$

This transformation remains valid if both coordinate systems are generic. Tensors transform like products of vectors, for example

$$C'^{\mu\nu} = \frac{\partial x'^\mu}{\partial x^\alpha} \frac{\partial x'^\nu}{\partial x^\beta} C^{\alpha\beta} . \tag{3.2}$$

Going from Minkowski coordinates x'^μ to generic coordinates x^μ, the spacetime interval ds^2 given by Eq. (2.3) becomes[1]

$$ds^2 = g_{\mu\nu}\, dx^\mu\, dx^\nu , \tag{3.3}$$

where the metric tensor is now

$$g_{\mu\nu} = \frac{\partial x'^\alpha}{\partial x^\mu} \frac{\partial x'^\beta}{\partial x^\nu} \eta_{\alpha\beta} . \tag{3.4}$$

Note that $g_{\mu\nu} = g_{\nu\mu}$. Expression (3.3) is called the **line element**, and it is a convenient way of defining the metric tensor in a given coordinate system.

If we now go from x^μ to *generic* coordinates x'^μ,

$$g'_{\mu\nu} = \frac{\partial x^\alpha}{\partial x'^\mu} \frac{\partial x^\beta}{\partial x'^\nu} g_{\alpha\beta} . \tag{3.5}$$

[1] We choose to go *from* primed *to* unprimed coordinates so that there is no prime in the final expression.

In generic coordinates, components are lowered using $g_{\mu\nu}$:

$$A_\mu = g_{\mu\nu} A^\nu. \tag{3.6}$$

The transformation of A_μ is

$$A'_\mu = \frac{\partial x^\nu}{\partial x'^\mu} A_\nu. \tag{3.7}$$

(This can be derived using Eqs. (3.1) and (3.6) or by the method given for Eq. (2.17). The latter procedure shows that the quotient theorem is valid with generic coordinates.) The transformation (3.5) of $g_{\mu\nu}$ is that of the product of two vectors with lowered indices, verifying that $g_{\mu\nu}$ is a tensor. The inverse of Eq. (3.6) is

$$A^\mu = g^{\mu\nu} A_\nu, \tag{3.8}$$

where $g^{\mu\nu}$ is the matrix inverse of $g_{\mu\nu}$,

$$g^{\mu\lambda} g_{\lambda\nu} = \delta^\mu_\nu. \tag{3.9}$$

In generic coordinates, the 4-volume element is defined as

$$dV^{(4)} = J dx^0 dx^1 dx^2 dx^3 \equiv J d^4x, \tag{3.10}$$

where J is the determinant of the transformation matrix $\partial x'^\nu / \partial x^\mu$ taking us from a Minkowski coordinate system x'^ν. This is a standard formula, used for any number of dimensions. It ensures that the integral $\int dV^{(4)}$ over a given region of spacetime is independent of the choice of coordinates. More generally, the integral $\int s dV^{(4)}$ of any scalar has that property.

Regarding Eq. (3.4) as the product of three matrices we have $J = \sqrt{-g}$, where g is the determinant of $g_{\mu\nu}$. We can therefore write

$$dV^{(4)} = \sqrt{-g}\, d^4x. \tag{3.11}$$

3.1.2 Covariant derivative

If f is a scalar (defined in a region of spacetime), $\partial_\mu f$ defines a 4-vector in any coordinate system because it has the correct transformation law Eq. (3.1). But if A^μ is a 4-vector, $\partial_\nu A^\mu$ does *not* define a tensor because it has the wrong transformation law. The **covariant derivative** $D_\nu A^\mu$ is defined as the tensor that reduces to $\partial_\nu A^\mu$ in an inertial frame. In a generic coordinate system x^μ,

$$D_\nu A^\mu \equiv \frac{\partial x'^\alpha}{\partial x^\nu} \frac{\partial x^\mu}{\partial x'^\beta} \frac{\partial A'^\beta}{\partial x'^\alpha} = \frac{\partial x^\mu}{\partial x'^\beta} \frac{\partial A'^\beta}{\partial x^\nu}, \tag{3.12}$$

where the prime denotes an arbitrarily chosen inertial frame. Using Eq. (3.1) one finds

$$\boxed{D_\nu A^\mu = \partial_\nu A^\mu + \Gamma^\mu_{\nu\alpha} A^\alpha}, \tag{3.13}$$

where the **metric connection** (Christoffel symbol) is defined by

$$\Gamma^{\mu}_{\nu\alpha} = \frac{\partial^2 x'^{\beta}}{\partial x^{\nu}\partial x^{\alpha}}\frac{\partial x^{\mu}}{\partial x'^{\beta}}. \tag{3.14}$$

It is symmetric in the lower indices, $\Gamma^{\mu}_{\nu\alpha} = \Gamma^{\mu}_{\alpha\nu}$.

The covariant derivative of a lower-index vector is defined in the same way, and repeating the derivation of Eq. (3.13) one finds

$$D_{\nu}A_{\mu} = \partial_{\nu}A_{\mu} - \Gamma^{\alpha}_{\mu\nu}A_{\alpha}. \tag{3.15}$$

Acting on a scalar, D_{μ} is equivalent to ∂_{μ}. From its definition, D_{μ} has the derivative property

$$D_{\mu}(XY) = (D_{\mu}X)Y + XD_{\mu}Y, \tag{3.16}$$

where X and Y are any two tensors, whose indices have been suppressed. This allows us to calculate the effect of D_{μ} on a tensor of any rank; for instance, the expressions

$$D_{\mu}(A_{\nu}B_{\lambda}) = \partial_{\mu}(A_{\nu}B_{\lambda}) - \Gamma^{\alpha}_{\mu\nu}A_{\alpha}B_{\lambda} - \Gamma^{\alpha}_{\mu\lambda}A_{\nu}B_{\alpha}, \tag{3.17}$$

$$D_{\mu}(A_{\nu}B^{\lambda}) = \partial_{\mu}(A_{\nu}B^{\lambda}) - \Gamma^{\alpha}_{\mu\nu}A_{\alpha}B^{\lambda} + \Gamma^{\lambda}_{\mu\alpha}A_{\nu}B^{\alpha}. \tag{3.18}$$

The covariant derivative of the metric tensor is

$$D_{\mu}g_{\nu\lambda} = \partial_{\mu}g_{\nu\lambda} - \Gamma^{\alpha}_{\mu\nu}g_{\alpha\lambda} - \Gamma^{\alpha}_{\mu\lambda}g_{\nu\alpha}. \tag{3.19}$$

However, in an inertial frame, $g_{\mu\nu}$ has the constant value $\eta_{\mu\nu}$, and $D_{\mu}g_{\nu\lambda}$ vanishes in that and hence all other frames, implying that

$$\partial_{\mu}g_{\nu\lambda} = \Gamma^{\alpha}_{\mu\nu}g_{\alpha\lambda} + \Gamma^{\alpha}_{\mu\lambda}g_{\nu\alpha}. \tag{3.20}$$

This relation makes no mention of the inertial coordinates x'_{μ}, and it can be inverted to give the connection in terms of the metric[2]

$$\boxed{\Gamma^{\gamma}_{\beta\mu} = \frac{1}{2}g^{\alpha\gamma}(\partial_{\mu}g_{\alpha\beta} + \partial_{\beta}g_{\alpha\mu} - \partial_{\alpha}g_{\beta\mu})}. \tag{3.21}$$

Note that the connection is *not* a tensor; its complicated transformation law, which may be derived from Eq. (3.21), is not the same as the one for a product of three vector components.

The d'Alembertian is defined as $\Box \equiv D^{\mu}D_{\mu}$. After some manipulation it can be written in the more convenient form

$$\boxed{\Box = \frac{1}{\sqrt{-g}}\partial_{\mu}\sqrt{-g}\; g^{\mu\nu}\partial_{\nu}}. \tag{3.22}$$

[2] To obtain this, cyclically permute $(\mu\nu\lambda)$ in Eq. (3.20) to give two more equations. Adding two of them and subtracting the third, remembering that $\Gamma^{\mu}_{\nu\alpha} = \Gamma^{\mu}_{\alpha\nu}$, gives $2g_{\alpha\nu}\Gamma^{\nu}_{\beta\mu} = \partial_{\mu}g_{\alpha\beta} + \partial_{\beta}g_{\alpha\mu} - \partial_{\alpha}g_{\beta\mu}$. Then, contracting with $g^{\gamma\alpha}$ gives the desired result.

Instead of tensors defined in a region of spacetime, we can consider tensors defined along a line through spacetime. Consider a line $x^\mu(\lambda)$, with λ any parameter and no restriction as to whether the interval along the line is spacelike, timelike or lightlike. Repeating the argument leading to Eq. (3.13), one finds that the object transforming as a vector, which reduces to $dA^\mu/d\lambda$ in an inertial frame is

$$\frac{DA^\mu}{D\lambda} \equiv \frac{dA^\mu}{d\lambda} + \Gamma^\mu_{\alpha\beta} A^\alpha \frac{dx^\beta}{d\lambda}. \tag{3.23}$$

This is called the directional derivative.

3.2 Special relativity with generic coordinates: laws of physics

Starting with covariant equations that are valid in an inertial frame, we obtain equations valid in a generic coordinate system by making the following substitutions:

$$\eta_{\mu\nu} \to g_{\mu\nu}, \qquad \partial_\mu \to D_\mu. \tag{3.24}$$

If ∂_μ acts on a scalar there is no need for D_μ, and the same is true for the 4-curl of a 4-vector:

$$D_\mu A_\nu - D_\nu A_\mu = \partial_\mu A_\nu - \partial_\nu A_\mu. \tag{3.25}$$

Let us see how this works for the equations of the previous chapter.

Transforming the geodesic condition $d^2x^\mu/d\lambda^2 = 0$, one finds

$$\frac{d^2x^\mu}{d\lambda^2} = -\Gamma^\mu_{\alpha\beta} \frac{dx^\alpha}{d\lambda}\frac{dx^\beta}{d\lambda}, \tag{3.26}$$

where λ is an affine parameter. For the 4-momentum of a particle this gives

$$\boxed{\frac{dp^\mu}{dx^0} = g^{\mu\nu}\left(\frac{1}{2}\partial_\nu g_{\alpha\beta} - \partial_\beta g_{\nu\alpha}\right)\frac{p^\alpha p^\beta}{p^0}}. \tag{3.27}$$

A quick derivation of this expression is to pretend that the worldline belongs to a fluid element. Then Eq. (2.26) for the 4-acceleration of a fluid element becomes in generic coordinates

$$a^\mu = u^\nu D_\nu u^\mu. \tag{3.28}$$

Equation (3.27) is then obtained by using Eqs. (3.13) and (3.21), and dividing by u^0.

The continuity equation (2.36) for a generic current becomes

$$D_\mu j^\mu = 0, \tag{3.29}$$

and in particular

$$D_\mu T^\mu_\nu = 0. \tag{3.30}$$

Equations (2.42) and (2.43) remain valid, defining the energy density, pressure and anisotropic stress.

At a given spacetime point, let us consider an inertial frame such that

$$g_{\mu\nu} = \eta_{\mu\nu} \quad \text{(at the spacetime point)}, \tag{3.31}$$

$$\partial_\lambda g_{\mu\nu} = 0 \quad \text{(at the spacetime point)}. \tag{3.32}$$

This is called a **locally inertial frame**. If the first condition is satisfied (but not necessarily the second) it is called a **locally orthonormal frame**.

In any locally inertial frame the geodesic equation (2.24) for a worldline becomes simply $du^\mu/dt = 0$. The rate of change of the geodesic's 4-velocity vanishes and so does that of its 3-velocity, just as it would in a truly inertial frame. Newton's second law, stating that the 3-acceleration of a free particle vanishes, is therefore valid in any locally inertial frame.

Remarkably, the entire discussion of fluids in Sections 2.4– 2.7 is also valid in any locally inertial frame, except for Eqs. (2.35), (2.47), (2.53), (2.54) and (2.56) which involve an integral over a finite region of space. (Even these are valid if the integral goes over an infinitesimal region.) The same applies to the field equations that we encounter in Part III. In all cases, the fact that the discussion is valid in a locally inertial frame derives from the fact that the equations can be written in covariant form, in which scalars are differentiated at most twice while vectors and tensors are generally differentiated at most once.[3]

3.3 Curved spacetime

The preceding discussion assumes the existence of Minkowski coordinates, defining an inertial frame. As we shall see, this amounts to ignoring gravity. To include gravity we need to consider a spacetime in which there are no Minkowski coordinates.

3.3.1 Metric tensor

Taking the spacetime interval ds^2 as a given concept, we postulate that in every coordinate system the line element is given by Eq. (3.3) with $g_{\mu\nu}$ some function of spacetime. Moreover, it is postulated that at any point in spacetime there are locally orthonormal coordinate systems satisfying Eq. (3.31). This requirement represents a mild restriction on the metric. At any point one can always make a coordinate transformation so that $g_{\mu\nu}$ is diagonal with each element ± 1. We are

[3] More differentiations are allowed if there is a cancellation as in Eq. (3.25). If scalars are differentiated at most once while vectors and tensors aren't differentiated at all, it is enough to use a locally orthonormal frame.

just postulating that three signs are positive and one (yielding a time direction) is negative.

All of this is analogous to the situation for a two-dimensional surface. If it is flat there exist Cartesian coordinates, such that the distance-squared between nearby points is $dl^2 = dx^2 + dy^2$. If the surface is curved, Cartesian coordinates generally don't exist, but there is still a metric such that $dl^2 = g_{ij} dx^i dx^j$ and at any point we can choose $g_{ij} = \delta_{ij}$. With this in mind, we call a spacetime flat if there are Minkowski coordinates and curved if there are not.

To handle curved spacetime, we can take over the machinery developed for flat spacetime with generic coordinates. Timelike, lightlike and spacelike lines are defined in the same way, and so is the lightcone at a given spacetime point. The transformation of a tensor in curved spacetime is exactly the same as in flat spacetime; vectors (first-rank tensors) transform like dx^μ and higher-rank tensors transform like a product of vectors. Contraction of indices reduces the rank of a tensor by two. Acting on a tensor defined in a region of spacetime, D_μ increases the rank by one. Acting on a tensor defined along a line with the directional derivative $u_\mu D^\mu$ leaves the rank unchanged.

3.3.2 Curvature tensor

The difference from flat spacetime comes when we consider repeated differentiation of a tensor. In flat spacetime differentiations commute, but that is not the case for curved spacetime. Taking a vector A^μ, we find from Eq. (3.18)

$$\boxed{D_\alpha(D_\beta A^\mu) - D_\beta(D_\alpha A^\mu) = R^\mu{}_{\nu\beta\alpha} A^\nu}, \tag{3.33}$$

where the **curvature tensor**, also known as the Riemann tensor, is defined by[4]

$$R^\mu{}_{\nu\beta\alpha} \equiv \partial_\alpha \Gamma^\mu_{\nu\beta} - \partial_\beta \Gamma^\mu_{\nu\alpha} + \Gamma^\mu_{\sigma\alpha}\Gamma^\sigma_{\nu\beta} - \Gamma^\mu_{\sigma\beta}\Gamma^\sigma_{\nu\alpha}. \tag{3.34}$$

According to the quotient theorem it is indeed a tensor. If spacetime is flat the curvature tensor vanishes, and it can be shown that the converse is true. The curvature tensor therefore characterizes the departure from flatness.

From the curvature tensor, we can form the symmetric Ricci tensor

$$R_{\mu\nu} \equiv R^\lambda{}_{\mu\lambda\nu}, \tag{3.35}$$

and the curvature (or Ricci) scalar

$$R \equiv R^\mu{}_\mu. \tag{3.36}$$

[4] We follow the conventions of Ref. [1].

3.3.3 Locally inertial frame

We postulated that, given any spacetime point, we can always find a locally orthonormal frame such that Eq. (3.31) holds. According to the **local flatness theorem** we may always go farther and find a locally inertial frame such that Eq. (3.32) holds. The theorem may be proved by construction as follows. Starting with a locally orthogonal frame, taken to be at the origin of coordinates, define new coordinates by

$$x'^{\mu} = x^{\mu} - \Gamma^{\mu}_{\alpha\beta} x^{\alpha} x^{\beta}, \tag{3.37}$$

where $\Gamma^{\mu}_{\alpha\beta}$ is the connection defined by Eq. (3.21). The new metric components $g'_{\mu\nu}$ are given by Eq. (3.4) and, differentiating this expression, we find that, at the origin,

$$\frac{\partial g'_{\mu\nu}}{\partial x'^{\lambda}} = \frac{\partial g_{\mu\nu}}{\partial x^{\lambda}} - \Gamma^{\alpha}_{\lambda\mu} g_{\alpha\nu} - \Gamma^{\alpha}_{\lambda\nu} g_{\mu\alpha}. \tag{3.38}$$

This vanishes according to Eq. (3.20), so that the new coordinates are indeed locally inertial.

Moving away from the chosen spacetime point, the derivatives $\partial g_{\mu\nu}/\partial x^{\alpha}$ will necessarily become non-zero in curved spacetime. The locally inertial coordinates remain useful as long as the derivatives are negligible. The size of the spacetime region within which this is true may be called the **curvature scale**, and is typically of order $|R|^{-1/2}$.

3.3.4 Local observers

It doesn't make sense to take linear combinations of vectors defined at different points in curved spacetime. The reason is that the transformation coefficients $\partial x'^{\mu}/\partial x^{\nu}$ are necessarily different at the two points, there being no inertial frame. As a result, the coefficients of the linear combination would change under a transformation and could have no physical meaning.

At a given point in spacetime, one should therefore think of an observer as measuring quantities only in the immediate vicinity of that point. For that purpose, the observer is equipped with a locally orthonormal or locally inertial frame, which is the rest frame of the observer. As an example of the use of such an observer one may reconsider Figure 2.2. With the restricted definition of 'observer', it defines simultaneity only for nearby events, providing a definition of the 'orthogonality' of a spacelike to a timelike line. Of course the same construction defines simultaneity even for distant events in curved spacetime, but that definition is not as useful as the local one.

3.4 Curved space and curved surfaces

The mathematics that we described in Section 3.3 goes through if the coordinates (x^0, x^1, x^2, x^3) are replaced by any set $(x^1, \ldots, x^N)$. Postulating only that $g_{\mu\nu}$ is symmetric, it can always be diagonalized at a given spacetime point and then the coordinates can be chosen so that each element is ± 1. If the signs are all the same, we are dealing with an N-dimensional **Riemannian manifold**. If instead both signs occur, as they do for spacetime, we are dealing with an N-dimensional **pseudo-Riemannian manifold**.[5] If it is possible to choose coordinates that make $g_{\mu\nu}$ everywhere diagonal with elements ± 1, the manifold is Euclidean (if all elements have the same sign) or pseudo-Euclidean (if both signs occur. In that case one may call the manifold flat, otherwise one may call it curved.

Let us apply these ideas to a slice of spacetime, which is a 3-dimensional Riemannian manifold. On a given slice of spacetime the interval ds^2 becomes the length-squared $d\ell^2$. It is given by

$$d\ell^2 \equiv g_{ij} dx^i dk^j , \tag{3.39}$$

where x^i are the spatial coordinates.

If we change the spatial coordinates, the transformation of the spatial metric is

$$g_{nm} = \frac{\partial x'^i}{\partial x^n} \frac{\partial x'^j}{\partial x^m} g'_{ij} . \tag{3.40}$$

In Cartesian coordinates the volume element is d^3x' and in generic coordinates it is $\sqrt{g}\, d^3x$, where now g is the determinant of g_{ij}.

Repeating the spacetime analysis, we arrive at the curvature tensor $R^{(3)}_{ijnm}$, the Ricci tensor $R^{(3)}_{ij}$, and the curvature scalar $R^{(3)}$ of the three-dimensional slice. In a generic coordinate system the spatial curvature is non-zero even if spacetime is flat.

Going down another dimension, one can repeat the analysis for a two-dimensional surface. Then the curvature tensor $R^{(2)}_{ijnm}$ specifies what is called the intrinsic curvature of the surface. A generic two-dimensional surface will have intrinsic curvature even if it is embedded in flat space. Indeed, the only intrinsically flat surfaces embedded in flat space are the plane, the cylinder and the cone.

Irrespectively of how it is embedded, a two-dimensional sphere may be characterized as a two-dimensional surface whose curvature tensor is homogeneous, $D_i R^{(2)}_{jkmn} = 0$ with $R^{(2)} > 0$. Embedding the sphere in flat space, one finds $R^{(2)} = 1/R^2$ where R is the radius of the sphere. The limit $R^{(2)} \to 0$ gives

[5] More generally, these terms apply if the manifold requires two or more sets of N coordinates to cover it, which are related by a coordinate transformation in each region where two or more members of the set overlap.

a flat surface and $R^{(2)} < 0$ corresponds to another surface. These are the only two-dimensional surfaces with homogeneous curvature.

3.5 The equivalence principle

When formulating the laws of physics with special relativity, one completely ignores gravity. To arrive at a simple form for the laws one uses Minkowski coordinates, corresponding to an inertial frame. But as we saw in Section 3.2, it turns out in the end to be enough to use a *locally* inertial frame, since fundamental laws of physics seem to take the same form in such a frame as in a truly inertial frame.

To fix the laws of physics in curved spacetime, one can adopt the **equivalence principle**. According to the equivalence principle, the laws of physics in a locally inertial frame have the same form that they would have in the absence of gravity (written in a truly inertial frame). In other words, gravity is abolished in a locally inertial frame.

To see how the equivalence principle might be violated, consider a law of physics which in flat spacetime takes the form $T = 0$, where T is some tensor with the indices suppressed, and suppose that in curved spacetime it takes the form '$(1 + aR)T = 0$' where a is a constant and R is the spacetime curvature scalar. Then the equivalence principle is satisfied if $a = 0$, but not otherwise.

Amongst the laws of physics that are supposed to apply in the absence of gravity is Newton's second law, stating that the worldline of any free particle is a geodesic. According to the equivalence principle, any particle which would be free in the absence of gravity (i.e. any free-falling particle) has constant velocity in a locally inertial frame, which means that all free-falling particles have the same acceleration in a generic frame.

A locally inertial frame is defined with reference to a point in spacetime, but as an approximation we can use it in a region around that point, which is small enough that $g_{\mu\nu}$ and $\partial_\lambda g_{\mu\nu}$ have negligible variation. In this way we can make sense of the integrals (2.35), (2.47), (2.53), (2.54) and (2.56) over a finite region of space.

3.6 Einstein gravity

3.6.1 Field equation

To define a theory of gravity, we have to specify how the sources of gravity determine the spacetime metric. According to Einstein's theory, the source of gravity is the energy–momentum tensor, which determines the metric through the field equation

$$\boxed{R_{\mu\nu} - \frac{1}{2} g_{\mu\nu} R = -8\pi G T_{\mu\nu}}, \tag{3.41}$$

where G is Newton's gravitational constant, which we are using here instead of the equivalent $M_{\rm Pl}^2 = (8\pi G)^{-1}$. Sometimes it is useful to write the equation in the following equivalent form:

$$R_{\mu\nu} = -8\pi G\left(T_{\mu\nu} - \frac{1}{2}g_{\mu\nu}T\right), \tag{3.42}$$

where $T \equiv T^\mu_\mu$.

The Einstein field equation implies the continuity equation (3.30), because there is an identity

$$D_\mu\left(R^{\mu\nu} - \frac{1}{2}g^{\mu\nu}R\right) \equiv 0. \tag{3.43}$$

To prove this identity, we focus on one point in spacetime, and work in a locally inertial frame, so that we need only prove

$$\partial_\mu\left(R^{\mu\nu} - \frac{1}{2}\eta^{\mu\nu}R\right) = 0. \tag{3.44}$$

(The more general expression is of the form 'tensor $= 0$', and reduces to Eq. (3.44) in a locally inertial frame, and so it is enough to prove the latter.) In the locally inertial frame, $D_\mu = \partial_\mu$ and $\Gamma^\mu_{\alpha\beta} = 0$. Differentiating Eq. (3.34) gives

$$\partial_\lambda R^\mu{}_{\nu\beta\alpha} = \partial_\lambda\partial_\alpha\Gamma^\mu_{\nu\beta} - \partial_\lambda\partial_\beta\Gamma^\mu_{\nu\alpha}, \tag{3.45}$$

and therefore

$$\partial_\lambda R^\mu{}_{\nu\alpha\beta} + \partial_\beta R^\mu{}_{\nu\lambda\alpha} + \partial_\alpha R^\mu{}_{\nu\beta\lambda} = 0. \tag{3.46}$$

This is called the Bianchi identity. Contracting μ with α, and λ with ν, gives

$$2\partial_\nu R^\nu{}_\mu - \partial_\mu R = 0. \tag{3.47}$$

Contracting with $\eta^{\alpha\mu}$ now gives the desired identity.

According to the equivalence principle, the worldline of a free-falling particle is a geodesic. By 'particle' we mean in this context an object small enough compared with the spacetime curvature scale, that it can be viewed in a locally inertial frame. The geodesic condition corresponds to momentum conservation. That, in turn, follows from the Einstein field equation (3.41), insofar as the interior of the 'particle' can be regarded as a fluid satisfying Einstein's field equation. As has often been remarked, the above situation may be summarized by saying that the field equation gives both the effect of matter on the spacetime curvature *and* the effect of spacetime curvature on matter. In other words, it gives the gravitational field due to matter *and* the motion of matter in a gravitational field. In Newtonian gravity, the Poisson equation that does the former job says nothing about the motion of matter.

Einstein gravity is known to be quite accurate on distance scales ranging from a millimetre or so to the size of the Solar System. It seems also to work well on cosmological scales. Going back in time, when successively smaller distances and higher energy densities are relevant, it seems to work back to at least the epoch of primordial nucleosynthesis. At smaller distances and/or earlier times, Einstein gravity might well receive corrections that eventually dominate, so that a completely different theory ultimately applies. Many kinds of modifications to the theory have been suggested, most of them satisfying general coordinate invariance and the equivalence principle.

3.6.2 Newtonian gravity

General relativity gives Newtonian gravity as an approximation. For a fluid, the Newtonian limit corresponds to the line element [2]

$$ds^2 = -(1+2\phi)dt^2 + (1-2\phi)\delta_{ij}dr^i dr^j, \tag{3.48}$$

$$|\phi| \ll 1, \qquad |\partial_0\phi| \ll |\nabla\phi|. \tag{3.49}$$

Putting this into the Einstein equation one finds that $T_{00} \simeq T^{00} \equiv \rho$ dominates the energy–momentum tensor. The 0–0 component of the Einstein equation gives the Newtonian Poisson equation

$$\boxed{\nabla^2\phi = 4\pi G\rho}. \tag{3.50}$$

Evaluating the geodesic equation for a worldline with velocity $v \ll 1$ we find the Newtonian acceleration equation for a free-falling particle:

$$\boxed{\frac{d\mathbf{v}}{dt} = -\boldsymbol{\nabla}\phi}. \tag{3.51}$$

These identify ρ as the mass density and ϕ as the Newtonian gravitational potential.

In using these equations one pretends that spacetime is flat. Since P is much less than ρ, the mass density satisfies Eq. (2.50) in a local rest frame. In any frame where $v \ll 1$ it becomes

$$\boxed{\frac{d\rho(\mathbf{x},t)}{dt} = -\left(\boldsymbol{\nabla}\cdot\mathbf{v}\right)\rho(\mathbf{x},t), \qquad \frac{d}{dt} \equiv \frac{\partial}{\partial t} + \mathbf{v}\cdot\boldsymbol{\nabla}}, \tag{3.52}$$

where $\mathbf{v}(\mathbf{x},t)$ is the fluid velocity field. Keeping the small but non-zero stress, assumed for simplicity to be isotropic, the acceleration of a fluid element is

$$\boxed{\frac{d\mathbf{v}}{dt} = -\frac{1}{\rho}\boldsymbol{\nabla}P - \boldsymbol{\nabla}\phi}. \tag{3.53}$$

This is the Newtonian Euler equation.

The definition of Newtonian gravity is completed by choosing the solution of Eq. (3.50) that corresponds to

$$\phi(\mathbf{x},t) = -G\int d^3y\frac{\rho(\mathbf{y},t)}{|\mathbf{y}-\mathbf{x}|} \tag{3.54}$$

$$-\nabla\phi(\mathbf{x},t) = G\int d^3y\rho(\mathbf{y},t)\frac{\mathbf{y}-\mathbf{x}}{|\mathbf{y}-\mathbf{x}|^2}. \tag{3.55}$$

This says that the acceleration of a free-falling particle is given by Newton's inverse square law.

Although it is only g_{00} that affects the motion of something moving with speed $v \ll 1$, the other components are of interest because they affect the motion of an object with speed $v \sim 1$. In particular they affect the motion of photons.

3.6.3 Gravitational waves

Extending the Newtonian analysis, we may consider [1, 2, 3] a metric perturbation $h_{\mu\nu}$ away from flat spacetime, defined by $g_{\mu\nu} = \eta_{\mu\nu} + h_{\mu\nu}$. It is required to satisfy $|h_{\mu\nu}| \ll 1$ and its spacetime derivatives are also required to be small, the exact requirement depending on the context. In this way one arrives at a theory of weak gravity which is the standard tool for comparing general relativity with observation.[6] The perturbation $h_{\mu\nu}$ is treated to first order, and is supposed to live in a fictitious flat spacetime. The contribution of $h_{\mu\nu}$ to the left-hand side of the Einstein equation is taken over to the right-hand side, and regarded as a contribution to the energy–momentum tensor.

One can choose coordinates such that $\partial_\mu h^{\mu\nu} = 0$. This is called a Lorenz gauge, by analogy with electromagnetism that we come to later. We shall need only the description of gravitational waves. For them one can choose a Lorenz gauge with $h_{00} = h_{0i} = 0$, and spatial components given by

$$h_{ij}(\mathbf{x},t) = \frac{1}{(2\pi)^3}\int\left[\left(h^+_{ij}(\mathbf{k}) + h^\times_{ij}(\mathbf{k})\right)e^{i(\mathbf{k}\cdot\mathbf{x}-kt)} + \text{c.c.}\right]d^3k, \tag{3.56}$$

$$h^+_{ij}(\mathbf{k}) \equiv \epsilon^+_{ij}h_+(\mathbf{k}), \qquad h^\times_{ij}(\mathbf{k}) \equiv \epsilon^\times_{ij}h_\times(\mathbf{k}). \tag{3.57}$$

Choosing the z-axis along $\mathbf{k}$, the polarization tensors have non-zero components

$$\boxed{e^+_{xx} = -e^+_{yy} = 1}, \qquad \boxed{e^\times_{xy} = e^\times_{yx} = 1}. \tag{3.58}$$

This defines a unique Lorenz gauge, up to a rotation about the $\mathbf{k}$-axis.

[6] In a similar way one can define small perturbations away from any particular solution of the Einstein field equation. A central theme of this book will be the study of perturbations away from the Robertson–Walker metric described in the next section.

3.7 The Robertson–Walker metric

The central premise of modern cosmology is that, at least on large scales, the Universe is almost homogeneous and isotropic.[7] This is borne out by a variety of observations, most spectacularly the nearly identical temperature of cosmic microwave background radiation coming from different parts of the sky. On the other hand, it is apparent that nearby regions of the Universe are at present highly inhomogeneous, with material clumped into stars, galaxies and galaxy clusters. It is believed that these structures have formed over time through gravitational attraction, from a distribution that was more homogeneous in the past.

It is convenient then to break up the dynamics of the Universe into two parts. The large-scale behaviour of the Universe can be described by assuming a homogeneous and isotropic background. On this background, we can superimpose the short-scale irregularities. For much of the evolution of the observable Universe, these irregularities can be considered to be small perturbations on the evolution of the background (unperturbed) Universe.

The metric of the unperturbed Universe is called the Robertson–Walker metric, after authors who emphasized that its form is independent of the theory of gravity. Models based upon it are also called Friedmann–Lemaître–Robertson–Walker (FLRW) models, Friedmann models, and FRW models.

3.7.1 Scale factor

To arrive at the form of the Robertson–Walker metric, we first choose the time coordinate so that spacetime slices of fixed t are homogeneous and isotropic. In other words, physical conditions on each slice are the same at every position, and in every direction. We choose the threading to be orthogonal to the slicing, corresponding to $g_{0i} = 0$. Isotropy requires that an observer moving with the threading measures zero velocity for the cosmic fluid, or in other words zero momentum density. The threading is therefore comoving (moving with the fluid flow).

Homogeneity demands that the proper time interval between slices is independent of position, which means that we can choose t as proper time corresponding to $g_{00} = -1$. Homogeneity and isotropy require that the distance between nearby threads is proportional to a universal **scale factor** $a(t)$. Putting all this together, the Robertson–Walker line element takes the form

$$ds^2 = -dt^2 + a^2(t)\tilde{g}_{ij}(x^1, x^2, x^3)dx^i dx^j. \qquad (3.59)$$

The Universe is observed to be expanding, corresponding to $a(t)$ increasing with

[7] The homogeneity and isotropy is verified by observation within a region around us (the observable Universe) and is usually supposed to hold in some bigger region which we call simply the Universe. We will use 'universe' without the capitalization to denote things different from the Universe, notably the separate universes of Section 5.3.2.

time. We will see shortly how homogeneity fixes the form of $\tilde{g}_{ij}$, but for the moment focus on the scale factor.

We normalize the scale factor to have present value $a_0 = 1$ This makes the coordinate distance between comoving worldlines equal to the present physical distance. We will usually refer to the coordinate distance as the **comoving distance**.

We need to be able to discuss the motion of a free-falling particle moving through the cosmic fluid. (The fluid itself is usually modelled as a gas, in which case the following discussion applies to each particle of the gas in between collisions.) The worldline of the particle will be a geodesic. Let us consider its momentum $p(t)$, measured at each instant by the comoving observer at its location. In an infinitesimal time dt it travels between nearby observers separated by a distance vdt, where v is its velocity measured by either of them. The relative velocity du of these observers is

$$du = \frac{\dot{a}}{a} v dt = \frac{da}{a} v. \tag{3.60}$$

The change in momentum is given by the Lorentz boost with velocity du, applied to the 4-momentum $(E, \mathbf{p})$; it is $dp = -E du$, which is equal to $-p du/v$. Using Eq. (3.60) we therefore find $dp/p = -da/a$. Finally, integrating this relation along the path of the particle we find

$$p(t) \propto \frac{1}{a(t)}. \tag{3.61}$$

The momentum, as measured by a sequence of observers, falls like $1/a$.

Applying this result to a photon, we learn that the wavelength λ of the corresponding electromagnetic radiation is proportional to $a(t)$. In this case, $E = p$ which means that the frequency is proportional to $1/a(t)$. This is the famous **redshift**, seen in the spectral lines of sources that are moving away from us with the cosmic expansion.

3.7.2 Geometry of space

We have yet to fix the form of $\tilde{g}_{ij}$, which determines the geometry of space. In any coordinate system, homogeneity requires that a given coordinate position (choice of all three spatial coordinates) can correspond to any position, and isotropy requires that a given coordinate line (choice of two of them) can correspond to any direction. It can be shown that this requires the existence of spatial coordinates (x, θ, ϕ) that bring the line element into the following form:

$$\boxed{ds^2 = -dt^2 + a^2(t)\left[\frac{dx^2}{1 - Kx^2} + x^2\left(d\theta^2 + \sin^2\theta \, d\phi^2\right)\right]}. \tag{3.62}$$

The proof of this statement is given for instance in Ref. [1]. One easily checks that it satisfies a couple of obvious requirements. First, the round bracket is the familiar line element on a sphere, which guarantees isotropy around the origin $x = 0$. Second, a straightforward calculation shows that the curvature scalar is $R^{(3)} = 2K/a^2(t)$, which is independent of position as is required by homogeneity.

The case $K = 0$ corresponds to Euclidean geometry and we can then choose comoving Cartesian coordinates $(x, y, z) \equiv \mathbf{x}$ to arrive at

$$\boxed{ds^2 = -dt^2 + a^2(t)\left(dx^2 + dy^2 + dz^2\right)}\,. \tag{3.63}$$

It is often convenient to define **conformal time** η, such that $dt = a d\eta$ and

$$\boxed{ds^2 = a^2(\eta)\left(-d\eta^2 + dx^2 + dy^2 + dz^2\right)}\,. \tag{3.64}$$

For a particle with velocity $c = 1$, the comoving distance travelled during a conformal time interval $d\eta$ is simply $d\eta$. (Recall that by comoving distance, we mean the coordinate distance.)

To understand the geometry for $K \neq 0$ we re-define the radial coordinate x so that it is the comoving distance from the origin. For K respectively positive and negative this gives

$$ds^2 = -dt^2 + a^2(t)\left[dx^2 + \frac{1}{K}\sin^2(K^{1/2}x)(d\theta^2 + \sin^2\theta\, d\phi^2)\right], \tag{3.65}$$

$$ds^2 = -dt^2 + a^2(t)\left[dx^2 + \frac{1}{|K|}\sinh^2(|K|^{1/2}x)(d\theta^2 + \sin^2\theta\, d\phi^2)\right]. \tag{3.66}$$

For a circle with comoving radius x, the distance around the circle subtended by angle θ is $x_{\rm d}\theta$, where

$$x_{\rm d} = K^{-1/2}\sin(K^{1/2}x) \qquad (K > 0), \tag{3.67}$$

$$x_{\rm d} = |K|^{-1/2}\sinh(|K|^{1/2}x) \quad (K < 0), \tag{3.68}$$

and $x_{\rm d}$ is called the **angular-diameter distance**. Space is flat (Euclidean) if $K = 0$, finite or closed if $K > 0$, and infinite or open if $K < 0$. It is also infinite if $K = 0$, but, by custom, one reserves the term open for the case of negative K. Euclidean geometry is recovered for $x \ll |K|^{-1/2}$.

Whether it is closed, flat or open, space might be periodic, and then its physical volume is finite even for $K \leq 0$. The period, if any, must be significantly bigger than the size of the observable Universe, or its effect on the cosmic microwave background would have been detected.

The value $K = 0$ corresponds to flat space. As we see later, this value is strongly favoured by the inflationary cosmology, and is in excellent agreement with observation.

A straightforward calculation shows that the value $K = -\dot{a}^2$ would make spacetime flat, corresponding to what is called the Milne universe. The spacetime metric could then be written in terms of Minkowski coordinates, at the expense of making the spatial slicing inhomogeneous. The Milne universe is strongly ruled out by observation.

3.8 Hubble parameter and horizons

At any epoch, the rate of expansion of the Universe is given by the **Hubble parameter** $H \equiv \dot{a}/a$. The Hubble time H^{-1} and the Hubble distance or length cH^{-1} (equal to H^{-1} with our chosen unit $c = 1$) are of crucial importance. It typically takes of order one Hubble time for the Universe to expand appreciably, and while that is happening light travels of order one Hubble length.

The number N of e-folds of expansion between epochs with scale factor a_1 and a_2 is

$$\boxed{N \equiv \ln \frac{a_2}{a_1}}. \tag{3.69}$$

It is also called the number of Hubble times, since $N = \int_{t_1}^{t_2} H dt$.

The relative velocity of a pair of nearby comoving observers, separated by distance $r \ll H^{-1}$, is $v = Hr \ll 1$. As we discussed in Section 3.7.1, v is equal (in the limit $v \to 0$) to the redshift $\delta\lambda/\lambda$ of light going between the two observers. Taking one of the observers as ourselves at the present epoch, we have

$$\frac{\delta\lambda}{\lambda} = H_0 r. \tag{3.70}$$

where H_0 is the present value of H, called the **Hubble constant**. This is Hubble's law.

We shall call the Hubble distance $H^{-1}(t)$ the **horizon**, because it provides an estimate of the distance that light can travel during a Hubble time.[8] In other words, it is an estimate of how far light can travel while the Universe expands appreciably. A given observer will see regions far beyond the horizon only after many Hubble times, if at all.

At a given time, a region with size $Ra(t)$ (comoving size R, defined say as the radius of a spherical region) is said to be outside or inside the horizon according to whether the ratio

$$\frac{Ra}{H^{-1}} \equiv RaH \equiv R\dot{a} \tag{3.71}$$

is bigger or less than unity. We will take R to be constant so that the region is

[8] Here 'light' indicates an idealized carrier of information, travelling at speed $c = 1$ without collisions.

comoving. If gravity attracts different regions of the cosmic fluid to each other, $\ddot{a}$ will be negative corresponding to a decelerating rate of expansion. Then the ratio Eq. (3.71) decreases with time and comoving regions are entering the horizon. If instead gravity is repulsive, $\ddot{a}$ will be positive corresponding to an accelerating rate of expansion. Then the ratio Eq. (3.71) increases with time and comoving regions are leaving the horizon. According to general relativity, gravity can be either repulsive or attractive depending on the nature of the cosmic fluid.

Consider a comoving region well inside the horizon, and a time interval $\ll H^{-1}$. The scale factor hardly varies, and setting it equal to 1 the comoving coordinates $\mathbf{x}$ become Cartesian coordinates in practically flat spacetime. According to the equivalence principle, the laws of special relativity then apply.

In addition to the Hubble distance, which we are calling simply the horizon, it can be useful to consider two other distances. The **particle horizon** is the distance that light could have travelled since the beginning of the Universe at $a = 0$. Specified as a comoving distance it is

$$\boxed{x_{\rm ph}(t) = \int_0^t \frac{dt}{a} = \eta(t) - \eta(0)}\,. \tag{3.72}$$

At any epoch, events separated by more than twice the particle horizon cannot have a common cause; they are said to be out of causal contact.

The **event horizon** is the distance that light will be able to travel in the future:

$$\boxed{x_{\rm eh}(t) = \int_t^\infty \frac{dt}{a} = \eta(\infty) - \eta(t)}\,. \tag{3.73}$$

An event occurring at a given epoch cannot be the cause of a future event which is outside the event horizon.

On the usual assumption that $\dot{a}(t)$ is positive, Eqs. (3.72) and (3.73) can be written

$$x_{\rm ph}(a) = \int_0^a \frac{da}{a^2 H}, \qquad x_{\rm eh}(a) = \int_a^\infty \frac{da}{a^2 H}. \tag{3.74}$$

Taken literally, these definitions make no sense because they invoke zero and infinite scale factors. In practice one considers what might be called effective horizons:

$$x_{\rm ph}(a_1, a) = \int_{a_1}^a \frac{da}{a^2 H}, \qquad x_{\rm eh}(a, a_2) = \int_a^{a_2} \frac{da}{a^2 H}. \tag{3.75}$$

Taking t as the reference time, the first quantity is the maximum comoving distance that light can have travelled since the scale factor was a_1, and the second is the maximum distance that light will be able to travel in the future until the scale factor is a_2.

3.9 Inflation and the Big Bang

At this point we describe briefly the behaviour of the scale factor, according to the cosmology of this book. There is an era of repulsive gravity, followed by an era of attractive gravity lasting almost to the present epoch. (Starting around the present epoch, an era of accelerated expansion is getting underway, which we forget about for the moment.)

The era of repulsive gravity, corresponding to accelerated expansion, is called **inflation**. Setting a_2 equal to the scale factor at the end of inflation, the effective event horizon $ax_{\rm eh}(a, a_2)$ is the maximum distance that light can travel during inflation, starting at a given epoch. It will typically be of order H^{-1} which we are calling simply the horizon. The particle horizon during inflation is not a useful concept.

After inflation comes the era of attractive gravity. In the simplest scenario, it begins with the creation of particles which are in thermal equilibrium at very high temperature. The gas cools and eventually the temperature becomes low enough for the formation of the first nuclei. Accepting this scenario, the era of attractive gravity begins with what is called the **Hot Big Bang**. In more general scenarios, the content of the cosmic fluid is more complicated and the onset of the Hot Big Bang is delayed. To allow for such scenarios, we will call the entire era of attractive gravity simply the Big Bang.[9]

Setting a_1 equal to its value at the beginning of the Big Bang, the effective particle horizon $ax_{\rm ph}(a_1, a)$ is the maximum distance that light can have travelled during the Big Bang up to a given epoch. It will typically be of order H^{-1} which we are calling simply the horizon. The event horizon during the Big Bang is not a useful concept.

Setting η to zero at the beginning of the Big Bang, $\eta = x_{\rm ph}(a_1, a)$. Its present value η_0 defines the radius of the **observable Universe**, and nothing from beyond the observable Universe (emitted during the Big Bang) has yet had time to reach us. At any epoch, the physical time since the beginning of the Big Bang is roughly of order H^{-1}. In other words, it is the age of the Universe born at the beginning of the Big Bang.

The crucial assumption of the inflationary cosmology is that *the observable Universe is far inside the horizon at the beginning of inflation, and far outside the horizon at the end of inflation.* As we shall see, this allows inflation to set the initial condition for the subsequent Big Bang.

[9] In an even more general scenario, the era of attractive gravity can be interrupted by a brief era of inflation, which we will still count as part of the Big Bang.

3.10 Continuity equation and Friedmann equation

3.10.1 Continuity equation

As the Universe is isotropic its energy–momentum tensor is specified by the energy density ρ and pressure P, measured at each spacetime position by a comoving observer. Independently of the theory of gravity, the rate of change of ρ is given by the continuity equation (2.51). In terms of the Hubble parameter. This reads

$$\boxed{\dot{\rho} = -3H\,(\rho + P)}\,. \tag{3.76}$$

This also can be written

$$\boxed{a\frac{d\rho}{da} = -3(\rho + P)}\,. \tag{3.77}$$

The continuity equation holds separately for any component of the cosmic fluid which is uncoupled from the rest, neither giving nor receiving energy.

If the Universe is gaseous, the pressure of each constituent is related to the mean-square random velocity of the particles. If a constituent has mean-square random velocity $v^2 \ll 1$, it has $P \ll \rho$ (remember that we are setting $c=1$). It is then called matter or, sometimes, dust. If instead the random velocity is close to 1, then $\rho = P/3$ and it is called radiation. For the contribution of matter, the continuity equation gives $\rho_{\mathrm{m}} \propto a^{-3}$, which expresses mass conservation. For the contribution of radiation, it gives $\rho_{\mathrm{r}} \propto a^{-4}$, the extra factor a^{-1} coming from the redshift of particle energy between collisions.

A given species of radiation will become matter if the redshift drives the random velocity below 1. Conversely, matter will become radiation if it decays into relativistic particles. If neither of these things happens, $\rho_{\mathrm{m}}/\rho_{\mathrm{r}} \propto a$. Given enough time, the matter will dominate over the radiation if it doesn't decay.

3.10.2 Friedmann equation

We have not yet invoked a theory of gravity. Now we assume the field equation of Einstein gravity. Because of the high degree of symmetry, it provides just one extra piece of information, which can be taken as the time–time component of the field equation. It reads

$$\boxed{H^2 = \frac{\rho}{3M_{\mathrm{Pl}}^2} - \frac{K}{a^2}}\,, \tag{3.78}$$

and is called the **Friedmann equation**.

If during some era ρ has a known dependence on the scale factor, $dt/da = (aH)^{-1}$ can be in integrated to give $t(a)$ and hence the dependence of the scale factor on t. We shall refer to this procedure as 'integrating the Friedman equation'.

Integrating the Friedmann equation assuming radiation or matter domination (with $K = 0$) gives

$$a \propto t^{1/2}, \qquad \eta = \frac{1}{aH} \propto a \qquad \text{(radiation domination)}, \tag{3.79}$$

$$a \propto t^{2/3}, \qquad \eta = \frac{2}{aH} \propto a^{1/2} \qquad \text{(matter domination)}. \tag{3.80}$$

In writing these expressions, we are pretending that there is radiation or matter domination right back to $a = 0$. At that epoch, we set $\eta = 0$ which makes $\eta(t)$ the particle horizon.

At the present epoch the energy density of matter is about 10^4 times the energy density of the radiation. One of the big surprises of recent years is that the matter density is itself outweighed by an even bigger contribution to the energy density, known as **dark energy**. Observation suggests that the dark energy density is homogeneous and time independent. Taking that to be the case we are dealing with what is called a **cosmological constant**, whose contribution to the energy–momentum tensor is of the form

$$(T_\Lambda)_{\mu\nu} = M_{\rm Pl}^2 \Lambda g_{\mu\nu}. \tag{3.81}$$

As the subscript suggests, T_Λ may be regarded as a property of the vacuum. This is because its components in a locally orthonormal frame are $(T_\Lambda)_{\mu\nu} = M_{\rm Pl}^2 \Lambda \eta_{\mu\nu}$, which is Lorentz invariant.

The energy density and pressure of the dark energy are given by

$$\rho_\Lambda = -P_\Lambda = M_{\rm Pl}^2 \Lambda. \tag{3.82}$$

Observation indicates that the cosmological constant accounts at the present epoch for about 70% of the energy density, leaving only about 30% for the matter density.

Differentiating the Friedmann equation with the aid of the continuity equation gives the acceleration of the expansion,

$$\boxed{\frac{\ddot{a}}{a} = -\frac{\rho + 3P}{6M_{\rm Pl}^2}}. \tag{3.83}$$

Written in terms of H this expression becomes

$$\dot{H} + H^2 = -\frac{\rho + 3P}{6M_{\rm Pl}^2}. \tag{3.84}$$

Radiation and matter give negative contributions to $\ddot{a}$, corresponding to attractive gravity which becomes Newtonian gravity in the matter-dominated case. The cosmological constant gives a positive contribution to $\ddot{a}$, corresponding to repulsive gravity.

3.10.3 Density parameter

It is useful to define

$$\boxed{\Omega - 1 \equiv \frac{K}{a^2H^2} \equiv \frac{K}{\dot{a}^2}}. \tag{3.85}$$

With the assumption of Einstein gravity Ω measures the energy density in units of H^2 and is therefore called the **density parameter**. Irrespectively of the theory of gravity, $\Omega - 1$ is (half of) the spatial curvature scalar in units of H^2.

During any era of repulsive gravity ($\ddot{a} > 0$) then Ω is driven towards 1, and during any era of attractive gravity ($\ddot{a} < 0$) it is driven away from 1. The value $\Omega = 1$ is time independent.

With Einstein gravity, the Friedmann equation gives

$$\boxed{\Omega(t) = \frac{\rho(t)}{\rho_{\rm crit}(t)}}, \tag{3.86}$$

where the **critical density** $\rho_{\rm crit}$ is defined as $3M_{\rm Pl}^2H^2(t)$, and we take ρ to include the dark energy density. The present value Ω_0 is constrained by observations relating to the geometry of the Universe, and with the assumption of Einstein gravity it is constrained also by observations of the energy density. It is close to the time-independent value $\Omega_0 = 1$, and from now on we set $\Omega = 1$ unless otherwise stated. Then the Friedman equation becomes

$$\boxed{3M_{\rm Pl}^2H^2 = \rho}, \tag{3.87}$$

and the Einstein field equation gives the curvature scalar as

$$R = M_{\rm Pl}^{-2}(3P - \rho). \tag{3.88}$$

Exercises

3.1 Verify Eq. (3.4) which gives the metric components in generic coordinates in terms of those in Minkowski coordinates.

3.2 Using both of the methods suggested in the text, verify Eq. (3.7) which gives the Lorentz transformation of a vector with lowered indices.

3.3 Using the chain rule and $\partial x^\mu/\partial x^\nu = \delta^\mu_\nu$, verify Eq. (3.13) which gives the covariant derivative of a vector.

3.4 Verify Eq. (3.18) which gives the covariant derivative of a tensor.

3.5 Verify Eq. (3.23) which gives the directional derivative of a vector field.

3.6 Verify Eq. (3.25) which is the 4-dimensional curl.

3.7 Show that in a local rest frame (comoving frame), Eq. (3.28) for the 4-acceleration reduces to $a^\mu = du^\mu/d\tau$.

3.8 Verify that the curvature tensor in Eq. (3.33) is given by Eq. (3.34).

3.9 Use Eq. (3.40) to calculate the spatial metric components in spherical polar coordinates.

3.10 Using Eq. (3.48), verify that the Newtonian Poisson equation (3.50) corresponds to the 00 component of the Einstein field equation, and that the Newtonian acceleration equation (3.51) corresponds to the geodesic equation (3.27).

3.11 Can an open universe evolve into a closed one?

3.12 Calculate the components of the metric connection for the flat-space Robertson–Walker metric (3.64). Hence, assuming Einstein's field equation, verify Eq. (3.88) for the spacetime curvature scalar.

References

[1] S. Weinberg. *Gravitation and Cosmology, Principles and Applications for General Relativity* (New York: John Wiley and Sons, 1972).

[2] P. J. E. Peebles. *Principles of Physical Cosmology* (Princeton: Princeton University Press, 1993).

[3] M. P. Hobson, G. P. Efstathiou and A. N. Lasenby. *General Relativity: An Introduction for Physicists* (Cambridge: Cambridge University Press, 2006).

Part II

The Universe after the first second

4
The unperturbed Universe

In Part II of this book, we consider the history of the Universe after it is roughly one second old. It is quite well understood, and is known to begin with an era of rapid expansion dominated by hot radiation. In contrast, we have no certain knowledge about the preceding era that is the subject of Part IV.

We assume general relativity, which seems to accord with observation. This chapter is devoted to the unperturbed Universe. We begin with a general discussion of the main possibilities for particle distribution functions in thermal equilibrium, applying whenever the Universe has a gaseous component. Then we discuss baryon and lepton number. After these preliminaries, we lay down initial conditions, that are assumed to hold when the Universe is slightly less than one second old. It is assumed that there exist, in thermal equilibrium, protons, neutrons, electrons, positrons and all three species of neutrino and antineutrino.

We then see how, as the Universe expands, the initial conditions determine the primordial abundance of the lightest nuclei, in apparent agreement with observation. The success of this **Big Bang Nucleosynthesis** (BBN) calculation is one of the cornerstones of modern cosmology.

Next we follow the subsequent evolution of the unperturbed Universe. For this evolution to agree with observation, two more components have to be added to the cosmic fluid, leading to what is called the ΛCDM model. We define the parameters of the ΛCDM model as it applies to the unperturbed Universe, and see how observation can determine the values of these parameters. We end by discussing the possible effect of neutrino mass.

4.1 Temperature and redshift

As we shall see, the temperature when the Universe is one second old is $T \sim 1\,\mathrm{MeV}$. Also, the photons have the blackbody distribution with no creation of photons once the temperature gets appreciably below $1\,\mathrm{MeV}$. As a result T is proportional to $1/a$. The present photon temperature is $T_0 = 2.73\,\mathrm{K}$, which in particle physics units is $2.36 \times 10^{-4}\,\mathrm{eV}$. The epoch $T \sim 1\,\mathrm{MeV}$ therefore corresponds to $1/a \sim 10^{10}$. When specifying an epoch in the early Universe the temperature is usually more convenient than the time.

To specify an epoch in the fairly recent past one often gives the redshift z of radiation emitted at that epoch. As explained in Section 3.7.1, it is given by

$$\boxed{1 + z \equiv \frac{\lambda_0}{\lambda(t)} = \frac{1}{a(t)}}\,, \tag{4.1}$$

where λ_0 is the measured wavelength and $\lambda(t)$ is the wavelength that would be measured by a comoving observer at the time of emission.

4.2 Thermal equilibrium in the early Universe

As described in Section 2.7, the term 'thermal equilibrium' refers to some chosen set of processes. Each process is either a collision, involving two or more initial particles, or a decay involving a single initial particle. The reverse of each process occurs with the same amplitude. In the early Universe, a process can be expected to be in thermal equilibrium if and only if each of the particles in the initial state interacts many times during one Hubble time. In other words, if the interaction rate per particle is much bigger than H.

In thermal equilibrium the distribution function of each species is given by Eq. (2.71). For every interaction that creates or destroys particles, the chemical potentials are subject to the constraint (2.73).

Assume now that μ_{a} and m_{a} are both negligible compared with the temperature T. Then Eq. (2.71) reduces to

$$\boxed{f_{\mathrm{a}}(p) = \frac{1}{e^{p/T} \pm 1}}\,. \tag{4.2}$$

For photons this is the blackbody distribution and we will call it the **generalized blackbody distribution**.

The generalized blackbody distribution gives number density

$$n_{\mathrm{a}} = \frac{g_{\mathrm{a}}}{2\pi^2} \int_0^\infty f_{\mathrm{a}}(p) p^2 dp = \frac{\zeta(3) g_{\mathrm{a}}}{\pi^2} T^3 \times \begin{cases} 1 & \text{(bosons)} \\ 3/4 & \text{(fermions)} \end{cases}\,, \tag{4.3}$$

where the zeta function evaluates to $\zeta(3) = 1.202\ldots$. It gives energy density

$$\rho_{\mathrm{a}} = \frac{g_{\mathrm{a}}}{2\pi^2} \int_0^\infty f_{\mathrm{a}}(p) p^3 dp = \frac{\pi^2 g_{\mathrm{a}}}{30} T^4 \times \begin{cases} 1 & \text{(bosons)} \\ 7/8 & \text{(fermions)} \end{cases}\,. \tag{4.4}$$

According to these expressions, the mean energy per particle is of order $T \gg m_{\mathrm{a}}$.

The species is therefore relativistic, corresponding to radiation as opposed to matter.

The total energy density of all species with the generalized blackbody distribution is

$$\rho_{\mathrm{r}} = \frac{\pi^2}{30} g_*(T)\, T^4, \tag{4.5}$$

where

$$g_*(T) \equiv \sum_{\text{bosons}} g_{\mathrm{a}} + \frac{7}{8} \sum_{\text{fermions}} g_{\mathrm{a}}. \tag{4.6}$$

If this energy density dominates the total we have $\rho_{\mathrm{r}} = 3M_{\mathrm{Pl}}^2 H^2$ with $H = 1/2t$. This gives the timescale

$$\frac{1}{4t^2} = \frac{\pi^2}{90}\, g_*(T)\, \frac{T^4}{M_{\mathrm{Pl}}^2}. \tag{4.7}$$

Substituting in numerical values,

$$\boxed{\frac{t}{1\,\mathrm{s}} \simeq 2.42\, g_*^{-1/2} \left(\frac{1\,\mathrm{MeV}}{T}\right)^2}. \tag{4.8}$$

Keeping the condition $\mu_{\mathrm{a}} \ll T$, let us now allow $T \lesssim m_{\mathrm{a}}$. Then Eq. (2.71) becomes

$$\boxed{f_{\mathrm{a}}(p) = \frac{1}{e^{E_{\mathrm{a}}(p)/T} \pm 1}}. \tag{4.9}$$

As T falls below m_{a}, the number density of the species falls exponentially and so does the energy density and entropy density. Usually the fall is caused by pairs of the particles annihilating. In any case, the exponential fall continues until the rate per particle for the relevant interactions falls below H, and interactions can no longer maintain thermal equilibrium. After equilibrium fails, a typical particle of the species travels freely. If the species is unstable with decay rate Γ, the number in a comoving volume falls like $e^{-\Gamma t}$. If instead it is stable, the number of particles in a comoving volume remains constant. That is called **freeze-out**, and is one mechanism for producing cold dark matter (CDM).

A species with the generalized blackbody distribution may fall out of thermal equilibrium while its mass is still negligible. This is called **decoupling**. After decoupling a typical particle of the species travels freely and has momentum $p \propto 1/a$ (redshift). The number of particles in a comoving volume $d\mathcal{V} \propto a^3$ with momentum in a range $d^3p \propto a^{-3}$ is constant. It follows from the definition (2.57) of f_{a} that $f_{\mathrm{a}}(p(a))$ is independent of a. Therefore, the generalized blackbody distribution function remains valid, with an effective temperature $T_{\mathrm{a}} \propto 1/a$.

If all particles in thermal equilibrium have $\mu_a \ll T$, and exchange no energy with the rest of the cosmic fluid, their total entropy density s depends only on the temperature and can be derived in the following way. For the particles in thermal equilibrium, the second law of thermodynamics is $dE = TdS - Pd\mathcal{V}$, where $\mathcal{V} \propto a^3$ is a comoving volume with energy $E = \rho\mathcal{V}$ and entropy $S = s\mathcal{V}$. This can be rewritten as

$$d\rho = (sT - \rho - P)\frac{d\mathcal{V}}{\mathcal{V}} + T\,ds\,. \tag{4.10}$$

Because ρ and P depend only on T, the solution of this equation is $s = (\rho + P)/T + f(\mathcal{V})$, with $df = -f d\mathcal{V}/\mathcal{V}$ leading to $f = \text{constant}/\mathcal{V}$. But the constant vanishes because entropy vanishes at $T = 0$, giving simply $s = (\rho + P)/T$.

Except when T is close to some mass m_a, ρ and P are dominated by the relativistic species satisfying Eq. (4.2), and we have

$$s = \frac{(\rho + P)}{T} = \frac{2\pi^2}{45} g_* T^3 \simeq 1.8\, g_* n_\gamma. \tag{4.11}$$

Because there is no heat transfer in an isotropic Universe, the entropy $S = a^3 s$ in a comoving volume is constant, and therefore $T \propto g_*^{-1/3}/a$. We see that $T \propto 1/a$ is a good approximation, being exact during any era when g_* is constant.

Sometimes we need to consider the effect of non-zero chemical potentials, which are equal and opposite for the particle and antiparticle and much less than T. In this case we are interested only in the regime $T \gg m_a$. Then we find, to first order in μ_a/T, that the number densities n_a and $\bar{n}_a$ of the particle and antiparticle are related by

$$n_a - \bar{n}_a = \frac{T^2}{6}\tilde{g}_a \mu_a. \tag{4.12}$$

Here μ_a is the chemical potential of the particle, while $\tilde{g}_a = g_a$ for fermions and $\tilde{g}_a = 2g_a$ for bosons. Dividing by the entropy density we get

$$\frac{n_a - \bar{n}_a}{s} = \frac{15\tilde{g}_a}{4\pi^2 g_*(T)}\frac{\mu_a}{T}. \tag{4.13}$$

4.3 Baryon and lepton number

Crucial to our discussion will be the existence of baryon number B and lepton number L. These are 'charges', analogous to electric charge Q. A definite amount of B, L and Q is carried by each particle, and the total 'charge' is the sum of the 'charges' of the individual particles.

The values of Q, B and L are given in Table 4.1, for particles existing at $T \lesssim 1\,\text{MeV}$. In each case the antiparticle carries opposite values for the 'charges'. A particle which is its own antiparticle, like the photon, cannot carry any 'charge'.

Table 4.1. *The values of Q, B and L for particles existing at $T \sim 1\,\mathrm{MeV}$*

	e	ν_{e}	ν_μ	ν_τ	p	n	γ
Q	-1	0	0	0	1	0	0
B	0	0	0	0	1	1	0
L	1	1	1	1	0	0	0

An important cosmological parameter is the baryon number per photon at the present epoch, usually denoted by η. An equivalent definition is $\eta = 3.6 n_B/s$, where s is the present entropy density (residing in the photons). As we see in Section 4.5, measurements of the CMB anisotropy and the galaxy distribution indicate a value

$$\eta = (6.1 \pm 0.2) \times 10^{-10}. \tag{4.14}$$

This corresponds to $(\rho_B/\rho_\gamma)_0 \sim 10^3$, which means that the baryons far outweigh the photons. As we will see, the neutrinos give an additional radiation contribution and CDM gives an additional matter contribution, but these don't change the orders of magnitude. We conclude that the matter far outweighs the radiation at the present epoch.

Now consider what happens as we go back in time, assuming that there is no creation of photons. The ratio of number densities remains the same, but the ratio of baryon density to photon density decreases like $1/a$, and the Universe is radiation dominated before redshift $z \sim (\rho_B/\rho_\gamma)_0$. With our assumed initial condition, we are going to find that there is indeed no photon creation at $T \ll 1\,\mathrm{MeV}$, and only modest creation at the epoch $T \sim 1\,\mathrm{MeV}$. Therefore, the Universe is radiation dominated at the beginning of the era that we are considering, and remains so until matter domination takes over.

The densities of the three 'charges' are related to the particle number densities by

$$n_Q = n_{\mathrm{p}} - (n_e - n_{\bar{e}}), \tag{4.15}$$
$$n_B = n_{\mathrm{p}} + n_{\mathrm{n}}, \tag{4.16}$$
$$n_L = (n_e - n_{\bar{e}}) + (n_{\nu_e} - n_{\bar{\nu}_e}) + (n_{\nu_\mu} - n_{\bar{\nu}_\mu}) + (n_{\nu_\tau} - n_{\bar{\nu}_\tau}). \tag{4.17}$$

All three 'charges' are conserved to high accuracy by the interactions that can take place at $T \lesssim 1\,\mathrm{MeV}$. Examples of such interactions are

$$\mathrm{n} \leftrightarrow \mathrm{p} + e + \bar{\nu}_e \qquad \mathrm{n} + \nu_e \leftrightarrow \mathrm{p} + e \tag{4.18}$$

$$e + \bar{e} \leftrightarrow \nu_\alpha + \bar{\nu}_\alpha \tag{4.19}$$

$$\nu_\alpha \leftrightarrow \nu_\beta \tag{4.20}$$

$$e \leftrightarrow e + \gamma \qquad \gamma \leftrightarrow e + \bar{e}. \tag{4.21}$$

In the second and third line, the subscripts α and β can take on any of the values e, μ and τ. In the fourth line the photon may be regarded as a virtual quantum object, or else as representing a classical electromagnetic field. The third line represents the important phenomenon of **neutrino mixing**. As described in Section 16.7, neutrino mixing occurs because the states ν_α that are created and absorbed are not mass eigenstates.

Electric charge is on a different footing from baryon and lepton number, because it represents the strength of an interaction (namely, the electromagnetic interaction). Particles carrying electric charge inevitably interact strongly with each other. Baryon and lepton number in contrast are merely book-keeping devices. This difference has two very important consequences.

First, the average electric charge density of the Universe must vanish to very high accuracy. If it did not, electrostatic repulsion would outweigh the observed effect of gravity. In contrast, there is no requirement that the densities of B and L should vanish. In fact, the density of baryon number definitely doesn't vanish in the present Universe, being equal to the number density of nucleons.

The second difference has to do with symmetries, which are discussed in Chapter 14. Electric charge conservation is a consequence of a gauge symmetry, and is therefore exact. In contrast, B and L conservation are consequences only of global symmetries and need not be exact. According to the Standard Model B and L are indeed violated by 'sphaleron' interactions that can take place at early times. As a result of these interactions *we expect that B and L will have roughly equal densities*. This expectation still holds if we take on board proposed extensions of the Standard Model that are expected to apply in the early Universe.

4.4 The evolution up to Big Bang Nucleosynthesis

4.4.1 Initial thermal equilibrium

The initial condition, assumed to hold when T is a few MeV, is that the processes (4.18)–(4.21) are in thermal equilibrium. The photon annihilation process in Eq. (4.21) makes $\mu_\gamma = 0$, so that photons have the blackbody distribution. Since the electron has mass $m_\mathrm{e} \simeq 0.5\,\mathrm{MeV}$, which is much less than T, it will have the generalized blackbody distribution (4.2) if its chemical potential is much less than T. As we will see the neutrino masses cannot be bigger than $1\,\mathrm{eV}$ or so, and are probably far smaller. We therefore set them equal to zero, which means that the neutrinos too will have the generalized blackbody distribution if their chemical potentials are much less than T.

To calculate the chemical potentials, note first that the electron–positron annihilation processes in Eq. (4.21) make $\mu_e = -\mu_{\bar{e}}$ and $\mu_\gamma = 0$. Then the neutrino–antineutrino annihilation process (4.19) makes $\mu_{\nu_\mathrm{a}} = -\mu_{\bar{\nu}_\mathrm{a}}$ for each neutrino species. In the same way, the neutrino mixing process (4.20) gives all three neutrino chemical potentials a common value $\mu_\nu = -\mu_{\bar{\nu}}$.

We saw in the last section that the baryon number per photon is of order 10^{-9}, and the proton number per photon is obviously no larger. Thus, electrical neutrality requires

$$n_e - n_{\bar{e}} \lesssim 10^{-9} n_\gamma. \tag{4.22}$$

Looking at Eq. (4.13) we see that $\mu_e/T \sim 10^{-9}$. In the same way, applying Eq. (4.13) to the neutrino chemical potential, we have $\mu_\nu/T \sim n_L/n_\gamma$. As we noted in the previous section it is expected that $n_L \sim n_B$, but all that we need for the present purpose is $n_L \ll n_\gamma$ corresponding to $|\mu_\nu/T| \ll 1$. This is ensured by the success of the BBN calculation which requires, with present observational uncertainties,

$$\left|\frac{n_L}{n_\gamma}\right| \lesssim 0.1. \tag{4.23}$$

We conclude that the electrons and neutrinos will indeed have the generalized blackbody distribution. The energy density is therefore given by Eq. (4.4) with

$$g_* = 2 + 5 \times 2 \times \frac{7}{8} = 10.75. \tag{4.24}$$

To arrive at this expression, we remembered that the electron and positron each have two spin states, and that neutrinos have left-handed spin and that antineutrinos have right-handed spin.[2] With this g_* Eq. (4.8) gives $T(t)$, confirming our earlier assertion that $T \sim 1\,\mathrm{MeV}$ corresponds to age $t \sim 1\,\mathrm{s}$.

We have yet to consider the nucleons. Either of the processes (4.18) gives the difference between the proton and neutron chemical potentials:

$$\mu_\mathrm{p} - \mu_\mathrm{n} = \mu_\nu - \mu_\mathrm{e}. \tag{4.25}$$

This is much smaller than T which in turn is much smaller than the nucleon mass

[2] The spin components along the direction of motion are respectively $\mp 1/2$. With neutrino mass taken into account these statements cannot be exact because they are not Lorentz invariant, but they are valid to high accuracy in the rest frame of the cosmic fluid in which the distribution functions are defined.

$m_N \simeq 1\,\mathrm{GeV}$. The equilibrium distribution (2.71) therefore reduces to

$$f_a(p) = \exp\left[-\frac{m_a + p^2/2m_a}{T}\right]. \tag{4.26}$$

This is the Maxwell–Boltzmann distribution, which gives number density

$$\boxed{n_a = g_a \left(\frac{m_a T}{2\pi}\right)^{3/2} \exp\left(-\frac{m_a}{T}\right)}. \tag{4.27}$$

The ratio of neutron to proton numbers is, to high accuracy

$$\boxed{\frac{n_n}{n_p} = e^{-(m_n - m_p)/T}}. \tag{4.28}$$

where $m_n - m_p = 1.3\,\mathrm{MeV}$.

Finally, we need to understand why none of the other known particles can be present at $T \lesssim 1\,\mathrm{MeV}$. For quarks and gluons it is because they cannot exist when their spacing is bigger than $(100\,\mathrm{MeV})^{-1}$, as described in Section 21.3.1. For the μ, τ and hadrons it is because they are too heavy to be created in a typical collision, and have a lifetime much less than one second.

4.4.2 Neutrino decoupling and electron–positron annihilation

We have been assuming thermal equilibrium. The condition for a process to be in thermal equilibrium is that the interaction rate per particle is much bigger than the Hubble parameter. The Standard Model gives the interaction rates, and one finds that they are indeed much bigger than H when T is a few MeV.

When T falls through $1\,\mathrm{MeV}$, three things happen in succession. First, the interaction rate per neutrino falls below H, after which a typical neutrino travels freely to the present epoch. This is **neutrino decoupling**. After neutrino decoupling, neutrino number is conserved and the momentum of each neutrino falls like $1/a$. As a result, the distribution of each neutrino species retains the generalized black-body form, with effective temperature $T_\nu \propto 1/a$, unless and until the temperature falls below its mass. For a short time each nucleon experiences the interaction in Eq. (4.18) with rate bigger than H, maintaining equilibrium. Soon though, these interactions practically cease, and the neutron to proton ratio freezes out except for the effect of neutron decay:

$$\frac{n_n}{n_p + n_n} = A e^{-\Gamma t}, \tag{4.29}$$

where $\Gamma = (886\,\mathrm{s})^{-1}$ is the neutron decay rate. In the approximation that the freeze-out occurs suddenly, the constant A is given by Eq. (4.28) with T equal to the freeze-out temperature. That isn't a good approximation though, and the actual value of $A \simeq 0.16$ has to be obtained from the relevant Boltzmann equations, either numerically or through an analytic approximation such as that described in Ref. [1].

The third event occurring at $T \sim 1\,\text{MeV}$ is that the electrons and positrons become non-relativistic and annihilate, except for the one electron per proton that maintains electrical neutrality. Ignoring the latter and applying entropy conservation to the electrons, positrons, and photons (the species in thermal equilibrium), we have $g_* = 2 + 7/2 = 11/2$ before annihilation (because the electron and the positron each have two spin states) and $g_* = 2$ after annihilation (because only the photons are now relativistic). The photon temperature therefore is boosted relative to the neutrino temperature by a factor $\sqrt[3]{11/4}$. As long as all three species of neutrino are relativistic, the radiation density is given by Eq. (4.5) with an effective g_*^{eff} given by

$$g_*^{\text{eff}} = 2 + \frac{7}{8} \times 6 \times \left(\frac{4}{11}\right)^{4/3} = 3.36. \tag{4.30}$$

This gives

$$R_\nu \equiv \frac{\rho_\nu}{\rho_\nu + \rho_\gamma} = 0.40. \tag{4.31}$$

4.4.3 Nucleosynthesis

When the temperature is about 10^{-1} MeV, practically all of the neutrons bind into ^{4}He nuclei. A small fraction of them go into deuterium and ^{3}He, and an even smaller fraction into ^{7}Li. The predicted abundances depend only on the assumed baryon number per photon η. Using the value in Eq. (4.14), there seems to be adequate agreement between the BBN calculation and observation.

The abundances are determined by the Boltzmann equations governing the relevant nuclear interactions. Accurate results are obtained by numerical integration. Analytic estimates exist [1], but they are in general quite complicated. The dominance of ^{4}He is easy to understand though. It comes from the fact that ^{4}He is the light nucleus with the biggest binding energy. According to the analytic estimate, practically all of the neutrons bind into ^{4}He, which happens at $t_{\text{bind}} \simeq 2.4 \times 10^2$ s [1, 2] (using the value of η in Eq. (4.14)). Putting this into Eq. (4.29) gives the ^{4}He fraction by mass as

$$X = 2Ae^{-\Gamma t_{\text{bind}}} = 0.25. \tag{4.32}$$

This agrees quite well with the numerical calculation.

The success of the BBN calculation confirms that the assumed initial condition at a temperature of a few MeV is correct with good accuracy. It also constrains the magnitude of possible departures from the initial condition. In particular, it constrains the abundance of any, as yet undiscovered, decaying particles that may be present when nucleosynthesis occurs.

4.5 The ΛCDM model

To explain the evolution of the Universe after BBN, including the cosmological perturbations, we need two more components of the cosmic fluid. First, we need some kind of non-baryonic matter, termed cold dark matter (CDM), which has more or less negligible interaction (with itself or anything else) and has more or less negligible random motion.[3] Second, we need a more or less homogeneous and time-independent 'dark energy' density, existing at redshift $z \lesssim 2$. Also, as seen in the next chapter, we need a primordial curvature perturbation.

The simplest ΛCDM model takes the interaction and random motion of the CDM to be completely negligible, and takes the dark energy density to come from the cosmological constant of Section 3.10.2. If we take for granted the accurately measured cosmic microwave background (CMB) temperature, the simplest ΛCDM model needs in principle just three parameters to describe the unperturbed Universe, which we can take as the present values of the Hubble parameter, the baryon density and the total matter density. To describe the primordial curvature perturbation we need two more parameters, specifying the spectrum and the spectral index.

Finally, we need in practice a sixth parameter to specify the effect of reionization of the cosmic medium. The reionization is caused by the formation of early objects and is in principle calculable if the other parameters of the ΛCDM model are known. In practice though, the astrophysical processes leading to reionization are not understood well enough to allow more than a crude estimate. A suitable parameter is the **optical depth** τ, such that $e^{-\tau}$ is the probability that a photon emitted before reionization (but after photon decoupling) rescatters.

With its six parameters, the simplest version of the ΛCDM model seems able at the time of writing to explain all relevant types of observation, some of which have order 1% accuracy. The most important observables are the CMB anisotropy and the inhomogeneity in the galaxy distribution. The parameter values that agree with these observations are shown in Table 4.2 (excluding the spectrum and spectral index that will be considered later). In this table, the Hubble constant is specified by $h \equiv H_0/100\,\mathrm{km\,s^{-1}\,Mpc^{-1}}$. In terms of h, the present Hubble time and Hubble

[3] By baryonic matter we mean ordinary matter, consisting of nucleons (baryons) and the electron. Since there is only one electron per proton, the baryon density is practically the same thing as the density of ordinary matter.

Table 4.2. *The values shown are those obtained at the time of writing from CMB and galaxy survey data, the uncertainties being roughly at the 2σ level. Assuming sudden reionization, the optical depth corresponds to* $z_{\text{ion}} = 9 \pm 2$

h	Ω_c	Ω_B	τ
0.70 ± 0.03	0.23 ± 0.02	0.046 ± 0.002	0.08 ± 0.03

distance are

$$H_0^{-1} = 9.78h^{-1}\,\text{Gyr} \simeq 14\,\text{Gyr}\,, \tag{4.33}$$

$$cH_0^{-1} = 3.00h^{-1}\,\text{Gpc} \simeq 4.3\,\text{Gpc}. \tag{4.34}$$

The present baryon and CDM densities are specified by Ω_B and Ω_c, their values in units of the present critical density $\rho_0 = 1.88h^2 \times 10^{-29}\ \text{g}\,\text{cm}^{-3}$. In astrophysical units,

$$\rho_0 = 2.78h^{-1} \times 10^{11} M_\odot/(h^{-1}\ \text{Mpc})^3, \tag{4.35}$$

where $M_\odot = 1.99 \times 10^{33}$ g is the Solar mass. In particle physics units it corresponds to

$$\rho_0 = (3.00 \times 10^{-3}\ \text{eV})^4 h^2 \simeq (2.5 \times 10^{-3}\ \text{eV})^4. \tag{4.36}$$

Since the total energy density is assumed to be critical, Ω_B and Ω_c are the fractional contributions to the present energy density, with the cosmological constant making up the remainder. The photon density is determined from the CMB temperature T_γ, whose measured value is (2.725 ± 0.001) K. Taking $T_\gamma = 2.73$ K we have $\Omega_\gamma h^2 = 2.47 \times 10^{-5}$, and $\Omega_B h^2 = 3.65\eta \times 10^7$.

When we come to relate Ω_c to scenarios of the early Universe, we will find the following relation useful:

$$\boxed{\frac{\Omega_c}{0.20} = 1.5 \times 10^9 \frac{\gamma n}{s} \frac{m}{\text{GeV}}}, \tag{4.37}$$

where m is the mass of the CDM particles, γn is their present number density, and s is the present entropy density. The factor γ is pulled out so that n can be evaluated in the early Universe with γ allowing for any subsequent entropy production, as described in Sections 21.1.2 and 21.6.

The parameter values in Table 4.2 are consistent with those found by direct observation. Most of the observations rely heavily on the existence of spectral lines, which allow an accurate determination of the redshift of a given source. Let us summarize the present situation.

We begin with the Hubble constant. For $z \ll 1$, the redshift is given by $z = H_0 r$

where r is the distance of the source. This is Hubble's law, which provides a direct determination of the Hubble constant H_0 if one knows the distance of the source or equivalently its power. In recent years, progress appears finally to have been made toward a consensus on the value of the Hubble constant. At the time of writing this gives something like $h = 0.7 \pm 0.1$ with the uncertainty dominated by systematic errors.

The best alternative information about the baryon density comes from the BBN calculation, but the uncertainty is still bigger than the one in Table 4.2. Determination of the baryon density by direct observation is difficult because most of it is dark, i.e. it neither emits nor absorbs radiation at a detectable level.

The total mass of matter in a galaxy or galaxy cluster can be estimated through its gravitational effect on the motion of its visible components. Adding the masses in a given volume gives an estimate of the matter density of the Universe. Alternatively, in the case of a rich galaxy cluster containing of order 1000 galaxies, most of the baryonic matter is X-ray luminous, allowing a direct estimate of the ratio of baryonic to total matter in the cluster. That in turn should be an estimate of the ratio of cosmological densities, because the content of a rich cluster is thought to be a fair sample of the primordial cosmic fluid.

Finally, observations of high-redshift supernovae probe the cosmological constant.

As one might expect in such a rapidly developing field, there are areas of tension between theory and observation when the simplest CDM model is adopted. But there are also proposals for making suitable small modifications of the model if the tension persists. It is therefore widely supposed that some version of the ΛCDM model is correct, which might well be the simplest version.

Among the modifications that one might contemplate are the following: (i) an additional particle species decaying before the present, (ii) random motion for the CDM, (iii) interactions for the CDM, (iv) a more complicated primordial perturbation, (v) neutrino mass, beyond the negligible level suggested by the simplest interpretation of the data, (vi) time dependence of the dark energy, (vii) departures from general relativity and (viii) non-critical density $\Omega \neq 1$.

The abundance of an additional unstable particle species may be constrained by its ability to spoil the success of BBN, by its contribution to the diffuse γ ray background, or by its ability to spoil the accurate blackbody distribution of the CMB.

A component of the CDM with random motion is called warm or hot dark matter. Hot dark matter has by definition significant relativistic motion when galaxy scales are entering the horizon. On its own it is ruled out because its free-streaming wipes out the perturbations. In particular, neutrinos on their own cannot account for the dark matter. Warm dark matter has significant random motion without being relativistic. Random motion of the CDM is constrained by the success of the ΛCDM model in describing the CMB anisotropy, the galaxy distribution and the properties of galaxies.

Time dependence of the dark energy and $\Omega \neq 1$ are constrained by various aspects of the success of the ΛCDM model. It is found that $|\Omega_0 - 1| \lesssim 10^{-2}$. For the dark energy it is found that $|P/\rho + 1| \lesssim 10^{-1}$.

4.6 Evolution of the scale factor

In this section we see how to calculate the evolution of the scale factor, from radiation domination to the present.

Assuming three relativistic species of neutrino, the present energy density in relativistic particles (radiation) is

$$\Omega_{\rm r} = 4.15 \times 10^{-5} h^{-2}, \tag{4.38}$$

and the redshift of **matter–radiation equality** is given by

$$1 + z_{\rm eq} = \frac{\Omega_{\rm m}}{\Omega_{\rm r}} = 2.41 \Omega_{\rm m} h^2 \times 10^4 \simeq 3.3 \times 10^3, \tag{4.39}$$

where $\Omega_{\rm m} \equiv \Omega_c + \Omega_B$ is the total matter density. The corresponding temperature is

$$T_{\rm eq} = T_0 (1 + z_{\rm eq}) = 6.57 \Omega_{\rm m} h^2 \times 10^4\,{\rm K} \simeq 6.2 \times 10^3\,{\rm K}. \tag{4.40}$$

The definition of the density parameter indicates that at any epoch the matter component obeys

$$\Omega_{\rm m}(z)(1+z)^{-3} H^2 = {\rm const} = \Omega_{\rm m} H_0^2. \tag{4.41}$$

By definition, at equality the radiation density matches the matter density, and the Friedmann equation gives the Hubble parameter at equality as

$$\frac{H_{\rm eq}}{H_0} = \sqrt{2}\, \Omega_{\rm m}^{1/2} (1 + z_{\rm eq})^{3/2} = 5.29 \times 10^6 h^3 \Omega_{\rm m}^2 \simeq 1.6 \times 10^5, \tag{4.42}$$

yielding

$$\frac{a_{\rm eq} H_{\rm eq}}{a_0 H_0} = \sqrt{2 \Omega_{\rm m}}\, z_{\rm eq}^{1/2} = 220\, \Omega_{\rm m} h \simeq 43. \tag{4.43}$$

The comoving Hubble length at matter–radiation equality was therefore

$$(a_{\rm eq} H_{\rm eq})^{-1} = 13.6 \Omega_{\rm m}^{-1} h^{-2}\ {\rm Mpc} \simeq 1.0 \times 10^2\,{\rm Mpc}. \tag{4.44}$$

Now let us consider the evolution of the scale factor between radiation domination and matter domination. Including both matter and radiation, the Friedmann equation can be written

$$H^2 = \frac{1}{2} H_{\rm eq}^2 \left[\left(\frac{a_{\rm eq}}{a} \right)^3 + \left(\frac{a_{\rm eq}}{a} \right)^4 \right], \tag{4.45}$$

where $a_{\rm eq}$ is the scale factor when the densities of matter and radiation are equal. Integrating the Friedmann equation using conformal time gives

$$\frac{a(\eta)}{a_{\rm eq}} = (2\sqrt{2} - 2) \left(\frac{\eta}{\eta_{\rm eq}} \right) + (1 - 2\sqrt{2} + 2) \left(\frac{\eta}{\eta_{\rm eq}} \right)^2 , \tag{4.46}$$

with

$$\eta_{\rm eq} = \frac{2\sqrt{2} - 2}{a_{\rm eq} H_{\rm eq}} = 11.3 \Omega_{\rm m}^{-1} h^{-2}\, {\rm Mpc} \simeq 83\, {\rm Mpc} \,. \tag{4.47}$$

We see a smooth rollover between the two characteristic behaviours of radiation domination and matter domination. There is no good way to rewrite this in terms of cosmic time.

Now we consider the cosmological constant. According to Table 4.2, the total matter density is $\Omega_{\rm m} = 0.28$ making $\Omega_\Lambda = 0.72$ and the baryon density is about 0.05. The latter value is within the relatively large range allowed by nucleosynthesis. Since $\rho_{\rm m}$ is proportional to a^{-3} while ρ_Λ is constant, those two densities become equal only at the recent epoch $1 + z = (\rho_\Lambda / \rho_{\rm m})^{1/3} = 1.3$, the cosmological constant being negligible at much earlier times.

Including both matter and the cosmological constant, the Friedmann equation becomes

$$H^2 = H_0^2 \left[\Omega_{\rm m} a^{-3} + (1 - \Omega_{\rm m}) \right] . \tag{4.48}$$

Using Eq. (4.41) this gives

$$\Omega_{\rm m}(z) = \Omega_{\rm m} \frac{(1+z)^3}{1 - \Omega_{\rm m} + (1+z)^3 \Omega_{\rm m}}, \tag{4.49}$$

where as usual $\Omega_{\rm m}$ without an argument denotes the present value. If nothing intervenes, the cosmological constant will completely dominate after another Hubble time or so, giving a constant Hubble parameter and an exponentially increasing scale factor.

Integrating the Friedmann equation gives the age of the Universe as

$$t_0 = \frac{2H_0^{-1}}{3} \frac{1}{\sqrt{1 - \Omega_{\rm m}}} \ln \left(\frac{1 + \sqrt{1 - \Omega_{\rm m}}}{\sqrt{\Omega_{\rm m}}} \right) \simeq 13.7\, {\rm Gyr}. \tag{4.50}$$

The last factor can equivalently be written as $\sinh^{-1}(\sqrt{(1 - \Omega_{\rm m})/\Omega_{\rm m}})$. For $\Omega_{\rm m} = 0.26$, the age equals the Hubble time.

The present conformal time η_0 is obtained by integrating the Friedmann equation. It defines the size of the observable Universe, and can be written

$$\eta_0 = H_0^{-1} \int_0^\infty \frac{dz}{\sqrt{1 + 3\Omega_m z + 3\Omega_m z^2 + \Omega_m z^3}} \simeq 14\,\mathrm{Gpc}. \tag{4.51}$$

4.7 Photon decoupling and reionization

For some time after matter–radiation equality, a typical photon undergoes frequent scattering off free electrons (Thomson scattering), but that ceases to be the case when atoms form (known as **recombination**), leading to **photon decoupling** after which a typical photon travels freely. One often calls the epoch of photon decoupling the **last-scattering epoch**. At some much later redshift, the Universe becomes ionized again, recreating a population of free electrons from which further Thomson scattering may occur. An understanding of the photon scattering history is crucial in determining the cosmic microwave anisotropies.

4.7.1 The optical depth and the visibility function

The probability per unit time for a photon to scatter is $n_e \sigma_T$, where n_e is the number density of free electrons and σ_T is the Thomson scattering cross-section. The optical depth $\tau(t)$ is defined as

$$\boxed{\tau(t) \equiv \sigma_T \int_t^{t_0} n_e(t)\,dt}. \tag{4.52}$$

The probability that a cosmic microwave background photon, now observed, has travelled freely since time t is $e^{-\tau(t)}$: to see this, note that $d\tau/dt = -\sigma_T n_e$, so that the probability $P(t)$ satisfies $dP/dt = -(d\tau/dt)P$, with $P(t_0) = 1$.

We define the ionization fraction as $\chi(z) \equiv n_e/n_p$, where n_p is the number density of protons. Because the electron number density grows with redshift as $(1+z)^3$, the optical depth is

$$\tau(z) = 0.88\, n_{B,0}\, \sigma_T \int_0^z (1+z')^2 \frac{dz'}{H(z')} \chi(z'), \tag{4.53}$$

where we used $dz = -(1+z)H\,dt$, and assumed $n_p = 0.88 n_B$ at present, corresponding to a 24% helium fraction by mass.

The **visibility function** g is defined by

$$\boxed{g(t) \equiv -\dot{\tau}(t) e^{-\tau(t)} = \sigma_T n_e(t) e^{-\tau(t)}}. \tag{4.54}$$

We see that $g(t)\,dt$ is the probability that a CMB photon, now observed, scattered in the time interval dt and has travelled freely since then. In equations where conformal time is being used, one defines

$$g(\eta) \equiv -\frac{d\tau(\eta)}{d\eta} e^{-\tau(\eta)} . \tag{4.55}$$

This makes $g(\eta) = a g(t)$ so that $g(\eta) d\eta = g(t) dt$.

4.7.2 Recombination and decoupling

As we go back in time through the decoupling era, τ rises sharply to become much bigger than 1. Correspondingly, g is peaked at the epoch of decoupling, its width corresponding to the thickness of the last-scattering surface.

At early times, there is full ionization, so that n_e is equal to the total electron number density. Then first ^{4}He and later H nuclei combine with electrons to form atoms; in the usual parlance they 'recombine', though that term isn't really appropriate because atoms never existed before. As a result, $n_e(t)$ falls abruptly to a small value. The time dependence of n_e during this fall is crucial for an accurate description of the CMB anisotropy.

To describe the recombination process for helium, it is a good approximation to use the thermal equilibrium distribution (4.2), which for atoms and nuclei leads to what is called the **Saha equation**. For hydrogen, we need to solve the coupled Boltzmann equations describing the transitions between different levels. Using an analytic approximation to their solution one finds [1, 2]

$$\chi \equiv \frac{n_e(t)}{n_p(t)} = 1.1 \times 10^8 z^{-1} e^{-14400/z} , \tag{4.56}$$

This indicates that g peaks at redshift $z_{ls} = 1050$, and Eqs. (4.46) and (4.47) give the conformal time of last scattering as $\eta_{ls} \simeq 207$ Mpc. The CMB originates on the sphere around us with comoving radius $\eta_0 - \eta_{ls} \simeq \eta_0$. One further finds that $g(\eta)$ is well approximated by a gaussian with dispersion $\Delta\eta = 13.7 (aH)_{dec}$, corresponding to $\Delta z \simeq 80$, which measures the thickness of the last-scattering surface.

Although the ionization fraction n_e/n_p falls sharply after the epoch $z \simeq 1000$, it is big enough to hold the baryons and photons at the same temperature until $z \sim 100$. As we see in Section 7.4.3, this residual ionization is important because it determines roughly the mass of the first baryonic objects in the Universe. More detail on these calculations can be found for instance in Refs. [1, 2].

4.7.3 Reionization

We now turn to reionization at low redshift. Using

$$\Omega_B = \frac{8\pi G}{3H_0^2} m_B n_{B,0}, \tag{4.57}$$

and Eq. (4.41) gives the optical depth to redshift z as

$$\tau(z) = \tau^* \int_0^z (1+z')^{1/2} \sqrt{\frac{\Omega_{\rm m}(z')}{\Omega_{\rm m}}}\, \chi(z')\, dz', \tag{4.58}$$

where the redshift dependence of Ω is given by Eq. (4.49) for a flat Universe and

$$\tau^* = 0.88\, \frac{3H_0 \Omega_B \sigma_{\rm T} c}{8\pi G m_{\rm b}} \simeq 0.061 h\, \Omega_B. \tag{4.59}$$

Well after photon decoupling but before the onset of reionization, $\tau(z)$ has a practically constant value denoted simply by τ. Its measured value is $\tau = 0.08 \pm 0.03$ at 2σ level. The probability that a photon was affected by late reionization is $1 - e^{-\tau} \simeq 0.08$.

A simple model for reionization is instantaneous reionization at redshift $z_{\rm ion}$, so that $\chi(z) = 1$ for $z < z_{\rm ion}$ and zero otherwise. Using Eq. (4.49) in Eq. (4.58) and integrating gives

$$\begin{aligned} \tau &\simeq \frac{2\tau^*}{3\Omega_{\rm m}} \left[\left(1 - \Omega_{\rm m} + \Omega_{\rm m} (1+z_{\rm ion})^3 \right)^{1/2} - 1 \right] && (4.60) \\ &\simeq 2.5 \times 10^{-3} (1+z_{\rm ion})^{3/2}. && (4.61) \end{aligned}$$

4.8 Neutrino mass

As described in Section 16.7, observation of neutrino mixing gives the differences between the neutrino masses-squared. At the time of writing the results are

$$|m_3^2 - m_2^2| = (2.5 \pm 0.5) \times 10^{-3}\,{\rm eV}^2 \qquad m_2^2 - m_1^2 = (8.0 \pm 0.3) \times 10^{-5}\,{\rm eV}^2. \tag{4.62}$$

An immediate conclusion is that one species has $m > 0.05\,{\rm eV}$ and another has $m > 0.009\,{\rm eV}$. The most straightforward possibility regarding the masses is the hierarchy $m_3 = 0.05\,{\rm eV}$, $m_2 = 0.009\,{\rm eV}$ and $m_1 \ll m_2$. Another possibility, known as the inverted hierarchy, is $m_1 \simeq m_2 \simeq 0.05\,{\rm eV}$ with $m_3 \ll m_1$. In either case, the heaviest neutrino has mass $\simeq 0.05\,{\rm eV}$. Much heavier neutrinos are possible only if all three have practically the same mass.

The observed CMB anisotropy and galaxy distribution together require that the total neutrino mass satisfies

$$\sum m_{\rm a} \lesssim 1\,{\rm eV}. \tag{4.63}$$

A given species of neutrino becomes non-relativistic if T_ν falls below its mass, which corresponds to the epoch

$$z_{\rm nr} = \frac{m_{\rm a}}{5\times 10^{-4}\,{\rm eV}} \lesssim 2000\,. \tag{4.64}$$

Unless all three neutrinos have practically the same mass, $z_{\rm nr} \simeq 100$ for the heaviest neutrino. The total matter density from all non-relativistic neutrino species is

$$\Omega_\nu = \frac{\sum_{\rm a} m_{\rm a}}{94\,{\rm eV}} h^{-2} \simeq \frac{\sum_{\rm a} m_{\rm a}}{46\,{\rm eV}} \lesssim 2\times 10^{-2}\,. \tag{4.65}$$

(That the density scales linearly with mass is easy to understand; because the particles are relativistic at decoupling, the final number density is independent of the particle mass.) Unless all three neutrinos have almost the same mass the actual value is $\Omega_\nu \simeq 10^{-3}$.

Exercises

4.1 Use Appendix B to verify Eq. (4.8), giving the timescale during radiation domination in terms of the energy density.

4.2 Use Eq. (4.28) to calculate the constant A in Eq. (4.29) for the neutrino abundance in terms of the temperature at freeze-out (taken to be instantaneous). What freeze-out temperature would reproduce the actual value $A = 0.16$?

4.3 Use Appendix B to verify Eq. (4.37), giving the density parameter of the CDM.

4.4 Calculate the maximum electric charge density that the Universe can possess, if the electrostatic repulsion isn't to dominate the force of Newtonian gravity. How many electrons per cubic meter could be added to the number required for electrical neutrality, while respecting your bound?

4.5 How many neutrino families would there have to be to make matter–radiation equality and decoupling coincide?

4.6 For a Universe containing only non-relativistic matter (and no cosmological constant), show that the Friedmann equation can be rewritten as

$$H(z) = H_0(1+z)(1+\Omega_{\rm m} z)^{1/2}.$$

Derive the equivalent of Eq. (4.49) for an open Universe.

4.7 Using Appendix B, compute the Hubble radius at $T = 0.1\,{\rm MeV}$. To what comoving scale does this correspond?

References

[1] V. Mukhanov. *Physical Foundations of Cosmology* (Cambridge: Cambridge University Press, 2006).

[2] S. Weinberg. *Cosmology* (Oxford: Oxford University Press, 2008).

5

The primordial density perturbation

In this chapter and the next, we consider the perturbation in the energy density of the Universe, as it exists when the run-up to nucleosynthesis begins at temperature $T \sim 1\,\mathrm{MeV}$. Although it is not essential, we will take T to actually be a bit below $1\,\mathrm{MeV}$ so that the positrons have annihilated leaving just the cold dark matter (CDM), baryons, photons and neutrinos, with the last decoupled.

Perturbations existing at this epoch may be called **primordial perturbations** because they provide a simple initial condition for the subsequent evolution of the perturbed Universe. That evolution and its contact with observation will occupy us for the rest of Part II.

We will also study what is called the **curvature perturbation**. It is a powerful quantity, because on superhorizon scales it is conserved provided that the pressure of the cosmic fluid depends only on its energy density. The curvature perturbation determines the perturbation in the total energy density, defined on a given slicing of spacetime. We consider also the isocurvature perturbations which determine the distribution of energy density between different components of the cosmic fluid.

The strategy will be to study the perturbations themselves in this chapter, and their stochastic properties in the next. The study is important both in its own right, and because it introduces basic concepts that will be used throughout the book.

5.1 A first look at the primordial perturbations

5.1.1 Epoch of horizon entry

Consider a spherically symmetric region with radius aR (comoving radius R), expanding with the Universe, which may later collapse to form a gravitationally bound object. Spherical collapse is at best an approximation but it serves to illustrate the concepts. The mass M of the object will be roughly the total mass of non-relativistic matter (including both baryonic matter and cold dark matter) contained within the initially expanding region. That mass is $M = (4\pi/3)R^3\Omega_{\mathrm{m}}\rho_{\mathrm{c},0}$, corresponding to

$$\frac{M}{10^{12}M_\odot} = 0.16 \left(\frac{R}{\text{Mpc}}\right)^3 . \tag{5.1}$$

The collapse may occur if the region is, on average, slightly denser than its surroundings; as a result it attracts more matter so that the fractional overdensity is increased. However, this process cannot be effective while the comoving scale R is far outside the horizon corresponding to $aR \gg H^{-1}$. Indeed, as we showed in Section 3.10, the quantity H^{-1} that we are calling the horizon is actually of order the particle horizon in a Universe dominated by matter and/or radiation. As the latter is the maximum distance that anything has had time to travel, there can be no significant movement of matter into a comoving region, until the region enters the horizon.

We will find that on very large scales ($R \gtrsim 10\,\text{Mpc}$ corresponding to $M \gtrsim 10^{15}M_\odot$) horizon entry is so late that there has not been time for the collapse to actually occur. This finding accords well with the observation that there are no gravitationally bound objects with such high masses, only expanding regions which are slightly overdense or underdense. These regions are actually of great interest, because their stochastic properties provide a direct window into the very early Universe. They are observed more or less directly through their effect on the cosmic microwave background, right up to the scale $R = \eta_0 \simeq 14\,\text{Gpc}$ which corresponds to the whole observable Universe.

What about the other extreme of very small scales? It turns out that even after horizon entry, gravitational collapse of the baryons can actually occur only on scales corresponding to $M \gtrsim 10^4 M_\odot$. This is the smallest scale of cosmological interest. Let us work out the temperature at which it enters the horizon. From Eq. (4.39), the temperature at matter–radiation equality is $T_{\text{eq}} = 3300T_0 = 0.77\,\text{eV}$. Using Eq. (4.46) with the advertized value of η_{ls}, and taking $\eta = 1/aH$, we find that the mass $M \sim 10^4 M_\odot$ enters the horizon when T is somewhat below $0.1\,\text{MeV}$ (note that this calculation refers to the mass in non-relativistic particles). This means that all scales of cosmological interest are outside the horizon when the run-up to nucleosynthesis begins at $T \sim 1\,\text{MeV}$. The properties of the perturbations at this time therefore provide a convenient 'initial condition' for the subsequent evolution.

5.1.2 Primordial density perturbations

At $T \sim 1\,\text{MeV}$ the main components of the Universe are neutrinos, photons, baryonic matter (which we call simply 'baryons') and CDM. We use subscripts ν, γ, B

and c to denote these components. In particular, the total energy density is

$$\rho = \rho_\nu + \rho_\gamma + \rho_B + \rho_{\rm c}. \tag{5.2}$$

We will denote a generic component by $\rho_{\rm a}$, where a= ν, γ, B or c. By the time that the temperature has fallen a bit below $1\,{\rm MeV}$, the neutrinos and the CDM are decoupled (no significant interactions) while the photons and baryons are still tightly coupled by the scattering of photons off electrons (Compton scattering).

Now consider a sphere whose comoving radius R may correspond to any cosmological scale. The centre of the sphere has position $\mathbf{x}$, and a time coordinate t. There is no need to specify the precise definition of t at the moment; it need not be the proper time τ measured by an ideal clock. At each point the actual quantity $\rho_{\rm a}(\mathbf{x}, t)$ is replaced by a smoothed one $\rho_{\rm a}(R, \mathbf{x}, t)$, defined as the average within the sphere. The smoothed perturbations $\delta\rho_{\rm a}(R, \mathbf{x}, t)$ are then defined as departures from the spatial average density, which we denote by $\rho_{\rm a}(t)$:

$$\rho_{\rm a}(R, \mathbf{x}, t) = \rho_{\rm a}(t) + \delta\rho_{\rm a}(R, \mathbf{x}, t). \tag{5.3}$$

Smoothed number densities $n_{\rm a}(R, \mathbf{x}, t)$ and their perturbations can be defined in the same way.

Particle flow into or out of the sphere is negligible well before horizon entry. This means that the local number densities are inversely proportional to the local comoving volume element $d\mathcal{V}$, just as if they lived in some unperturbed universe. If we consider two or more non-overlapping spheres, each of them in this sense represents a 'separate universe'. We can define a local scale factor $a(\mathbf{x}, t)$, such that $d\mathcal{V} \propto a^3$. Then the local number densities are proportional to a^{-3} just as in an unperturbed universe. Also, the matter densities evolve as a^{-3} while the radiation densities evolve as a^{-4}.

5.1.3 Main features of the primordial density perturbation

Now we are ready to ask three crucial questions about the nature of the primordial density perturbation on cosmological scales. First, does the composition of the Universe alter as we move from one location to another? The answer, within present observational uncertainties, seems to be that it doesn't. Using say the (smoothed) total energy density ρ as a time coordinate, each individual energy density and number density is a unique function of ρ, the same throughout spacetime;

$$\rho_{\rm a} = \rho_{\rm a}(\rho), \qquad n_{\rm a} = n_{\rm a}(\rho). \tag{5.4}$$

This is called the **adiabatic condition** for the quantity $\rho_{\rm a}$ or $n_{\rm a}$. The term is justified because the relation (5.4) holds in particular for the evolution with time at a given location, and that evolution is indeed adiabatic before horizon entry since there is no particle flow and hence no heat flow. Quite generally, any cosmological quantity is said to satisfy the adiabatic condition if it is a unique function of ρ. Current observation shows that the density perturbation is adiabatic to within roughly ten percent, with no detection of non-adiabaticity.

Our second and third questions relate to the smoothed total energy density perturbation $\delta\rho(R, \mathbf{x}, t)$. The second question relates to the possible correlation between the Fourier components of $\delta\rho_{\mathbf{k}}$. If we measure one component, does that tell us anything about the likely outcome of measuring another one? The answer, again within observational uncertainties, is that there is no correlation except for the reality condition $\delta\rho_{\mathbf{k}}^* = \delta\rho_{-\mathbf{k}}$. We say that $\delta\rho$ is **gaussian**, because as we discuss later the lack of correlation implies that the probability distribution of $\delta\rho(\mathbf{x})$ is gaussian. Current observation shows that the level of non-gaussianity (measured say by the skewness of the probability distribution) is less than 10^{-3}. As we discuss in Section 6.7.2, the possibility that non-gaussianity may be detected in the future is of great interest.

Finally, we ask how the *rms* value of $\delta\rho(R, \mathbf{x})$, evaluated around the epoch of horizon entry for a random location, depends on the smoothing scale. The answer is that it is almost independent of the smoothing scale. One says that the primordial density perturbation is almost **scale invariant**. The *rms* value of $\delta\rho/\rho$ (or rather a closely related quantity denoted by $\mathcal{P}_\zeta^{1/2}$) is one of the fundamental parameters of cosmology. Observations determines it to be about 5×10^{-5}, and detects a small departure from scale invariance characterized by what is called the spectral index. These matters are further discussed in Section 6.7.1.

5.1.4 Other types of primordial perturbation

There may be other types of primordial perturbation, which have nothing to do with the density perturbation. One is the **primordial tensor perturbation**. It determines the initial amplitude of gravitational waves, whose oscillation begins at horizon entry. There may also be a primordial magnetic field, and going beyond smooth perturbations there may be cosmic strings or other topological defects. None have been detected at the time of writing.

Whatever they are, the primordial perturbations are of crucial importance, because they provide the initial condition for the subsequent evolution of all perturbations. The task of theory is to explore scenarios of the early Universe that may generate primordial perturbations, and to confront those scenarios with observation.

5.2 Cosmological perturbations

5.2.1 *Defining the perturbations*

To describe spacetime it is necessary to lay down coordinates, which define a slicing and threading of spacetime. Let us begin by considering an unperturbed universe. In that case there is a uniquely preferred slicing and threading. The preferred threading consists of *comoving* worldlines; observers on such worldlines move with the expansion. These worldlines are *geodesics* (free-falling), and they correspond to uniform expansion which in technical terms means that they have zero shear and zero vorticity. The preferred slicing is orthogonal to the preferred threading. On the preferred slicing, all quantities are homogeneous. In particular both the energy density and the spatial curvature are homogeneous. In accordance with observation (and the expectation from inflation) the curvature is taken to vanish. We may summarize this by saying that the preferred slicing is *comoving* (i.e. orthogonal to comoving worldlines), is *flat*, and has *uniform energy density*. To label the slices one uses proper time t or conformal time η, and to label the threads one uses comoving Cartesian coordinates $\mathbf{x} = (x, y, z)$ or comoving spherical polar coordinates (r, θ, ϕ). Choosing say (t, x, y, z), the line element is given by Eq. (3.63).

In the presence of perturbations, it is impossible to find coordinates with all of the properties listed above for an unperturbed universe. In order to implement cosmological perturbation theory, one requires that Eq. (3.63) is recovered in the limit of zero perturbations, but there is in principle no other restriction. A coordinate choice satisfying this requirement is called a **gauge**. Several gauges have been considered in the literature, each of them corresponding to a threading and slicing that has some subset of the properties listed for an unperturbed universe.

Once a gauge has been chosen, a perturbed universe is described by some set of functions, which depend on position. In an unperturbed universe, the corresponding functions would either vanish or have spatially homogeneous values. The perturbation $\delta f(\mathbf{x}, t)$ of a given function is defined as the difference between its value $f(\mathbf{x}, t)$ in the perturbed universe, and its value $f(t)$ in some unperturbed ('background') universe:

$$f(t, \mathbf{x}) = f(t) + \delta f(t, \mathbf{x}). \tag{5.5}$$

We used the symbol $\mathbf{x}$ to indicate the space coordinates x^i. These will be taken to be Cartesian coordinates in the background. Referring to the perturbed universe they are generic coordinates. The choice of the background is arbitrary, as long as it makes the perturbation small within the observable Universe.

We shall meet many examples of perturbations. Sticking to Newtonian perturbations, we can consider the perturbation in the mass density $\rho(\mathbf{x}, t)$. This is a single function. There is also the peculiar velocity v_i of the cosmic fluid (its departure from the uniform Hubble flow). This vanishes in the background so that v_i is itself the perturbation.

Moving on to relativistic perturbations, we might consider the spacetime metric $g_{\mu\nu}$. In that case the perturbation is labelled by the pair of indices $\mu\nu$. Finally, we might consider the perturbation in the photon distribution function $f(\mathbf{x}, \mathbf{p}, t)$. It is observed at the present time as the CMB anisotropy. In this case Eq. (5.5) applies at each fixed $\mathbf{p}$. The perturbation now consists of a continuous set of functions, labelled by $\mathbf{p}$. In writing Eq. (5.5) we have suppressed any indices and/or continuous parameters, that may label f.

The different perturbations are not independent. Rather, their evolution with time is given by a set of coupled partial differential equations. According to the simplest possibility, which is consistent with observation at the time of writing, the sole initial condition needed to solve that system is provided by just one 'primordial curvature perturbation'. This perturbation, which we denote by $\zeta(\mathbf{x})$, will occupy much of our attention. We will also look at some other possible 'primordial' perturbations which might be needed to complete the specification of the initial condition.

5.2.2 Smoothing

In our overview we introduced smoothed perturbations, $\delta\rho(R, \mathbf{x}, t)$ and so on. At a given point, the smoothed quantity is the average of the original quantity, within a comoving sphere of radius aR. Smoothing is frequently necessary to handle cosmological perturbations, and it introduces some tricky issues. One needs to be careful about which quantities are smoothed. For example, it would not be a good idea to smooth away the small-scale oscillations of the electromagnetic field that correspond to the photon energy density. In practice smoothing is generally used only for the metric perturbations, the energy–momentum tensor, and (sometimes) scalar fields.

A more serious issue, in principle, is to understand how the equations of a non-linear theory like general relativity can apply to smoothed quantities. However, since general relativity is formulated from the beginning by postulating a source of gravity that is smooth at some level, one usually assumes that additional smoothing will do no harm.

In terms of Fourier components, smoothing on a scale R corresponds to dropping components with wavenumber less than $k \equiv 1/R$.[1] With this in mind, one generally denotes the smoothing scale by $1/k$ instead of by R.

[1] A detailed discussion of this correspondence is given in Section 9.1.

5.3 The evolution of cosmological perturbations

5.3.1 *Cosmological perturbation theory*

Perturbations can be handled using cosmological perturbation theory. This follows the usual procedure, familiar from quantum mechanics. In some set of exact equations, the physical perturbations are multiplied by a common parameter so that the exact equations become a power series in this parameter. First-order perturbation theory, also called linear perturbation theory, keeps only the linear term of the power series, so that all equations are linear in the perturbations. Keeping also the quadratic term one arrives at second-order perturbation theory, and so on.

An equivalent procedure is that of iteration. Having evaluated the first-order perturbation, we can include it as part of the 'unperturbed' quantity. The 'first-order' perturbation of that quantity will then be the second-order perturbation of the original quantity.

Cosmological perturbation theory assumes that the perturbations are suitably small. Take for instance the energy density contrast $\delta \equiv \delta\rho/\rho$. To describe it using cosmological perturbation theory one needs $|\delta(\mathbf{x})| \ll 1$. In the early Universe this may be valid without any smoothing, but after galaxies form we will certainly need to smooth on a scale containing many galaxies.

5.3.2 *Separate universe assumption*

While scales of interest are outside the horizon, an alternative to cosmological perturbation theory is provided by what is called the **separate universe assumption**.

The separate universe assumption is a powerful tool for dealing with perturbations in the very early Universe. In this section we see what the separate universe assumption entails, and in the next section we use it to describe what is called the curvature perturbation. After that we will mostly forget about the separate universe assumption until Part IV, where it plays a central role.

The separate universe assumption refers to the behaviour of the Universe after smoothing on a specified comoving scale k^{-1}, during the super-horizon era $k \ll aH$. It may actually be regarded as a pair of assumptions. The first assumption is that the spatial gradients, at most of order k/a, are negligible. This amounts to assuming that k^{-1} is the biggest relevant comoving distance scale, throughout the superhorizon era. By virtue of this assumption, the Universe at each position will evolve as if it were homogeneous. The second assumption is that the Universe at each position is locally isotropic. Given both assumptions, the smoothed universe at each position evolves like some unperturbed (homogeneous and isotropic) universe throughout the super-horizon regime $aH \gg k$. In other words, the whole Universe can be regarded as a collection of unperturbed separate universes.

Let us see to what extent the separate universe assumption can be justified. Within the inflationary scenario, isotropy is expected. This is because (i) the initial era of inflation is expected to make each comoving region both homogeneous and isotropic at the classical level (Section 18.3) and (ii) at the quantum level, the vacuum fluctuation of scalar fields generates inhomogeneity (Chapters 24, 25 and 26) but it retains the isotropy.

Taking isotropy for granted, the separate universe assumption is equivalent to assuming that k^{-1} is the biggest relevant distance scale, throughout the super-horizon regime. If the smoothing scale isn't far below the shortest cosmological scale, one expects that to be the case for the following reason. On the very largest cosmological scale of order H_0^{-1}, the smoothed universe will be smooth on all observable scales. It should then serve as the unperturbed universe that provides the background upon which cosmological perturbations are superimposed. In other words, we expect the separate universe assumption to be valid on the largest cosmological scale. Going down in scale, even the smallest cosmological scale is still of order $10^{-6}H_0^{-1}$, and it seems quite reasonable to expect the separate universe assumption to remain valid over this modest range of scales.

The separate universe assumption has profound implications. It means that after smoothing, the evolution described in Chapters 4 and 18–22 applies at each comoving location throughout the super-horizon era. This way of thinking about perturbations on super-horizon scales gives results that are valid to all orders of cosmological perturbation theory.

Given the content of the cosmic fluid, the separate universe assumption can and should be verified by using cosmological perturbation theory. We occasionally refer to such calculations, but in general we take the separate universe assumption for granted.

Finally, we mention that there is an extension of the separate universe assumption, called the gradient expansion (see for instance Ref. [1]). In this scheme, each spatial gradient in some set of exact equations is multiplied by a common parameter, and one keeps only a finite number of terms in the series. The separate universe assumption keeps only the zero-order term in the expansion, and is the only version that we shall invoke.

5.3.3 Strategy for this book

In this book, we handle the evolution of cosmological perturbations in the three stages: (i) their origin at horizon exit during inflation is described using first-order or second-order perturbation theory, the latter being required if the theory predicts rather small non-gaussianity for ζ, (ii) their evolution while outside the horizon is described by the separate universe assumption, (iii) their subsequent evolution is described using first-order cosmological perturbation theory.

Our decision to exclude higher-order perturbation theory at stage (iii) should not be taken to mean that it is irrelevant. On the contrary, we see in Chapter 9 that first-order theory will fail at late times, on scales corresponding to the formation of gravitationally bound objects. Higher-order theory in that context is well developed, and provides a useful supplement to numerical simulation.

Even at early times, second-order theory might be needed to handle the evolution of non-gaussianity. Roughly speaking, that will happen if non-gaussianity is below the level 10^{-4} (to be compared with the present observational upper bound of 10^{-3}). A more or less equivalent statement is that second-order theory will be needed for stage (iii) if it is needed for stage (i). At the time of writing, second-order theory in this context is under active development.

5.4 Primordial curvature perturbation

We now discuss a quantity ζ, called the primordial curvature perturbation. After smoothing the energy–momentum tensor and the metric on the shortest cosmological scale, the curvature perturbation has a constant value $\zeta(\mathbf{x})$ while that scale approaches horizon entry (in other words, at the epoch $T \sim 10^{-1}\,\mathrm{MeV}$). This constant value determines the total energy density perturbation that we discussed in Section 5.1, and it provides the main (possibly the sole) initial condition for the subsequent evolution of all perturbations. As a result it is directly accessible to observation.

5.4.1 Definition

In Part IV we see how $\zeta(\mathbf{x})$ may be generated within the inflationary scenario. To permit that discussion, we define ζ in a very general way, allowing for possible time-dependence. Our only demand is that the energy–momentum tensor and the metric tensor are smoothed on some comoving scale, which is well outside the horizon to that the separate universe assumption can be invoked.

Without invoking cosmological perturbation theory, we consider a gauge in which the threads are comoving and the slices have uniform energy density. Then we define ζ by writing the spatial metric in the form

$$\boxed{g_{ij} = a^2(\mathbf{x},t)\gamma_{ij}(\mathbf{x})}\,, \tag{5.6}$$

with

$$a(\mathbf{x},t) \equiv a(t)e^{\zeta(\mathbf{x},t)}, \qquad \gamma_{ij}(\mathbf{x}) \equiv \left(Ie^{2h}\right)_{ij}. \tag{5.7}$$

Here I is the unit matrix, and γ has unit determinant or, equivalently, h is traceless.

The separate universe assumption requires that at a given location we can choose the space coordinates to make $\gamma_{ij} = \delta_{ij}$. To make that possible, γ_{ij} and hence h_{ij} must be time-independent. We see later that h_{ij} defines at first order the primordial tensor perturbation.

The functions ζ and h obviously determine the perturbation in the intrinsic curvature of the slices. If the tensor perturbation is negligible, ζ alone determines the curvature perturbation. In practice the term 'curvature perturbation' is taken to refer only to ζ.

In accordance with the separate universe assumption, let us assume that the local evolution of the energy–momentum tensor is the same as that of some homogeneous universe (without actually requiring isotropy as well). Then the threads must be almost orthogonal to the slices. This means that the expansion rate of an infinitesimal volume element $\mathcal{V}$, seen by a comoving observer, is given by

$$\frac{1}{\mathcal{V}}\frac{d\mathcal{V}}{dt} = \frac{3}{a}\frac{da(\mathbf{x},t)}{dt}. \tag{5.8}$$

(It is important to realize that t is the coordinate time labelling the slices, which won't in general be the same as proper time.) From this, we see that $a(\mathbf{x},t)$ is a locally defined scale factor. The curvature perturbation defines its perturbation, $\zeta = \delta(\ln a)$, without any assumption that the perturbation is actually small.

5.4.2 Conservation

By virtue of the separate universe assumption, the change in the energy $\rho\mathcal{V}$ within a given comoving volume element is equal to $-Pd\mathcal{V}$ with P is the pressure. With our chosen threading and slicing, this is equivalent to the continuity equation

$$\dot\rho(t) = -3\frac{\dot a(\mathbf{x},t)}{a(\mathbf{x},t)}\left[\rho(t) + P(\mathbf{x},t)\right] \tag{5.9}$$

$$= -3\left[\frac{\dot a(t)}{a(t)} + \dot\zeta(\mathbf{x},t)\right]\left[\rho(t) + P(\mathbf{x},t)\right]. \tag{5.10}$$

Because we are working on uniform-density slices, the energy density has no spatial dependence.

Now suppose that during some era P is a unique function of ρ, the same throughout spacetime:

$$P = P(\rho). \tag{5.11}$$

Then the spatial dependence of P also vanishes, making $\dot\zeta$ independent of position. Without loss of generality, we choose the background scale factor so that $\dot\zeta$ vanishes.

Without reference to a theory of gravity, and without using cosmological perturbation theory, we have found that ζ is conserved if and only if Eq. (5.11) is satisfied. According to the terminology established after Eq. (5.4), this is the adiabatic condition for the pressure, and when it is satisfied we will simply say that the pressure is adiabatic.

The pressure is adiabatic during any era where there is sufficiently complete radiation domination ($P = \rho/3$) or matter domination ($P = 0$), irrespective of the composition of the cosmic fluid. More generally, the pressure is adiabatic if the cosmic fluid consists of several components, each of them either radiation or matter, which satisfy the adiabatic condition (5.4).

At the primordial epoch $T \sim 1\,\mathrm{MeV}$, we know that the Universe is radiation dominated to very high accuracy, making ζ practically constant. From now, $\zeta(\mathbf{x})$ will denote this constant value.

5.4.3 The δN formula

In Eqs. (5.6) and (5.7) we defined ζ in terms of the spatial metric on the uniform-density slicing. On any slicing we can write the spatial metric in a similar form;

$$\tilde{g}_{ij} = \tilde{a}^2(\mathbf{x}, t)\tilde{\gamma}_{ij}(\mathbf{x}, t), \tag{5.12}$$

with

$$\tilde{a}(\mathbf{x}, t) \equiv a(t)e^{\psi(\mathbf{x},t)}, \qquad \tilde{\gamma}_{ij}(\mathbf{x}) \equiv \left(Ie^{2\tilde{h}}\right)_{ij}. \tag{5.13}$$

We still suppose that the metric is smoothed on a comoving scale well outside the horizon, and we require that the displacement between the new slicing and the uniform-density slicing is smooth on the same scale. The slicing will then be practically orthogonal to the comoving threading, just like the uniform-density slicing. As a result, $\tilde{h}_{ij}(\mathbf{x})$ will be independent of t by the same argument as before, and $\tilde{a}$ will be the local scale factor. At a given value of t, the local scale factor $\tilde{a}$ is different from the original one only because of the time displacement between the slices of a given t. In view of this, there is no need for the tilde and we will drop it.

The perturbation in the local number of Hubble times between two generic slices is

$$\boxed{\delta N_{12}(\mathbf{x}) = \delta \int_{t_1}^{t_2} \frac{1}{a}\frac{da}{dt}dt = \psi(\mathbf{x}, t_2) - \psi(\mathbf{x}, t_1)}\,. \tag{5.14}$$

A particularly interesting choice is the slicing with $\psi = 0$, which we will call the flat slicing. Its time displacement from the uniform-density slicing is obviously smooth on the same scale as the latter.

We define $N(\mathbf{x},t)$ as the local number of Hubble times starting from a fixed flat slice, and ending at a uniform-density slice labelled by time t. Then

$$\boxed{\zeta(\mathbf{x},t) = \delta N(\mathbf{x},t)}\,. \tag{5.15}$$

This is the δN formula, which proves very useful when considering how the curvature perturbation may originate in the very early Universe.

5.5 Linear density perturbations

The uniform-density slicing provides a clear definition of ζ but it has limitations. In particular, exactly because it does correspond to uniform density, this slicing cannot be used to define the density perturbation that was our starting point in Section 5.1. In this section, we see how to define the density perturbation by using a different slicing, which allows us to relate the density perturbation to ζ. To that end, we first calculate the effect of going from one gauge to another.

5.5.1 Gauge transformation

Let us first consider a quantity f, which at a given spacetime point is defined independently of the coordinates. Consider a change of gauge affecting only the time coordinate (hence changing only the slicing);

$$\tilde{t}(t,\mathbf{x}) = t + \delta t(t,\mathbf{x}). \tag{5.16}$$

This is just a relabelling of points in the perturbed spacetime. The relabelling doesn't change f at a given point, but it does change its separation into a background plus a perturbation. Making the separation first in the old gauge and then in the new one, we have

$$f(t) + \delta f(\mathbf{x},t) = f(\tilde{t}) + \widetilde{\delta f}(\mathbf{x},\tilde{t}), \tag{5.17}$$

and hence

$$\widetilde{\delta f}(\mathbf{x},\tilde{t}) - \delta f(\mathbf{x},t) = f(t) - f(\tilde{t}). \tag{5.18}$$

The right-hand side is the difference between the unperturbed values, which to first order is $-\dot{f}\delta t$. We conclude that the gauge transformation for the perturbation is

$$\boxed{\widetilde{\delta f}(\mathbf{x},\tilde{t}) - \delta f(\mathbf{x},t) = -\dot{f}(t)\delta t(\mathbf{x},t)}\,. \tag{5.19}$$

If we consider instead a change of threading, there is no change in δf to first order, because the background value $f(t)$ is independent of position.

The above analysis applies to any quantity whose definition at a given point is independent of the coordinate choice. It applies to energy density and pressure, since according to Section 3.2 they are defined at each point once and for all in the local rest frame (Eqs. (2.39) and (2.40)). It applies also to number density n, if that is defined in the local rest frame.

In contrast, the components of a 4-vector or 4-tensor, evaluated at a given spacetime position, depend on the coordinates. We will need the effect of a gauge transformation on a second-rank tensor, allowing the space coordinates to change as well as the time coordinate;

$$\widetilde{x}^{\mu} = x^{\mu} + \delta x^{\mu}(x^0, x^1, x^2, x^3). \tag{5.20}$$

As in the previous case the gauge transformation shifts the slicing of spacetime, and now it shifts the threading as well. The effect of those shifts is given by the generalization of Eq. (5.19). But because we now deal with a tensor, we also need to take into account the effect (3.2), that the coordinate transformation would have even if there were no shifts. Including everything we find to first order

$$\boxed{\widetilde{\delta B}_{\mu\nu} - \delta B_{\mu\nu} = -B_{\alpha\nu}\partial_\mu \delta x^\alpha - B_{\mu\alpha}\partial_\nu \delta x^\alpha - \delta x^\lambda \partial_\lambda B_{\mu\nu}}\,. \tag{5.21}$$

5.5.2 Primordial density perturbation

Starting from the uniform-density slicing, let us go to a generic slicing with time coordinate $\tilde{t} = t + \delta t(\mathbf{x}, t)$. To first order, $\tilde{t}(\mathbf{x}, t - \delta t(\mathbf{x}, t)) = t$. Therefore, the time shift going from a uniform density slice at time t to a generic slice at time $\tilde{t} = t$ is $-\delta t$. Using Eq. (5.14) with $t_2 - t_1 = \delta t$, we learn that the first-order curvature perturbation on the generic slicing is given by[2]

$$\psi = \zeta - H\delta t. \tag{5.22}$$

From Eq. (5.19) the density perturbation is given by

$$\delta\rho(\mathbf{x}, t) = -\dot{\rho}(t)\delta t(\mathbf{x}, t). \tag{5.23}$$

(From now on we are dropping the tilde on the time coordinate of the generic slicing.) We conclude that the density perturbation $\delta\rho$ and the curvature perturbation ψ are related by

$$\zeta = \psi - H\frac{\delta\rho}{\dot{\rho}} \tag{5.24}$$

$$= \psi + \frac{1}{3}\frac{\delta\rho}{\rho + P}. \tag{5.25}$$

This expression may be regarded as a gauge-invariant definition of ζ, valid to

[2] This formula can also be derived from Eq. (5.21) and is actually valid on all scales.

first order in the cosmological perturbations on scales outside the horizon. An important slicing is the flat slicing with $\psi = 0$. On that slicing

$$\boxed{\zeta = -H\frac{\delta\rho}{\dot{\rho}} = \frac{1}{3}\frac{\delta\rho}{\rho + P}}. \tag{5.26}$$

For future reference we note the following extension of this result. Suppose that the cosmic fluid has separate components which don't exchange energy. Suppose also that each component has a unique $P_{\rm a}(\rho_{\rm a})$. Then there are separately conserved quantities

$$\zeta_{\rm a} \equiv -\frac{H\delta\rho_{\rm a}}{\dot{\rho}_{\rm a}} = \frac{1}{3}\frac{\delta\rho_{\rm a}}{\rho_{\rm a} + P_{\rm a}}, \tag{5.27}$$

where $\delta\rho_{\rm a}$ are defined on flat slices. This leads to

$$\boxed{\zeta(t) = \frac{\sum(\rho_{\rm a} + P_{\rm a})\zeta_{\rm a}}{\rho + P}}. \tag{5.28}$$

Finally, we consider the perturbations $\delta\rho_{\rm a}$ of the fluid components, with a $= \gamma$, ν, c or B. If the adiabatic condition Eq. (5.4) is satisfied, the separate perturbations $\delta\rho_{\rm a}$ vanish on the slicing where $\delta\rho$ vanishes. On any other slicing, they are then given by $\delta\rho_{\rm a} = -\dot{\rho}_{\rm a}\delta t$. Using the continuity equation (3.76) to evaluate $\dot{\rho}_{\rm a}$ we conclude that the adiabatic condition implies

$$\boxed{\frac{1}{3}\delta_B = \frac{1}{3}\delta_{\rm c} = \frac{1}{4}\delta_\gamma = \frac{1}{4}\delta_\nu}, \tag{5.29}$$

where $\delta_{\rm a} \equiv \delta\rho_{\rm a}/\rho_{\rm a}$ are the density contrasts. The adiabatic condition thus determines each $\delta\rho_{\rm a}$ in terms of the total $\delta\rho$, which in turn is related to the curvature perturbation ψ by Eq. (5.25).

Exercises

5.1 Show that the adiabatic condition $\rho_{\rm a}(\rho)$ is equivalent to the adiabatic condition $n_{\rm a}(\rho)$, at the epoch $T \sim 1\,\text{MeV}$ when Eq. (5.4) is supposed to apply.

5.2 Verify Eq. (5.21) for the gauge transformation of a tensor.

5.3 (a) Use Eq. (5.26) to calculate $\dot{\zeta}$ to first order in the perturbations δP and $\delta\rho$. (b) Show that $\dot{\zeta} = 0$ if the pressure is adiabatic (i.e. is a unique function of energy density, the same at every location).

5.4 Use Eqs. (5.26) and (5.29) to obtain the separate density contrasts $\delta_{\rm a}$ in terms of ζ.

Reference

[1] D. H. Lyth, K. A. Malik and M. Sasaki. A general proof of the conservation of the curvature perturbation. *JCAP*, **0505** (2005) 004.

6
Stochastic properties

The time dependence of each perturbation is well defined, being determined by laws of physics. Viewed instead as a function of position at fixed time, the perturbations have random distributions. It is the statistical properties of these distribution that we wish to uncover via observation, and relate to fundamental physics models for the origin of perturbations. Those are usually referred to as **stochastic** properties.

The inherent randomness means that one shouldn't aim to predict things like the precise location of particular galaxies. Questions should refer to stochastic properties only. Don't ask 'how far is it to the nearest large galaxy?'; instead ask 'what is the typical separation between large galaxies?'. This randomness echoes simple quantum mechanics, e.g. one shouldn't hope to predict the precise position of a single particle in a closed box, but could compute the typical distance of the particle from its centre averaged over many such boxes. Indeed, we will see that in the inflationary cosmology the randomness of cosmological perturbations does have its origin in quantum uncertainty.

To describe the stochastic properties of the perturbations, one invokes the mathematical concept of a random field. In this chapter we describe the relevant aspects of that concept, without tying ourselves at this stage to any particular perturbation.

6.1 Random fields

Consider just one perturbation, evaluated at some instant, which we denote by $g(\mathbf{x})$. We take $g(\mathbf{x})$ to be associated with what is called a **random field**. This mathematical term denotes a set of functions $g_n(\mathbf{x})$ each coming with a probability $\mathbf{P}_n$.[1] The set of functions is referred to as the **ensemble**, and each individual function is called a **realization** of the ensemble. In practical applications, including

[1] We are taking the functions to depend only on three space coordinates, as is appropriate for cosmology. In many other applications of the theory of random fields, the functions are taken to depend only on time.

the cosmological one, the set of functions is continuous, so that instead of discrete probabilities there is a continuous probability distribution.

We will discuss the relevant aspects of random fields at the informal level that is usual in physics. A rigorous discussion can be found in mathematics texts, such as Ref. [1] for the elementary aspects or Ref. [2] for an advanced discussion.

Instead of dealing directly with the probability functions, one usually works with the correlators. The two-point correlator is

$$\langle g(\mathbf{x})g(\mathbf{x}')\rangle \equiv \sum_n \mathbf{P}_n g_n(\mathbf{x})g_n(\mathbf{x}'), \tag{6.1}$$

and the N-point correlator is defined similarly.

The random field is usually supposed to be statistically homogeneous and isotropic. In other words, the probabilities attached to the realizations are supposed to be invariant under translations and rotations. They are also taken to be invariant under the parity transformation which reverses the handedness of the coordinate system.

Translation invariance (homogeneity) means that the probability attached to a realization $g_n(\mathbf{x})$ is the same as the one attached to the realization $g_n(\mathbf{x}+\mathbf{X})$, for each fixed $\mathbf{X}$. Starting with *one* realization, it can be shown that the entire random field is generated by allowing $\mathbf{X}$ to take on arbitrary values. This is referred to as the **ergodic** property of the field, and follows from the homogeneity with weak assumptions [3]. The word 'ergodic' was coined in the context of statistical mechanics, where one deals with a random field that depends on time as opposed to position. According to the ergodic theorem, the ensemble average (two-point correlator) $\langle g(\mathbf{x})g(\mathbf{x}')\rangle$ can be regarded as a spatial average at fixed $\mathbf{x}'-\mathbf{x}$ for a single realization of the ensemble, and similarly for higher correlators. We prove this property of the correlators in Section 6.5.

Rotational invariance (isotropy) means that the probability attached to a realization $g_n(\mathbf{x})$ is the same as the one attached to the realization $g_n(\tilde{\mathbf{x}})$ where $\tilde{\mathbf{x}}$ are rotated coordinates.[2] Dealing with a set of observations made on a sufficiently small patch of the sky, rotational invariance allows us to replace the ensemble average by an average with respect to the location of the patch on the sky for a single realization.

The parity transformation reverses the handedness of the coordinates. It can be achieved, for instance, by reversing just one Cartesian coordinate. The parity transformation relates pairs of the realizations, which as we will see may reduce the number of independent correlators that need be considered.

From now on, we dispense with the notation $g_n(\mathbf{x})$ and use a single function

[2] At this point we are taking g to be a single number, as opposed to a vector or tensor whose components would undergo a linear transformation if a rotation were performed. As we see in connection with the cosmic microwave background (CMB) anisotropy, it is easy to take such a transformation into account.

$g(\mathbf{x})$ which, according to the context, will be a generic member of the ensemble or the particular realization that corresponds to the perturbation of our Universe.

6.2 Fourier expansion

The Fourier expansion is a powerful tool for analyzing stochastic properties. One can work with a Fourier series, defined in a finite box L much bigger than the region of interest, so that the wave-vectors $\mathbf{k}_n$ form a cubic lattice with spacing $2\pi/L$. Alternatively, one takes the box size to infinity so that there are continuous wave-vectors $\mathbf{k}$ and the Fourier series becomes a Fourier integral. The Fourier series generally leads to a discussion which is conceptually simple, but too messy for everyday use, where one employs instead the Fourier integral.

Our approach will be to use the Fourier series where necessary to explain the concepts. Then we allow the box to be much bigger than the observable Universe, so that the wave-vectors can be regarded as continuous. In that way we arrive at the Fourier integral, which will normally be used. We will see that, in principle, the box size should remain finite. In particular, we sometimes need a box such that $\ln(H_0 L)$ is only of order a few, so that it can be set equal to 1 for rough estimates. Since cosmological scales don't extend over very many orders of magnitude, $\ln(kL)$ will also be only of order a few over the whole range of cosmological scales. We will call such a box a **minimal box**. Usually though, there is no need to specify the box size.

We consider a generic perturbation $g(\mathbf{x})$, evaluated at some particular time. Working in a box of comoving size L, the Fourier series and its inverse may be written as

$$g(\mathbf{x}) = \frac{1}{L^3} \sum_n g_n e^{i\mathbf{k}_n \cdot \mathbf{x}}, \qquad g_n = \int g(\mathbf{x}) e^{-i\mathbf{k}_n \cdot \mathbf{x}} d^3x. \tag{6.2}$$

The momenta $\mathbf{k}_n$ form a cubic lattice with spacing $2\pi/L$, and they can be labelled by a single index n. If we allow the index to have either sign and choose $\mathbf{k}_{-n} = -\mathbf{k}_n$, the reality of $g(\mathbf{x})$ is equivalent to $g_n^* = g_{-n}$. We have

$$\int e^{i(\mathbf{k}_n - \mathbf{k}_m)\cdot \mathbf{x}} d^3x = L^3 \delta_{nm}. \tag{6.3}$$

If the $k = 0$ mode (spatial average) is non-zero, it is ignored in the context of stochastic properties. It can be absorbed into the background unless it corresponds to anisotropy.

The Fourier integral corresponds to the limit $L \to \infty$ and may be written

$$g(\mathbf{x}) = \frac{1}{(2\pi)^3} \int g(\mathbf{k}) e^{i\mathbf{k}\cdot\mathbf{x}} d^3k, \qquad g(\mathbf{k}) = \int g(\mathbf{x}) e^{-i\mathbf{k}\cdot\mathbf{x}} d^3x. \tag{6.4}$$

Then the reality condition is $g(-\mathbf{k}) = g^*(\mathbf{k})$ and

$$\int e^{i(\mathbf{k}-\mathbf{k}')\cdot\mathbf{x}} d^3x = (2\pi)^3 \delta^3(\mathbf{k}-\mathbf{k}'). \tag{6.5}$$

Following a common practice in cosmology, we shall generally call $\mathbf{k}$ and k the momentum even though that is correct only in the quantum regime with g a field.

The Fourier component of a product is

$$(fg)_{\mathbf{k}} = \frac{1}{(2\pi)^3} \int d^3q f_{\mathbf{q}} g_{\mathbf{k}-\mathbf{q}}. \tag{6.6}$$

This is the convolution theorem. Conversely, the Fourier component of a convolution is the product of Fourier components:

$$\left[\int f(\mathbf{x}-\mathbf{y}) g(\mathbf{y}) d^3y\right]_{\mathbf{k}} = f_{\mathbf{k}} g_{\mathbf{k}}. \tag{6.7}$$

As in these expressions, we will sometimes write the argument of a Fourier component as a subscript. Where it causes no confusion we will drop the argument altogether.

In going between the Fourier sum (6.2) and Fourier integral (6.4), the correspondences are

$$\left(\frac{2\pi}{L}\right)^3 \sum_n \rightarrow \int d^3k, \tag{6.8}$$

$$g_n \rightarrow g(\mathbf{k}), \tag{6.9}$$

$$\left(\frac{L}{2\pi}\right)^3 \delta_{nn'} \rightarrow \delta^3(\mathbf{k}-\mathbf{k}'). \tag{6.10}$$

The last relation gives the rule

$$[\delta^3(\mathbf{k}-\mathbf{k}')]^2 = (L/2\pi)^3 \delta^3(\mathbf{k}-\mathbf{k}'), \tag{6.11}$$

which is useful when considering volume averages.

6.3 Gaussian perturbations

6.3.1 Momentum space

Now we describe the simplest type of random field, called a gaussian random field. We deal with a generic perturbation denoted by $g(\mathbf{x})$, the argument t being suppressed because the stochastic properties of the perturbations are defined at fixed t.

A gaussian random field may be defined as one whose Fourier coefficients have no correlation except for the reality condition. The two-point correlator in Fourier space is of the form

$$\langle g_n g_m^* \rangle = \delta_{nm} P_{g_n}, \tag{6.12}$$

where $P_{g_n} \equiv \langle |g_n|^2 \rangle$. If we allow the index n to have either sign and choose $\mathbf{k}_{-n} = -\mathbf{k}_n$ this can be written

$$\langle g_n g_m \rangle = \delta_{n,-m} P_{g_n}, \tag{6.13}$$

The two-point correlator determines all higher correlators. The odd-n correlators all vanish:[3]

$$\langle g_n \rangle = \langle g_{n1} g_{n2} g_{n3} \rangle = \ldots = 0, \tag{6.14}$$

and the four-point correlator is given by

$$\langle g_{n1} g_{n2} g_{n3} g_{n4} \rangle = \langle g_{n1} g_{n2} \rangle \langle g_{n3} g_{n4} \rangle + \langle g_{n1} g_{n3} \rangle \langle g_{n2} g_{n4} \rangle + \langle g_{n1} g_{n4} \rangle \langle g_{n2} g_{n3} \rangle, \tag{6.15}$$

with similar expressions for higher correlators.

Now we go to the continuum limit of these expressions, defining the **spectrum** $P_g(k)$ by

$$L^{-3} P_{g_n} \to P_g(k). \tag{6.16}$$

The spectrum is taken to depend only on the magnitude of $\mathbf{k}$, corresponding to rotational invariance.

The first three correlators are now

$$\boxed{\langle g_{\mathbf{k}} \rangle = 0}, \tag{6.17}$$

$$\boxed{\langle g_{\mathbf{k}} g_{\mathbf{k}'} \rangle = (2\pi)^3 \delta^3_{\mathbf{k}+\mathbf{k}'} P_g(k)}, \tag{6.18}$$

$$\boxed{\langle g_{\mathbf{k}_1} g_{\mathbf{k}_2} g_{\mathbf{k}_3} \rangle = 0}. \tag{6.19}$$

The four-point correlator is[4]

$$\langle g_{\mathbf{k}_1} g_{\mathbf{k}_2} g_{\mathbf{k}_3} g_{\mathbf{k}_4} \rangle = \langle g_{\mathbf{k}_1} g_{\mathbf{k}_2} \rangle \langle g_{\mathbf{k}_3} g_{\mathbf{k}_4} \rangle + \langle g_{\mathbf{k}_1} g_{\mathbf{k}_3} \rangle \langle g_{\mathbf{k}_2} g_{\mathbf{k}_4} \rangle + \langle g_{\mathbf{k}_1} g_{\mathbf{k}_4} \rangle \langle g_{\mathbf{k}_2} g_{\mathbf{k}_3} \rangle, \tag{6.20}$$

which can be written

$$\boxed{\langle g_{\mathbf{k}_1} g_{\mathbf{k}_2} g_{\mathbf{k}_3} g_{\mathbf{k}_4} \rangle = (2\pi)^6 \delta^3_{\mathbf{k}_1+\mathbf{k}_2} \delta^3_{\mathbf{k}_3+\mathbf{k}_4} P_g(k_1) P_g(k_3) + \text{ two permutations}}. \tag{6.21}$$

Similar expressions hold for higher correlators.

[3] Note that $\langle g_n \rangle$ isn't a correlator, but from translation invariance it could be non-zero only for the zero mode (the one with $\mathbf{k}_n = 0$) which can be absorbed into the unperturbed (background) universe.

[4] We are using a subscript to distinguish different momenta, but they are continuous quantities and not the discrete momenta $\mathbf{k}_n$ that we considered earlier for a finite box.

Although these expressions are motivated by the discrete case, they can be taken as the definition of a gaussian perturbation, with Eq. (6.18) defining the spectrum. It is often convenient to define a quantity $\mathcal{P}_g \equiv (k^3/2\pi^2)P_g$, also called the spectrum. The prefactor is designed to give Eq. (6.22) below its simple form.

Because of the delta function in Eq. (6.18), the Fourier components $g_{\mathbf{k}}$ don't have numerical values and neither does the function $g(\mathbf{x})$. This may be regarded as a pathology, associated with the process of taking the infinite limit for the finite box.

6.3.2 Position space

The position-space quantity $g(\mathbf{x})$ is a superposition of Fourier modes. Gaussianity of g means that there is no correlation between the modes; drawing them randomly from the ensemble, each of them has an independent probability distribution. But the central limit theorem states, under very general conditions, that the sum of uncorrelated quantities has a gaussian probability distribution independently of the probability distributions of the original quantities. We conclude that, for a gaussian random field, *the probability distribution of $g(\mathbf{x})$ at a given point is gaussian.* Using the definitions of P and $\mathcal{P}$, the mean-square (ensemble average) is

$$\boxed{\sigma_g^2(\mathbf{x}) \equiv \langle g^2(\mathbf{x})\rangle = \frac{1}{(2\pi)^3}\int P_g(k) d^3k = \int_0^\infty \mathcal{P}_g(k)\frac{dk}{k}}. \tag{6.22}$$

We see that the spectrum $\mathcal{P}_g(k)$ is the contribution to σ_g^2 per unit logarithmic interval of k. Notice that σ_g^2 is independent of position, by virtue of the delta function in Eq. (6.18).

What about the convergence of this integral? Some quantities, such as the matter density perturbation, have $\mathcal{P}_g(k)$ increasing strongly with k, up to a cutoff which corresponds to a **coherence length** typically far smaller than cosmological scales. Before one can apply cosmological perturbation theory, such quantities typically need to be smoothed on some scale bigger than the actual coherence length. This is equivalent to imposing a cutoff on the spectrum at the inverse of the smoothing scale. From Eq. (9.3), the smoothed quantity has mean-square

$$\langle g^2(R, \mathbf{x})\rangle \equiv \sigma_g^2(R) \simeq \int_0^{R^{-1}} \mathcal{P}_g(k)\frac{dk}{k} \sim \mathcal{P}_g(R^{-1}). \tag{6.23}$$

In words, the mean-square is of order the spectrum, evaluated at the cutoff scale.

Other quantities, such as the primordial curvature perturbation, have $\mathcal{P}_g$ almost scale-independent, at least on cosmological scales. If $\mathcal{P}_g$ is scale-independent, the integral (6.22) is logarithmically divergent at small and large k. Large k is again handled with a smoothing scale R, while small k by working in a box with comoving size L. Then

$$\sigma_g^2(R) \simeq \mathcal{P}_g \int_{L^{-1}}^{R^{-1}} \frac{dk}{k} = \mathcal{P}_g \ln \frac{L}{R}. \tag{6.24}$$

Choosing a minimal box, such that $\ln(kL)$ is only a few, the mean-square is again roughly of order the spectrum.

Using Eq. (6.21) we can also find $\langle g^4(\mathbf{x})\rangle = 3\sigma_g^4$, and using the analogous expression for higher products we can find

$$\langle g^{2n}(\mathbf{x})\rangle = (2n-1)!!\,\sigma_g^{2n} \equiv (2n-1)!!\langle g^2\rangle^n . \tag{6.25}$$

The double exclamation is multiplication over every second integer up to the argument. This is equivalent to the statement that the probability distribution of $g(\mathbf{x})$ at a given point is gaussian, as one can check by calculating the moments under that assumption. Indeed, the gaussian probability distribution $\mathbf{P}(g)$, giving the probability of finding g within a given interval dg, is

$$\boxed{\mathbf{P}(g) = \frac{1}{\sqrt{2\pi}\sigma} \exp\left(-\frac{g^2}{2\sigma_g^2}\right)}, \tag{6.26}$$

and the moments are

$$\langle g^{2n}(\mathbf{x})\rangle = \frac{1}{\sqrt{2\pi}\sigma_g} \int_{-\infty}^{\infty} \mathbf{P}(g) g^{2n} dg \, , \tag{6.27}$$

leading to Eq. (6.25).

We also can work out the two-point correlator, using Eq. (6.18):

$$\langle g(\mathbf{y}) g(\mathbf{x}+\mathbf{y})\rangle = \frac{1}{(2\pi)^3} \int d^3k P_g(k) e^{i\mathbf{k}\cdot\mathbf{x}} \tag{6.28}$$

$$= \int_0^\infty \mathcal{P}_g(k) \frac{\sin(kx)}{kx} \frac{dk}{k}. \tag{6.29}$$

It is invariant under translations because of the delta function in Eq. (6.18), and invariant under rotations because the spectrum was assumed to be invariant under rotations. Since this correlator completely specifies the stochastic properties of the perturbation, its invariance under translations is equivalent to statistical homogeneity of the perturbation, and its invariance under rotations is equivalent to statistical isotropy. Notice that the delta function is a consequence of gaussianity, which means that *gaussianity implies statistical homogeneity*.

6.4 Non-gaussian perturbations

6.4.1 Correlators

The Fourier coefficients of a gaussian perturbation have only the minimal correlation, demanded by the reality condition. As a result the stochastic properties of the perturbation are completely defined by its spectrum. In particular the non-zero correlators of the Fourier coefficients are given by Eqs. (6.17)–(6.21). The Fourier coefficients of a non-gaussian perturbation have additional correlation, not specified by the spectrum.

At the time of writing no additional correlation has been observed, for a cosmological perturbation described by linear perturbation theory. As the observations are quite accurate, this means that the observed perturbations are gaussian to quite high accuracy. As we shall see though, scenarios for the origin of perturbations can produce non-gaussianity at a level that could be observed in the future, or even at a level that should already have been observed.

Let us consider the form that additional correlation might take, assuming invariance under rotations. Translation invariance demands that each correlator vanishes when the sum of the momenta $\mathbf{k}_i$ vanishes, and we are interested only in non-zero momenta. The one-point correlator $\langle g_{\mathbf{k}} \rangle$ would be proportional to $\delta^3_{\mathbf{k}}$ which means that it can be absorbed into the unperturbed (background) quantity. The form of the two-point correlator is the same as in the gaussian case, since the only other two-point correlator consistent with translation invariance would be proportional to $\delta^3_{\mathbf{k}_1}\delta^3_{\mathbf{k}_2}$ which also vanishes for non-zero momenta. In the following, contributions to correlators proportional to a product of $\delta_{\mathbf{k}_i}$ factors will be dropped without comment.

The three-point correlator vanishes in the gaussian case. Its possible form in the non-gaussian case is

$$\boxed{\langle g_{\mathbf{k}_1} g_{\mathbf{k}_2} g_{\mathbf{k}_3} \rangle = (2\pi)^3 \delta^3_{\mathbf{k}_1+\mathbf{k}_2+\mathbf{k}_3} B_g(k_1, k_2, k_3)}\,, \tag{6.30}$$

where B_g is called the **bispectrum**. The delta function corresponds to invariance under translations, and the fact that the bispectrum depends only on the lengths of the three sides of the triangle formed by the momenta corresponds to invariance under rotations. It is often convenient to define the **reduced bispectrum** $\mathcal{B}_g$ by

$$\boxed{B_g(k_1, k_2, k_3) = \mathcal{B}_g(k_1, k_2, k_3)\left[P_g(k_1)P_g(k_2) + \text{cyclic permutations}\right]}\,. \tag{6.31}$$

Going to position space we have

$$\langle g(\mathbf{z}+\mathbf{x})g(\mathbf{z}+\mathbf{y})g(\mathbf{z})\rangle = \frac{1}{(2\pi)^6}\int d^3k_1 d^3k_2 B_g(k_1,k_2,k_3)e^{i(\mathbf{k}_1\cdot\mathbf{x}+\mathbf{k}_2\cdot\mathbf{y})}, \quad (6.32)$$

$$\langle g^3(\mathbf{x})\rangle = \frac{1}{(2\pi)^6}\int d^3k_1 d^3k_2 B_g(k_1,k_2,k_3). \quad (6.33)$$

The second relation gives the skewness of the probability distribution defined by $S_g \equiv \langle g^3\rangle/\langle g^2\rangle^{3/2}$. If $\mathcal{P}_g$ and $\mathcal{B}_g$ are scale independent

$$\boxed{S_g \sim \mathcal{B}_g\mathcal{P}_g^{1/2}}. \quad (6.34)$$

where as usual we work in a minimal box. To the extent that S_g is a measure of non-gaussianity, an almost-gaussian perturbation will have $\mathcal{B}_g \ll \mathcal{P}_g^{-1/2}$.

The four-point correlator is non-vanishing in the gaussian case, with the form (6.21) which relates it to the power spectrum. This contribution to the correlator is always present, but in the non-gaussian case there can be an additional contribution of the form

$$\boxed{\langle g_{\mathbf{k}_1}g_{\mathbf{k}_2}g_{\mathbf{k}_3}g_{\mathbf{k}_4}\rangle = (2\pi)^3\delta^3_{\mathbf{k}_1+\mathbf{k}_2+\mathbf{k}_3+\mathbf{k}_4}T_g}. \quad (6.35)$$

The **trispectrum** T_g will depend on the six scalars specifying the quadrilateral formed by the wave-vectors. The reduced trispectrum $\mathcal{T}_g$ is defined by

$$\boxed{T_g(\mathbf{k}_1,\mathbf{k}_2,\mathbf{k}_3,\mathbf{k}_4) = \mathcal{T}_g\left[P_g(k_1)P_g(k_2)P_g(k_{14}) + 23\,\text{permutations}\right]}, \quad (6.36)$$

with $k_{14} = |\mathbf{k}_1 - \mathbf{k}_4|$.

In position space the contribution (6.21) has the form

$$\langle g(\mathbf{x}_1)g(\mathbf{x}_2)g(\mathbf{x}_3)g(\mathbf{x}_4)\rangle = \langle g(\mathbf{x}_1)g(\mathbf{x}_2)\rangle\langle g(\mathbf{x}_3)g(\mathbf{x}_4)\rangle + 2\,\text{permutations}. \quad (6.37)$$

It doesn't go to zero as the separation between (say) the pair of points $(\mathbf{x}_1,\mathbf{x}_2)$ and the pair $(\mathbf{x}_3,\mathbf{x}_4)$ goes to infinity. It is called the **disconnected contribution** to the correlator. In contrast, the new contribution (6.35) will go to zero in this case, and is significantly different from zero only if all four points are sufficiently close to each other. It is called the **connected contribution** to the correlator. The connected contributions give

$$\boxed{\langle g^4(\mathbf{x})\rangle - 3\langle g^2(\mathbf{x})\rangle^2 \equiv \langle g^4(\mathbf{x})\rangle_c = (2\pi)^{-9}\int d^3k_1 d^3k_2 d^3k_3 T_g}, \quad (6.38)$$

from which follows the kurtosis $K_g \equiv \langle g^4\rangle_c/\langle g^2\rangle^2$. If the reduced spectrum is scale independent $K_g \sim \mathcal{T}_g\mathcal{P}_g$, which provides a measure of the non-gaussianity associated with the trispectrum.

One can now see how things will go for higher correlators. The N-point correlator will have disconnected contributions proportional to products of delta functions, plus a possible connected contribution proportional to the overall delta function. The disconnected contributions are inevitable, determined by lower-order correlators. The connected contribution is optional, and if present requires a new function to specify it analogous to the bispectrum and the trispectrum. Within the inflationary cosmology, higher connected correlators of the primordial perturbations are generally expected to be small.

6.5 Ergodic theorem and cosmic variance

6.5.1 Ergodic theorem

Now we are in a position to prove the ergodic property of correlators, that the ensemble average can be regarded as a spatial average. Consider the two-point correlator. We work in a finite box, and take the spatial average, to find

$$\begin{aligned} \overline{g(\mathbf{y})g(\mathbf{x}+\mathbf{y})} &= L^{-3}\int g(\mathbf{y})g(\mathbf{x}+\mathbf{y})d^3y \\ &= L^{-3}(2\pi)^{-6}\int g_{\mathbf{k}}g_{\mathbf{k}'}e^{i[\mathbf{k}\cdot\mathbf{y}+\mathbf{k}'\cdot(\mathbf{x}+\mathbf{y})]}d^3y d^3k d^3k' \\ &= L^{-3}(2\pi)^{-3}\int \delta^3(\mathbf{k}+\mathbf{k}')g_{\mathbf{k}}g_{\mathbf{k}'}e^{i\mathbf{k}\cdot\mathbf{x}}d^3k d^3k'. \end{aligned} \tag{6.39}$$

We took the continuous limit of the Fourier series, which is justified if kL is sufficiently large or, in other words, if the box size is sufficiently bigger than the scales under consideration.

Within each volume element d^3k, the Fourier coefficients are uncorrelated, because of the delta functions in Eqs. (6.18), (6.30) and (6.35) and so on. Therefore in the limit where d^3kd^3k' contains an infinite number of points, we can replace $g_{\mathbf{k}}g_{\mathbf{k}'}$ by its ensemble average $\langle g_{\mathbf{k}}g_{\mathbf{k}'}\rangle$ to find

$$\overline{g(\mathbf{y})g(\mathbf{x}+\mathbf{y})} = L^{-3}(2\pi)^{-3}\int \delta^3(\mathbf{k}+\mathbf{k}')\langle g_{\mathbf{k}}g_{\mathbf{k}'}\rangle e^{i\mathbf{k}\cdot\mathbf{x}}d^3k d^3k' \tag{6.40}$$

$$= L^{-3}\int [\delta^3(\mathbf{k}+\mathbf{k}')]^2 P_g(k)e^{i\mathbf{k}\cdot\mathbf{x}}d^3k d^3k' \tag{6.41}$$

$$= (2\pi)^{-3}\int P_g(k)e^{i\mathbf{k}\cdot\mathbf{x}}d^3k, \tag{6.42}$$

where we used Eq. (6.11) for the last step. Comparing with Eq. (6.28), we see that the spatial average is indeed equal to the ensemble average. The same thing works for higher correlators.

6.5.2 Cosmic variance

If the box isn't big enough, the spatial average of a correlator will be significantly different from the ensemble expectation value. Equivalently, the average of the momentum-space correlator within a cell of momentum space will be significantly different from the ensemble average. Similarly, the angular average mentioned on page 86 will be significantly different from the ensemble average if the patch of sky isn't sufficiently small.

The mean-square difference between the ensemble expectation value of a quantity that is predicted by theory and the observed average that approximates it is called the **cosmic variance**. If it isn't negligible, it has to be taken into account along with the error bars of the data and the uncertainty of the theoretical prediction. If X is the quantity and $\bar{X}$ is the observed average, the cosmic variance is

$$(\Delta X)^2 \equiv \langle (\bar{X} - \langle X \rangle)^2 \rangle = \langle \bar{X}^2 \rangle - \langle X \rangle^2. \tag{6.43}$$

One usually says that this is 'the cosmic variance of $\langle X \rangle$'.

Cosmic variance of the spectrum

Let us consider the cosmic variance of the spectrum of a perturbation g. Working in a finite box the spectrum P_n is defined as $\langle |g_n|^2 \rangle$. To estimate it from observation one will measure a 'pseudo-spectrum'[5]

$$\widetilde{P}_n \equiv \overline{|g_m|^2}, \tag{6.44}$$

where the average is over a cell of $\mathbf{k}$-space around $\mathbf{k}_n$. The pseudo-spectrum is calculated by summing a large number of uncorrelated Fourier components. Running over the ensemble, its probability distribution will be gaussian by virtue of the central limit theorem. In fact, we can go further and declare that each individual Fourier component has a gaussian distribution. This is because an individual Fourier component within the box that we are considering would correspond to an average over many components were we to consider a much bigger box (corresponding to much smaller spacing of the cubic lattice of discrete momenta). The cosmic variance of P_n is

$$(\Delta P_n)^2 \equiv \left\langle \left(\widetilde{P}_n - P_n \right)^2 \right\rangle = \langle \widetilde{P}_n \rangle^2 - P_n^2, \tag{6.45}$$

Suppose first that we just measure the real part x, of a single Fourier coefficient within the cell around $\mathbf{k}_n$. Then

$$(\Delta P_n)^2 = \left\langle \left(x^2 - P_n \right)^2 \right\rangle = \langle x^4 \rangle - P_n^2 = 2P_n^2. \tag{6.46}$$

[5] In this subsection we drop the subscript g, writing P_n instead of P_{g_n} and so on.

Measuring both the real and imaginary parts of N Fourier coefficients will reduce $(\Delta P_n)^2$ by a factor $2N$. This gives $\Delta P_n/P_n = 1/\sqrt{N}$.

Suppose we demand a resolution $\delta \equiv dk/k$. The number of independent real and imaginary parts in this interval is

$$N = 4\pi k^2 \left(\frac{L}{2\pi}\right)^3 dk \sim (kL)^3 \delta. \tag{6.47}$$

Going to the continuous limit, so as to consider the spectrum $P(k)$, we conclude that the cosmic variance $(\Delta P)^2$ of the spectrum is given by

$$\frac{\Delta P(k)}{P(k)} \left(= \frac{\Delta P_n}{P_n}\right) \sim \frac{1}{\delta^{1/2}(kL)^{3/2}}. \tag{6.48}$$

The box size L should be taken as the size of the region observed, and the cosmic variance of the spectrum will be very small if the scale being probed is much smaller than L.

Cosmic variance of the bispectrum

The situation is very different if we consider the bispectrum, because the observed cosmological perturbations are almost gaussian making the bispectrum very small. Let us define a finite-box version of the bispectrum by

$$B_{12} \equiv \langle g_1 g_2 g_3 \rangle, \tag{6.49}$$

where $\mathbf{k}_1 + \mathbf{k}_2 + \mathbf{k}_3 = 0$.[6] Define the pseudo-bispectrum by

$$\tilde{B}_{12} \equiv \overline{g_1 g_2 g_3}, \tag{6.50}$$

where the average is over cells of $\mathbf{k}$-space centred on the specified values, subject to the constraint that the sum of the three momenta vanishes. Define the cosmic variance as

$$(\Delta B_{12})^2 \equiv \left\langle \left(\tilde{B}_{12} - B_{12}\right)^2 \right\rangle. \tag{6.51}$$

Suppose first that we measure just the real parts, x, y and z, of just a single Fourier component within each cell. Then

$$(\Delta B_{12})^2 = \left\langle \left(x^2 y^2 z^2 - B_{12}\right)^2 \right\rangle \tag{6.52}$$

$$= \langle x^2 y^2 z^2 \rangle - B_{12}^2 \tag{6.53}$$

$$\simeq P_1 P_2 P_3 - B_{12}^2 \simeq P_1 P_2 P_3, \tag{6.54}$$

[6] We are choosing a triplet $(\mathbf{k}_1, \mathbf{k}_2, \mathbf{k}_3)$ from the discrete momenta available, and ordering the infinite sequence k_n so that the chosen triplet corresponds to the first three of the sequence. When we go to the continuum limit, the triplet $(\mathbf{k}_1, \mathbf{k}_2, \mathbf{k}_3)$ is chosen from the continuous momenta and there is no question of a sequence.

where in the last line we took g to be almost gaussian. If N independent sets of Fourier components are measured, the cosmic variance will be reduced by a factor $2N$.

The continuous limit for the bispectrum is

$$L^{-3} b_{123} \to B(k_1, k_2, k_3). \tag{6.55}$$

Invoking also the continuous limit (6.16) of the spectrum, and assuming $k_1 \sim k_2 \sim k_3$, we conclude that the cosmic variance of the bispectrum is

$$[\Delta B(k)]^2 \sim \frac{L^3 P^3(k)}{N}. \tag{6.56}$$

If we demand a resolution $\delta = dk/k$, the number of momenta available after taking account of rotational invariance is $N \sim (kL)^6 \delta$. Putting this into Eq. (6.56) we find for the reduced bispectrum

$$\boxed{\Delta \mathcal{B}(k) \sim \frac{1}{\mathcal{P}^{1/2}(k)} \frac{1}{\delta^{1/2} (kL)^{3/2}}}. \tag{6.57}$$

In Section 6.7.2 we apply this formula to the bispectrum of the curvature perturbation ζ.

6.6 Spherical expansion

Instead of the Fourier expansion, which refers to Cartesian coordinates, we can use the expansion into radial functions. Each of these is the product of a spherical Bessel function and a spherical harmonic. The spherical expansion provides a good way of understanding the CMB anisotropy, because the last-scattering surface is a sphere. Also, in contrast with the Fourier series, it can be generalized readily to the case of an open or closed Universe; all that happens is that the radial functions become different. Here we give the flat-Universe version, which according to observation is the relevant one.

The expansion is of the form

$$g(\mathbf{x}) = \int_0^\infty dk \sum_{\ell m} g_{\ell m}(k) Z_{k\ell m}(x, \theta, \phi), \tag{6.58}$$

where $g_{\ell m}(k)$ are the expansion coefficients and

$$Z_{k\ell m}(x, \theta, \phi) \equiv \sqrt{\frac{2}{\pi}}\, k\, j_\ell(kx) Y_{\ell m}(\theta, \phi). \tag{6.59}$$

Here, j_ℓ is the spherical Bessel function, (θ, ϕ) is the direction of $\mathbf{x}$, and $Y_{\ell m}$

is the spherical harmonic. The main properties of these functions are given in Appendix A. The basis functions $Z_{k\ell m}$ are orthonormal,

$$\int Z^*_{k\ell m} Z_{k'\ell' m'} d^3x = \delta(k-k')\delta_{\ell\ell'}\delta_{mm'}, \tag{6.60}$$

where $d^3x = x^2 \sin\theta \, d\theta \, d\phi \, dx$.

The spherical expansion is equivalent to the Fourier integral of Eq. (6.4) because of the identity (A.19). Substituting the latter into Eq. (6.4) and reading off the coefficients $g_{\ell m}$ we find

$$g_{\ell m}(k) = (2\pi)^{-3/2} k \, i^\ell \int g(\mathbf{k}) Y^*_{\ell m}(\hat{\mathbf{k}}) \, d\Omega_{\mathbf{k}}. \tag{6.61}$$

Using Eq. (A.9),

$$\sum_{\ell m} |g_{\ell m}(k)|^2 = (2\pi)^{-3} \int |g(\mathbf{k})|^2 d\Omega_{\mathbf{k}}. \tag{6.62}$$

From Eqs. (6.18), (6.61) and (A.8), the two-point correlator of the multipoles $g_{\ell m}$ is given by

$$\langle g^*_{\ell m}(k) g_{\ell' m'}(k') \rangle = (2\pi)^3 P_g(k) \delta(k-k')\delta_{\ell\ell'}\delta_{mm'}. \tag{6.63}$$

If g is gaussian, the correlator of an odd number of multipoles then vanishes, while the correlator of an even number is completely determined by the two-point correlator.

In the gaussian case, each multipole has a gaussian probability distribution, with no correlation except for the reality condition $g^*_{\ell m} = g_{\ell,-m}$. This follows from the fact that the transformation from the Fourier expansion to the spherical expansion is unitary (up to an overall factor) which doesn't mix different k. To see why, return to the finite box. For a gaussian perturbation, the real and imaginary parts of the Fourier components of g_n (call them R_n and I_n) have independent gaussian probability distributions.[7] The probability of finding them in given intervals is $\mathbf{P} \prod_n dR_n \, dI_n$, where

$$\mathbf{P} = \prod_n \frac{1}{2\pi\sigma_n^2} \exp\left(-\frac{1}{2}\frac{R_n^2}{\sigma_n^2}\right) \exp\left(-\frac{1}{2}\frac{I_n^2}{\sigma_n^2}\right), \tag{6.64}$$

and σ_n is equal to the spectrum P_{g_n} defined in Eq. (6.12). Let us take the box size large enough for the continuum limit to apply, so that P_{g_n} becomes the spectrum $P_g(k)$. Then σ_n is practically constant within a shell k to $k + dk$ and can be taken

[7] We can ignore the reality condition in this context.

outside the sum. Inside the sum we can use Eq. (6.9) to replace $|g_n|^2$ by $|g(\mathbf{k})|^2$ so that the probability (6.64) becomes

$$\mathbf{P} = \text{const } \times \exp\left(-\int |g(\mathbf{k})|^2 d\Omega_{\mathbf{k}}\right). \tag{6.65}$$

Going to the spherical expansion and using Eq. (6.62) this becomes (with a different prefactor)

$$\mathbf{P} = \text{const } \times \exp\left(-\sum_{\ell m} |g_{\ell m}(k)|^2\right), \tag{6.66}$$

showing that indeed the real and imaginary parts of the multipoles have independent gaussian probability distributions.

6.7 Correlators of the curvature perturbation

6.7.1 Spectrum and spectral index

Observation gives information about the spectrum $\mathcal{P}_\zeta(k)$ of the curvature perturbation. At present the most important data come from the CMB anisotropy and the galaxy distribution. Other types of observation will become important in the future. In later chapters we see in some detail how to compare theory with the CMB and galaxy observations.

According to observation, the spectrum $\mathcal{P}_\zeta(k)$ of the curvature perturbation is almost scale independent, with a value of order 10^{-9}. The scale dependence of the spectrum is characterized by the **spectral index** $n(k)$, defined by

$$\boxed{n - 1 \equiv \frac{d\ln \mathcal{P}_\zeta(k)}{d\ln k}}. \tag{6.67}$$

Scale-independence, also referred to as a flat spectrum, corresponds to $n = 1$, and $n - 1$ is often called the **tilt** of the spectrum. If $n(k)$ is constant, $\mathcal{P}_\zeta(k) \propto k^{n-1}$.

If n depends on k one says that the spectral index is **running**. In that case one usually assumes that n can be approximated as a linear function of $\ln k$, so that the running is defined by $n' \equiv dn/d\ln k$.

Observation is compatible with the hypothesis that ζ is the only primordial perturbation and that $n' = 0$. Adopting this hypothesis, a fit of the ΛCDM model to the current CMB anisotropy and the galaxy distribution gives

$$\mathcal{P}_\zeta^{1/2}(k_0) = (4.9 \pm 0.2) \times 10^{-5}, \tag{6.68}$$

$$n = 0.96 \pm 0.03. \tag{6.69}$$

The spectrum is specified at a 'pivot' scale $k_0 \equiv 0.002\,\text{Mpc}^{-1}$, and the quoted uncertainties are at something like the 2σ level. If n' is allowed to float (taking it to be a constant) its allowed range is something like

$$-0.07 < n' < 0.01, \quad (6.70)$$

and $n(k_0) \simeq 1.0 \pm 0.1$. (With n' floating, the best-fit value of $n(k_0)$ becomes strongly dependent on the pivot scale k_0.)

The observed spectral amplitude is commonly referred to as the CMB normalization (also sometimes as the COBE normalization, having first been accurately measured by the COBE satellite).

6.7.2 Non-gaussianity

Observation shows that ζ is gaussian to high accuracy, with significant bounds on both the bispectrum and trispectrum.

The bispectrum is conventionally written in the form

$$\boxed{B_\zeta(k_1, k_2, k_3) = \frac{6}{5} f_{\rm NL}(k_1, k_2, k_3) \left[P_\zeta(k_1) P_\zeta(k_2) + \text{cyclic permutations.} \right]}. \quad (6.71)$$

We see that $f_{\rm NL}$ is the reduced bispectrum as defined in Section 6.4, except for a conventional factor $6/5$ whose origin is explained in Section 8.5.

Local non-gaussianity

The simplest possibility for the form of the non-gaussianity is

$$\boxed{\zeta(\mathbf{x}) = \zeta_{\rm g.} + \zeta_{\rm n.g.} \equiv g(\mathbf{x}) + b g^2(\mathbf{x})}, \quad (6.72)$$

where g is a gaussian perturbation. This is known as local non-gaussianity. Since ζ is gaussian to good accuracy, the first term must dominate at a typical position. Its magnitude at a typical position is $|\zeta(\mathbf{x})| \sim \mathcal{P}_\zeta^{1/2} \sim 10^{-4}$. The fractional contribution of the second term at a typical position is

$$\left| \frac{\zeta_{\rm n.g.}(\mathbf{x})}{\zeta_{\rm g.}(\mathbf{x})} \right| \sim |b\zeta(\mathbf{x})| \sim {\mathcal{P}_\zeta}^{1/2} |b|. \quad (6.73)$$

This non-gaussian fraction is also, according to Eq. (6.34), an estimate of the skewness.

Treating the non-gaussian part to first order, the bispectrum is

$$\langle \zeta_{\mathbf{k}_1} \zeta_{\mathbf{k}_2} \zeta_{\mathbf{k}_3} \rangle \simeq b \left[\langle g_{\mathbf{k}_1} g_{\mathbf{k}_2} (g^2)_{\mathbf{k}_3} \rangle + \text{cyclic permutations} \right]. \quad (6.74)$$

To work this out we use the convolution theorem to find

$$\langle \zeta_{\mathbf{k}_1} \zeta_{\mathbf{k}_2} \zeta_{\mathbf{k}_3} \rangle = b \left[\int \langle g_{\mathbf{k}_1} g_{\mathbf{k}_2} g_{\mathbf{q}} g_{\mathbf{k}_3 - \mathbf{q}} \rangle d^3 q + \text{cyclic permutations} \right]. \quad (6.75)$$

Using Eq. (6.21), and demanding that all three momenta are non-zero, gives the result $3f_{\rm NL}/5 = b$. In contrast with the general case, $f_{\rm NL}$ is independent of the momenta. A similar calculation for the trispectrum gives $2\mathcal{T}_\zeta = (6f_{\rm NL}/5)^2$.

Observational bounds on non-gaussianity

At the time of writing, the strongest bounds on the bispectrum and trispectrum come from the CMB anisotropy. If $f_{\rm NL}$ is taken to be momentum independent (corresponding to local non-gaussianity) the current bound is $-10 < f_{\rm NL} < 110$ at something like 95% confidence level. This constraint comes mostly from the 'squeezed' configuration where the triangle of momenta collapses to a line (say $k_1 \ll k_2 \simeq k_3$). If instead we consider the equilateral case (equal momenta), the bound is something like $-150 < f_{\rm NL} < 250$. As we see in Chapters 25 and 26, the inflationary cosmology suggests that $f_{\rm NL}$ will either be local, or will practically vanish in the squeezed configuration.

What about future measurements of $f_{\rm NL}$? Although we see in Section 9.6.2 that it might be avoided, cosmic variance is a serious limitation for most methods of detection. For observation of the galaxy distribution, we can see this from the discussion of Section 6.5. Taking resolution $\delta \sim 1$, the estimate (6.57) says that observation can go down only to

$$|f_{\rm NL}| \simeq {\mathcal{P}_\zeta}^{-1/2}(kL)^{-3/2} \simeq 2 \times 10^4 (kL)^{-3/2}, \qquad (6.76)$$

where L is the size of the observed region and k^{-1} is the smallest scale observed. Taking the biggest possible value $L \sim H_0^{-1} \sim 10^4\,{\rm Mpc}$, this suggests that we need $k^{-1} \lesssim 10\,{\rm Mpc}$ to get down to $|f_{\rm NL}| \sim 1$ and $k^{-1} \lesssim 1\,{\rm Mpc}$ to get down to $|f_{\rm NL}| \sim 10^{-2}$.

The estimates of the previous paragraph are quite crude, and take account only cosmic variance. Using realistic estimates, it seems that in the foreseeable future, combined observation of the CMB anisotropy and the galaxy distribution is expected to be capable of detecting $f_{\rm NL}$ down to something like $|f_{\rm NL}| \sim 5$, and it has been suggested that an observation of the anisotropy of the 21 cm background may go even lower. We see from Eq. (6.73) that if $f_{\rm NL}$ is so small, the non-gaussian fraction of ζ will be no bigger than ${\mathcal{P}_\zeta}^{1/2}$, which is the typical magnitude ζ itself.[8] In other words, the non-gaussian part of ζ will be of order ζ^2, which means that second-order perturbation theory will be required to handle it.

[8] Equation (6.73) was derived assuming Eq. (6.72), but a similar result will usually hold provided only that the non-gaussian part of ζ is fully correlated with the gaussian part. Uncorrelated non-gaussianity is discussed in Section 26.3.2.

Exercises

6.1 Use Eq. (6.18) to show that the variance of a generic perturbation $g(\mathbf{x})$ is given by Eq. (6.22), independently of $\mathbf{x}$. For $\mathbf{x} = 0$, check that the same result is obtained from the spherical expansion Eq. (6.58), with the definition Eq. (6.63) of the spectrum.

6.2 Extending the derivation of Eq. (6.29), calculate the spatial four-point correlator corresponding to the momentum-space expression Eq. (6.21). Verify that it is invariant under translations and rotations.

6.3 Verify Eq. (6.34), giving the skewness of a perturbation with a scale-invariant spectrum and reduced bispectrum

6.4 Using the method described in the text, verify that the local non-gaussianity expression (6.72) leads to $3f_{\rm NL}/5 = b$.

6.5 Show that $\delta^3(\mathbf{k}-\mathbf{k}') = k^{-2}\delta(k-k')\delta^2(\hat{\mathbf{k}}-\hat{\mathbf{k}}')$, and hence prove Eq. (6.63).

References

[1] S. Karlin and H. M. Taylor. *A First Course on Stochastic Processes* (New York: Academic Press, 1975).

[2] R. J. Adler. *The Geometry of Random Fields* (Chichester: John Wiley & Sons, 1981).

[3] S. Weinberg. *Cosmology* (Oxford: Oxford University Press, 2008).

7

Newtonian perturbations

In the previous two chapters we considered the cosmological perturbations before horizon entry. As no causal processes can then operate, the description of these perturbations is simple, but they are not directly observable, except on the very largest scales via the cosmic microwave anisotropy. Over the next few chapters we follow the evolution of the cosmological perturbations after horizon entry, under the influence of causal processes.

The first section of this chapter is devoted to an overview of the evolution. In the rest of the chapter we see how to calculate the evolution in the regime of Newtonian gravity.

7.1 Free-streaming, oscillation, and collapse

As summarized in Table 7.1, the cosmic fluid at $T < 1\,\mathrm{MeV}$ has four components: baryons, cold dark matter (CDM), photons, and neutrinos. The CDM and neutrinos have negligible interaction. Until the temperature falls below its mass each neutrino species has relativistic motion, so that it behaves as radiation rather than matter up to that point.

The essence of what happens after horizon entry can be stated very simply. For each component, there is a competition between gravity, which tries to increase the density perturbation by attracting more particles to the overdense regions, and random particle motion which drives particles away from those regions. The relative importance of these effects depends on which component we are considering, and it may depend also on what era we are talking about.

For **CDM**, random motion is by definition negligible and gravity wins. The growth of the CDM density contrast is only logarithmic during radiation domination, but is proportional to $a(t)$ during matter domination.

For **neutrinos** the random motion is relativistic at horizon entry, except on very large scales (how large depends on the neutrino mass). With relativistic random

Table 7.1. *The four fluids*

Type of fluid	Matter	Radiation
Non-interacting	CDM	Neutrinos
Interacting	Baryons	Photons

motion, the neutrinos travel unimpeded with the speed of light (free-stream), which completely washes out the neutrino density perturbation.

For **baryons** and **photons** the story is more complicated. By baryons, in this context we mean nuclei *and* electrons, because the Coulomb interaction ensures that the number densities of electrons and protons are practically equal at each point in space. Consider first photon decoupling. There are free electrons and nuclei, with frequent scattering of photons off electrons (Compton scattering). The tightly coupled baryon–photon fluid oscillates after horizon entry, as a standing **acoustic wave**. However, on small scales the amplitude of the oscillation is strongly damped by photon diffusion, the photons carrying the baryons with them. The comoving scale below which this occurs is called the **Silk scale**, and the corresponding mass given by Eq. (5.1) is called the Silk mass. As we shall see, these quantities grow with time, so that by the epoch of decoupling the Silk mass is of order $10^{15} M_\odot$ and the Silk scale is of order $10\,\mathrm{Mpc}$.

At photon decoupling two things happen. Firstly, most of the electrons bind into atoms, allowing the photons to travel freely to become the cosmic microwave background (CMB). On scales longer than the Silk scale, the CMB anisotropy provides a snapshot of the acoustic oscillation on the last-scattering sphere, as it existed at the epoch of decoupling.

Secondly, the baryons are no longer carried along with the photons (except on rather small scales). Instead, they are free to fall into the gravitational potential wells already created by the CDM density perturbation. As a result, gravitationally bound objects with successively bigger mass form. At the present epoch the process is ceasing because of the rapid expansion caused by the dark energy. The largest structures that have had time to form are rich galaxy clusters with mass of order $10^{15} M_\odot$.

In Table 7.2 we give the values of various quantities at the epoch of horizon entry, for a sample of cosmological scales from the smallest to the biggest. The first column shows conformal time η, specified as a length so that it is the comoving particle horizon. This in turn is about equal to the comoving horizon $1/aH$, which means that the scale k^{-1} entering the horizon at a given epoch is about equal to η. The range of cosmologically interesting scales is $10^{-3}\,\mathrm{Mpc} \lesssim k^{-1} \lesssim 10^4\,\mathrm{Mpc}$,

Table 7.2. *Various quantities as a function of the time of horizon entry*

	η/ Mpc	$M/M_\odot$	ℓ	T/ eV	z
	1.0×10^{-3}	$\simeq 10^3$		7.9×10^4	3.3×10^8
	1.0×10^{-2}	$\simeq 10^6$		7.9×10^3	3.3×10^7
small scales	0.10	$\simeq 10^9$		7.9×10^2	3.2×10^6
	1.0	$\simeq 10^{12}$		79	3.3×10^5
	10	$\simeq 10^{15}$	1400	7.9	3.3×10^4
(equality)	89		157	0.77	3.3×10^3
(decoupling)	2.1×10^2		63	0.25	1050
large scales	1.0×10^3		14	0.022	94
	1.4×10^4		~ 2	2.348×10^{-4}	present

spanning a range $\Delta \ln k \simeq 16$. The second column shows the mass $M(k)$ enclosed by the scale $k^{-1} = \eta$, given by Eq. (5.1). The third column shows the multipole $\ell = k\eta_0$ corresponding to this scale, which we explain in Chapter 10. Finally we show the temperature and redshift at horizon entry.

We show M only in the regime where gravitational collapse to baryonic structures occurs, and show ℓ only in the regime where the primary CMB anisotropy is observable. The overlap between these regimes happens to be very small. Horizon entry is well *before* matter–radiation equality for the small scales that lead to the formation of baryonic structures. Also, horizon entry is well *after* matter–radiation equality for the large scales that give the low multipoles of the CMB anisotropy.

7.2 Newtonian perturbations: total mass density

For the rest of this chapter we adopt Newtonian gravity. Roughly speaking, Newtonian gravity is the non-relativistic approximation to Einstein gravity. We derived this approximation in Section 3.6, for the case of a fluid which is the case of interest for cosmology. According to the conditions specified there, Newtonian gravity will apply during matter domination, on scales well inside the horizon. That means that it will apply towards the end of the story that we just outlined.

7.2.1 Newtonian cosmology

Let us derive the Friedmann equation (3.78) with ρ the mass density. Consider a comoving fluid element at distance $a(t)x$ from the origin. According to a famous result, the Newtonian gravitational potential due to everything inside the sphere of radius $a(t)x$ is

$$\phi_{\rm gr}^{(0)}(\mathbf{x},t) = -\frac{GM}{ax} = -\frac{4\pi G}{3}(ax)^2\rho(t), \tag{7.1}$$

where M is the mass inside the sphere. Also, the potential due to everything between this sphere and a larger one is zero. Taking the second sphere to be much larger than the first one, the matter outside the second sphere can be ignored because it gives the same acceleration to everything inside the small sphere, and we are setting that acceleration to zero by a choice of the reference frame. We conclude that $\Phi_{\rm gr}^{(0)}$ is the complete gravitational potential. If m is the mass of the fluid element, its potential energy is $m\Phi_{\rm gr}^{(0)}$ and its kinetic energy is $m(axH)^2/2$. Adding these and setting them equal to a constant $E_{\rm tot}$, we arrive at the Friedmann equation with $K = -2E_{\rm total}/mx^2$.

In the perturbed Universe we write

$$\boxed{\phi_{\rm gr}(\mathbf{x},t) = \phi_{\rm gr}^{(0)}(x,t) + \Phi(\mathbf{x},t)}. \tag{7.2}$$

We have denoted the perturbation of the gravitational potential by Φ because that is the usual notation for the corresponding relativistic quantity that we encounter later. We call this perturbation the **peculiar gravitational potential.**

We also need the perturbation $\mathbf{v}$ in the fluid velocity:

$$\boxed{\mathbf{u}(\mathbf{x},t) = H(t)\mathbf{r} + \mathbf{v}(\mathbf{x},t)}. \tag{7.3}$$

It is called the **peculiar velocity**. Finally, we need to define perturbations in the mass density and pressure of the matter;

$$\rho(\mathbf{x},t) = \rho(t) + \delta\rho(\mathbf{x},t) \equiv \rho(t)\left[1 + \delta(\mathbf{x},t)\right], \tag{7.4}$$
$$P(\mathbf{x},t) = P(t) + \delta P(\mathbf{x},t). \tag{7.5}$$

We are using the same symbol to denote the perturbed and the unperturbed quantities. This compact notation works, provided that the argument (t) or $(\mathbf{x},t)$ is displayed when that is necessary to avoid ambiguity. Also, we introduced the density contrast $\delta \equiv \delta\rho/\rho$.

We will treat $\delta\rho$, δP and $\mathbf{v}$ to first order, which requires $|\delta| \ll 1$. To write down the first-order equations we need the time derivative along the trajectory of a fluid element:

$$\frac{d}{dt} = \frac{\partial}{\partial t} + \frac{dx_i}{dt}\frac{\partial}{\partial x_i}, \tag{7.6}$$

with summation over x_i understood.

In Section 3.6.2 we wrote down the Newtonian continuity, Euler, and Poisson equations. In Fourier space, the first-order perturbations of these equations take the form

$$a\dot{\delta}_{\mathbf{k}} = -i\mathbf{k}\cdot\mathbf{v}_{\mathbf{k}} \quad \text{(continuity)}, \tag{7.7}$$

$$a\dot{\mathbf{v}}_{\mathbf{k}} + aH\mathbf{v}_{\mathbf{k}} + i\mathbf{k}\Phi_{\mathbf{k}} = -i\mathbf{k}\delta P_{\mathbf{k}}/\rho \quad \text{(Euler)}, \tag{7.8}$$

$$-\frac{k^2}{a^2}\Phi_{\mathbf{k}} = 4\pi G\rho\delta_{\mathbf{k}} \quad \text{(Poisson)}. \tag{7.9}$$

7.2.2 Scalar and vector modes

For the peculiar velocity we write :

$$\boxed{\mathbf{v}_{\mathbf{k}} = \mathbf{v}_{\mathbf{k}}^{\text{sc}} + \mathbf{v}_{\mathbf{k}}^{\text{vec}}}, \tag{7.10}$$

where the scalar part $\mathbf{v}_{\mathbf{k}}^{\text{sc}}$ is parallel to $\mathbf{k}$ and the vector part $\mathbf{v}_{\mathbf{k}}^{\text{vec}}$ is perpendicular to it. The scalar part can be written

$$\boxed{\mathbf{v}_{\mathbf{k}}^{\text{sc}} = -\frac{i\mathbf{k}}{k}V_{\mathbf{k}}}. \tag{7.11}$$

Then $|V_{\mathbf{k}}|$ is the magnitude of $\mathbf{v}_{\mathbf{k}}^{\text{sc}}$. The vector part satisfies

$$\boxed{\mathbf{k}\cdot\mathbf{v}^{\text{vec}} = 0}. \tag{7.12}$$

The notation 'scalar' and 'vector' part is usual in relativistic cosmological perturbation theory, and adopt it here. More generally, the scalar and vector parts of any vector field often are called 'longitudinal' and 'transverse' (to the $\mathbf{k}$ direction).

The perturbation in the expansion rate depends only on the scalar part of $\mathbf{v}$:

$$\boxed{(\nabla\cdot\mathbf{v})_{\mathbf{k}} = (\nabla\cdot\mathbf{v}^{\text{sc}})_{\mathbf{k}} = \frac{k}{a}V_{\mathbf{k}}}. \tag{7.13}$$

Inserting the scalar–vector decomposition of $\mathbf{v}_{\mathbf{k}}$ into Eqs. (7.7)–(7.9), we find that these equations break into two uncoupled sets, called modes. The vector mode involves just $\mathbf{v}_{\mathbf{k}}^{\text{vec}}$ and consists of a single equation:

$$\boxed{a\dot{\mathbf{v}}_{\mathbf{k}}^{\text{vec}} + aH\mathbf{v}_{\mathbf{k}}^{\text{vec}} = 0} \quad \text{(Euler)}. \tag{7.14}$$

The solution decays like $1/a$, corresponding to the conservation of the angular momentum of a fluid element.

In cosmology one usually drops decaying modes, on the ground that they would lead to a very perturbed universe at early times. Within the inflationary cosmology one can usually check explicitly that they are negligible. Accordingly, we drop $\mathbf{v}_{\mathbf{k}}^{\text{vec}}$.

Writing $\mathbf{v}_\mathbf{k}^{\rm sc}$ in terms of $V_\mathbf{k}$, the equations of the scalar mode are

$$\boxed{\dot{\delta}_\mathbf{k} + \frac{k}{a} V_\mathbf{k} = 0} \qquad \text{(continuity)}, \tag{7.15}$$

$$\boxed{\dot{V}_\mathbf{k} + H V_\mathbf{k} - \frac{k}{a}\Phi_\mathbf{k} = \frac{k}{a}\frac{\delta P_\mathbf{k}}{\rho}} \qquad \text{(Euler)}, \tag{7.16}$$

$$\boxed{-\frac{k^2}{a^2}\Phi_\mathbf{k} = 4\pi G\rho\delta_\mathbf{k} = \frac{3}{2}H^2\Omega_{\rm m}(t)\delta_\mathbf{k}} \qquad \text{(Poisson)}. \tag{7.17}$$

In the final equality we expressed the mass density ρ as a fraction $\Omega_{\rm m}(t)$ of the critical density.

7.2.3 Solution for the scalar mode

Eliminating $V_\mathbf{k}$ from Eqs. (7.15) and (7.16) gives

$$\ddot{\delta}_\mathbf{k} + 2H\dot{\delta}_\mathbf{k} - \frac{3}{2}H^2\Omega_{\rm m}(t)\delta_\mathbf{k} = -\left(\frac{k}{a}\right)^2 \frac{\delta P_\mathbf{k}}{\rho}. \tag{7.18}$$

As we see later, $\delta P_\mathbf{k}$ is negligible on cosmological scales, simplifying Eq. (7.18) to

$$\boxed{\ddot{\delta} + 2H\dot{\delta} - \frac{3}{2}H^2\Omega_{\rm m}(t)\delta = 0}. \tag{7.19}$$

As k doesn't actually appear in this equation, it is valid also for the perturbations $\delta(\mathbf{x})$ etc. at each comoving position. For that reason, we dropped the subscript $\mathbf{k}$. The equation has a growing solution, usually denoted by $D_1(t)$, and a decaying solution denoted by $D_2(t)$.

Before the dark energy becomes significant the matter has critical density, corresponding to $H = 2/3t$ and $3H^2 = 8\pi G\rho$. Then $D_1 \propto t^{2/3}$ and $D_2 \propto t^{-1}$. The most general solution is a linear combination

$$\delta = f_1 D_1 + f_2 D_2. \tag{7.20}$$

Dropping the decaying mode, we find from Eq. (7.17) that the peculiar gravitational potential $\Phi_\mathbf{k}$ is time independent:

$$\boxed{\delta_\mathbf{k}(t) = -\frac{2}{3}\left(\frac{k}{aH}\right)^2 \Phi_\mathbf{k}}. \tag{7.21}$$

From Eq. (7.16), the corresponding peculiar velocity is given by

$$\boxed{V_\mathbf{k}(t) = \frac{2}{3}\frac{k}{aH}\Phi_\mathbf{k} = -\frac{aH}{k}\delta_\mathbf{k}(t)}. \tag{7.22}$$

In position space this corresponds to $\mathbf{v}(t) = -t\boldsymbol{\nabla}\Phi$.

Whether or not Φ is time dependent, it is given by

$$\Phi(\mathbf{x},t) = -Ga^2(t)\int d^3y \frac{\delta\rho(\mathbf{x},t)}{|\mathbf{y}-\mathbf{x}|}. \tag{7.23}$$

There is an extra factor a^2 compared with Eq. (3.54) because we are now using comoving coordinates.

7.3 Effect of the cosmological constant

At very late times, the cosmological constant contribution to the energy density and pressure becomes significant. The contributions are related by $P_\Lambda = -\rho_\Lambda$, which violates the condition $|P| \ll \rho$ that we specified for the validity of Newtonian gravity. Nevertheless, it can be shown that the Newtonian Poisson equation (3.50), geodesic equation (3.51) and Euler equation (3.53) all remain valid in the presence of a cosmological constant, if ρ and P in those equations refer to just the matter [1]. The only effect of the cosmological constant is to add to the Newtonian gravitational potential (3.54) a contribution $\phi_\Lambda(\mathbf{x},t) = \Lambda x^2/6$, which adds a repulsive term $\Lambda\mathbf{x}/3$ to Newton's inverse square law (3.55).

The additional contribution to the acceleration has zero divergence, in accordance with the fact that the Newtonian Poisson equation remains valid with ρ the matter density. Its effect is negligible within our galaxy. Going out to bigger distances, it gives each object an additional outward acceleration, which gives the correct Friedmann equation.

Coming to the perturbations, the only effect of the additional term is to alter the time dependence of the scale factor. For any $a(t)$, the two independent solutions of Eq. (7.19) are

$$D_1 \equiv H\int_0^a \frac{da}{(aH)^3}, \tag{7.24}$$

$$D_2 \equiv H, \tag{7.25}$$

where

$$H^2 = H_0^2\left(\frac{\Omega_\mathrm{m}}{a^3} + \Omega_\Lambda\right), \tag{7.26}$$

with Ω_m and Ω_Λ taking on their present values.

As usual we drop the decaying solution D_2. Requiring that the growing solution reduces to Eq. (7.21) at early times we have

$$\delta_\mathbf{k}(t) = -\frac{5}{3}\Phi_\mathbf{k}k^2H\int_0^a \frac{da}{(aH)^3}, \tag{7.27}$$

where $\Phi_{\mathbf{k}}$ is the constant early-time value of $\Phi_{\mathbf{k}}(t)$. The latter turns out to decrease with time and it is usual to define a suppression factor $g \equiv \Phi_{\mathbf{k}}(t)/\Phi_{\mathbf{k}}$. Then we can write

$$\boxed{\delta_{\mathbf{k}}(t) = -\frac{2}{3}\left(\frac{k}{H_0}\right)^2 \frac{1}{\Omega_{\mathrm{m}}}\Phi_{\mathbf{k}} g(t) a(t)} . \tag{7.28}$$

The integral in Eq. (7.27) can be written in terms of elliptic functions, but is more commonly performed numerically or parameterized. Once the integral is done we can read off the suppression factor. An accurate parameterization is

$$g(\Omega) = \frac{5}{2}\Omega\left(\frac{1}{70} + \frac{209\Omega}{140} - \frac{\Omega^2}{140} + \Omega^{4/7}\right)^{-1}, \tag{7.29}$$

where $\Omega \equiv \Omega_{\mathrm{m}}(z)$ which is related to $a^{-1} = 1 + z$ by Eq. (4.49). By the present epoch $g(t)$ has fallen to around 0.75.

For the peculiar velocity, it is useful to introduce a suppression factor f into Eq. (7.22),

$$\boxed{V = -\frac{aH}{k} f\delta} . \tag{7.30}$$

From Eq. (7.15),

$$f = \frac{a}{D_1}\frac{dD_1}{da} . \tag{7.31}$$

A good approximation is $f(t) = \Omega_{\mathrm{m}}^{0.6}(t)$. At the present epoch $f \simeq 0.5$.

7.4 Baryon density perturbation

7.4.1 Separate equations

We are taking the CDM to have negligible random motion, corresponding to negligible pressure. To handle the baryon pressure we replace Eq. (7.18) by a pair of equations describing the separate evolution of the CDM and the baryonic matter. As we are dealing with Newtonian gravity such a procedure makes sense only after decoupling, since before that the baryons are tightly coupled to the photons requiring the use of Einstein gravity.

For the separate quantities, the Euler and Poisson equations become

$$\begin{aligned}
\dot{\delta}_{\rm c} + \frac{k}{a} V_{\rm c} &= 0, &(7.32)\\
\dot{V}_{\rm c} + H V_{\rm c} - \frac{k}{a}\Phi &= 0, &(7.33)\\
\dot{\delta}_B + \frac{k}{a} V_B &= 0, &(7.34)\\
\dot{V}_B + H V_B - \frac{k}{a}\Phi &= -\frac{k}{a}\frac{\delta P_B}{\rho_B}. &(7.35)
\end{aligned}$$

Each species feels its *own* pressure gradient, but the *total* gravitational acceleration. Eliminating the velocities, and using Eq. (7.21) for Φ, gives for the baryons

$$\boxed{\ddot{\delta}_B + 2H\dot{\delta}_B - \frac{3}{2}H^2\delta + \left(\frac{k}{a}\right)^2 c_{\rm s}^2 \delta_B = 0}, \tag{7.36}$$

where $c_{\rm s}^2 \equiv \delta P_B/\delta\rho_B$. For the CDM we find

$$\boxed{\ddot{\delta}_{\rm c} + 2H\dot{\delta}_{\rm c} - \frac{3}{2}H^2\delta = 0}. \tag{7.37}$$

In each equation δ is the total matter density contrast, representing the effect of gravity on the evolution of δ_B or $\delta_{\rm c}$.

7.4.2 Growth and oscillation

If the last term of Eq. (7.36) is negligible compared with the previous one, the baryon pressure is negligible and the baryon density contrast is free to grow along with the CDM density contrast. To understand this growth, note first that δ_B and $\delta_{\rm c}$ satisfy the same equation. The difference $S_{\rm cb} \equiv \delta_{\rm c} - \delta_B$ satisfies

$$\ddot{S}_{\rm cb} + 2H\dot{S}_{\rm cb} = 0. \tag{7.38}$$

The solution of this equation is $S_{\rm cb} = A + Bt^{-1/3}$. On the other hand, as we saw earlier the total density contrast is given by $\delta = Ct^{2/3} + Dt^{-1}$, and clearly

$$\delta = f_B\delta_B + f_{\rm c}\delta_{\rm c}, \tag{7.39}$$

with $f_B \simeq 1/6$ the fraction of baryonic matter and $f_{\rm c} \simeq 5/6$ the fraction of CDM. We expect the non-decaying modes to dominate after a few Hubble times, leading to

$$\boxed{\delta_B = \delta_{\rm c} = \delta = Ct^{2/3}}. \tag{7.40}$$

As we see later the CDM density contrast dominates at the beginning of the Newtonian regime. According to the above analysis, the baryons then fall into the potential wells already created by the CDM.

If the last term of Eq. (7.36) dominates the previous one, the baryon density contrast doesn't grow. Instead it undergoes slowly damped standing-wave oscillations with angular frequency $c_s k/a$. Before decoupling, the tightly coupled baryon–photon fluid undergoes a similar oscillation, which we study later using general relativity. Such oscillations are referred to in cosmology as acoustic oscillations, and the speed c_s with which they would propagate (were they travelling waves) is called the sound speed.

7.4.3 Baryon Jeans mass

The last term of Eq. (7.36) dominates the first one if k is bigger than some time-dependent value $k_J(t)$, called the Jeans wavenumber. The corresponding wavelength λ_J is called the Jeans length. The **Jeans mass** M_J is conventionally defined as the mass of matter in a sphere of radius $\lambda_J/2$.

To estimate M_J, it is good enough to assume that the density contrasts are equal as in Eq. (7.40). This gives

$$\frac{\pi a}{k_J} \equiv \frac{\lambda_J}{2} = \pi c_s (4\pi G\rho)^{-1/2}, \tag{7.41}$$

and

$$M_J = \frac{\pi^{5/2}}{6} \frac{c_s^3}{G^{3/2}\rho^{1/2}}. \tag{7.42}$$

Let us calculate c_s. It can be shown that well before $z = 140$, the Compton scattering of photons off residual free electrons keeps the baryon temperature T_B close to the uniform photon temperature T_γ. Then

$$c_s^2 \equiv \frac{\delta P_B}{\delta \rho_B} = T_\gamma \frac{\delta n_B}{\delta \rho_B} = \frac{T_\gamma}{\mu m_p} \propto \frac{1}{a}, \tag{7.43}$$

where $\mu = 1.22$ is the mean molecular weight of the cosmic gas in atomic units and m_p is the proton mass. This gives a time-independent Jeans mass,

$$M_J = 1.4 \times 10^5 M_\odot. \tag{7.44}$$

Well after $z = 140$, Compton scattering ceases to be effective and the baryon fluid expands adiabatically. We are dealing with a monatomic gas (because there is only helium and hydrogen with the latter monatomic at the relevant epochs) which corresponds to $c_s^2 = 5T_B/3\mu m_p$. The temperature T_B is proportional to the mean-square baryon velocity which is in turn proportional to the mean-square baryon momentum (both measured in the rest frame of the baryon fluid). But the momentum of each baryon redshifts like a^{-1} which means that T_B is proportional to a^{-2}. The Jeans mass is therefore a constant times $(1+z)^{3/2}$. The constant can be estimated by matching this result to the earlier one, taking both to be exact at $z = 140$. A proper calculation following the evolution through $z = 140$ gives at $z \ll 140$ [2]

$$\boxed{M_{\mathrm{J}}(z) = 6 \times 10^3 \left(\frac{1+z}{10}\right)^{3/2} M_\odot} . \tag{7.45}$$

The Jeans mass becomes irrelevant when the evolution on the Jeans scale ceases to be linear. As we see later, this happens at $z \sim 10$. We conclude that $M_{\mathrm{J}}(t)$ falls from around $10^5 M_\odot$ to around $10^4 M_\odot$.

At any epoch, on scales bigger than the Jeans length, the baryons fall into the CDM potential wells. We will see in Section 9.2 how this leads to to the formation of baryonic objects. On scales below the Jeans length, the last term of Eq. (7.36) dominates causing the baryon fluid to oscillate, but this oscillation doesn't have any observable consequence beyond the fact that it places a lower limit on the mass of the first baryonic objects.

We end by noting that there is a cruder way of estimating the Jeans mass, which has wider applicability. Consider a spherically symmetric density enhancement with radius R (defined, for example, as the distance at which the density contrast has fallen by one-half). At a distance r from the centre, the pressure gives an outward acceleration $-[\partial(\delta P)/\partial r]/\rho$, and the inward peculiar gravitational acceleration is $G\,\delta M/r^2$, where δM is the excess mass within radius r. For an estimate, we can set $r \sim R$, $\delta M \sim \delta\rho R^3$, and $\partial(\delta P)/\partial r \sim \delta P/R$. An estimate of the Jeans length is provided by the value of R for which pressure and gravity balance:

$$\lambda_{\mathrm{J}} \sim \sqrt{\frac{1}{G\rho}\frac{\delta P}{\delta\rho}} . \tag{7.46}$$

Inserting $\delta P = \delta\rho\,\overline{v^2}/3$, where $\overline{v^2}$ is the mean-square velocity of the particles, gives

$$\boxed{\lambda_{\mathrm{J}} \sim G^{-1/2}\rho^{-1/2}\left(\overline{v^2}\right)^{1/2}} . \tag{7.47}$$

This formula applies even if diffusion or free-streaming causes the stress to be anisotropic, because the components of the stress are still of order $\rho\overline{v^2}$, and the criterion is still that the acceleration due to the stress should balance that due to the gravity. In the case of free-streaming, one can regard the formula as giving the velocity required to escape from the overdense region. We use it later when considering massive neutrinos.

Exercises

7.1 Use the formulas in Section 4.6 to verify the entries in Table 7.2.

7.2 Verify Eq. (7.18), describing the Newtonian evolution of the matter density contrast.

7.3 Verify Eqs. (7.21) and (7.22), relating in **k**-space the Fourier components of the peculiar gravitational potential and the peculiar velocity potential to those of the density contrast. Show that Eq. (7.21) is just equivalent to Newton's inverse-square law of gravity. Derive from Eq. (7.22) an expression relating $\delta(\mathbf{x})$ to the gradient of the peculiar velocity field.

7.4 By considering a single Fourier coefficient, show that the vector part of a velocity field is determined by its vorticity.

7.5 Derive Eqs. (7.32)–(7.35) by mimicking the derivation of the corresponding equation for the total density contrast.

References

[1] P. J. E. Peebles. *Large Scale Structure of the Universe* (Princeton: Princeton University Press, 1980).

[2] R. Barkana and A. Loeb. In the beginning: the first sources of light and the reionization of the Universe. *Phys. Rept.*, **349** (2001) 125.

8

General relativistic perturbations

In Chapters 2 and 3 we studied the relativistic continuity and Euler equations. We also gave the field equation defining Einstein's theory of gravity.

Now we study the first-order perturbation of these equations. After a rather general treatment, we specialize to the evolution on cosmological scales. The initial condition for such scales may be specified just after neutrino decoupling, as described in Chapter 4. We are dealing with four fluids, but we treat the baryon–photon fluid only in the tight-coupling approximation. Using that approximation we follow the evolution of the matter density until the Newtonian description takes over. We also follow the acoustic oscillation of the baryon–photon fluid until photon decoupling. In both cases, the adiabatic initial condition is adopted.

8.1 Scalar, vector, and tensor modes

In Newtonian perturbation theory, we found that the Fourier components decouple. In other words, there is a separate set of equations for each $\mathbf{k}$. We also found that there is further decoupling, so that for each $\mathbf{k}$ there is a 'scalar' mode and a 'vector' mode, with no coupling between them. As we now see, these are generic features of first-order cosmological perturbation theory, and tensor modes are also possible.

The cosmological perturbations satisfy coupled equations, containing partial derivatives with respect to time and comoving position x^i. At first order, any one of these equations will be of the form $G = 0$, with G a linear combination of the perturbations[1]

$$G(\mathbf{x}, t) \equiv a_1(t)g_1(\mathbf{x}, t) + a_2(t)g_2(\mathbf{x}, t) + \ldots = 0. \tag{8.1}$$

The coefficients a_n are independent of $\mathbf{x}$ because the unperturbed Universe is

[1] In this equation, $g_n(\mathbf{x}, t)$ denotes a perturbation which may have been differentiated any number of times with respect to time and/or position coordinates. In the following equation $g_n(\mathbf{k}, t)$ denotes a perturbation which may have been differentiated any number of times with respect to t, and multiplied any number of times by components of $\mathbf{k}$.

invariant under translations. Taking the Fourier transform of both sides we have

$$G(\mathbf{k},t) \equiv a_1(t)g_1(\mathbf{k},t) + a_2(t)g_2(\mathbf{k},t) + \ldots = 0, \tag{8.2}$$

As advertised, we have a separate equation for each $\mathbf{k}$.

The Newtonian continuity, Euler and Poisson equations, Eqs. (7.15), (7.16) and (7.17), are of this form. In the continuity and Poisson equations, G and the g_n are 3-scalars while in the Euler equation they are 3-vectors. In the relativistic case we will find that G and the g_n can also be second-rank tensors. (G and the g_n must be of the same rank, because the coefficients a_n are 3-scalars by virtue of the rotational invariance of the unperturbed Universe.)

In the relativistic case, the equation $G = 0$ is obtained by perturbing an exact equation of the form '4-tensor= 0'. In practice the 4-tensor is at most of rank two, which means that G and the g_n can be 3-tensors of at most rank two. This happens for two reasons. First, the Einstein and Maxwell field equations both involve at most second-rank 4-tensors. Second, within the inflationary cosmology that we study in Part IV, all cosmological perturbations originate from the perturbations of one or more classical fields. Such fields are 4-scalars, 4-vectors or 4-tensors, and are expected to be at most rank two because tensors of higher rank are difficult to accommodate within quantum field theory, and would correspond to fundamental particles with spin bigger than two which are not observed.

The decoupling into scalar, vector and tensor modes comes when we consider a rotation by an angle ϕ about the $\mathbf{k}$ direction, which changes G to some $G(\phi)$. The rotation acts on each index of G with the rotation matrix (2.11). Suppressing those indices we can write

$$G(\phi) = \sum_m G_m e^{2\pi i m\phi}, \qquad g_1(\phi) = \sum_m (g_1)_m e^{2\pi i m\phi}, \tag{8.3}$$

and so on, where m is *at most* equal to the rank of the 3-tensor G.

Rotational invariance requires that $G(\phi) = 0$ independently of ϕ, which means that we deal with uncoupled equations:

$$G_m(\mathbf{k},t) \equiv a_1(t)g_{1m}(\mathbf{k},t) + a_2(t)g_{2m}(\mathbf{k},t) + \ldots = 0. \tag{8.4}$$

Assuming that the relevant laws of physics are invariant under the parity transformation, we can choose $G_m = G_{-m}$, so that we need consider only $m \geq 0$. The modes $m = 0$, 1 and 2 are called respectively the scalar, vector, and tensor modes. The terms 'scalar', 'vector' and 'tensor' may be taken to refer to the transformation properties of G, regarded as a 2-tensor whose indices take on only two values (corresponding to components in the plane orthogonal to $\mathbf{k}$).

To make the discussion more concrete, we have to restore the indices. Consider first the scalar mode corresponding to $m = 0$. Each perturbation in this mode is invariant under rotations about $\mathbf{k}$. A 3-scalar perturbation g obviously belongs to this mode.[2] A vector perturbation g_i belongs to the mode if it is the gradient of a scalar, so that it is of the form $g_i = gk_i$. Indeed, choosing the z-axis along $\mathbf{k}$ we find that the perturbation is then invariant under rotations about $\mathbf{k}$. Adopting the notation introduced for the Newtonian peculiar velocity, we may say that a longitudinal vector perturbation belongs to the scalar mode.

Analogously, a tensor perturbation g_{ij} belongs to the scalar mode if it is obtained from a scalar g by differentiation, corresponding to $g_{ij} = k_i k_j g$. It also belongs to the scalar mode if it is of the form $g_{ij} = g\delta_{ij}$, because the components of the tensor δ_{ij} are not affected by any rotation. Any symmetric second-rank tensor belonging to the scalar mode is a combination of these two. We met examples of both in the Newtonian case. An antisymmetric tensor of the form $g_{ij} = \epsilon_{ijn} k_n g$ would belong to the scalar mode, but isn't encountered in practice. Higher-rank tensors of the scalar mode would have to be built from scalars using k_i, δ_{ij} and ϵ_{ijk} but are also not encountered. In summary, all scalar perturbations belong to the scalar mode, and so do those vectors and tensors that can be built from scalars using k_i, δ_{ij}, and ϵ_{ijk}.

Now we come to the vector mode. A vector g_i belongs to the vector mode if, with the z-axis along $\mathbf{k}$, it is of the form $g_i = (g_x, g_y, 0)$. Indeed, from Eq. (2.11), a rotation about the z-axis has the effect

$$(g_x \pm i g_y) \to e^{\pm i\phi}(g_x \pm i g_y). \tag{8.5}$$

Adopting the terminology used for the Newtonian peculiar velocity, we see that a vector belongs to the vector mode if it is transverse. We also see that the most general vector is the sum of a scalar and a vector mode.

Tensors belonging to the vector mode have to be constructed from transverse vectors using k_i, δ_{ij} and ϵ_{ijk}. From a transverse vector g_i we get second-rank tensors in the vector mode by writing $g_{ij} = k_i g_j$ or $g_{ij} = \epsilon_{ijk} g_k$. Counting the independent components, we see that the most general antisymmetric second-rank tensor is the sum of a scalar and a vector mode.

Finally we come to the tensor mode. A tensor g_{ij} belongs to the tensor mode if it is symmetric, traceless ($g_{ii} = 0$) and transverse ($k_i g_{ij} = 0$). This means that it is of the form

$$g_{ij}(\mathbf{k}) = g_+(\mathbf{k}) e^+_{ij} + g_\times(\mathbf{k}) e^\times_{ij}, \tag{8.6}$$

where the polarization tensors $e^{+,\times}_{ij}$ are defined by Eq. (3.58). This tensor indeed belongs to the tensor mode, because Eqs. (2.11) and (2.21) imply that under a

[2] From now one we suppress the index distinguishing the different perturbations $g_1, g_2 \ldots$. The discussion can refer to any one of them. We restore the suppressed spatial indices, so that g denotes a scalar, g_i a 3-vector and g_{ij} a 3-tensor.

rotation

$$(g_+ \pm i g_\times) \to e^{\pm 2i\phi}(g_+ \pm g_\times). \tag{8.7}$$

(We are taking the polarization tensors to be fixed as discussed in footnote 1 of Chapter 11.) If the handedness of the coordinate system is reversed, $g_+(\mathbf{k})$ is unchanged but $g_\times(\mathbf{k})$ reverses sign (i.e. they are respectively even and odd under the parity transformation). By counting the number of independent components, one sees that the most general symmetric second-rank tensor is the sum of scalar, vector and tensor modes.

8.2 Perturbing the metric and energy–momentum tensors

Now we begin our study of the evolution of the perturbations. As is usual for relativistic perturbations, we will use conformal time η instead of physical time t, and an overdot will indicate $d/d\eta$ instead of d/dt. Care is needed to avoid confusion; one piece of guidance is that equations featuring the combination (k/a) will be using physical time t, whereas those featuring k alone will be using conformal time.

8.2.1 Metric

For the moment we leave open the gauge choice, and consider the most general first-order perturbation:

$$\boxed{ds^2 = a^2(\eta)\{-(1+2A)d\eta^2 - 2B_i\, d\eta\, dx^i + [(1+2D)\,\delta_{ij} + 2E_{ij}]dx^i dx^j\}}\,. \tag{8.8}$$

The term in square brackets specifies the spatial metric perturbation $2(D\delta_{ij} + E_{ij})$, and E_{ij} is taken to be traceless so that the separation of the two terms is unique. The term B_i is the **shift function**; as we see in a moment, it specifies the relative velocity between the threading and the worldlines orthogonal to the slicing. The term A is the **lapse function**, which specifies the relation between η and the proper time τ along the threading. To first order,

$$\frac{1}{a(\eta)}\frac{d\tau}{d\eta} = \sqrt{1+2A} \simeq 1 + A. \tag{8.9}$$

The chosen coordinate system $x^\mu = (\eta, x^i)$ applies throughout spacetime; it is a global coordinate system. In addition, at each spacetime point we need a locally orthonormal coordinate system (t, r^i) (locally orthonormal frame) with the following properties: the time directions of the global and local coordinate systems

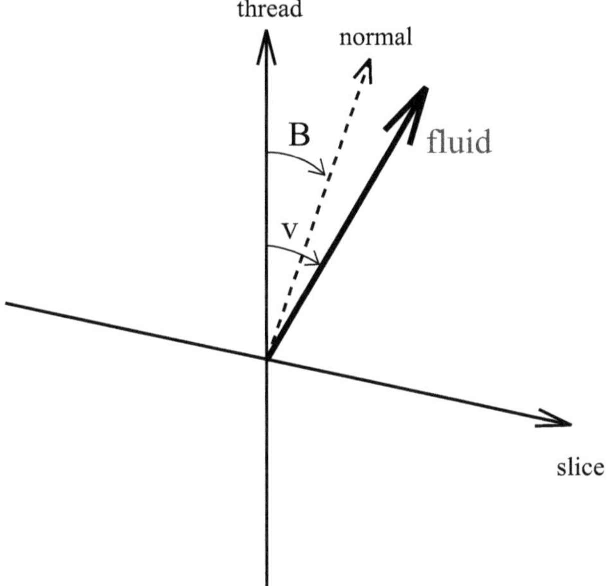

Fig. 8.1. The coordinates (η, x^i) define a threading and a slicing (corresponding, respectively, to fixed x^i and fixed η). As indicated, the slicing typically isn't orthogonal to the threading. The time direction of the locally orthonormal frame considered in the text is lined up with the time coordinate line (thread). The space directions of the locally orthonormal frame, not shown, are orthogonal to the time direction, and they coincide with the space coordinate lines to zero order in the perturbation. In the locally orthonormal frame, the velocity of a comoving observer (fluid velocity) is $\mathbf{v}$, and the velocity of the worldline that is normal (orthogonal) to the slices is $\mathbf{B}$, where B_i is the shift function defined by Eq. (8.8).

are lined up to first order, and the spatial directions are lined up to zero order. From the perturbed metric Eq. (8.8), this is equivalent to

$$dt = a(1 + A)\, d\eta \qquad \text{(first order)}, \tag{8.10}$$
$$dr^i = a\, dx^i \qquad \text{(zero order)}. \tag{8.11}$$

In the locally orthonormal frame, the velocity of the worldline orthogonal to the fixed-η slice is B^i. This is illustrated in Figure 8.1, and the proof is very simple. Working with the coordinates (η, x^i), the 4-velocity of the worldline with velocity B_i is to first order of the form $B^\mu = B^0(1, B^i)$. On the other hand, any 4-vector lying in the fixed-η slice is of the form $e^\mu = (0, e^i)$. One can check easily that indeed $g_{\mu\nu} B^\mu e^\nu$ vanishes to first order, for all choices of e^i.

8.2.2 Energy–momentum tensor

As illustrated in Figure 8.1, the fluid velocity in the locally orthonormal frame is $v^i = dr^i/dt$. Reverting to the global coordinates, the components $u^\mu = dx^\mu/d\tau$ of the fluid 4-velocity are

$$
\begin{aligned}
au^0 &= 1 && \text{(zero order)}, && (8.12)\\
au^i &= v^i \equiv v_i && \text{(first order)}. && (8.13)
\end{aligned}
$$

There is no distinction between upper and lower indices for the 3-velocity v^i, but there is for the 4-velocity. Indeed, for $u_\mu = g_{\mu\nu}u^\nu$, we find

$$
\begin{aligned}
\frac{1}{a}u_0 &= -1 && \text{(zero order)}, && (8.14)\\
\frac{1}{a}u_i &= v_i - B_i && \text{(first order)}. && (8.15)
\end{aligned}
$$

Using these expressions, we can work out Eq. (2.43) to first order:

$$
\begin{aligned}
T^0_0 &= -(\rho + \delta\rho)\,, && (8.16)\\
T^0_i &= (\rho + P)(v_i - B_i)\,, && (8.17)\\
T^i_0 &= -(\rho + P)v^i\,, && (8.18)\\
T^i_j &= (P + \delta P)\delta^i_j + \Sigma^i{}_j. && (8.19)
\end{aligned}
$$

Raising and lowering indices has no effect on either v_i or Σ_{ij} because they are spatial components defined in a locally orthonormal frame. It is usual to define a dimensionless version of the anisotropic stress by $\Pi_{ij} \equiv \Sigma_{ij}/P$.

8.2.3 Scalar mode

In this chapter we deal only with the scalar mode, and write

$$
\begin{aligned}
B_i &\equiv -\frac{ik_i}{k}B, && (8.20)\\
E_{ij} &\equiv \left(-\frac{k_ik_j}{k^2} + \frac{1}{3}\delta_{ij}\right)E, && (8.21)\\
v_i &\equiv -\frac{ik_i}{k}V, && (8.22)\\
\Pi_{ij} &= \left(-\frac{k_ik_j}{k^2} + \frac{1}{3}\delta_{ij}\right)\Pi. && (8.23)
\end{aligned}
$$

In the scalar mode we can write $\delta x_i = -i(k^i/k)\delta x$. Applying Eq. (5.21) to the metric tensor gives

$$\begin{aligned}\tilde{A} &= A - (\delta\eta)^{\cdot} - aH\delta\eta, &(8.24)\\ \tilde{B} &= B + (\delta x)^{\cdot} + k\delta\eta, &(8.25)\\ \tilde{D} &= D - \frac{k}{3}\delta x - aH\delta\eta, &(8.26)\\ \tilde{E} &= E + k\delta x. &(8.27)\end{aligned}$$

Applying it to the energy–momentum tensor gives

$$\begin{aligned}\tilde{V} &= V + (\delta x)^{\cdot}, &(8.28)\\ \tilde{\delta} &= \delta + 3(1+w)aH\delta\eta, &(8.29)\\ \widetilde{\delta P} &= \delta P - \dot{P}\delta\eta, &(8.30)\\ \tilde{\Pi} &= \Pi. &(8.31)\end{aligned}$$

In the second equation, we used $\dot{\rho} = -3aH(\rho + P)$ along with the definitions $\delta \equiv \delta\rho/\rho$ and $w \equiv P/\rho$.

We see that a slicing is needed to define the perturbations $\delta\rho$ and δP, a threading is needed to define the velocity perturbation V, and the anisotropic stress Π is gauge invariant.

8.3 Evolution of the scalar mode perturbations

8.3.1 Conformal Newtonian gauge

We will use the **conformal Newtonian gauge** which is determined uniquely by the line element

$$\boxed{ds^2 = a^2(\eta)\left[-(1+2\Psi)d\eta^2 + (1-2\Phi)\delta_{ij}dx^i dx^j\right]}. \tag{8.32}$$

We will call Φ and Ψ the **gravitational potentials**. Starting with an arbitrary gauge, we can indeed uniquely choose δx so that E vanishes, Eq. (8.27), and then uniquely choose $\delta\eta$ so that B vanishes, Eq. (8.25). In the conformal Newtonian gauge, slicing and threading are orthogonal, and the worldlines defining the latter have zero shear because the spatial part of the metric perturbation is isotropic.

8.3.2 Evolution equations

Although one can use the Einstein field equations directly, it is helpful to start with the continuity and Euler equations which follow from them. Using Eqs. (2.39), (2.40) and (2.41) for the energy–momentum tensor, we can find the first-order perturbations in the continuity equation $D_\mu T^\mu{}_0 = 0$ and also in the Euler equation

$D_\mu T^\mu_{\ i}=0$. These perturbations give, respectively, the two equations

$$\boxed{\dot{\delta} = -(1+w)(kV - 3\dot{\Phi}) + 3aHw\left(\delta - \frac{\delta P}{P}\right)}, \tag{8.33}$$

$$\boxed{\dot{V} = -aH(1-3w)V - \frac{\dot{w}}{1+w}V + k\frac{\delta P}{\rho+P} - \frac{2}{3}k\frac{w}{1+w}\Pi + k\Psi}, \tag{8.34}$$

where $w \equiv P/\rho$. The field equations give in addition

$$\boxed{\delta + 3\frac{aH}{k}(1+w)V = -\frac{2}{3}\left(\frac{k}{aH}\right)^2\Phi}, \tag{8.35}$$

$$\boxed{\Pi = \left(\frac{k}{aH}\right)^2(\Psi - \Phi)}. \tag{8.36}$$

These are constraint equations (i.e. equations involving no time derivatives).

Well after horizon entry, one expects the first constraint equation to become simply

$$\boxed{\delta = -\frac{2}{3}\left(\frac{k}{aH}\right)^2\Phi}. \tag{8.37}$$

(This expectation can be verified for the growing mode after neutrino decoupling, which is our concern here.) During matter domination $\Phi = \Psi$, and on scales well inside the horizon we recover Newtonian gravity with Φ the Newtonian gravitational potential and Eq. (8.37) the Poisson equation.[3] We will in any case call Eq. (8.37) the Poisson equation.

8.3.3 Initial conditions

For each scale k, we need initial conditions to be imposed well before horizon entry. In general, that means during radiation domination. For the very long wavelength modes that enter the horizon after matter domination, we can instead take the initial epoch to be during matter domination at least for an approximate solution.

Well before horizon entry, we expect the comoving threads to have practically zero shear. Then Eq. (5.25) will apply with $\psi = -\Phi$:

$$\boxed{\frac{\delta}{3(1+w)} = \zeta + \Phi}. \tag{8.38}$$

[3] Some authors define Φ and/or Ψ with a minus sign relative to our definition.

We also expect the right-hand sides of Eqs. (8.35) and (8.36) to be negligible. Then, taking k/aH to be negligible in Eqs. (8.33) and (8.44) (or more simply, in the 0–0 component of the Einstein equation) we find $\delta = -2(\Psi + (aH)^{-1}\dot{\Phi})$ leading to

$$\boxed{-\zeta = \Phi + \frac{2}{3}\frac{\Psi + (aH)^{-1}\dot{\Phi}}{1+w}}. \tag{8.39}$$

8.4 Separate fluids

If the cosmic fluid has a number of components, the total perturbations are

$$\delta\rho = \sum_i \delta\rho_i, \tag{8.40}$$

$$(\rho + P)V = \sum_i (\rho_i + P_i)V_i, \tag{8.41}$$

$$\delta P = \sum_i \delta P_i, \tag{8.42}$$

$$P\Pi = \sum_i P_i\Pi_i. \tag{8.43}$$

If a component of the cosmic fluid doesn't exchange energy with its surroundings it will satisfy the continuity equation (8.33), and if it doesn't exchange momentum it will satisfy the Euler equation (8.34). *We use these facts without comment in the following.*

We are interested in the epoch after $T \sim 10^{-1}$ MeV and assume that the cosmic fluid has just the four components i =CDM, B, γ and ν. The cold dark matter (CDM) doesn't interact and has negligible pressure. It therefore satisfies the continuity and Euler equations

$$\boxed{\dot{\delta}_{\rm c} = -k\, V_{\rm c} + 3\dot{\Phi}} \qquad \boxed{\dot{V}_{\rm c} = -aHV_{\rm c} + k\,\Psi}. \tag{8.44}$$

Thomson scattering doesn't alter the photon energy, so that the photons and baryons satisfy the continuity equation;

$$\boxed{\dot{\delta}_\gamma = -\frac{4}{3}kV_\gamma + 4\dot{\Phi}} \qquad \boxed{\dot{\delta}_B = -k\,V_B + 3\dot{\Phi}}. \tag{8.45}$$

In this chapter we will adopt the tight-coupling approximation for the baryons and photons. In this approximation, Thomson scattering of photons off electrons is taken to completely prevent photon diffusion. Since the electron number density is everywhere equal to the proton number density, this gives

$$\boxed{V_B = V_\gamma} \qquad \boxed{\frac{\delta(n_B/n_\gamma)}{n_B/n_\gamma} = \delta_B - \frac{3}{4}\delta_\gamma = \text{const}}. \tag{8.46}$$

With the adiabatic initial condition, the constant vanishes. The baryon–photon fluid is isotropic in a local rest frame, and its anisotropic stress vanishes.

In the tight-coupling approximation, the Euler equation for the baryon–photon fluid is

$$\boxed{\dot{V}_\gamma = -aH\left(1-3\tilde{w}\right)V_\gamma - \frac{\dot{\tilde{w}}}{1+\tilde{w}}V_\gamma + \frac{k}{3}\frac{\rho_\gamma}{\rho_B+\frac{4}{3}\rho_\gamma}\delta_\gamma + k\Psi} . \tag{8.47}$$

Here the tilde refers to the baryon–photon fluid, and to a good approximation $\tilde{w} = 3\rho_\gamma/(\rho_\gamma+\rho_B)$. In writing this equation we took $\delta P = \delta P_\gamma$, valid for scales above the Jeans scale.

If we ignore the neutrino perturbation (which otherwise would contribute an anisotropic stress), the tight-coupling approximation gives $\Phi = \Psi$. Then a closed set of equations is provided by the tight-coupling approximation (8.46)–(8.47), together with the exact equations (8.35), (8.40)–(8.42) and (8.44). In the rest of this chapter, we use this set to arrive at some rough approximations which encapsulate the essential physics. In order to do that we need initial conditions, holding well before horizon entry.

Until Chapter 12 we adopt the adiabatic initial condition for the density perturbation. Then ζ is constant, and Eq. (8.39) gives during any era when w is constant

$$\boxed{\Phi = \Psi = -\frac{3+3w}{5+3w}\zeta} \qquad \text{(before horizon entry).} \tag{8.48}$$

(As usual we drop a decaying solution.)

We will normally lay down the initial condition during radiation domination, so that

$$\boxed{\Phi = \Psi = -\frac{2}{3}\zeta} \qquad \text{(before horizon entry).} \tag{8.49}$$

During radiation domination, Eq. (8.38) gives the density contrast as

$$\boxed{\delta = -2\Phi = \frac{4}{3}\zeta} \qquad \text{(before horizon entry)} . \tag{8.50}$$

8.5 Matter density transfer function

In this section we consider the matter-dominated era, after photon decoupling. We ignore the radiation completely so that $\Phi = \Psi$. We first show that the gravitational potential $\Phi_{\mathbf{k}}$ is time independent. Then we see how to calculate a transfer function which gives its **k**-dependence. The calculation is vital because $\Phi_{\mathbf{k}}$ has several important effects. It determines the course of bottom-up structure formation (Section 9.2), it affects the acoustic oscillation of the baryon–photon fluid (Section

8.6), and it gives the dominant contribution to the cosmic microwave background (CMB) anisotropy on large angular scales (Section 10.6).

Let us first pretend that there are no baryons. From Eqs. (8.44) and (8.35), or more simply from the ij component of the Einstein equation, we find

$$\ddot{\Phi}_{\mathbf{k}} + 6\eta^{-1}\dot{\Phi}_{\mathbf{k}} = 0. \tag{8.51}$$

It indeed has a time-independent solution, and a decaying solution which we drop as usual.

To include the baryons we have to wait until after photon decoupling, when the baryons cease to be tightly coupled to the photons. Then, on scales in excess of the baryon Jeans scale, we have a single pressureless fluid and arrive again at Eq. (8.51).

To calculate the time-independent $\Phi_{\mathbf{k}}$, we have to follow the evolution of the density contrast between the initial epoch and photon decoupling. Consider first those scales which are still far outside the horizon when matter domination first becomes a good approximation. Then $\zeta_{\mathbf{k}}$ is constant, and Eq. (8.48) evaluated during matter domination gives $\Phi_{\mathbf{k}} = -(3/5)\zeta_{\mathbf{k}}$.[4]

We therefore define a **matter transfer function** by

$$\boxed{\Phi_{\mathbf{k}} = -\frac{3}{5}T(k)\zeta_{\mathbf{k}}}\,. \tag{8.52}$$

The transfer function is time independent after photon decoupling and is close to 1 on very large scales.

For an approximate determination of $T(k)$ it is enough to keep only the CDM, and to consider only scales which are well inside the horizon at decoupling. The evolution of the CDM is given in terms of the gravitational potentials by Eq. (8.44). The gravitational potentials depend on the perturbations of all four components of the cosmic fluid. Well before horizon entry, $V_i = 0$ while the δ_i and gravitational potentials have time-independent values given by the adiabatic initial condition. As horizon entry approaches, all of those quantities start to evolve. Around the time of horizon entry, all relevant perturbations have the same order of magnitude and

$$\dot{\delta}_{\mathrm{c}} \sim aH\delta_{\mathrm{c}}. \tag{8.53}$$

The order of magnitude of the right-hand side is determined by the Hubble parameter, which is the only scale in the problem. The precise coefficient including the sign can be calculated using either the exact equations or the tight-coupling approximation.

[4] The factor $6/5$ in Eq. (6.71) occurs because f_{NL} was first defined using Φ instead of ζ, with a sign convention for Φ that is opposite to ours, and with f_{NL} normalized so that on large scales during matter domination $\Phi = g + bg^2$ corresponds to $f_{\mathrm{NL}} = b$.

Well after horizon entry Eq. (8.37) applies, with $\delta\rho$ practically equal to the photon density perturbation because the neutrino density perturbation has free-streamed away. The photon density contrast undergoes the decaying oscillation of the baryon–photon fluid, which means that Φ falls faster than $(aH/k)^2$ and soon becomes negligible compared with δ_c. Combining the continuity and Euler equations for δ_c one then finds

$$\ddot{\delta}_c + aH\dot{\delta}_c = 0. \tag{8.54}$$

With the initial conditions (8.50) and (8.53), the solution of this equation is $\delta_c \simeq \zeta \ln(k/aH)$. This holds until matter–radiation equality, by which time

$$\delta_c \simeq \zeta \ln(k\eta_{eq}). \tag{8.55}$$

At this stage δ_B/δ_c is well below its primordial value of unity, because the numerator has been undergoing a decaying oscillation while the denominator has been slowly growing. Setting $\delta_B = 0$ the density contrast is $(\rho_c/\rho_m)\delta_c$, but we can drop the prefactor within the accuracy of the present discussion. Applying Eq. (8.37) and again dropping the numerical factor, we conclude that $\Phi_{\mathbf{k}}$ is of order $-(k_{eq}/k)^2 \ln(k/k_{eq})\zeta_{\mathbf{k}}$ when $k_{eq} = (aH)_{eq}$. This gives

$$\boxed{T(k) \simeq \frac{k_{eq}^2}{k^2} \ln\frac{k}{k_{eq}}} \qquad (k \gtrsim k_{eq}). \tag{8.56}$$

The small non-zero value of δ_B at decoupling is an oscillating function of k, being a snapshot of the acoustic oscillation at that epoch. This gives a small oscillating contribution to $T(k)$, known as baryon acoustic oscillations (BAO), which may ultimately prove to be a powerful standard ruler for exploring the expansion history of the Universe and hence the properties of dark energy.

We have so far ignored the fact that neutrinos contribute a non-zero amount Ω_ν to the present matter density Ω_m. At least one neutrino species has mass $m > 0.05\,\text{eV}$. If m isn't too big the species becomes non-relativistic during matter domination. When that happens, the scale entering the horizon is $250\,\text{Mpc}\,(1\,\text{eV}/m)^{1/2}$, from Eqs. (4.46) and (4.64). On much smaller scales the density of this species free-streams away, reducing the matter density contrast by a fractional amount of order Ω_ν/Ω_m where Ω_ν is the contribution of just this species. On much larger scales there is no significant reduction. For this effect to be observable the mass has to be much bigger than $0.05\,\text{eV}$, requiring the mass of two or all three species to be practically the same.

The present observational constraint, $\sum m_i \lesssim 1\,\text{eV}$, is obtained by fitting both galaxy distribution data, which is sensitive to the above reduction, and CMB anisotropy data that we consider in the next chapter. With future improvements in both

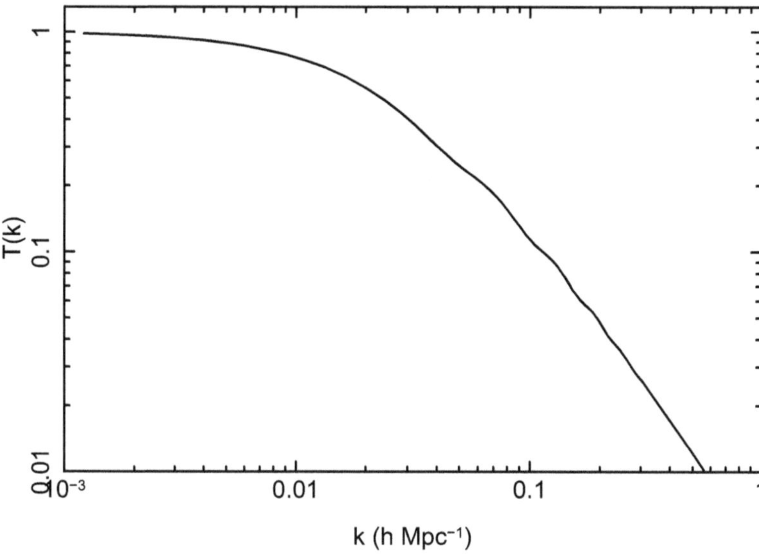

Fig. 8.2. The adiabatic transfer function for the standard cosmology.

types of data, it will probably be possible to infer at least the heaviest neutrino mass.

To obtain an accurate transfer function over the full range of scales, we have to solve the full equations of Chapter 11 numerically. Taking the neutrino mass to be negligible, this gives the result shown in Figure 8.2, obtained from the CMB-fast code.[5] The baryon oscillations can just be identified, and the approach to the asymptotic regime $k \gg aH$ is clearly seen.

8.6 Acoustic oscillation

Until photon decoupling, the baryons and photons form a tightly coupled fluid, supporting a standing-wave acoustic oscillation. The photons then travel freely, but imprinted on their distribution is a snapshot of the oscillation as it existed just before decoupling. As we see in Chapters 10 and 11, this gives rise to the peak structure in the CMB anisotropy. The same peak structure gives the small oscillation in the matter transfer function that we noted earlier.

We base our discussion of the acoustic oscillation on the tight-coupling approximation. To make the formulae useful for the CMB discussion, we shall sometimes translate the comoving wavenumber k into the multipole $\ell = k\eta_0$ of the CMB anisotropy, as was done in Table 7.2.

Combining the continuity and Euler equations gives an equation governing the

[5] Seljak and Zaldarriaga, http://cfa-www.harvard.edu/~mzaldarr/CMBFAST/cmbfast.html.

acoustic oscillation of the baryon–photon fluid. It is

$$\boxed{\frac{1}{4}\ddot{\delta}_{\gamma\mathbf{k}} + \frac{1}{4}\frac{R}{1+R}\dot{\delta}_{\gamma\mathbf{k}} + \frac{1}{4}k^2 c_s^2 \delta_{\gamma\mathbf{k}} = F_{\mathbf{k}}(\eta)}\,. \tag{8.57}$$

In this equation,

$$F_{\mathbf{k}}(\eta) \equiv -\frac{k^2}{3}\Psi_{\mathbf{k}}(\eta) + \frac{\dot{R}(\eta)}{1+R(\eta)}\dot{\Phi}_{\mathbf{k}}(\eta) + \ddot{\Phi}_{\mathbf{k}}(\eta), \tag{8.58}$$

$$c_s^2(\eta) \simeq \frac{\dot{P}_\gamma}{\dot{\rho}_\gamma + \dot{\rho}_B} = \frac{1}{3(1+R(\eta))}, \tag{8.59}$$

$$R(\eta) \equiv \frac{3}{4}\frac{\rho_B}{\rho_\gamma}. \tag{8.60}$$

In the regime where c_s is slowly varying, Eq. (8.57) is the equation of a forced oscillator, with a driving term F and speed of sound c_s. We will ignore the small damping of this oscillation provided by the second term of Eq. (8.57) because it is negligible compared with Silk damping that we come to later.

The two constraint equations determine F in terms of the perturbations of the neutrinos, the CDM, and the baryon–photon fluid itself. As the first two vary slowly, the angular frequency of the driving term is kc_s. This is also the frequency of the oscillation, which has the form

$$\boxed{\frac{1}{4}\delta_{\gamma\mathbf{k}}(\eta) = A_{\mathbf{k}}(\eta) + B_{\mathbf{k}}(\eta)\cos(kr_s(\eta)) + C_{\mathbf{k}}(\eta)\sin(kr_s(\eta))}\,, \tag{8.61}$$

with slowly varying coefficients $A_{\mathbf{k}}$, $B_{\mathbf{k}}$ and $C_{\mathbf{k}}$.

The coefficient $A_{\mathbf{k}}$ corresponds to the oscillation being off-centre, owing to the effect of the non-oscillating contributions to the driving term. After matter domination, it corresponds to the non-oscillating solution of Eq. (8.57),

$$\boxed{A_{\mathbf{k}}(\eta) = -[1+R(\eta)]\Phi_{\mathbf{k}}}\,. \tag{8.62}$$

The slowly varying quantity r_s is the distance that sound has had time to travel since a much earlier time, idealized as $\eta = 0$. It is called the sound horizon and is given by

$$\boxed{r_s(\eta) = \int_0^\eta c_s(\eta)d\eta}\,. \tag{8.63}$$

At early times $R \simeq 0$, $c_s = 1/\sqrt{3}$ and $r_s = c_s\eta$. The baryon fraction increases with time,

$$R(\eta) = \frac{3}{4} f_B \left(1 - R_\nu\right)^{-1} \frac{z_{\rm eq}}{z(\eta)}, \tag{8.64}$$

where $f_B \equiv \rho_B/\rho_{\rm m}$ and $R_\nu \equiv \rho_\nu/\rho_{\rm r} = 0.40$. The sound horizon also increases with time. Using Eqs. (4.46), (8.59) and (8.63),

$$r_{\rm s}(\eta) = \frac{2}{3k_{\rm eq}}\sqrt{\frac{6}{R_{\rm eq}}}\ln\left(\frac{\sqrt{1+R(\eta)}+\sqrt{R(\eta)+R_{\rm eq}}}{1+\sqrt{R_{\rm eq}}}\right). \tag{8.65}$$

At decoupling R and $r_{\rm s}$ are given by

$$\boxed{R = \frac{3}{4}f_B(1-f_\nu)^{-1}\frac{z_{\rm eq}}{z_{\rm ls}} = 29\Omega_B h^2 \simeq 0.65} \qquad \boxed{r_{\rm s} = 1.5\times 10^2\,{\rm Mpc}}. \tag{8.66}$$

We need to check if $c_{\rm s}$ is indeed slowly varying on the timescale of the oscillation, corresponding to $\dot c_{\rm s}/c_{\rm s} \ll k c_{\rm s}$. If this requirement is satisfied at decoupling it is satisfied at earlier times too. Remembering that $R_{\rm ls} \sim 1$, it corresponds to $k\eta_{\rm ls} \gg 1$ or $\ell \gg 70$.

8.7 Silk damping

We have so far been ignoring Silk damping, coming from the diffusion of the photons which carry the baryons with them. As long as the time $t_{\rm c}$ between collisions is short compared with the Hubble time, the effect of diffusion can be calculated using standard kinetic theory. It multiplies the amplitude by some factor $\exp\left(-k^2/k_{\rm D}^2(\eta)\right)$. The Silk scale $k_{\rm D}^{-1}$ is roughly the comoving distance that a photon has had time to travel since some much earlier epoch (idealized as $\eta = 0$). Let us see how to calculate the Silk scale.

8.7.1 Thomson scattering

The only significant interaction experienced by the photons is the elastic scattering process $\gamma\,{\rm e} \to \gamma\,{\rm e}$. In general this is Compton scattering. But we are interested in the regime $T \lesssim 10^{-1}$ MeV. In this regime the typical photon energy is much less than the electron mass. Also, the electron kinetic energy, $m_{\rm e}v^2/2 \sim T$ is much less than $m_{\rm e}$ so that the electrons are non-relativistic.

In this regime, Compton scattering becomes Thomson scattering. For an isolated electron, Thomson scattering is a classical process. Taking the electron to be initially at rest an incoming plane wave with given frequency causes it to oscillate with that frequency so that it emits dipole radiation with the same frequency. From this we see that in the electron rest frame, (i) Thomson scattering doesn't change the photon energy and (ii) the scattered photon has equal probability of emerging in the forward and backward directions so that on average the photon loses all of its momentum. Going to a frame where the electron has speed $v \ll 1$, it remains true to first order in v that the photon energy is unchanged.

In our situation there are many electrons, and there are photons moving in all directions. The effect of Thomson scattering on photons with a given momentum will therefore be given by Eq. (2.69), with $\mathcal{M}$ the Thomson scattering amplitude. The quantum effects of Pauli blocking and stimulated emission seem to appear in the square bracket, and we have to check that they are actually absent before we can treat Thomson scattering as an isolated event.

One easily checks that the electron occupation number is much less than 1 so that there is no Pauli blocking factor. The photon occupation number is typically of order 1 since the photons have a blackbody distribution. Here, the absence of stimulated emission corresponds to a cancellation between two terms in the square bracket of Eq. (2.69). To see that the cancellation occurs, we work in the local rest frame of the baryons. The typical electron momentum, $m_{\rm e} v \sim \sqrt{T m_{\rm e}}$, is then much less than the typical photon momentum T. We can therefore take the electrons to be practically at rest, which means that the scattering doesn't alter the electron or the photon energy. For the electrons, this means that the initial and final occupation numbers ($f_{\rm a}$ and $f_{\rm c}$ in Eq. (2.69)) are the same, since the Coulomb interaction keeps them in good equilibrium with the photons so that their occupation number depends only on their energy. For the photons that isn't the case because they are falling out of thermal equilibrium. But the equality of the initial and final electron occupation numbers leads to a cancellation of factors $f_{\rm c} f_{\rm d} f_{\rm b}$ and $f_{\rm a} f_{\rm b} f_{\rm d}$ in Eq. (2.69), so that the square bracket becomes simply $f_{\rm c} f_{\rm d} - f_{\rm a} f_{\rm b}$ which is the same as if there were no stimulated emission.

We can therefore treat each Thomson scattering event as if it occurred in isolation. In the local baryon rest frame, the photons perform a random walk as they fall out of thermal equilibrium. We will describe Thomson scattering in detail in Section 11.4 but for the moment we just need the rate. If there are $n_{\rm e}$ free electrons per unit volume, the probability per unit time of a photon undergoing Thomson scattering is $n_{\rm e} \sigma_{\rm T}$, where $\sigma_{\rm T}$ is the Thomson scattering cross-section. It is given by $\sigma_{\rm T} = (8\pi/3)(\alpha/m_{\rm e})^2$, where $\alpha = e^2/4\pi = 1/137$ is the fine-structure constant and $m_{\rm e}$ is the electron mass. With c restored, the rate is $n_{\rm e} c \sigma_{\rm T}$, and $\sigma_{\rm T} = 6.652 \times 10^{-25}\,{\rm cm}^2$.

8.7.2 Silk scale estimate

Now we can estimate the Silk scale. In the local baryon rest frame the photon performs a random walk. The mean time between collisions is $t_{\rm c} \sim (n_{\rm e}\sigma_{\rm T})^{-1}$. The average number of steps in time t is $N = t/t_{\rm c}$, and in that time a photon diffuses a distance $d \sim \sqrt{N} t_c \sim (t t_c)^{1/2}$. This gives the estimate

$$\boxed{a k_D^{-1} \simeq \left(\frac{t}{n_e \sigma_T} \right)^{1/2}}. \tag{8.67}$$

Before decoupling, k_D^{-1} is proportional to $a^{5/4}$ during matter domination, and to $a^{3/2}$ during radiation domination. Each scale k^{-1} starts out bigger than the Silk scale at horizon entry, photon diffusion setting in only at the later epoch when $k_S(\eta) = k$. During decoupling the electrons bind into atoms so that n_e falls sharply. Taking as a representative epoch the peak of the visibility function, and using Eq. (8.67), one finds $k_D^{-1} \simeq 8\,\mathrm{Mpc}$. Unfortunately this scale is about the same as the photon mean free path, which means that $t_c \sim H^{-1}$ and kinetic theory is failing. As a result the Silk scale at decoupling isn't a well-defined concept, though it is good enough for use in the approximate calculation of the CMB anisotropy that we describe in the next chapter.

8.7.3 Acoustic oscillation at decoupling

Putting everything together, we have the desired snapshot of the acoustic oscillation at decoupling:

$$\boxed{\frac{1}{4}\delta_{\gamma\mathbf{k}} = A_{\mathbf{k}} + e^{-k^2/k_D^2} \left[B_{\mathbf{k}} \cos(k r_s) + C_{\mathbf{k}} \sin(k r_s) \right]}, \tag{8.68}$$

$$\boxed{\frac{1}{4}\dot{\delta}_{\gamma\mathbf{k}} = -\frac{k}{3} V_{\gamma\mathbf{k}} = k c_s e^{-k^2/k_D^2} \left[-B_{\mathbf{k}} \sin(k r_s) + C_{\mathbf{k}} \cos(k r_s) \right]}. \tag{8.69}$$

All quantities are evaluated at decoupling, so that $R \simeq 0.70$, $r_s \simeq 150\,\mathrm{Mpc}$ and $k_D^{-1} \simeq 8\,\mathrm{Mpc}$. In the second equation we used Eq. (8.45), with Φ dropped because it is suppressed by the factor $(aH/k)^2$ appearing in the Poisson equation.

Adopting say the adiabatic initial condition, the coefficients $A_{\mathbf{k}}$, $B_{\mathbf{k}}$ and $C_{\mathbf{k}}$ can be evaluated by solving the wave equation (8.57), taking the gravitational potentials from either an exact calculation or one using the closed system of equations that ignores the neutrino perturbation. We will instead use Eq. (8.62) for the first coefficient, and assume that the other coefficients are given by the initial condition. The latter approximation isn't completely unreasonable, because Silk damping allows only a few oscillations before decoupling.

Using the adiabatic initial condition (8.50), we find

$$\boxed{\frac{1}{4}\delta_{\gamma\mathbf{k}} \simeq -(1+R)\Phi_{\mathbf{k}} + \frac{1}{3}e^{-k^2/k_{\rm D}^2}\cos(kr_{\rm s})\zeta_{\mathbf{k}}}\,, \tag{8.70}$$

$$\boxed{\frac{1}{4}\dot{\delta}_{\gamma\mathbf{k}} = -\frac{k}{3}V_{\gamma\mathbf{k}} \simeq -\frac{1}{3}kc_{\rm s}e^{-k^2/k_{\rm D}^2}\sin(kr_{\rm s})\zeta_{\mathbf{k}}}\,, \tag{8.71}$$

with $\Phi_{\mathbf{k}} = -(3/5)T(k)\zeta_{\mathbf{k}}$. We shall use this expression to understand the peak structure of the CMB anisotropy.

8.8 Synchronous gauge

With an appropriate gauge transformation, we can write the evolution and constraint equations in any gauge. The most commonly used alternative gauge is the synchronous gauge. It corresponds to $A = B = 0$, so that the only non-zero metric perturbations are D and E. The condition $A = 0$ says that the threading consists of geodesics, and $B = 0$ says that the slicing is orthogonal to them.

In contrast with the conformal Newtonian gauge, the specification of the metric doesn't completely define the gauge. The definition is best completed by demanding that the threading becomes comoving in the super-horizon regime. This is equivalent to saying that the threads follow the motion of the CDM.

If one applies the Einstein equations using only the metric, one obtains solutions that correspond to changes in the choice of threading. These are referred to as gauge modes, and are dropped. After dropping the gauge modes, the synchronous gauge provides a useful alternative to the conformal Newtonian gauge. It is especially convenient for numerical work.

Exercises

8.1 Choosing the z-axis along the $\mathbf{k}$ direction, write down the matrix specifying a symmetric second-rank tensor as the sum of the scalar, vector and tensor modes described in the text. Verify that it is the most general such tensor.

8.2 For the metric tensor, show that Eq. (5.21) leads to Eqs. (8.24) and (8.27). For the energy–momentum tensor, verify that Eq. (8.28) is the gauge transformation of a 3-vector, whereas each of Eqs. (8.29) and (8.30) is just the gauge transformation Eq. (5.19) of a scalar.

8.3 Verify that Eq. (8.62) is a solution of Eq. (8.57), during matter domination and after horizon entry.

8.4 Write down the closed system of equations, existing if the neutrino perturbation is ignored and the baryon–photon fluid is taken to be tightly coupled. Show that they are indeed closed.

9
The matter distribution

In this chapter, we describe one ultimate consequence of the evolution of the primordial perturbation, namely the observable matter distribution in the Universe. It is not our intention to provide a detailed account of how observers probe the matter distribution, something which is now carried out with impressive precision. Instead, we focus on the outcome of those observations and their comparison with theory. We will explain the fundamental theoretical strategy, and highlight some simple physical arguments which account for the broad features of the data.

To use linear perturbation theory after galaxies begin to form we need to smooth the density contrast as described in Section 5.1.2. After looking in more detail at the description of smoothing, we will see in a qualitative way how the theory describes 'bottom-up' structure formation. We go on to give a more quantitative description, which applies to the formation of objects with a given mass as long as they are rare. In this way we obtain an estimate of the abundance of such objects, which may be compared with observation.

Next we come to the issue of comparing the calculated density perturbation directly with observation. More precisely, we seek to compare with observation the spectrum of the density contrast, and higher correlators which may signal non-gaussianity. For this purpose we have to remember that the galaxy number density won't precisely trace the matter density, since the latter includes dark matter (both baryonic and cold dark matter (CDM)). We therefore consider direct probes of the matter density, as well as the more traditional tool of galaxy surveys. We end the chapter by looking at the possibility that objects may form before (maybe long before) matter domination.

9.1 Smoothing

Irrespectively of whether smoothing is needed to validate linear theory, it will be needed because observations will have a finite resolution corresponding to a smoothing of the perturbation.

We consider a smoothed quantity

$$g(R, \mathbf{x}) = \frac{1}{V} \int W(|\mathbf{x}' - \mathbf{x}|/R)\, g(\mathbf{x}')\, d^3x' \,, \tag{9.1}$$

where the **window function** $W(y)$ that governs the weighting is to fall off rapidly for $y > 1$, and V is its volume:

$$V \equiv \int d^3x\, W(x/R) = 4\pi R^3 \int y^2 W(y)\, dy. \tag{9.2}$$

The exact choice of window function is a matter of convenience. We have taken it to be spherically symmetric, with $x = |\mathbf{x}|$, though this isn't mandatory. Our introductory discussion employed the **top hat**, defined by $W = 1$ for $y \leq 1$ and $W = 0$ for $y > 1$, which gives equal weight to all material within radius R. Another simple possibility is the **gaussian**, $W(y) = \exp(-y^2/2)$.

Using Eq. (6.7)

$$g_{\mathbf{k}}(R) = \widetilde{W}(kR)\, g_{\mathbf{k}}, \tag{9.3}$$

where

$$\widetilde{W}(kR) \equiv \frac{\int d^3y W(y) e^{iR\mathbf{k}\cdot\mathbf{y}}}{\int d^3y W(y)}. \tag{9.4}$$

We see that $\widetilde{W}(kR)$ is equal to 1 at $k = 0$ and falls off rapidly at $kR > 1$. For the top hat, we can perform the angular integration in Eq. (9.4) to find

$$\widetilde{W}(kR) = 3\left[\frac{\sin(kR)}{(kR)^3} - \frac{\cos(kR)}{(kR)^2}\right]. \tag{9.5}$$

For the gaussian, we can use Cartesian coordinates to find

$$\widetilde{W}(kR) = \exp\left(-\frac{k^2R^2}{2}\right). \tag{9.6}$$

These two filters are shown in Figure 9.1.

For each window function, it is useful to define a mass $M = V\rho_{c,0}$, which is the mass of matter enclosed in comoving volume V. For the top hat

$$M = 1.16 \times 10^{12} h^{-1} \left(\frac{R}{h^{-1}\,\mathrm{Mpc}}\right)^3 M_\odot, \tag{9.7}$$

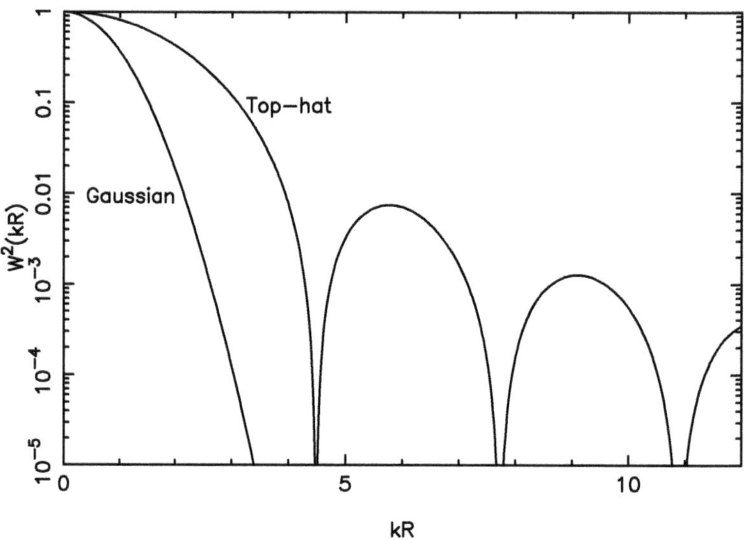

Fig. 9.1. Top-hat and gaussian filters in Fourier space.

and for the gaussian

$$M = 4.37 \times 10^{12} h^{-1} \left(\frac{R}{h^{-1}\,\mathrm{Mpc}} \right)^3 M_\odot . \tag{9.8}$$

The smoothing removes structure on scales $\lesssim R$, without affecting structure on much bigger scales. Correspondingly, it filters out the Fourier components with $kR \gtrsim 1$ without significantly affecting those with $kR \ll 1$.

9.2 Bottom-up structure formation

In this section we see that the formation of gravitationally bound structures from the primordial density perturbation proceeds in a bottom-up fashion, starting with small objects and ending with the biggest galaxy clusters.

According to linear theory, the density contrast smoothed on scale R is

$$\boxed{\delta_{\mathbf{k}}(z) = \frac{2}{5} \widetilde{W}(kR) \left(\frac{k}{H_0} \right)^2 \frac{T(k)}{\Omega_\mathrm{m}} \frac{g(z)}{1+z} \zeta_{\mathbf{k}}} . \tag{9.9}$$

This expression corresponds to Eq. (7.28), with smoothing and using Eq. (8.52) for

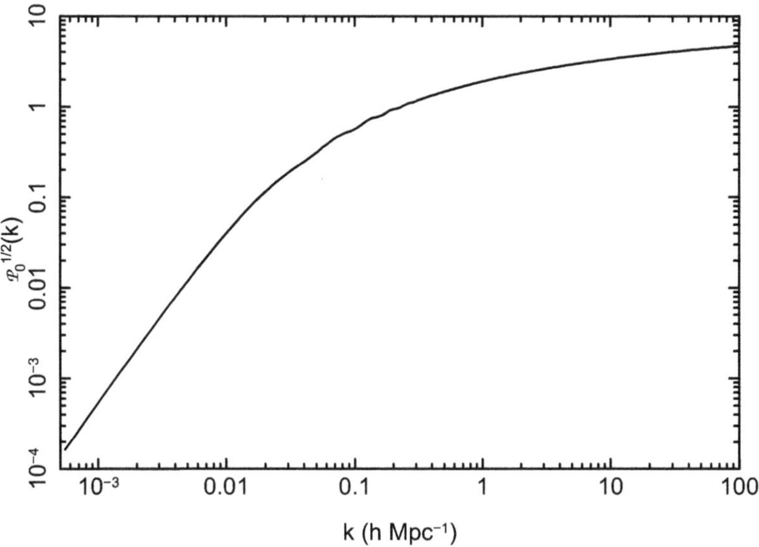

Fig. 9.2. Spectrum of the density contrast during matter domination, according to the ΛCDM model.

$\Phi_{\mathbf{k}}$. The spectrum of δ is conveniently written

$$\mathcal{P}_\delta^{1/2}(k, R, z) = \widetilde{W}(kR)\frac{g(z)}{1+z}\mathcal{P}_0^{1/2}(k), \tag{9.10}$$

$$\mathcal{P}_0^{1/2}(k) = \frac{2}{5}\left(\frac{k}{H_0}\right)^2 \frac{T(k)}{\Omega_{\mathrm{m}}}\mathcal{P}_\zeta^{1/2}(k), \tag{9.11}$$

where $\mathcal{P}_0$ is $\mathcal{P}_\delta$, calculated at the present epoch *using linear theory without smoothing*, and not including the suppression factor $g(z)$ which instead is in the first line. From now on we drop the suppression factor, which fell below 1 only recently.

The linear regime ends at roughly the epoch z_{nl} given by $\sigma(R, z_{\mathrm{nl}}) = 1$, where $\sigma(R, z)$ is the *rms* density contrast. This can be written

$$\boxed{1 + z_{\mathrm{nl}} = \sigma_0(R)}\,, \tag{9.12}$$

where $\sigma_0(R)$ is calculated at the present epoch using linear theory:

$$\boxed{\sigma_0^2(R) = \int \widetilde{W}^2(kR)\mathcal{P}_0(k)\frac{dk}{k} \sim \mathcal{P}_0(R^{-1})}\,. \tag{9.13}$$

Figure 9.2 shows $\mathcal{P}_0^{1/2}(k)$, and Figure 9.3 shows $\sigma_0(R)$. Figure 9.4 shows $\sigma_0(R)$ as a function of the enclosed mass $M(R)$ given by Eq. (9.7).

Since the spectrum $\mathcal{P}_0$ is fairly flat to large k, and the window function falls off rapidly for $k > 1/R$, the spectrum $\tilde{W}^2(kR)\mathcal{P}_0(k)$ of the smoothed density

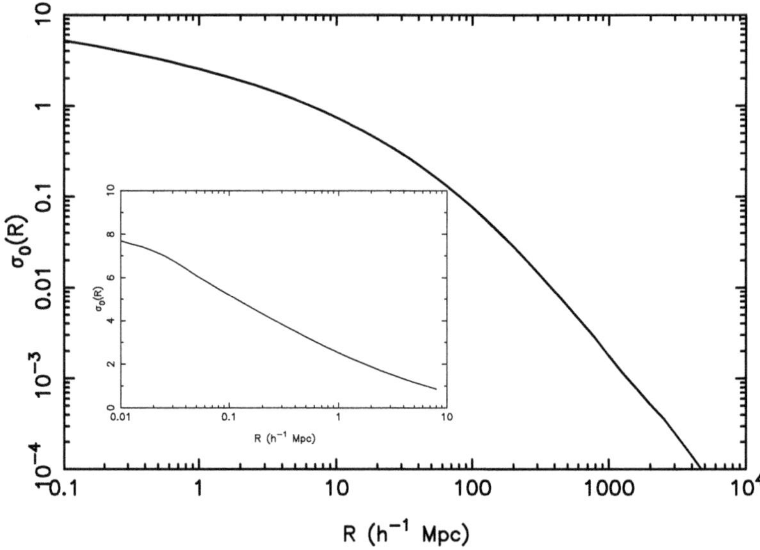

Fig. 9.3. Dispersion of the top-hat filtered density contrast according to the ΛCDM model.

contrast peaks at $k \sim 1/R$, where it is roughly equal to the mean-square $\sigma_0^2(R)$. The position of the peak means that an overdense region is typically of size R. When the linear regime ends such a region typically collapses, because the inward peculiar velocity at its edge, given by Eq. (7.22), becomes of the same order as the Hubble velocity aRH. Since the overdense regions account for half of the volume, it follows that when the density contrast smoothed on scale R ceases to evolve linearly, a significant fraction of the matter in the Universe collapses to form structures with mass of order $M(R)$.

Meanwhile, the density contrast smoothed on significantly bigger scales is still evolving linearly. In other words, while regions of a given size are collapsing, significantly bigger ones are still expanding with the Universe. This is the bottom-up picture of structure formation. From Figure 9.3, we see that at the present epoch scales $\gtrsim 10h^{-1}$ Mpc are still evolving linearly. On smaller scales, linear evolution has ceased, and the linear theory prediction $\sigma_0(R)$ is invalid, even though it is the one that should be used in Eq. (9.12).

Let us see how the bottom-up picture refers to the various objects in the Universe. At $z \sim 10$ to 30, the Jeans mass is of order $10^4 M_\odot$. This provides a very crude estimate of the mass of the first baryonic objects to form, though arguably the true value might be an order of magnitude or two lower when non-linear collapse is described properly. Looking at Figure 9.4 we see that a large fraction of the mass of the Universe collapses into baryonic objects at $z \simeq 10$.

We can see from Figure 9.4 that, after the first baryonic objects form, heavier

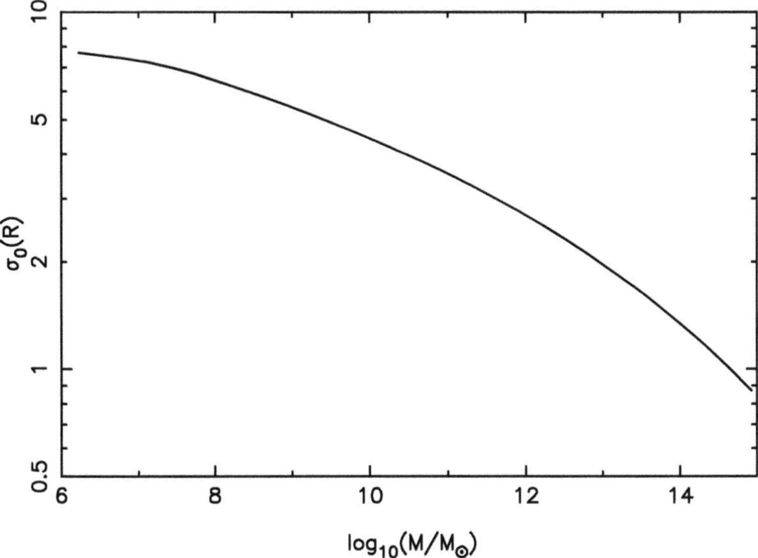

Fig. 9.4. Dispersion of the top-hat filtered density contrast according to the ΛCDM model, as a function of the filter mass M.

ones form not much later, because $\sigma_0(M)$ is quite flat on small scales. Objects with mass $M \sim 10^{11} M_\odot$ or so will form at a redshift of a few. After the first galaxies form, they can merge and split, leading to a complicated evolution, which prevents us from saying much about present-day galaxies on the basis of linear evolution.

The situation is different for the scales $M \sim 10^{13} M_\odot$ to $10^{15} M_\odot$, which correspond to galaxy groups and clusters. In this range, $\sigma(M)$ is decreasing more strongly, and we can expect a more sharply defined bottom-up picture, with the heavier structures forming quite a bit later than the lighter ones. This is consistent with high-redshift observations.

9.3 Critical density for collapse

In this section and the next, we see how the evolution of the density in a collapsing region can be followed analytically if it is assumed to be perfectly spherical. Our treatment ignores the cosmological constant, which is effective only at very late times.

For simplicity we postulate, at some initial epoch, a perturbation with a top-hat profile of radius R. (The evolution equations that we develop actually apply to each spherical shell of a generic profile.) According to the two famous theorems of Newton regarding spherical mass distributions, the Universe outside the perturbation continues to evolve as a critical-density Universe, and the perturbed region evolves independently of the behaviour of the exterior region, behaving as a closed universe with density parameter $\Omega(z)$ Writing the radius as $r = Ra_{\rm loc}(t)$ where $a_{\rm loc}$ is the local scale factor, the solution of the Friedmann equation with $K = 1$ may be written as

$$\frac{a_{\rm loc}(t)}{a_{\rm max}} = \frac{1}{2}(1 - \cos\theta), \qquad \frac{t}{t_{\rm max}} = \frac{1}{\pi}(\theta - \sin\theta), \tag{9.14}$$

where the development angle θ runs from 0 to 2π. The subscript max indicates the epoch of maximum expansion for the perturbed region.

To study the linear regime, we need the small-parameter expansions of these to second order, giving

$$\frac{a_{\rm loc}(t)}{a_{\rm max}} \simeq \frac{\theta^2}{4} - \frac{\theta^4}{48}, \qquad \frac{t}{t_{\rm max}} \simeq \frac{1}{\pi}\left(\frac{\theta^3}{6} - \frac{\theta^5}{120}\right). \tag{9.15}$$

Combining these gives what we call the linearized scale factor,

$$\frac{a_{\rm lin}(t)}{a_{\rm max}} = \frac{1}{4}\left(6\pi\frac{t}{t_{\rm max}}\right)^{2/3}\left[1 - \frac{1}{20}\left(6\pi\frac{t}{t_{\rm max}}\right)^{2/3}\right]. \tag{9.16}$$

This last expression is the important one. If we ignore the square-bracketed term, it gives the expansion of the background spatially flat Universe. If we include both terms, that gives us the *linear theory* expression for the growth of a perturbation. The full non-linear evolution is given by the preceding parametric solution. Note that $a_{\rm max}$ is the maximum scale factor of the full solution, and that all three models are normalized to the same scale factor at early times when the perturbation is small. The three different scale factors are shown in Figure 9.5.

The first interesting event in the evolution of the full perturbation is **turnaround**, when it reaches maximum expansion at $\theta = \pi$. Up until that point the general expansion of the Universe has been dominating over the gravitational collapse and the physical size has continued to grow (though of course the comoving size is always decreasing). Suppose we had continued to use the linear-theory expression. Because we are assuming matter domination throughout, the energy density in our various model universes always goes as a^{-3}. The linear density contrast (which, in this subsection, we subscript for extra clarity) therefore is given by

$$1 + \delta_{\rm lin} = \frac{a_{\rm back}^3}{a_{\rm lin}^3}, \tag{9.17}$$

where $a_{\rm back}$ is the background evolution given by the lowest-order truncation of

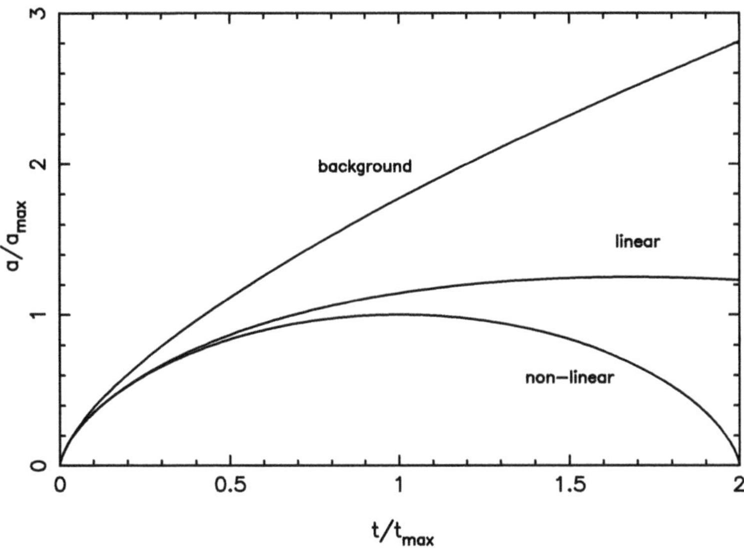

Fig. 9.5. The evolution of the three scale factors, namely the background evolution, the full non-linear collapse, and the linearized evolution.

Eq. (9.16). Substituting in the preceding expressions and linearizing again gives

$$\delta_{\rm lin} = \frac{3}{20}\left(6\pi\frac{t}{t_{\rm max}}\right)^{2/3} . \tag{9.18}$$

So, at turnaround, $t = t_{\rm max}$, the linear density contrast is

$$\boxed{\delta_{\rm lin}^{\rm turn} = \frac{3}{20}(6\pi)^{2/3} = 1.06} . \tag{9.19}$$

Theoretically, it is always simplest to deal with linear theory; what this result tells us is that the end of the linear regime, where $\delta_{\rm lin}$ reaches unity, corresponds roughly to the time that structures break away from the general expansion, but at that time, gravitationally bound structures have yet to form.

After turnaround, collapse proceeds symmetrically to the expansion phase, and the object collapses to a point at $t = 2\,t_{\rm max}$. At this time, the linear-theory density contrast has become

$$\boxed{\delta_{\rm lin}^{\rm coll} = \frac{3}{20}(12\pi)^{2/3} = 1.686} . \tag{9.20}$$

So, a linear density contrast of about 1.7 corresponds to the epoch of complete gravitational collapse of a spherically symmetric perturbation.

9.4 Virialization

In the real Universe, perfect spherical symmetry won't be respected; asphericities grow during the collapse of a spheroid from rest, so in fact we generically expect pancakes to form rather than collapse to a point. Ultimately, we expect the object to reach a state of virial equilibrium, where its radius has shrunk by a factor of 2 from that at turnaround. Numerical simulations show that this does indeed happen, and that $\delta_{\rm lin}^{\rm coll}$ provides a good estimate of the epoch of virialization. This is when gravitationally bound objects really can be said to have formed. This connection between linear and non-linear evolution plays a crucial part in the Press–Schechter theory that we discuss in the next section.

We also can use the spherical collapse model to study the densities at various stages. The actual non-linear density contrast at turnaround is just

$$\boxed{1+\delta_{\rm nonlin}^{\rm turn}=\frac{a_{\rm back}^3}{a_{\rm max}^3}=\frac{(6\pi)^2}{4^3}=5.55}\,. \tag{9.21}$$

In the spherical collapse model, the density goes infinite at the collapse time. However, if we assume that, instead, the collapsing object virializes at half the radius, its density will have gone up by a factor of 8. Meanwhile, the background density will have continued to fall, through expansion, by a further factor of 4. In combination then, we expect the overdensity at virialization to be

$$\boxed{1+\delta_{\rm nonlin}^{\rm vir}\simeq 178}\,. \tag{9.22}$$

This number is well verified by numerical simulations as dividing virialized regions from those where matter is still infalling, but needs some adjustment in the presence of a cosmological constant.

The final information that we extract from the spherical collapse model is the virial velocity of the bound objects. The virial theorem states that

$$v^2=\frac{GM}{R_{\rm g}}\,, \tag{9.23}$$

where M is the mass of the system and $R_{\rm g}$ the gravitational radius defined by the requirement that the gravitational potential energy be $-GM^2/R_{\rm g}$. The mass within an initial comoving radius $R_{\rm com}$ before collapse is

$$M=\frac{4\pi}{3}\rho_0 R_{\rm com}^3\,. \tag{9.24}$$

The virial theorem tells us that the system will collapse by a factor of 2 from the radius at which it is at rest (assuming no dissipation), and so, $R_{\rm g}=R_{\rm turn}/2$. From

Eq. (9.21), the turnaround radius is given in physical units by

$$R_{\rm turn}^3 = \frac{1}{5.55} R_{\rm phys}^3 = \frac{1}{5.55} \frac{1}{(1+z_{\rm turn})^3} R_{\rm com}^3. \tag{9.25}$$

Because the virialization time is twice the time to maximum and we assume matter domination, we have $1 + z_{\rm turn} = 2^{2/3}(1 + z_{\rm vir})$, leading to

$$R_{\rm g}^3 = \frac{1}{178} \frac{1}{(1+z_{\rm vir})^3} R_{\rm com}^3. \tag{9.26}$$

Combining Eqs. (9.23), (9.24) and (9.26) and substituting in values gives the scaling relation

$$\boxed{\left(\frac{v}{130\,{\rm km\,s^{-1}}}\right)^2 = \left(\frac{M}{10^{12}h^{-1}M_\odot}\right)^{2/3} (1+z_{\rm vir})}. \tag{9.27}$$

The scalings in this expression are well verified by numerical simulations, though the precise normalization really needs to be fixed by them rather than by the approximate analysis above. Note that among objects of the same mass, those that form earlier have a higher virial velocity because they are more compact through forming when the Universe was smaller and denser. Galaxies have virial velocities of order $100\,{\rm km\,s^{-1}}$, whereas those of clusters are of order $1000\,{\rm km\,s^{-1}}$.

This account of gravitational collapse assumes that the collapsing object loses no energy, which means that it applies to the CDM. The baryons can radiate energy, to become more tightly bound. For a galaxy, the end product is a more or less spherical and static distribution of CDM (dark halo), within which is a more tightly bound distribution of baryons of which a significant fraction is in stars. The angular momentum of the CDM and of the baryons is individually conserved during gravitational collapse. In the case of spiral galaxies like our own, this gives the baryons significant rotation, resulting in the formation of the galactic disk.

There is a scaling relation for a bound gas in hydrostatic equilibrium. Because the temperature measures the kinetic energy of the gas, the appropriate scaling law comes from $T \propto v^2$. The most interesting application of this is to rich clusters. For a rich cluster there is relatively little energy loss and most of the baryonic matter is in the form of intergalactic gas. The gas is observed in X-rays, with energies of a few kilo-electron volts, and one finds the relation normalized as

$$\boxed{\frac{T}{5\,{\rm keV}} = \left(\frac{M_{500}}{3\times 10^{14}h^{-1}M_\odot}\right)^{2/3} (1+z_{\rm vir})}. \tag{9.28}$$

where M_{500} is the mass within a sphere enclosing an overdensity of 500.

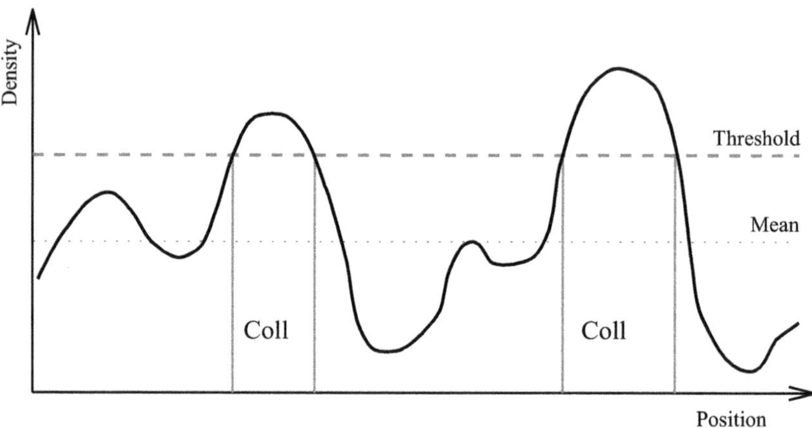

Fig. 9.6. Schematic of Press–Schechter theory applied to the density field smoothed on some scale. The volume in regions above the threshold, indicated by 'Coll' for collapsed, is identified with objects of the smoothing mass and above. If we smoothed further, more regions would drop below the threshold, giving a smaller total mass in objects above the higher smoothing scale.

9.5 Abundance of premature objects

The spherical collapse model is in general a very crude approximation to reality, but there is one situation in which it may be fairly accurate. This is when the collapse occurs long before the end of the linear regime, for the relevant smoothing scale $R(M)$. In that case the collapsing region, where the density contrast is of order 1, represents an exceptionally high peak of the density, and it can be shown that the density contrast near the peak becomes spherically symmetric in the limit where the background density goes to zero. This means that spherical collapse may be a reasonable approximation for premature objects of a given mass, forming well before the bulk of such objects appear at the end of the linear era.

We learned earlier that collapse occurs if the density contrast exceeds some threshold δ_c, which the spherical collapse model estimates as $\delta_c = 1.69$. The fraction of the volume of the Universe with $\delta(R, z, \mathbf{x}) > \delta_c$ at a given epoch, provides a rough estimate of the fraction f of the mass of the Universe which collapses into objects with mass $M(R)$ at the same epoch. This is known as Press–Schechter theory, illustrated schematically in Figure 9.6. To allow for the transfer of material from underdense to overdense regions, usually the fraction of mass is estimated as twice the volume

$$\boxed{f(> M(R), z) = \mathrm{erfc}\left(\frac{\delta_c}{\sqrt{2}\,\sigma(R, z)}\right)}\,. \tag{9.29}$$

Here, erfc is the complementary error function, defined by

$$\operatorname{erfc}(x) = \frac{2}{\sqrt{\pi}} \int_x^\infty \exp(-u^2)\, du. \tag{9.30}$$

If we shift the filtering scale from M to $M + dM$, the dispersion σ is reduced and hence so is the fraction of space above the threshold; we associate the change in the fraction of the mass above the threshold with objects of mass between M and $M + dM$. To obtain the comoving number density of objects of mass M, per mass interval dM, at redshift z, we do the following. Differentiating Eq. (9.29) with respect to the filtering mass gives the change in the volume fraction of the Universe, brought about by increasing the filter mass by dM. Multiplying this by ρ_0 gives the change in comoving mass density[1] above the threshold, and because that mass density is to be assigned to objects of mass M, we divide by M to get the comoving number density. Carrying this out yields

$$\frac{\partial n(M,z)}{\partial M}\, dM = -\sqrt{\frac{2}{\pi}}\, \frac{\rho_0}{M}\, \frac{\delta_{\rm c}}{\sigma^2(M,z)}\, \frac{\partial \sigma(M,z)}{\partial M} \exp\left(-\frac{\delta_{\rm c}^2}{2\sigma^2(M,z)}\right) dM, \tag{9.31}$$

from which the number density of objects of mass above M is

$$\boxed{n(>M,z) = \int_M^\infty \frac{\partial n(M,z)}{\partial M}\, dM}\,. \tag{9.32}$$

At a given redshift, dn/dM falls steeply with M, which means that $n(> M, z)$ is dominated by objects with mass not far below M. At any epoch, the expression for $n(> M)$ is valid in the linear regime, which corresponds to $\sigma_0(M) \ll (1 + z)$. From Figure 9.4 we see that it is valid at the present epoch only for the very biggest objects, which are galaxy clusters with $M \sim 10^{15} M_\odot$. Going back to higher redshift, it gives the number density of premature objects with successively smaller mass; lighter galaxy clusters, early galaxies, quasars, and the very first baryonic objects which were responsible for reionizing the Universe. In all of these cases, the prediction of the ΛCDM model is compatible with observation.

The Press–Schechter mass function works with surprising accuracy. However, it does significantly underestimate the number density of the rarest objects. More accurate mass functions, which prove necessary for precision applications, are given in for instance in Ref. [1].

[1] It is the *present* density that appears here, because we are using comoving units and, during matter domination, the comoving mass density is a constant. We could equally well have used the physical mass density $\rho(z)$ and included a factor $(1 + z)^{-3}$ to convert the density to comoving.

9.6 The observed mass density perturbation

9.6.1 Early matter distribution

At sufficiently-high redshift, the density perturbation on a given scale is evolving linearly. As we noticed at the end of the previous section, the density perturbation in this regime may be probed by observing the number density of rare objects. Direct detection of the density perturbation is also possible.

The so-called Lyman alpha forest consists of lines in the Lyman alpha spectrum, coming from the absorbtion of light from distant quasars by clouds of neutral hydrogen [2]. The clouds represent modest over-densities that are in the linear regime.

Ultimately the total mass density perturbation (dominated by that of the CDM) may be accurately mapped using weak gravitational lensing, whereby the images of distant galaxies are distorted by the gravitational effects of matter close to the line of sight. The path of a light ray is calculated in terms of the Newtonian gravitational potential using the line element (3.48). Depending on the redshift of the galaxy and the scale of the observation, the mass density perturbation may be in the linear or mildly non-linear regime.[2] As we write this promising field is still in its infancy.

Further down the line is the possible observation of the anisotropy in the 21 cm radiation emitted by the spin-flip transition of atomic hydrogen. This radiation probes the distribution of neutral hydrogen gas, and in the future should be able to do so to very high redshift. An important application of this will be to probe the inhomogeneity structure of the Universe before reionization, and to explore the mechanism of reionization itself. Due to the narrowness of the line, its redshifting allows receivers to be tuned to radiation emitted in narrow redshift intervals, which can then be assembled into a three-dimensional view.

9.6.2 Galaxy distribution

On large enough scales, the CDM density perturbation is still evolving according to linear theory. On smaller scales, its evolution can be determined quite accurately using purely gravitational N-body simulations. At the time of writing, the largest simulations follow the evolution of ten billion individual particles. An example is shown in Figure 9.7.

Most directly observable are the baryons which have formed galaxies, the stellar mass comprising around ten percent of the total baryonic mass. One can attempt to follow the detailed evolution of the baryons using computer simulations which include gas dynamics, so-called hydrodynamical N-body simulations, but including all physical processes relevant to galaxy formation itself is currently impossible. Instead, various semi-analytic prescriptions can be applied to model this process.

[2] Similarly, the matter distribution will lens the cosmic microwave background anisotropy, but as those photons come from a single redshift there is less information available to be extracted.

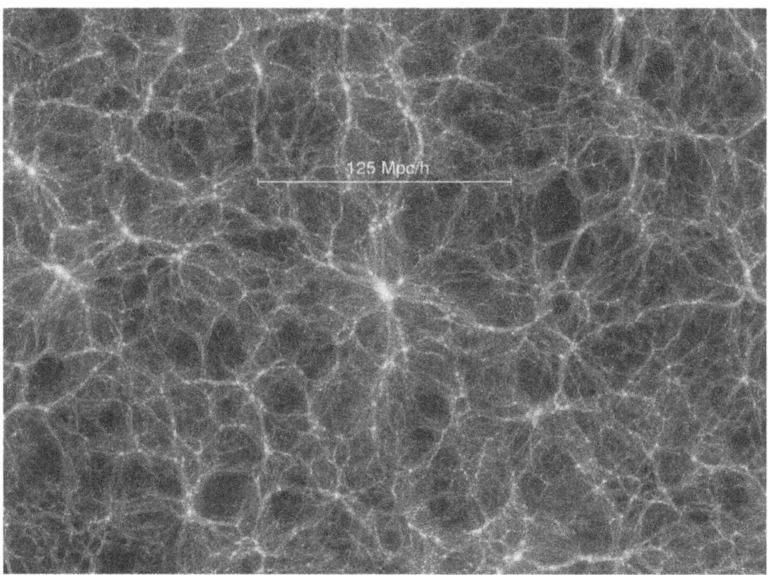

Fig. 9.7. A slice from a large N-body simulation, the Millennium Simulation, showing the large-scale dark matter distribution in the standard cosmology. The hierarchical nature of dark matter structures is plainly seen. Image courtesy Volker Springel and the Virgo Consortium.

Without worrying about the detailed properties of individual galaxies, one can use their clustering to trace the large-scale distribution of dark matter, in whose halos the galaxies form. This is one of the most important cosmological probes.

Gaussian primordial perturbation

Let us first consider the galaxy clustering on the assumption that the primordial density perturbation is gaussian. Then the power spectrum of galaxies is taken to be related to the power spectrum of matter by a bias factor b such that

$$\mathcal{P}_g(k, z) = b^2 \mathcal{P}_\delta(k, z) \,. \tag{9.33}$$

At fixed epoch and for a well-defined class of galaxies, the bias parameter is expected to be constant, except on small scales where non-linear evolution has become important. However, even on large scales the value of the bias is expected to be different for galaxies selected in different ways (e.g. optical versus infra-red selection) or with different luminosities. It is also predicted to evolve with redshift, being large at an epoch where the object class is rare and decreasing subsequently.

In cosmological parameter fitting, usual practice is to compare a theoretical prediction for the matter power spectrum with an observed galaxy power spectrum, the latter restricted to scales where the bias is expected to be constant. The bias is treated as an independent parameter to be fit from the data and then marginalized over, rather than derived from first principles.

While early surveys were typically interpreted in terms of the galaxy correlation function, and demonstrated that the correlation length, where it crossed unity, was only around $5\,h^{-1}\,\mathrm{Mpc}$, modern treatments almost invariably estimate the power spectrum directly from data. The analysis of a typical survey is rather complicated, due to the selection procedures of the original catalogue and the need to determine accurate uncertainties. Essential features of modelling including allowing for survey geometry, accounting for survey depth as a function of angle, survey completeness including omissions due to 'fibre collisions' which prevent multi-fibre spectrographs measuring very close objects, and allowing for the effect of redshift-space distortions due to galaxy peculiar velocities. Luminosity dependence of the bias is typically also included. The spectrum is estimated by comparing the clustering properties of the real catalogue and randomly generated unclustered mock catalogues, as for example in Ref. [3].

The galaxy power spectra measured by two large galaxy redshift surveys are shown in Figure 9.8. There are some discrepancies between the two power spectra, which is attributed to the different selection criteria for the galaxy samples (e.g. red galaxy subsamples give better agreement than the full catalogues).

Effect of primordial non-gaussianity

Although the galaxy distribution on short scales has evolved to become non-gaussian, it is consistent with the hypothesis that the primordial density perturbation is gaussian. To consider the effect of primordial non-gaussianity, let us suppose that the primordial density perturbation is adiabatic, with local non-gaussianity specified by the single parameter f_{NL} that specifies the primordial bispectrum. It can be shown that this leads in the linear regime to scale-dependent bias. Focussing on a fixed redshift the dependence is of the form

$$\mathcal{P}_g(k) = \left(b^2 + \Delta b(k) f_{\mathrm{NL}}\right) \mathcal{P}_\delta(k), \tag{9.34}$$

with [4]

$$\Delta b(k) = 2(b-p)\delta_{\mathrm{c}} \frac{\Phi_{\mathbf{k}}}{\delta_{\mathbf{k}}} = \frac{3(b-p)\delta_{\mathrm{c}}\Omega_{\mathrm{m}} H_0^2 (1+z)}{k^2 T(k) g(z)}, \tag{9.35}$$

where in the simplest case $p = 1$. Through this expression, the *spectrum* of the galaxy distribution probes the *bispectrum* of the primordial density perturbation. In addition, one can consider the bispectrum of the galaxy distribution which in the linear regime will be proportional to the primordial bispectrum.

By methods such as these, the galaxy distribution may give constraints on primordial non-gaussianity that are of comparable strength to those from the cosmic

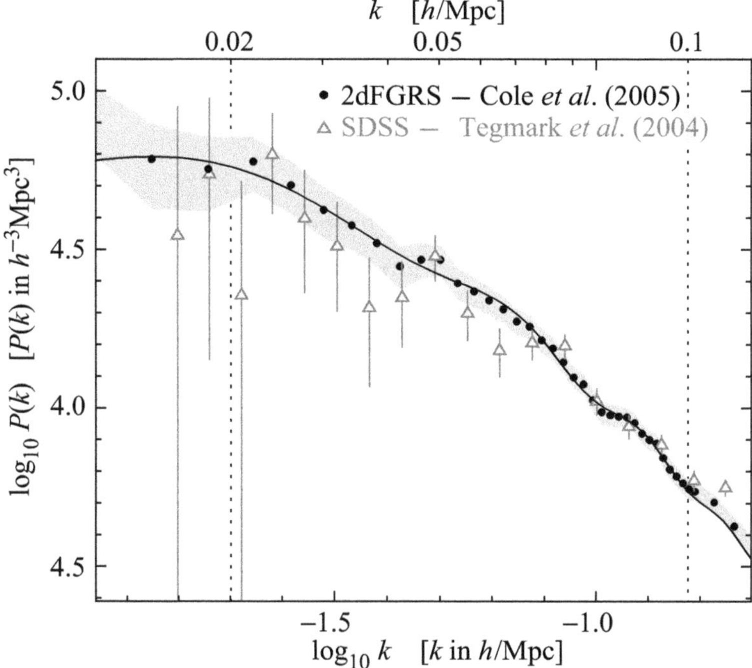

Fig. 9.8. The galaxy power spectrum $P(k)$ as measured by the 2dF galaxy redshift survey and by the Sloan Digital Sky Survey. The shaded area shows the uncertainty of the 2dF points. The solid curve shows a fit using the ΛCDM model with $\Omega_{\rm m}h = 0.168$, $\Omega_B/\Omega_{\rm m} = 0.17$, $h = 0.72$, $n = 1$ and normalization matched to the 2dFGRS power spectrum. The dotted vertical lines indicate the range over which the best-fit model was evaluated. Figure courtesy Shaun Cole and Will Percival.

microwave background anisotropy, but probing a rather different length scale. To reach $f_{\rm NL}$ values of order unity would, however, require a massive galaxy redshift survey extending out to redshift one.

9.6.3 Inter-galactic baryons

While the baryons incorporated into galaxies and undergoing nuclear fusion in stars are the most visible ones, they comprise only around ten percent of the total baryons. The remainder are much harder to detect directly. We noticed in Section 9.6.1 how unbound baryons might be detected at high redshift. One can also look at hot baryonic gas in galaxy clusters.

This gas is shock heated during gravitational collapse, and in the deep cluster potential wells it reaches temperatures where it emits in X-rays which are readily detectable. The hot gas is held in hydrostatic equilibrium by the gravitational attraction due primarily to the dark matter. Galaxy clusters can be used as cosmological probes both by studying the dark matter to baryon density ratio on a cluster-by-cluster basis, and by using evolution of the cluster number density to probe the growth of cosmic structure.

9.6.4 CDM halos and primordial black holes

The bottom-up picture as we have described it deals with the formation of objects containing baryons (baryonic objects), which can radiate energy and collapse in the manner that we have described. The process begins with the formation of the lightest baryonic objects, with mass of order $10^4 M_\odot$. Lighter objects will form earlier, but they will be entirely of CDM because they are below the baryon Jeans mass.

These 'CDM halos' are less tightly bound than galaxies. Nevertheless, some of them may still be around when the first baryonic objects form and if so the resulting clumpiness of the CDM could have a significant effect on way in which the baryonic objects collapse. In the extreme case of axion dark matter (Section 23.1), CDM mini-halos could still be around today with important implications for CDM detection.

Going back to very early times there is the more radical possibility of primordial black hole (PBH) formation, described in Section 23.4. These would come from the collapse of everything within a horizon volume, making no distinction between radiation, baryonic matter and non-baryonic matter.

Exercises

9.1 Verify Eq. (9.5) giving the top-hat window function in Fourier space, and Eq. (9.6) giving the gaussian window function.

9.2 Verify Eq. (9.9), giving the smoothed density contrast with the adiabatic initial condition.

9.3 Verify Eqs. (9.7) and (9.8). For this and the following question, you need

$$\int_0^\infty x^n e^{-x^2/2}\,dx = 2^{(n-1)/2}\,\Gamma\left(\frac{n+1}{2}\right) \qquad (n > -1).$$

9.4 Suppose wavenumbers and scales are related by $k_R = \alpha/R$, where α is a constant. Assuming that $\mathcal{P}_g(k) \propto k^m$, and using a gaussian filter, find α such that Eq. (9.13) is exact. For what range of m is your calculation valid? Over roughly what range of k must the power-law approximation be maintained to obtain an approximately correct result?

9.5 Verify Eq. (9.16), giving the linearized scale factor for spherical collapse.

References

[1] A. Jenkins *et al*. The mass function of dark matter halos. *Mon. Not. Roy. Astr. Soc.*, **321** (2001) 372.

[2] S. Weinberg. *Cosmology* (Oxford: Oxford University Press, 2008).

[3] H. A. Feldman, N. Kaiser and J. A. Peacock. Power spectrum analysis of three-dimensional redshift surveys. *Astrophys. J.*, **426** (1994) 23.

[4] U. Seljak. Measuring primordial non-gaussianity without cosmic variance. arXiv:0807.1770 [astro-ph].

10

Cosmic microwave background anisotropy

We have seen how the *inhomogeneities* of the matter and photon densities evolve, from early times to the present epoch. Now we will do the same thing for the *anisotropy* of the photon distribution function. This anisotropy is observed today as the cosmic microwave background (CMB) anisotropy.

Photons in the early Universe are in thermal equilibrium, with the blackbody distribution of momenta. As the epoch of photon decoupling is approached, the distribution begins to fall out of equilibrium, developing anisotropy which is different for the two polarization states. After decoupling at $z \sim 1000$, the redshifting of the photons through the inhomogeneous gravitational field generates more anisotropy, without affecting the polarization. Then reionization at $z \sim 10$ generates further anisotropy and polarization. Finally, the photons are observed at the present epoch, as the CMB anisotropy. The anisotropy is characterized by a perturbation in the intensity, which corresponds to a perturbation in the temperature of the blackbody distribution, and by two polarization parameters. In this chapter we study the temperature perturbation.

We are going to see that the CMB anisotropy is observable only on rather large scales, corresponding to comoving wavenumber $k \lesssim (10\,\text{Mpc})^{-1}$. In this regime, first-order (linear) cosmological perturbation theory is almost always a good approximation, failing only on the smallest scales and only around the present epoch. In the latter regime, the dominant effect is the thermal Sunyaev–Zel'dovich effect, which occurs when a galaxy cluster is in the line of sight. This effect comes from photons gaining energy by scattering off hot gas on their way through the cluster. It occurs on scales of order one arc-minute and is identified readily for individual clusters. The cumulative contribution of distant clusters on the CMB spectrum is believed small on the main scales where the CMB anisotropy is observed, but starts to dominate on the small scales where the primary anisotropies are damped away.

We assume first-order perturbation theory except where stated. Since the theory is to be used right up to the present epoch, it is understood that the perturbations have been smoothed on a scale at least of order 10 Mpc.

10.1 CMB multipoles

Throughout this chapter we use conformal time η. We start the discussion at an 'initial' epoch soon after neutrino decoupling. In the perturbed Universe we define the photon distribution function in the locally orthonormal frame lined up with the coordinates, as in Section 8.2. The evolution of δf is governed by the Boltzmann equation (2.68), with Thomson scattering as the collision term. In the rest frame of the electron, Thomson scattering doesn't change the photon momentum. As we see in the next chapter, this means that to first order, the perturbed f retains the black-body form with an effective temperature perturbation that is *independent of the magnitude p of the photon momentum*. This perturbation is called the **brightness function**, and we denote it by $\Theta(t, \mathbf{x}, \mathbf{n})$:

$$\boxed{\Theta(\eta, \mathbf{x}, \mathbf{n}) \equiv \frac{\delta T(\eta, \mathbf{x}, \mathbf{n})}{T(\eta)}}, \tag{10.1}$$

where $\mathbf{n} = \mathbf{p}/p$ is the direction of $\mathbf{p}$. The photons with momentum $\mathbf{p}$ in a given range d^3p have intensity I proportional to $T^4(t, \mathbf{x}, \mathbf{n})$. The fractional perturbation in this intensity is therefore $\delta I/I = 4\Theta$. Some authors include the factor 4 in the definition of the brightness function.

The brightness function depends on the direction $\mathbf{n}$ of the photon momentum or, equivalently, on the direction $\mathbf{e} = -\mathbf{n}$ in which the photon is seen. We expand it into multipoles at each point in spacetime:

$$\boxed{\Theta(\eta, \mathbf{x}, \mathbf{n}) = \sum_{\ell m} (-1)^\ell \Theta_{\ell m}(\eta, \mathbf{x}) Y_{\ell m}(\mathbf{n}) = \sum_{\ell m} \Theta_{\ell m}(\eta, \mathbf{x}) Y_{\ell m}(\mathbf{e})}. \tag{10.2}$$

From Eq. (4.4) we see that the monopole ($\ell = 0$) is related to the photon energy density contrast by

$$\boxed{\Theta_{00}(\eta, \mathbf{x}) = \frac{1}{4}\delta_\gamma(\eta, \mathbf{x})}. \tag{10.3}$$

This means that the monopole needs a slicing to define it, but it is independent of the threading since it is a scalar. It also means that we cannot determine the monopole, because there is no way of measuring the photon energy density at positions other than our own.

The dipole ($\ell = 1$) is the Doppler shift caused by the motion of the photon fluid relative to the observer,

$$\boxed{\sum_m \Theta_{1m} Y_{1m}(\mathbf{e}) = -\mathbf{v}_\gamma \cdot \mathbf{e}}\,. \tag{10.4}$$

We have evaluated the Doppler shift to first order in $\mathbf{v}_\gamma$ (non-relativistic formula).

At this point we have to distinguish an observer who moves with the threading of the gauge used to implement cosmological perturbation theory, and an observer who moves with the Earth's motion. Considering the observer moving with the threading, we learn that the dipole depends on the threading of the gauge, but not on the slicing. For this observer, $\mathbf{v}_\gamma$ is a cosmological perturbation and the evaluation of the Doppler shift to first order just corresponds to first-order cosmological perturbation theory.

On the other hand, it is the observer moving with the Earth's motion that corresponds to what is actually measured. The diurnal average of the observed CMB dipole gives the velocity of the Sun relative to the CMB rest frame as

$$v = 371\,\text{km s}^{-1} = 1.2 \times 10^{-3}, \tag{10.5}$$

the final figure in our chosen unit $c = 1$. The use of the non-relativistic formula is clearly justified in this case also.

Coming to the higher multipoles, the use of the relativistic Doppler shift would change each one by an amount of order v_γ^ℓ. In the context of cosmological perturbation theory, this means that they are independent of the threading at first order. They are also independent of the slicing because they vanish in the background, which means that they are gauge-invariant.

Considering instead the observer moving with the Earth, it is easy to check that the Earth's motion has only a small effect on the higher multipoles. The Earth's motion changes the quadrupole ($\ell = 2$) by an amount of order $v_\gamma^2 \sim 10^{-6}$, while its observed value is of order 10^{-5}; this 'kinematic quadrupole' is enough that it should be subtracted in accurate analyses. The observed higher multipoles have about the same 10^{-5} magnitude, which means that the effect of the Earth's motion is indeed small.

The observed multipoles with $\ell \geq 2$ therefore represent the intrinsic anisotropy of the CMB. They are denoted by $a_{\ell m}$:

$$\boxed{a_{\ell m} \equiv \Theta_{\ell m}(\eta_0, \mathbf{x}_0)}\,, \tag{10.6}$$

where $\mathbf{x}_0$ is our position, conveniently taken to be the origin of coordinates.

10.2 Spectrum of the CMB anisotropy

We are interested in the stochastic properties of the CMB multipoles $a_{\ell m}$. Invariance of the stochastic properties under rotations demands that $\langle a_{\ell m}\rangle = 0$, and that the two-point correlator of the multipoles is of the form

$$\boxed{\langle a_{\ell m} a^*_{\ell' m'}\rangle = \delta_{\ell\ell'}\delta_{mm'} C_\ell} . \tag{10.7}$$

The quantity $C_\ell = \langle |a_{\ell m}|^2\rangle$ is called the spectrum of the CMB anisotropy.

Using Eq. (A.14), an equivalent definition of C_ℓ is provided by the two-point correlation function:

$$C(\theta) \equiv \langle \Theta(\mathbf{e}_1)\Theta(\mathbf{e}_2)\rangle = \sum_\ell \frac{2\ell+1}{4\pi} C_\ell P_\ell(\cos\theta), \tag{10.8}$$

where θ is the angle between the directions $\mathbf{e}_1$ and $\mathbf{e}_2$.

Given a model of the early Universe one can calculate C_ℓ. In comparing such a prediction with observation, we have to remember that $C_\ell \equiv \langle |a_{\ell m}|^2\rangle$ is an ensemble average, whereas we make a single measurement of the multipoles. On the usual assumption that we observe a typical member of the ensemble, the typical difference between C_ℓ and an actually measured multipole will be of order the *rms* deviation of $|a_{\ell m}|^2$ from C_ℓ. The latter is given by

$$(\Delta C_\ell)^2 \equiv \left\langle \left(|a_{\ell m}|^2 - C_\ell\right)^2 \right\rangle = \langle |a_{\ell m}|^4\rangle - C_\ell^2 . \tag{10.9}$$

This is the cosmic variance of C_ℓ, similar to the cosmic variance of correlators of Fourier components that we discussed in Section 6.5.

As we see in Sections 10.4 and 12.5, the real and imaginary parts of the multipoles will have independent gaussian probability distributions, provided that the primordial perturbations are gaussian. If a single real or imaginary part is measured the cosmic variance is $2C_\ell^2$. If all $(2\ell+1)$ independent components of $a_{\ell m}$ are measured, the cosmic variance is reduced by a factor $(2\ell+1)$. If the data are further binned by averaging over a range $\Delta\ell$ about ℓ, this reduces the cosmic variance by an additional factor $\Delta\ell$, giving

$$\boxed{(\Delta C_\ell)^2 = \frac{2}{(2\ell+1)\Delta\ell} C_\ell^2} . \tag{10.10}$$

The cosmic variance of C_ℓ is a serious limitation at low ℓ, but becomes negligible at higher ℓ.

It is worth remarking that direct observational constraints on non-gaussianity become very weak on large scales, corresponding to very low multipoles of CMB. Indeed, in this regime they come only from the CMB itself, and they depend on the assumed form of the non-gaussianity. However, at least within the inflationary

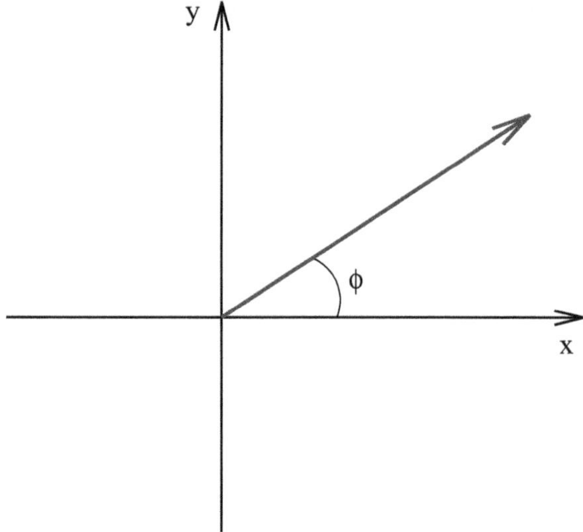

Fig. 10.1. In a small patch of sky near the pole, we can lay down almost-Cartesian coordinates $\vec{\theta} = (x, y) = (\theta \cos\phi, \theta \sin\phi)$. A typical wave-vector $\mathbf{k}$ points well away from the patch, so that there is a unique Cartesian vector $\vec{\theta}_{\mathbf{k}}$ pointing toward it. The contribution of $\zeta(\mathbf{k})$ to the CMB anisotropy $\delta T(\vec{\theta})/T$ is constant on the lines perpendicular to this vector.

scenario, one doesn't expect an abrupt change from gaussianity to non-gaussianity going down in ℓ, and the estimate (10.10) of cosmic variance is generally taken to hold even for low multipoles.

10.3 Flat-sky approximation

To describe the CMB anisotropy on small angular scales, there is no need to consider the whole sky simultaneously. Instead, we can focus on a small patch of sky. In such a patch, taken without loss of generality to be near the pole of spherical coordinates, we can lay down almost-Cartesian coordinates as in Figure 10.1:

$$\vec{\theta} = (x, y) = (\theta \cos\phi, \theta \sin\phi). \tag{10.11}$$

On angular scales much smaller than the patch, it makes sense to write

$$\boxed{\Theta(\vec{\theta}) = \frac{1}{2\pi} \int d^2\vec{\ell}\, a(\vec{\ell}) e^{i\vec{\theta}\cdot\vec{\ell}}}. \tag{10.12}$$

One pretends that the Fourier integral goes over the whole $\vec{\ell}$ plane, so that the functions $\exp(i\vec{\theta} \cdot \vec{\ell})/2\pi$ are orthonormal.

Let us introduce polar coordinates $(\ell, \phi_{\vec{\ell}})$ in the $\vec{\ell}$ plane, as shown in Figure 10.2.

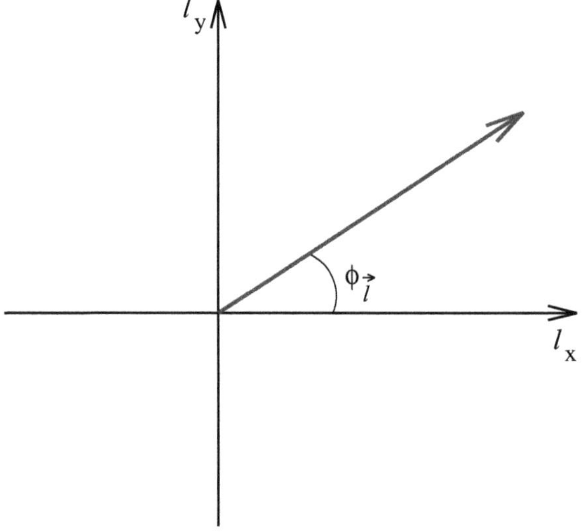

Fig. 10.2. The CMB anisotropy within the patch may be written as a two-dimensional Fourier integral. The position vector in Fourier space is $\vec{\ell} = (\ell_x, \ell_y) = (\ell \cos \phi_{\vec{\ell}}, \ell \sin \phi_{\vec{\ell}})$.

They are defined by $\vec{\ell} = (\ell_x, \ell_y) = (\ell \cos \phi_{\vec{\ell}}, \ell \sin \phi_{\vec{\ell}})$, so that $\ell = |\vec{\ell}|$. As the notation suggests, ℓ is the continuous limit of the integer labelling the multipoles $a_{\ell m}$. To see this, one can use the multipole expansion (10.2), and the approximation $P_\ell^m(\theta) \simeq \ell^m J_{-m}(\ell\theta)$ (valid for $\theta \to 0$ with $\ell\theta$ constant). This is equivalent to

$$Y_{\ell m}(\theta, \phi) = \sqrt{\frac{\ell}{2\pi}} J_{-m}(\ell\theta) e^{im\phi}. \tag{10.13}$$

Then, using the relation

$$J_n(x) = \frac{i^{-n}}{\pi} \int_0^\pi d\theta e^{ix\cos\theta} \cos(n\theta), \tag{10.14}$$

one has

$$e^{im\phi} J_{-m}(\ell\theta) = \frac{i^m}{2\pi} \int_0^{2\pi} e^{i\vec{\ell}\cdot\vec{\theta}} e^{im\phi_{\vec{\ell}}} d\phi_{\vec{\ell}}, \tag{10.15}$$

leading to

$$a(\vec{\ell}) \equiv \left(\frac{1}{2\pi\ell}\right)^{1/2} \sum_m a_{\ell m} i^{-m} e^{im\phi_{\vec{\ell}}}. \tag{10.16}$$

Since $\vec{\ell}$ and $\vec{\theta}$ are related by the Fourier transform, it follows that a multipole ℓ corresponds to Fourier modes with wavelength

$$\Delta\theta = 2\pi/\ell. \tag{10.17}$$

This is the angular scale in the sky explored by a given ℓ.

In the flat-sky approximation we are dealing with a two-dimensional random field, which is gaussian if the primordial perturbations are. Its spectrum is defined by

$$\boxed{\langle a^*(\vec{\ell})a(\vec{\ell}')\rangle = C(\ell)\,\delta^2(\vec{\ell}-\vec{\ell}')}\,. \tag{10.18}$$

The transformation from the expansion (10.12) to the multipole expansion is unitary (because both use orthonormal functions), so that $C(\ell) = C_\ell$ for large integer ℓ.

In the flat-sky approximation, the ensemble average can be replaced by an average over the single observed sky. Cosmic variance of C_ℓ is negligible since we are dealing with high ℓ. Rotational invariance of the stochastic properties requires that the stochastic properties observed in a sufficiently small patch of sky are independent of the location of that patch. In particular, the observed spectrum should be independent of the location.

10.4 Scalar mode

Now consider a Fourier component $\Theta(\eta, \mathbf{k}, \mathbf{n})$ of the brightness function. We can define its scalar, vector, and tensor modes through Eq. (8.3) with $G \equiv \Theta$. The definition is made with reference to the transformation properties of $\Theta(\eta, \mathbf{k}, \mathbf{n})$ under a rotation about the $\mathbf{k}$-axis. The only difference from the cases we discussed there is that $\Theta(\eta, \mathbf{k})$ isn't labelled by discrete indices but by the continuous label $\mathbf{n}$. This corresponds to the fact that Eq. (8.3) for Θ could in principle contain an infinite number of terms, corresponding to tensor modes of arbitrarily high rank. But as we discussed in Section 8.1, modes with $m > 2$ are not expected.

In this chapter we will be focusing on the scalar mode. Then the brightness function is independent of the azimuthal angle about the $\mathbf{k}$-direction, and we can define multipoles Θ_ℓ by

$$\boxed{\Theta(\eta, \mathbf{k}, \mathbf{n}) = \sum_\ell (-i)^\ell \sqrt{4\pi(2\ell+1)}\, Y_{\ell 0}(\mathbf{n})\Theta_\ell(\eta, \mathbf{k})}\,. \tag{10.19}$$

Note that $\sqrt{4\pi}Y_{\ell 0}(\mathbf{n}) = \sqrt{2\ell+1}P_\ell(\mu)$ where $\mu \equiv \cos(\hat{\mathbf{k}}\cdot\mathbf{n})$. The inverse is

$$\begin{aligned}
\Theta_\ell(\eta) &= -i^\ell\,[4\pi\,(2\ell+1)]^{-1/2}\int Y_{\ell 0}(\mathbf{n})\Theta(\eta, \mathbf{k}, \mathbf{n})d^2\mathbf{n} \\
&= \frac{1}{2}(-i)^\ell \int_{-1}^{1} d\mu P_\ell(\mu)\Theta(\eta, \mu).
\end{aligned} \tag{10.20}$$

Using Eq. (A.14) we find

$$\Theta_{\ell m}(\eta,\mathbf{x}) = \frac{4\pi}{(2\pi)^3} i^\ell \int \Theta_\ell(\eta,\mathbf{k}) Y^*_{\ell m}(\hat{\mathbf{k}}) e^{i\mathbf{k}\cdot\mathbf{x}} d^3\mathbf{k}. \tag{10.21}$$

Taking our position to be $\mathbf{x}=0$, the observed CMB multipoles are

$$a_{\ell m} = \frac{4\pi}{(2\pi)^3} i^\ell \int \Theta_\ell(\mathbf{k}) Y^*_{\ell m}(\hat{\mathbf{k}}) d^3\mathbf{k}, \tag{10.22}$$

where $\Theta_\ell(\mathbf{k}) \equiv \Theta_\ell(\eta_0,\mathbf{k})$.

The orthonormality of the spherical harmonics gives

$$C_\ell = 4\pi \int_0^\infty \mathcal{P}_{\Theta_\ell}(\eta_0,k)\frac{dk}{k}. \tag{10.23}$$

Keeping just the adiabatic mode, it is useful to define a transfer function by

$$\boxed{\Theta_\ell(\mathbf{k}) = T_\ell(k)\zeta_\mathbf{k}}\,, \tag{10.24}$$

so that

$$\boxed{C_\ell = 4\pi \int_0^\infty T_\ell^2(k)\mathcal{P}_\zeta(k)\frac{dk}{k}}\,. \tag{10.25}$$

An equivalent definition of the transfer function is

$$a_{\ell m} = \frac{4\pi}{(2\pi)^{3/2}} \int_0^\infty T_\ell(k)\,\zeta_{\ell m}(k)\,\frac{dk}{k}. \tag{10.26}$$

We saw in Section 6.6 that if ζ is gaussian (i.e. no correlation of the $\zeta_\mathbf{k}$ except for the reality condition) then each $\zeta_{\ell m}(k)$ has a gaussian probability distribution, with no correlations except for the reality condition. It follows that the same is then true of the $a_{\ell m}$.

In Figure 10.3 we show some currently available datapoints. The error bars include cosmic variance, which dominates at small ℓ. Also shown is a theoretical curve corresponding to the ΛCDM model. We are now going to explain the physical effects that give the various features labelled.

10.5 Sudden-decoupling approximation

In Chapter 11 we give the full first-order calculation of the CMB anisotropy for the scalar mode. Here we deal with the sudden-decoupling approximation described in Section 8.4. In this approximation, the baryon–photon fluid decouples suddenly, at a conformal time that we denote by $\eta_{\rm ls}$ (for last scattering). Last-scattering takes place on the sphere around us with radius $\eta_0 - \eta_{\rm ls} \simeq \eta_0$. Before decoupling, the frequent photon collisions keep the photon distribution isotropic in its rest frame.

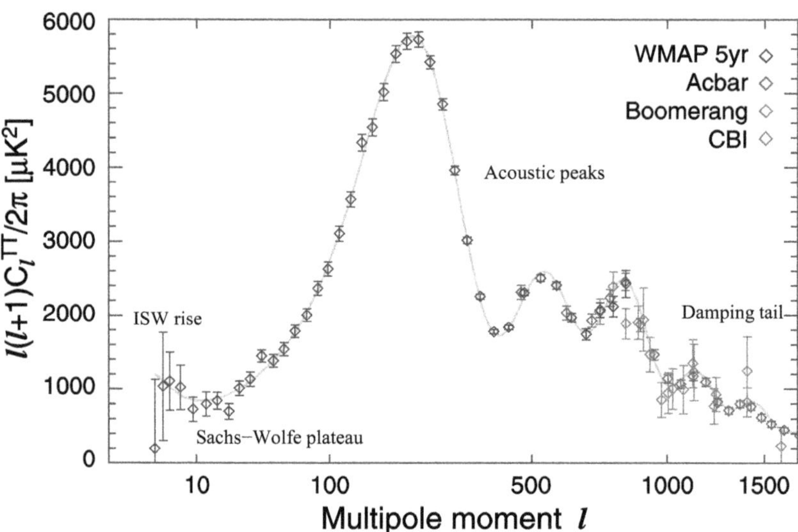

Fig. 10.3. The spectrum of the CMB anisotropy. The curve is the prediction of the ΛCDM model with parameters giving a best fit to CMB and galaxy survey data. The features have various physical origins as labelled. At low ℓ the error bars are dominated by cosmic variance. The figure is taken from the WMAP website [1].

As a result, an observer just after last-scattering sees no anisotropy in the CMB, except for the monopole and a dipole corresponding to the velocity $\mathbf{v}_\gamma$ of the CMB with respect to the observer:

$$\Theta_{\text{ls}}(\mathbf{e}) = \left(\frac{1}{4}\delta_\gamma + \mathbf{e}\cdot\mathbf{v}_\gamma\right)_{\text{ls}} . \tag{10.27}$$

The right-hand side is to be evaluated at decoupling (last scattering), corresponding to $(\eta_{\text{ls}}, \mathbf{x}_{\text{ls}})$, where

$$\mathbf{x}_{\text{ls}} = (\eta_0 - \eta_{\text{ls}})\,\mathbf{e} \simeq \eta_0\mathbf{e}. \tag{10.28}$$

On their journey from last scattering to our location, the CMB photons acquire a redshift, which depends on the direction $\mathbf{e}$. This causes an additional contribution $\Theta_{\text{SW}}(\mathbf{e})$, called the Sachs–Wolfe contribution after the authors who first considered it in 1967.

In total, the CMB anisotropy in the sudden-decoupling approximation is

$$\boxed{\Theta(\mathbf{e}) = \left(\frac{1}{4}\delta_\gamma + \mathbf{e}\cdot\mathbf{v}_\gamma\right)_{\text{ls}} + \Theta_{\text{SW}}(\mathbf{e})} . \tag{10.29}$$

The $\ell \geq 2$ multipoles calculated from this expression are gauge independent, but the same isn't true for the three separate terms. The first term (monopole) depends on the slicing, and the second (dipole) depends on the threading. The third (Sachs–Wolfe) contribution depends on both the slicing and the threading, which along each photon trajectory defines the locally orthonormal frames of the observers who measure the successive redshifts.

To calculate the Sachs–Wolfe contribution we denote the photon energy in the locally orthonormal frame of Section 8.2 by p, and define $q(\mathbf{x},t) \equiv a(t)p(\mathbf{x},t)$. In the unperturbed Universe q would be a constant. The redshift between nearby points on the photon trajectory is therefore to first order

$$\frac{dp}{p} = -\frac{da}{a} + \frac{dq}{q}. \tag{10.30}$$

The Sachs–Wolfe effect is the integral of the second term:

$$\Theta_{\mathrm{SW}}(\mathbf{e}) = \int_{\eta_{\mathrm{ls}}}^{\eta_0} \frac{d\eta}{q}\frac{dq}{d\eta}. \tag{10.31}$$

In the locally orthonormal frame of Section 8.2 the components of the photon four-momentum are $aq(-1,n^i)$. In the conformal Newtonian gauge they are

$$p^0 = (1+\Psi)q, \tag{10.32}$$

$$p^i = (1-\Phi)qn^i. \tag{10.33}$$

Taking the first-order perturbation of the time-component of the geodesic equation (3.27) we find

$$\frac{1}{q}\frac{dq}{d\eta} = \frac{\partial\Phi}{\partial\eta} - n_i\frac{\partial\Psi}{\partial x^i}. \tag{10.34}$$

Since

$$\frac{d\Psi}{d\eta} \equiv \frac{\partial\Psi}{\partial\eta} + \frac{dx^i}{d\eta}\frac{\partial\Psi}{\partial x^i}, \tag{10.35}$$

and $dx^i/d\eta = n_i$ this leads to

$$\boxed{\Theta_{\mathrm{SW}}(\mathbf{e}) = \Psi_{\mathrm{ls}} + \int_{\eta_{\mathrm{ls}}}^{\eta_0} \frac{\partial}{\partial\eta}(\Phi+\Psi)\,d\eta}, \tag{10.36}$$

where the integral is along the line of sight corresponding to position $\mathbf{x} = (\eta_0 - \eta)\mathbf{e}$. The contribution from the integral is called the **Integrated Sachs–Wolfe (ISW) effect**. We dropped a term $-\Psi_0$ (the potential at our position) which affects only the monopole.

Using this result, the sudden-decoupling approximation becomes

$$\boxed{\Theta(\mathbf{e}) = \left[\left(\frac{1}{4}\delta_\gamma + \Psi\right) + \mathbf{e}\cdot\mathbf{v}_\gamma\right]_{\mathrm{ls}} + \int_{\eta_{\mathrm{ls}}}^{\eta_0} \frac{\partial}{\partial\eta}(\Phi+\Psi)}. \tag{10.37}$$

To a reasonable approximation we can pretend that the Universe is completely matter-dominated from decoupling to the present. Then $\Phi = \Psi$ is time independent and there is no ISW effect, so that

$$\Theta_{\rm SW}(\mathbf{e}) = \Phi_{\rm ls}. \tag{10.38}$$

As the ISW effect and the corrections to the sudden-decoupling approximation are both fairly small, Eq. (10.29) implies that the CMB anisotropy is determined almost entirely by the perturbations on the last-scattering surface. As a result, each angular scale explores the linear scale that it subtends at the last-scattering surface. This surface lies practically at the particle horizon, whose comoving distance is $\eta_0 = 14\,000\,\mathrm{Mpc}$. An angle θ (in radians) subtends a comoving distance $x = \eta_0\theta$, corresponding to

$$\boxed{x = \eta_0\theta \sim 100\,\mathrm{Mpc}\,\frac{\theta}{1^\circ}}\,. \tag{10.39}$$

This gives the linear scale explored by the CMB anisotropy on a given angular scale.

Dropping the ISW effect, let us use Eq. (10.37) to calculate $\Theta_\ell(\mathbf{k})$. We first express the right-hand side as a Fourier integral, setting $\eta_0 - \eta_{\rm ls} \simeq \eta_0$:

$$\Theta(\mathbf{e}) = \frac{1}{(2\pi)^3}\int \left[\left(\frac{1}{4}\delta_\gamma(\mathbf{k}) + \Phi(\mathbf{k})\right) + \mathbf{e}\cdot\mathbf{v}_\gamma(\mathbf{k})\right]_{\rm ls} e^{i\mathbf{k}\cdot\mathbf{e}\,\eta_0}\, d^3k. \tag{10.40}$$

Then we use the identity (A.27) to obtain

$$\boxed{\Theta_\ell(\mathbf{k}) = \left[\frac{1}{4}\delta_\gamma(\eta_{\rm ls},\mathbf{k}) + \Phi(\eta_{\rm ls},\mathbf{k})\right] j_\ell(k\eta_0) + V_\gamma(\eta_{\rm ls},\mathbf{k})\, j'_\ell(k\eta_0)}\,, \tag{10.41}$$

where j' is the derivative of j with respect to its argument. From $\Theta_\ell(\mathbf{k})$ we obtain the observed multipoles through Eq. (10.22), and the CMB spectrum through Eq. (10.23).

For $\ell \gg 1$, j_ℓ and j'_ℓ peak at roughly $k\eta_0 \simeq \ell$. This is the same peaking of j_ℓ that, in quantum mechanics, yields the classical relation $\ell = rp$ for the angular momentum ℓ of a particle with momentum p, rotating at a distance r from the centre. Using Eq. (10.17), it reproduces the linear scale (10.39) associated with a given angular scale.

10.6 Sachs–Wolfe plateau

10.6.1 Height of the plateau

Now we study the prediction of Eq. (10.41), starting with the regime $\ell \lesssim 30$ which is called the Sachs–Wolfe plateau. The corresponding scales are well outside the

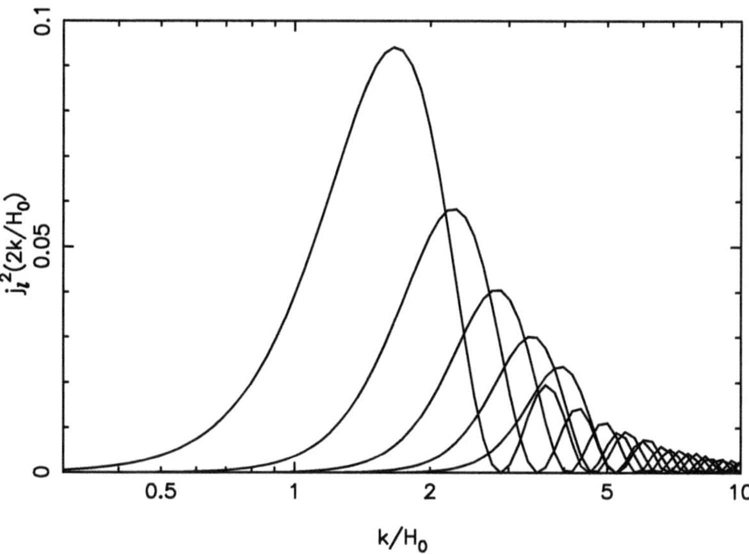

Fig. 10.4. Spherical Bessel functions $j_\ell^2(2k/H_0)$ for $\ell = 2, 3, 4, 5$, and 6, showing how the different C_ℓ sample scales k. For a scale-invariant $\mathcal{P}_\Phi(k)$, the C_ℓ is just the area under the curve.

horizon at the time of last scattering, which means that the sudden-decoupling approximation is essentially perfect.

We are assuming complete matter domination at last scattering. Using Eqs. (8.38) and (8.48) we find $\delta_{\rm m} = -2\Phi$, and the adiabatic condition then gives $\delta_\gamma = -(8/3)\Phi$. Since δ_γ is constant, V_γ vanishes and Eq. (10.37) gives

$$\boxed{\Theta(\mathbf{e}) = \left(\frac{1}{4}\delta_\gamma + \Phi\right) = \frac{1}{3}\Phi_{\rm ls} = -\frac{1}{5}\zeta_{\rm ls}}\,. \tag{10.42}$$

This result is called the Sachs–Wolfe effect.

The spectrum is

$$C_\ell = \frac{4\pi}{25}\int_0^\infty \frac{dk}{k} j_\ell^2(k\eta_0)\mathcal{P}_\zeta(k). \tag{10.43}$$

The factor j_ℓ^2 is sharply peaked at $k\eta_0 = \ell$ (shown in Figure 10.4), which means that we can take $\mathcal{P}_\Phi$ outside the integral with its argument $k = \ell/\eta_0$. This leaves a standard integral

$$\int_0^\infty \frac{dx}{x} j_\ell^2(x) = \frac{1}{2\ell(\ell+1)}, \tag{10.44}$$

which gives

$$\boxed{\ell(\ell+1)C_\ell = \frac{2\pi}{25}\mathcal{P}_\zeta(\ell/\eta_0)}. \tag{10.45}$$

This prediction for the Sachs–Wolfe plateau is flat if the primordial curvature perturbation is scale-invariant.

We have yet to consider the ISW effect. There is an early-time ISW contribution coming from radiation still present at decoupling, and a late-time one coming from dark energy. Only the late-time ISW effect is significant on the Sachs–Wolfe plateau. As one sees from Figure 10.3 it gives the predicted C_ℓ a downward slope, which, however, isn't very significant because of cosmic variance. A more important consequence of the late-time ISW effect, which we shall not describe, is to produce an observable correlation between the CMB anisotropy and the matter density perturbation.

10.6.2 Very large scale contribution

The spectrum $\mathcal{P}_\zeta(k)$ is observed to be almost scale-independent on cosmological scales, but it might in principle rise sharply as we go to scales $k^{-1} \gg H_0^{-1}$ (super-large scales). Grishchuk and Zel'dovich in 1978 pointed out that such an effect would enhance the mean-square of the quadrupole. The quadrupole dominates because the integrand in Eq. (10.43) is proportional to $k^{2\ell-1}$ for small k.

To quantify the Grishchuk–Zel'dovich effect, let us represent the very large scale contribution by a delta function:

$$\mathcal{P}_\zeta(k) = \langle\zeta^2\rangle\delta(\ln k - \ln k_{\rm vl}). \tag{10.46}$$

We are supposing that its contribution to the mean-square curvature perturbation dominates the contribution of smaller scales. (Recall that the latter is of order $10^{-10}\ln(k_{\rm vl}/H_0)$, assuming a flat spectrum for $k > k_{\rm vl}$.) Using the small-argument limit $j_2(x) = x^2/15$, and taking the observational limit on the quadrupole to be $C_2 \lesssim 10^{-9}$, gives the constraint

$$\frac{H_0}{k_{\rm vl}} \gtrsim 10^2\langle\zeta^2\rangle^{1/4}. \tag{10.47}$$

Perturbation theory requires $\langle\zeta^2\rangle \lesssim 1$. This bound will be roughly saturated if there is an era of eternal inflation (Section 18.3), ending when the scale $k_{\rm vl}$ leaves the horizon. The constraint Eq. (10.47) therefore suggests that eternal inflation must end more than a few e-folds ($\ln 100 \simeq 5$) before our Universe leaves the horizon. In typical models this requirement is amply satisfied.

More generally, the absence of the effect perhaps suggests that the observable

Universe is part of a smooth patch whose size is $\gtrsim 100 H_0^{-1}$. However, as the original authors emphasized, the effect is a stochastic one that applies to our Universe on the assumption that it is a fairly typical realization of an ensemble whose stochastic properties are invariant under translations and rotations. The only known origin for such an ensemble is a vacuum fluctuation during inflation, which is why we emphasized the connection with models of inflation. The Grishchuk–Zel'dovich effect does *not* give model-independent knowledge of what lies beyond the edge of the observable Universe. The existence of such knowledge would be a violation of causality.

10.7 Acoustic peaks and Silk damping

The Sachs–Wolfe plateau extends to $\ell \sim 30$. As ℓ increases further, C_ℓ has a peak structure caused by the acoustic oscillation of the photon fluid. Let us see how the peak structure arises, again within the context of the sudden-decoupling approximation (10.41). We will drop the ISW effect for simplicity, though both the early- and late-time effects need to be included for an accurate calculation.

The peaks represent a snapshot of the acoustic oscillation at decoupling. With the adiabatic initial condition we can use Eqs. (8.70) and (8.71) for a rough approximation. This gives the transfer function

$$T_\ell(k) = \left[A_1 + A_2 e^{-k^2/k_{\rm D}^2}\cos(kr_{\rm s})\right] j_\ell(k\eta_0) - kc_{\rm s} A_2 e^{-k^2/k_{\rm D}^2}\sin(kr_{\rm s}) j_\ell'(k\eta_0), \tag{10.48}$$

with $A_1 = (3/5)RT(k) \simeq 0.4T(k)$ and $A_2 \simeq -1/3$. As seen earlier, $r_{\rm s} \simeq 150\,\text{Mpc}$, $\eta_0 \simeq 14{,}000\,\text{Mpc}$ and $k_{\rm D}^{-1} \simeq 8\,\text{Mpc}$.

The spectrum is given by Eq. (10.25). The factor T_ℓ^2 gives a cross-term proportional to $j_\ell j_\ell'$, which oscillates rapidly about zero and gives a negligible contribution to the integral. There is also a term proportional $j_\ell'^2$, which can be shown to give a smooth and very small contribution to the integral [2, 3]. As a result of these features, the dipole contribution to the CMB that is present just after last scattering gives a smooth and very small contribution to the observed anisotropy. Dropping this contribution we arrive at

$$\boxed{C_\ell = 4\pi \mathcal{P}_\zeta \int_0^\infty \left[A_1(k) + A_2\cos(kr_{\rm s}) e^{-k^2/k_{\rm D}^2}\right]^2 j_\ell^2(k\eta_0)\,\frac{dk}{k}}\,. \tag{10.49}$$

To estimate C_ℓ we adopt the approximations of Refs. [2, 3]. For large ℓ, the Bessel function is practically zero for $x < \ell$, while for $x > \ell$ it oscillates rapidly. For the purpose of evaluating the integral we may average over the oscillation.

Then

$$j_\ell^2(x) \simeq \frac{1}{2}\frac{1}{x\sqrt{x^2-\ell^2}}. \tag{10.50}$$

This expression peaks at $k\eta_0 = \ell$, but the peaking is not sharp enough to take the square bracket of Eq. (10.49) outside the integral because the cosine factor oscillates too rapidly. To handle the oscillation of the cosine one can use

$$\int_1^\infty \frac{f(x)\cos(ax)}{\sqrt{x^2-1}}dx = f(1)\sqrt{\frac{\pi}{a}}\cos\left(a+\frac{\pi}{4}\right), \tag{10.51}$$

which is valid for sufficiently large a and is therefore approximately valid in our case. This gives

$$\ell^2 C_\ell = B_1 + B_2\cos\left(\pi\frac{\ell}{\ell_1}+\frac{\pi}{4}\right) + B_3\cos\left(2\pi\frac{\ell}{\ell_1}+\frac{\pi}{4}\right), \tag{10.52}$$

$$B_1 = \frac{1}{2}A_1^2 + \frac{1}{4}A_2^2\, e^{-(\ell/\ell_{\rm D})^2}, \tag{10.53}$$

$$B_2 = A_1 A_2\sqrt{\frac{\ell_1}{\ell}}\, e^{-\frac{1}{2}(\ell/\ell_{\rm D})^2}, \tag{10.54}$$

$$B_3 = \frac{1}{2}A_2^2\sqrt{\frac{2\ell_1}{\ell}}\, e^{-(\ell/\ell_{\rm D})^2}. \tag{10.55}$$

with $\ell_1 = \pi\eta_0/r_{\rm s} \simeq 300$ and $\ell_{\rm D} = \eta_0 k_{\rm D}/\sqrt{2} \simeq 1200$.

With estimates for A_1 and A_2 that are somewhat more sophisticated than ours, it is found [2, 3] that Eq. (10.52) is in quite good agreement, up to $\ell \simeq 1000$ or so, with the outcome of the exact calculation that we describe in the next chapter, and which is shown in Figure 10.3. At higher ℓ it falls off too slowly, and that remains true if we replace Eq. (10.49) by the actual solution of the tight-coupling equations. A much more correct result is obtained by taking into account the gradual decoupling described in Section 4.7.2. As most of the CMB photons come from around the peak of the visibility function, where photon scattering is still frequent enough that the CMB is isotropic in the rest frame, it may seem reasonable to assume that the sudden-decoupling approximation applies individually to those photons emerging at a given redshift. (We will see in Section 11.8 that such intuition is justified.) Then Eq. (10.41) becomes

$$\Theta_\ell(\mathbf{k}) = \int_0^{\eta_0} d\eta\, g(\eta)\left\{\left[\frac{1}{4}\delta_\gamma(\eta,\mathbf{k}) + \Phi(\eta,\mathbf{k})\right] j_\ell + V(\eta,\mathbf{k})j_\ell'\right\}, \tag{10.56}$$

where the argument of j_ℓ is $k(\eta_0-\eta)$. Except for $g(\eta)$, the most important η-dependence comes from the Silk damping factor. Keeping just this corresponds to making the following replacement in the sudden-decoupling approximation:

$$\boxed{e^{-k^2/k_{\rm D}^2(\eta_{\rm ls})} \to \int_0^\infty e^{-k^2/k_{\rm D}^2(\eta)} g(\eta) d\eta}. \tag{10.57}$$

With this replacement, the sudden-decoupling approximation gives quite a good approximation to the exact result. For sufficiently large ℓ it would leave only the Sachs–Wolfe contribution, coming from the term of Eq. (10.49) proportional to A_1, but the primordial CMB anisotropy is unlikely to be observable in that regime.

10.8 Reionization

As we saw in Section 4.7, the probability that a CMB photon, now observed, scattered at least once after decoupling is given by

$$P_{\rm scatt} = 1 - \exp(-\tau) \simeq 0.07. \tag{10.58}$$

What we see looking out along a given direction is therefore a superposition, consisting of a fraction $1 - P_{\rm scatt}$ of photons from the original last-scattering surface, plus a fraction $P_{\rm scatt}$ that have been scattered into the line of sight from a variety of different directions. The rescattered photons have originated from a surface whose radius is the distance between the original last-scattering surface and the rescattering point, in the case of a single rescattering, or from within that sphere if multiple scatterings have occurred. This is shown in Figure 10.5.

The fraction of the photons that reach us directly from the last-scattering surface without rescattering give the usual contribution to the anisotropy because they have been unaffected by reionization. Those that did scatter won't give the usual contribution, because their point of origin was somewhere on the surface of the small circle shown in Figure 10.5 (or within it in the case of multiple rescattering), which has a different anisotropy pattern than the original last-scattering surface. When all rescattered photons are taken into account, we observe an averaging of the temperature over their possible points of origin.

The effect on the anisotropy depends on the scale being examined. On large angular scales, larger than the horizon size at the time of rescattering (the small circle in Figure 10.5), the anisotropies are unaffected by rescattering because the rescattering occurs among photons within a large region sharing the same large-scale temperature contrast. This is simply a statement of causality; rescattering cannot affect physics on scales larger than the horizon size at that time. On small scales the situation is different; the temperature on the original last-scattering surface is uncorrelated with the mean temperature on the sphere[1] from which the rescattered

[1] In general, this has to be weighted both for multiple scatterings and for the anisotropy of Thomson scattering, but neither affect this argument.

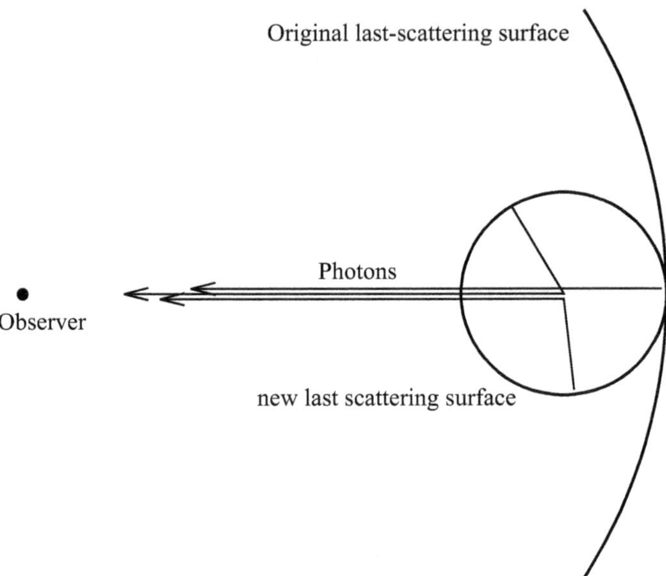

Fig. 10.5. Schematic representation of CMB photons scattering from ionized electrons. Photons that scatter a single time after last scattering originate on the last-scattering sphere surrounding the rescattering point. As seen by a distant observer, that sphere just touches the original last-scattering surface. Photons that scatter more than once originate from inside the sphere. The plot is in comoving coordinates $\mathbf{x}$.

photons originated (the small sphere in Figure 10.5). This means that when considering the small-scale anisotropy, the rescattered photons can be considered to have the *mean* CMB temperature, so that they contribute no anisotropy at all.

We consequently have two limiting behaviours:

$$C_\ell^{\rm obs} = C_\ell^{\rm int} \qquad \text{(Small } \ell\text{)}, \tag{10.59}$$

$$C_\ell^{\rm obs} = \exp\left(-2\tau\right) C_\ell^{\rm int} \simeq 0.85 \qquad \text{(Large } \ell\text{)}, \tag{10.60}$$

where $C_\ell^{\rm int}$ is the intrinsic CMB spectrum on the last-scattering surface at $z \simeq 1\,000$, and $C_\ell^{\rm obs}$ is that which we actually see. Equation (4.61) gives τ, and the factor 2 is simply because the C_ℓ is given by the square of the temperature contrast. The break between these two behaviours can be estimated from the angular size of the region in which the rescattered photons originate, that is, from the horizon size at rescattering, as being at ℓ of a few tens.

10.9 Non-gaussianity of the CMB anisotropy

If the CMB is gaussian, the correlator of an odd number of multipoles vanishes, while the correlator of an even number is completely determined by the two-point correlator. Non-gaussianity is characterized by additional correlations, as described for instance in Ref. [4].

Rotational invariance implies that the stochastic properties of each correlator are independent of the orientation of the polar coordinates $\{\theta, \phi\}$. To see how this works it will be enough to consider the three-point correlator. Rotational invariance requires that it be of the form

$$\langle a_{\ell_1 m_1} a_{\ell_2 m_2} a_{\ell_3 m_3} \rangle = \begin{pmatrix} \ell_1 & \ell_2 & \ell_3 \\ m_1 & m_2 & m_3 \end{pmatrix} B_{\ell_1 \ell_2 \ell_3}. \tag{10.61}$$

Here $B_{\ell_1\ell_2\ell_3}$, called the bispectrum of the CMB anisotropy, is invariant under rotations. The other factor, called the Wigner 3-j symbol, has the same transformation under rotations as the left-hand side. We won't give the definition and properties of the Wigner 3-j symbol.

Going to the flat-sky approximation, we can define a bispectrum $b(l_1, l_2, l_3)$ by

$$\langle a(\vec{\ell}_1) a(\vec{\ell}_2) a(\vec{\ell}_3) \rangle = (2\pi)^2 \delta^2(\vec{\ell}_1 + \vec{\ell}_2 + \vec{\ell}_3) b(\ell_1, \ell_2, \ell_3). \tag{10.62}$$

It can be shown [4] that $b(l_1, l_2, l_3)$ is related to the bispectrum of the multipoles by

$$B_{\ell_1\ell_2\ell_3} = \sqrt{\frac{(2\ell_1+1)(2\ell_2+1)(2\ell_3+1)}{4\pi}} \begin{pmatrix} \ell_1 & \ell_2 & \ell_3 \\ 0 & 0 & 0 \end{pmatrix} b_{\ell_1\ell_2\ell_3}. \tag{10.63}$$

Assuming that $a(\vec{\ell})$ is gaussian, the cosmic variance of b can be calculated along the lines indicated in Section 6.5.2 (using two instead of three dimensions). Alternatively, assuming that each $a_{\ell m}$ has an independent gaussian distribution the cosmic variance of $B_{\ell_1\ell_2\ell_3}$ is a numerical factor times $C_{\ell_1} C_{\ell_2} C_{\ell_3}$.

Extending this treatment, one can define a connected four-point correlator of the multipoles and the trispectrum of the CMB anisotropy, and also connected higher correlators though the latter are expected to be negligible.

Assuming first-order perturbation theory, each multipole is linearly related to the primordial perturbations. This allows the bispectrum (say) of the CMB anisotropy to be calculated in terms of the bispectrum of the primordial perturbations.

Exercises

10.1 Show that the relativistic Doppler shift changes each of the multipoles by an amount of order v^ℓ.

10.2 Verify Eq. (10.21) giving $\Theta_{\ell m}$ in terms of Θ_ℓ.

10.3 Use Eq. (6.61) to show that Eqs. (10.24) and (10.26) provide equivalent definitions of the transfer function T_ℓ.

10.4 Use the transformation (A.12) to show that the form of Eq. (10.7) is a consequence of rotational invariance.

10.5 Use Eqs. (10.24) and (10.22), and the orthogonality of the spherical harmonics, to show that there will be no correlation between different multipoles if Θ_ℓ is gaussian.

10.6 Verify the geodesic equation (10.34), for a massless particle in the perturbed Universe.

References

[1] *WMAP Satellite*, home page at http://wmap.gsfc.nasa.gov/.

[2] V. F. Mukhanov. CMB-slow, or how to estimate cosmological parameters by hand. *Int. J. Theor. Phys.*, **43** (2004) 623.

[3] V. Mukhanov. *Physical Foundations of Cosmology* (Cambridge: Cambridge University Press, 2006).

[4] E. Komatsu. The pursuit of non-gaussian fluctuations in the cosmic microwave background. arXiv:astro-ph/0206039.

11

Boltzmann hierarchy and polarization

We have thus far given only an approximate account of the evolution of the four components of the cosmic fluid. We relied on the tight-coupling approximation for the baryons and photons before photon decoupling, and more or less ignored the neutrino perturbation. Now we give the exact first-order theory, using the Boltzmann equations which describe the evolution of the photon and neutrino distribution functions. The exact theory is required if the calculations of the matter density perturbation and cosmic microwave background (CMB) anisotropy are to match the accuracy of the observations. Further, the relaxation of the tight-coupling approximation will allow us to calculate the polarization of the CMB.

11.1 Perturbed Boltzmann equation

As in Section 8.2, we define the photon distribution function in the locally orthonormal frame lined up with the coordinates. For the argument in Section 2.7 leading to the collisionless Boltzmann equation, we need the corresponding locally inertial frame. We also need that frame to work out the collision term appearing in the Boltzmann equation (2.68). As in Eq. (10.30), we use $q \equiv a(\eta)p(\eta, \mathbf{x})$ instead of the physical energy p. With all this in place we have a distribution function $f(\eta, \mathbf{x}, \mathbf{n}, q)$, where $\mathbf{n}$ points along the photon direction. It satisfies

$$\frac{\partial f}{\partial \eta} + \frac{\partial f}{\partial x^i}\frac{dx^i}{d\eta} + \frac{\partial f}{\partial q}\frac{dq}{d\eta} + \frac{\partial f}{\partial n^i}\frac{dn^i}{d\eta} = \frac{df}{d\eta} \equiv C[f]. \tag{11.1}$$

We start the discussion at an 'initial' epoch soon after neutrino decoupling. In the unperturbed Universe the photons are in thermal equilibrium. The photon temperature in the unperturbed Universe falls like $1/a$ and we can write the blackbody distribution as

$$f(q) = \frac{1}{e^{q/T_0} - 1}. \tag{11.2}$$

The two sides of Eq. (11.1) vanish in the unperturbed Universe. At first order, the two sides will be linear in the perturbation δf, defined by

$$f(\eta, \mathbf{x}, \mathbf{n}, q) = f(q) + \delta f(\eta, \mathbf{x}, \mathbf{n}, q). \tag{11.3}$$

The last term on the left-hand side of Eq. (11.1) is of second order, and so we drop it. In the second term, $\partial f/\partial x^i$ is of first order. We therefore can identify $dx^i/d\eta$ with n_i. In the third term, $dq/d\eta$ is given by Eq. (10.34). It is of first order, and so, we can replace $\partial f/\partial q$ by $df(q)/dq$. Putting all this into Eq. (11.1) gives

$$\dot{\delta f} + n^i \partial_i(\delta f) + q\frac{df}{dq}\left(\dot{\Phi} - n^i \partial_i \Psi\right) = C[\delta f]. \tag{11.4}$$

We are using dot to denote $\partial/\partial\eta$.

Now we relate δf to the brightness function $\Theta \equiv \delta T/T$. By writing

$$f(\eta, \mathbf{x}, \mathbf{n}, q) = \frac{1}{e^{q/T_0(1+\Theta)} - 1}, \tag{11.5}$$

we have to first order

$$\delta f(\eta, \mathbf{x}, \mathbf{n}, q) = -q\frac{df(q)}{dq}\Theta(\eta, \mathbf{x}, \mathbf{n}), \tag{11.6}$$

where $f(q)$ is the unperturbed quantity. The Boltzmann equation for $\Theta(\eta, \mathbf{k}, \mathbf{n})$ is

$$\boxed{\frac{\partial\Theta}{\partial\eta} + ik\mu\Theta - \frac{\partial\Phi}{\partial\eta} + ik\mu\Psi = C[\Theta]}, \tag{11.7}$$

with $\mu \equiv \hat{\mathbf{k}} \cdot \mathbf{n}$. The left-hand side of this equation doesn't involve q, and we see soon that neither does the collision term. This justifies the assumption that Θ has no q dependence.

A different expression for Θ follows from the second equality of Eq. (4.4), with f and T now depending on position $\mathbf{x}$ and direction $\mathbf{n}$. Using a locally orthonormal frame and $\mathbf{q} \equiv \mathbf{p}a(\eta)$ it gives

$$a^4(\eta) \int_0^\infty f(\eta, \mathbf{x}, \mathbf{n}, q) q^3 dq = (\text{constant})\, T^4(\eta, \mathbf{x}, \mathbf{n}). \tag{11.8}$$

Since $\Theta = \delta T/T$ this gives

$$4\Theta = \frac{\int_0^\infty \delta f q^3 dq}{\int_0^\infty f q^3 dq}. \tag{11.9}$$

By comparing this expression with the first-order perturbations of Eqs. (2.63), (2.64) and (2.66), and paying attention to the dependence on $\mathbf{n}$, we see that Θ_0 determines the energy density perturbation of the photon fluid, that Θ_1 determines the fluid velocity, and that Θ_2 determines the anisotropic stress. For the scalar mode the relations are

$$\Theta_0 = \frac{\delta_\gamma}{4}, \qquad \Theta_1 = \frac{V_\gamma}{3}, \qquad \Theta_2 = \frac{\Pi_\gamma}{12}, \tag{11.10}$$

where V and Π are defined by Eqs. (7.11) and (8.23).

We encountered the first relation in Chapter 10, and the second is just the scalar part of the Doppler shift seen by an observer moving with the threading.

11.2 Boltzmann hierarchy

Now we consider the $\Theta_\ell(\eta, \mathbf{k})$ defined by Eq. (10.19). Using Eq. (A.2), the Boltzmann equation becomes

$$\dot{\Theta}_0 = -k\Theta_1 + \dot{\Phi} + C[\Theta_0], \tag{11.11}$$

$$\dot{\Theta}_1 = \frac{k}{3}(\Theta_0 - 2\Theta_2) + \frac{k}{3}\Psi + C[\Theta_1], \tag{11.12}$$

$$\dot{\Theta}_\ell = \frac{k}{2\ell+1}\left[\ell\Theta_{\ell-1} - (\ell+1)\Theta_{\ell+1}\right] + C[\Theta_\ell] \qquad (\ell \geq 2). \tag{11.13}$$

This set of equations is called the Boltzmann hierarchy.

Without the collision terms, the first two equations of the Boltzmann hierarchy would be respectively the photon continuity and Euler equations. The continuity equation is actually valid, because Thomson scattering doesn't alter the photon energy. Thus the first equation of the hierarchy can be written

$$\boxed{\dot{\delta}_\gamma = -\frac{4}{3}kV_\gamma + 4\dot{\Phi}}. \tag{11.14}$$

Now we come to the neutrinos, ignoring for the moment their mass. As they do not interact, the neutrino component of the cosmic fluid satisfies the continuity and Euler equations

$$\boxed{\dot{\delta}_\nu = -\frac{4}{3}kV_\nu + 4\dot{\Phi}}, \tag{11.15}$$

$$\boxed{\dot{V}_\nu = k\left(\frac{1}{4}\delta_\nu - \frac{1}{6}\Pi_\nu + \Psi\right)}. \tag{11.16}$$

The neutrino brightness function may be defined as the fractional perturbation in T_ν, or equivalently by

$$\delta f_\nu(\eta, \mathbf{x}, \mathbf{n}, q) = -q\frac{df_\nu(q)}{dq}\Theta_\nu(\eta, \mathbf{x}, \mathbf{n}), \tag{11.17}$$

where f_ν is the unperturbed quantity

$$f_\nu(q) = \frac{1}{e^{q/T_{\nu 0}} + 1}. \tag{11.18}$$

It satisfies the same hierarchy as the photon brightness function, without the collision term. The first two equations of the hierarchy are the continuity and Euler equations, while the rest are

$$\boxed{\dot{\Theta}_{\nu\ell} = \frac{k}{2\ell+1}\left[\ell\Theta_{\nu(\ell-1)} - (\ell+1)\Theta_{\nu(\ell+1)}\right]} \qquad (\ell \geq 2). \tag{11.19}$$

The expressions in Eq. (11.10) are valid also for the neutrinos:

$$\Theta_{\nu 0} = \frac{\delta_\nu}{4}, \qquad \Theta_{\nu 1} = \frac{V_\nu}{3}, \qquad \Theta_{\nu 2} = \frac{\Pi_\nu}{12}. \tag{11.20}$$

If we consider Eq. (11.19) for $\ell = 2$ well before horizon entry, the first term of the square bracket dominates and using Eq. (11.10) we find

$$\boxed{\dot{\Pi}_\nu = \frac{4}{5}kV_\nu} \qquad (k \ll aH). \tag{11.21}$$

Neutrino mass becomes significant at late times. To include it one has to treat the three species separately, keeping track of the energy $E = \sqrt{m_\nu^2 + p^2}$. No new principle is involved and we shall not give the equations.

11.3 Collision term without polarization

The collision term comes from Thomson scattering, which is the scattering of classical electromagnetic waves off free electrons. In this section we give an approximate treatment which ignores polarization, that is accurate to a few percent.

11.3.1 Unpolarized Thomson scattering

Ignoring polarization, the differential cross-section $d\sigma_{\rm T}/d\Omega'$ for the Thomson scattering is given by

$$\boxed{\frac{d\sigma_{\rm T}(\mathbf{n}, \mathbf{n}')}{d\Omega'} = \frac{3}{16\pi}\sigma_{\rm T}\left(1 + \cos^2\theta\right)}, \tag{11.22}$$

where $\sigma_{\rm T} = (8\pi/3)\alpha^2/m_{\rm e}$ and θ is the angle between $\mathbf{n}$ and $\mathbf{n}'$. Integrating over $d\Omega'$ gives the total cross-section for Thomson scattering, $\sigma_{\rm T}$.

In Section 11.4.2 we derive this formula from first principles, along with its generalization to include polarization. More important is the meaning of the cross-section. It refers to the rest frame of the electron. If there are electrons with number density n_e and negligible motion, the cross-section is usually defined by saying that $n_e(d\sigma_T/d\Omega')d\Omega'$ is the probability per unit time, that a photon moving in the $\mathbf{n}$ direction will scatter into a solid angle $d\Omega'$ whose direction $\mathbf{n}'$ makes an angle θ with the incoming direction $\mathbf{n}$. For a plane wave, this is the fraction (assumed to be small) of energy lost by the wave per unit time, due to scattering into the same solid angle. The total fraction of energy lost per unit time is $n_e\sigma_T$, which means that each electron acts as if it were a total absorber with geometric cross-section σ_T.

For our purpose an equivalent statement is equally important; $n_e(d\sigma_T/d\Omega')d\Omega'$ is the probability per unit time that a photon whose initial momentum is within the solid angle $d\Omega'$ will scattered to become a photon with momentum $\mathbf{n}$. Taken literally, this statement makes sense only if we work in a finite box so that the momenta are discrete, because otherwise the probability for scattering into a single momentum state would be zero. To fix that we could consider states within a solid angle $d\Omega$ in the $\mathbf{n}$ direction but to keep things simple we take the statement at face value.

11.3.2 Collision term

The collision term is by the general formula Eq. (2.69), with the transition amplitude corresponding to Thomson scattering. We saw in Section 8.7.1 that the square bracket in Eq. (2.69) reduces to the difference between the final and initial photon distribution functions, and that in consequence each scattering process can be treated as if it occurred in isolation. Ignoring polarization, the effect of the collisions is described by the Thomson scattering differential cross-section (11.22). Let us see how to work out the collision term using that formula.

The electrons move with the baryons, and Eq. (11.22) assumes that they are at rest. We eventually want to work in the conformal Newtonian gauge where the baryons have velocity V_B, but for the moment let us work in the baryon rest frame. Then the collision term is

$$C[\delta f(\mathbf{n})] \equiv \frac{d\delta f}{d\eta} = a n_e \int d\Omega' \frac{d\sigma_T(\mathbf{n},\mathbf{n}')}{d\Omega'} \delta f(\mathbf{n}') - a n_e \sigma_T \delta f(\mathbf{n}). \qquad (11.23)$$

To understand this formula, recall the two interpretations of the differential cross-section offered in the previous subsection. The first of these interpretations immediately explains the second term. It is the rate at which $\delta f(\mathbf{n})$ *decreases* due to scattering out of the state with momentum in the direction $\mathbf{n}$. More interesting is the second of the two interpretations, which explains the first term of Eq. (11.23). It is the rate at which $\delta f(\mathbf{n})$ *increases*, due to the scattering of photons with direction in the volume element $d\Omega'$.

Going to the brightness function, let us define

$$\frac{d\Theta(\mathbf{n},\mathbf{n}')}{d\Omega'} \equiv \frac{3}{16\pi}\sigma_{\rm T}\left(1+\cos^2\theta\right)\Theta(\mathbf{n}') = \frac{d\sigma_{\rm T}(\mathbf{n},\mathbf{n}')}{d\Omega'}\Theta(\mathbf{n}'). \tag{11.24}$$

Then

$$C[\Theta(\mathbf{n})] = an_{\rm e}\int d\Omega'\frac{d\Theta(\mathbf{n}',\mathbf{n})}{d\Omega'} - an_{\rm e}\sigma_{\rm T}\Theta(\mathbf{n}). \tag{11.25}$$

The interpretation of the two terms is as before, with Θ instead of δf.

Using Eqs. (A.2) and (A.14) we find

$$\frac{d\Theta(\mathbf{n},\mathbf{n}')}{d\Omega'} = \sigma_{\rm T}\left(\frac{1}{4\pi}+\frac{1}{10}Y_{20}(\mathbf{n})Y_{20}(\mathbf{n}')\right)\Theta(\mathbf{n}'). \tag{11.26}$$

By putting this into Eq. (11.25) we find

$$C[\Theta_\ell] = \dot\tau\left(-\Theta_\ell+\delta_{\ell 0}\Theta_0+\frac{1}{10}\delta_{\ell 2}\Theta_2\right). \tag{11.27}$$

In writing this equation we made the replacement $\dot\tau = an_{\rm e}\sigma_{\rm T}$, which will be used from now on.

We find $C[\Theta_0] = 0$ in accordance with the earlier expectation. For $\ell = 1$, in terms of $V_\gamma = \Theta_1/3$, we find

$$\dot V_\gamma = k\left(\frac{1}{4}-2\Pi_\gamma+\Psi\right) - \dot\tau V_\gamma. \tag{11.28}$$

We have yet to allow for the fact that the electrons share the velocity of the baryons, specified by V_B. As we discussed in Section 10.1, a shift in the velocity of the observer affects only the $\ell = 1$ multipole. Therefore, the inclusion of the electron velocity changes only $C[V_\gamma]$, the effect being to replace V_γ by $V_\gamma - V_B$. We conclude that the only effect of the baryon velocity is to change Eq. (11.28) to

$$\boxed{\dot V_\gamma = k\left(\frac{1}{4}-2\Pi_\gamma+\Psi\right) - \dot\tau\left(V_\gamma - V_B\right)}\,. \tag{11.29}$$

The photon fluid doesn't satisfy the Euler equation because it exchanges momentum with the baryon fluid. But the two fluids taken together will still satisfy the Euler equation. The Euler equation of the baryon fluid is Eq. (8.34), with $V = V_B$, $w = 0$ and $\delta P/\rho = c_{\rm s}^2\delta_B$ where $c_{\rm s}$ is the speed of sound discussed in Section 7.4.3. Remembering that the Euler equation really refers to the momentum density given by Eq. (8.41), the rate of change of the baryon velocity must therefore satisfy

$$\boxed{\dot V_B = -aHV_B + c_{\rm s}^2k\delta_B + k\Psi + \frac{4\rho_\gamma}{3\rho_B}\dot\tau(V_\gamma - V_B)}\,. \tag{11.30}$$

11.4 Polarization and Thomson scattering

11.4.1 Stokes parameters

We first recall the classical description of polarized electromagnetic radiation. Consider a plane wave which is a superposition of different frequencies, arriving at the observer's position from the $+z$ direction. At this position, the sum of frequences within a narrow interval $d\omega$ around some angular frequency ω will give a contribution

$$\mathbf{E}(t) = \mathrm{Re}\left[\mathbf{E}e^{i\omega t}\right] = \frac{1}{2}\left(\mathbf{E}e^{i\omega t} + \mathbf{E}^* e^{-i\omega t}\right), \tag{11.31}$$

where the complex amplitude $\mathbf{E}$ varies slowly on the timescale ω^{-1}. We use the argument t to distinguish between the physical electric field $\mathbf{E}(t)$ and the complex amplitude $\mathbf{E}$.

We will assume that the average intensity in a given time interval $\gg \omega^{-1}$ is independent of the interval, because that is the case for the CMB. We will just call it the intensity. Let us calculate the intensity measured by a detector of linearly polarized radiation, in a plane with azimuthal angle ϕ. The component of $\mathbf{E}$ in this plane is

$$E_\phi = E_x \cos\phi + E_y \sin\phi. \tag{11.32}$$

The time average kills the cross-terms coming from Eq. (11.31), so that the intensity measured by the detector is

$$\boxed{\frac{dI}{d\omega} = \overline{|E_\phi^2|} = I + Q\cos 2\phi + U \sin 2\phi}, \tag{11.33}$$

where

$$\begin{aligned} I &\equiv \overline{|E_x|^2} + \overline{|E_y|^2}, \\ Q &\equiv \overline{|E_x|^2} - \overline{|E_y|^2}, \\ U &\equiv 2\,\mathrm{Re}\,\overline{E_x^* E_y}, \end{aligned} \tag{11.34}$$

and the bars denote time averages. The unpolarized intensity (equal to the average over the angle ϕ) is I, and Q and U are the **Stokes parameters** specifying the plane polarization.

We can also calculate the intensity of radiation measured by a detector of circular polarization. It is specified by a third Stokes parameter:

$$V = 2\,\mathrm{Im}\,\overline{E_x^* E_y}. \tag{11.35}$$

It is not expected in the CMB and we shall not need to consider it.

Returning to the other two Stokes parameters, let us see how they are affected if the x and y axes are rotated. The rotation is specified by Eq. (2.11), and we denote the angle appearing there by ψ. The rotation increases the azimuthal angle $\phi \equiv \tan^{-1}(x/y)$ by an amount ψ. Requiring that the physical quantity $\overline{|E_\phi|^2}$ is unchanged, we find

$$\begin{pmatrix} Q \\ U \end{pmatrix} \rightarrow \begin{pmatrix} \cos 2\psi & -\sin 2\psi \\ \sin 2\psi & \cos 2\psi \end{pmatrix} \begin{pmatrix} Q \\ U \end{pmatrix}. \tag{11.36}$$

It is convenient to use the combinations $Q_\pm \equiv Q \pm iU$. Then,

$$\boxed{Q_\pm \rightarrow \exp(\pm 2i\psi) Q_\pm}. \tag{11.37}$$

There is a preferred choice for the orientation of the x–y plane that makes U vanish, and this defines the 'plane of polarization' of the radiation. The quantity $(Q^2 + U^2)$ is independent of the orientation and satisfies

$$0 \leq \frac{\sqrt{Q^2 + U^2}}{I} \leq 1. \tag{11.38}$$

This is the degree of linear polarization.

To arrive at Eq. (11.37) in a more elegant way, we can define a traceless symmetric polarization tensor P_{ij} by

$$I P_{ij} \equiv \overline{E_i^* E_j} + \overline{E_j^* E_i} - I\delta_{ij}. \tag{11.39}$$

It can be written

$$P_{ij} = Q\epsilon_{ij}^{+} + U\epsilon_{ij}^{\times}, \tag{11.40}$$

where the polarization tensors $\epsilon_{ij}^{+,\times}$ are defined in Eq. (3.58). (The term polarization tensor in both cases is the usual one.) Applying the rotation matrix to each of the indices of P_{ij}, we arrive at Eq. (11.37). This is Eq. (8.7), valid for any traceless symmetric 2-tensor.[1]

The appearance of traceless symmetric 2-tensors is ubiquitous in cosmology. They describe a plane-polarized gravitational wave, and they also describe the shear field in the sky produced by weak gravitational lensing.

11.4.2 Polarized Thomson scattering

Now we consider the Thomson scattering of a polarized wave, travelling in some direction $\mathbf{n}$ with angular frequency in a narrow interval around some value ω as described in the previous subsection. We suppose that there is a single free electron located at the origin, and denote the field at the origin by $\mathbf{E}(t)$. The electron

[1] Note that Eq. (8.7) is derived with the convention that the polarization tensors $\epsilon_{ij}^{+,\times}$ are *not* transformed by the rotation, despite their name and appearance. An entirely equivalent procedure would be to have $\epsilon_{ij}^{+,\times}$ transform as tensors, in which case $g^{+,\times}$ in Eq. (8.6) would be scalars.

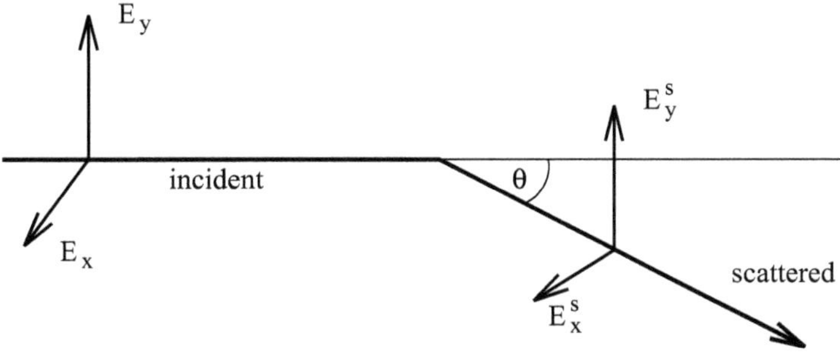

Fig. 11.1. Schematic of Thomson scattering. The scattering angle is aligned with the x-axis.

oscillates, with displacement $\mathbf{r}(t)$ and acceleration $\ddot{\mathbf{r}} = -e\mathbf{E}(t)/m_e$. The corresponding dipole moment is $\mathbf{d}(t) = -e\mathbf{r}(t)$, and it generates an outgoing 'scattered' spherical wave whose electric field at position $\mathbf{r} = r\mathbf{n}'$ is

$$\mathbf{E}(\mathbf{r}, t) = \frac{\left[\ddot{\mathbf{d}}(t-r) \times \mathbf{n}'\right] \times \mathbf{n}'}{4\pi r}. \tag{11.41}$$

Considering a scattering direction in the x–z plane and adopting the notation in Figure 11.1, we find for the scattered wave

$$E'_x = \frac{\alpha}{m_e r} E_x \cos\theta, \qquad E'_y = \frac{\alpha}{m_e r} E_y, \tag{11.42}$$

where $\alpha = e^2/4\pi$ is the fine-structure constant. The field on the left-hand side is evaluated at time t and that on the right-hand side is evaluated at the retarded time $t - r$.

As these formulas do not depend on angular frequency, they apply also to the Fourier components, and taking the time averages we relate the outgoing Stokes parameters to the ingoing ones;

$$I' + Q' = \frac{3\sigma_T}{8\pi r^2}(I + Q)\cos^2\theta, \tag{11.43}$$

$$I' - Q' = \frac{3\sigma_T}{8\pi r^2}(I - Q), \tag{11.44}$$

$$U' = \frac{3\sigma_T}{8\pi r^2} U \cos\theta, \tag{11.45}$$

where $\sigma_T \equiv (8\pi/3)(\alpha/m_e)^2$. Going to $Q_\pm$ we have

$$\boxed{I' = \frac{3\sigma_T}{8\pi r^2}\left[2(\cos^2\theta + 1)I + (\cos^2\theta - 1)Q_+ + (\cos^2\theta - 1)Q_-\right]}, \quad (11.46)$$

$$\boxed{Q'_\pm = \frac{3\sigma_T}{8\pi r^2}\left[2(\cos^2\theta - 1)I + (\cos\theta \pm 1)^2 Q_+ + (\cos\theta \mp 1)^2 Q_-\right]}. \quad (11.47)$$

The first expression gives the fraction of incoming power scattered into solid angle $d\Omega'$ in the θ direction. If the incoming radiation is unpolarized this gives the differential cross-section (11.22). But the second expression shows that even for unpolarized incoming radiation, the outgoing radiation is polarized. We cannot ignore polarization when there are repeated scatterings.

We are going to use Eqs. (11.46) and (11.47) to work out the collision terms for the CMB, including polarization. For that purpose, we will need to consider also the inverse scattering process; an incoming polarized wave encounters the electron at the origin, causing it to emit a beam of radiation in the $-\mathbf{n}$ direction. The inverse process is still described by Eqs. (11.46) and (11.47).

11.5 CMB polarization

The above description applies to CMB radiation coming from a given direction $\mathbf{e} = -\mathbf{n}$, with frequency in a given range $d\omega$. We already considered the total intensity, which is given by the blackbody distribution Eq. (11.2) with the fractional temperature perturbation $\Theta(\mathbf{e})$ independent of frequency. Since the intensity I in a given interval $d\omega$ is proportional to T^4, we have $\Theta(\mathbf{e}) = \delta I(\mathbf{e})/4I$. The Stokes parameters vanish in the unperturbed Universe, since there can be no polarization in the absence of a preferred direction. It is therefore convenient to define 'cosmological' Stokes parameters by $Q_\pm \to Q_\pm/4I$. As the unperturbed radiation is unpolarized we have $|Q_\pm| < |\Theta|$, and as we see later the inequality is actually very strong. Also, we see later that $Q_\pm$ is predicted to be independent of photon energy, just like Θ.

For a given direction of observation $\mathbf{e}$, we take the z-axis used to define the Stokes parameters to point along $\mathbf{e}$. We take the x direction used for that purpose to point in the direction of increasing θ at fixed ϕ, and the y direction to point along the direction of increasing ϕ at fixed θ, where θ and ϕ are the usual polar angles.

A convenient spherical expansion for $Q_\pm(\mathbf{e})$ uses the spin-weighted spherical harmonics ${}_sY_{\ell m}$, defined in Appendix A. We use ${}_{\pm 2}Y_{\ell m} \equiv Y^\pm_{\ell m}$, and write

$$\boxed{Q_\pm(\mathbf{e}) = \sum_{\ell=2}^{\infty}\sum_{m=-\ell}^{\ell}(-1)^\ell Q^\pm_{\ell m} Y^\mp_{\ell m}(\mathbf{n}) = \sum_{\ell=2}^{\infty}\sum_{m=-\ell}^{\ell} Q^\pm_{\ell m} Y^\pm_{\ell m}(\mathbf{e})}. \quad (11.48)$$

The ordinary spherical harmonics have weight zero so that $Y_{\ell m} = {}_0Y_{\ell m}$. A spin-weighted spherical harmonic of weight s transforms under rotations in the same way as $Y_{\ell m}$, except for a phase factor $e^{is\psi}$, where $\psi(\mathbf{e})$ is the change in the azimuthal angle induced by the rotation. (This is the angle $\tan^{-1}(x/y)$ where the directions x and y were define in the previous paragraph. Equivalently, $-\psi$ is the rotation of those directions.) As a result, $Q^{\pm}_{\ell m}$ transform under rotations in the same way as the temperature multipoles $\Theta_{\ell m}$.

Polarization multipoles $E_{\ell m}$ and $B_{\ell m}$ are defined by

$$\boxed{Q^{\pm}_{\ell m} = E_{\ell m} \pm iB_{\ell m}}. \tag{11.49}$$

Under a parity transformation, $E_{\ell m} \to (-1)^{\ell} E_{\ell m}$, the same as for $\Theta_{\ell m}$, while $B_{\ell m}$ picks up an extra factor -1. Inserting these into Eq. (11.48) we could write $Q_{\pm}(\mathbf{e}) = E \pm iB$, but we shall not need E and B.[2]

The invariance of the stochastic properties under rotations and the parity transformation allows the following correlators between pairs of multipoles;

$$\boxed{\langle a^*_{\ell m} a_{\ell' m'} \rangle = C_\ell \, \delta_{\ell\ell'} \delta_{mm'}}, \qquad \boxed{\langle a^*_{\ell m} E_{\ell' m'} \rangle = C^{TE}_\ell \, \delta_{\ell\ell'} \delta_{mm'}},$$

$$\boxed{\langle E^*_{\ell m} E_{\ell' m'} \rangle = C^{EE}_\ell \, \delta_{\ell\ell'} \delta_{mm'}}, \qquad \boxed{\langle B^*_{\ell m} B_{\ell' m'} \rangle = C^{BB}_\ell \, \delta_{\ell\ell'} \delta_{mm'}}, \tag{11.50}$$

where for completeness we reproduced the correlation between temperature multipoles.

The predictions of the ΛCDM model are shown in Figure 11.2. Now we see how to obtain them.

11.6 Boltzmann hierarchy with polarization

Now we consider the Boltzmann equations for Θ and $Q_{\pm}$. Keeping things at a classical level, we should consider plane electromagnetic waves instead of photons, each of them with some definite amount of I and $Q_{\pm}$. To work with photons, we should really attach a polarization matrix to each photon, which is altered at each scattering. Loosely though, we might as well think of each photon as carrying some definite amount of Θ and of $Q_{\pm}$.

We define polarization multipoles E_ℓ and B_ℓ by analogy to Eq. (10.19), by writ-

[2] The notation E and B for the polarization components has nothing to do with the electric and magnetic fields of the incident radiation. It is used because the polarization tensor $P_{ij}(\theta, \phi)$ can be written as a gradient plus a curl, corresponding respectively to the contributions of $E_{\ell m}$ and $B_{\ell m}$ to Eq. (11.48). A similar decomposition can be made for the gravitational wave amplitude.

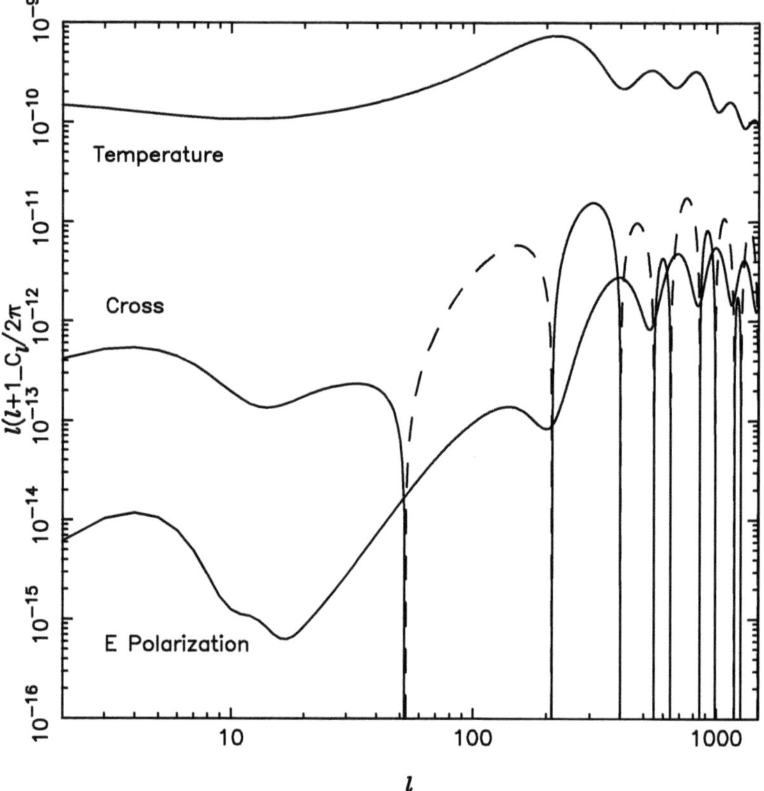

Fig. 11.2. Predicted polarization of the CMB.

ing

$$\boxed{Q_{\mp}(\eta, \mathbf{k}, \mathbf{n}) = 4\pi \sum_{\ell} (-i)^{\ell} \sqrt{\frac{4\pi}{(2\ell+1)}} Y_{\ell 0}^{\pm}(\mathbf{n}) \left[E_{\ell}(\eta, \mathbf{k}) \pm i B_{\ell}(\eta, \mathbf{k}) \right]}, \tag{11.51}$$

and note that

$$Y_{\ell 0}^{\pm}(\mathbf{n}) = \sqrt{\frac{2\ell+1}{4\pi} \frac{(\ell-2)!}{(\ell+2)!}} P_{\ell}^{2}(\hat{\mathbf{k}} \cdot \mathbf{n}), \tag{11.52}$$

where P_{ℓ}^{2} is the associated Legendre function (*not* the Legendre function squared). We wrote $Q_{\mp}$ so that this object has the transformation (11.37) under rotations. This swapping of $+$ and $-$ would matter for the stochastic properties only if these failed to be invariant under the parity transformation.

The initial condition for the scalar mode, set by ζ in the adiabatic case, doesn't distinguish between right- and left-handedness.[3] Neither does the subsequent evolution, as the electromagnetic interaction and Einstein gravity are both invariant under the parity transformation. As a result B_ℓ vanishes in the scalar mode.

For $\Theta(\eta, \mathbf{k}, \mathbf{n})$, Eq. (11.7) remains valid with some collision term that we are going to work out. It just describes the free-streaming of photons under the influence of the gravitational potentials. As gravity affects both polarization states equally, the gravitational potentials do not appear in the Boltzmann equation for $Q_\pm(\eta, \mathbf{k}, \mathbf{n})$, which reads

$$\frac{\partial Q_\pm}{\partial \eta} + ik\mu Q_\pm = C[Q_\pm]. \tag{11.53}$$

To write the Boltzmann hierarchy for E_ℓ we use (A.25) to find

$$(2\ell+1)\dot{E}_\ell = k\left[{}_2\kappa_\ell^m E_{\ell-1} - {}_2\kappa_{\ell+1}^m E_{\ell+1}\right] + C[E_\ell]. \tag{11.54}$$

For convenience we give again the corresponding equation for the $\ell \geq 2$ temperature multipoles:

$$\dot{\Theta}_\ell = \frac{k}{2\ell+1}\left[\ell\Theta_{\ell-1} - (\ell+1)\Theta_{\ell+1}\right] + C[\Theta_\ell]. \tag{11.55}$$

We need the collision terms $C[\Theta(\mathbf{n})]$ and $C[Q_\pm(\mathbf{n})]$. As before, we pretend that the electron is at rest, allowing for the electron motion at the end by changing V_γ to $V_\gamma - V_B$. The collision terms will be of the form Eq. (11.25);

$$C[\Theta(\mathbf{n})] = an_\mathrm{e}\int d\Omega' \frac{d\Theta(\mathbf{n}, \mathbf{n}')}{d\Omega'} - an_\mathrm{e}\sigma_\mathrm{T}\Theta(\mathbf{n}), \tag{11.56}$$

$$C[Q_\pm(\mathbf{n})] = an_\mathrm{e}\int d\Omega' \frac{dQ_\pm(\mathbf{n}, \mathbf{n}')}{d\Omega'} - an_\mathrm{e}\sigma_\mathrm{T}Q_\pm(\mathbf{n}). \tag{11.57}$$

The second term gives the rate at which $\Theta(\mathbf{n})$ or $Q_\pm(\mathbf{n})$ decrease, due to scattering out of the state with momentum $\mathbf{n}$. Omitting the integration, the first term gives the rate of increase of I or Q, due to the scattering of photons with initial momentum in a solid angle $d\Theta'$ in a direction $\mathbf{n}'$.

We first suppose that $\mathbf{n}$ is along the $-z$ direction and that $\mathbf{n}'$ has azimuthal angle $\phi = 0$. If the photons with momentum in the direction $\mathbf{n}'$ happen to be unpolarized, $d\Theta(\mathbf{n}, \mathbf{n}')/d\Omega'$ will be given by Eq. (11.24). Keeping polarization, $d\Theta(\mathbf{n}, \mathbf{n}')/d\Omega'$ and $dQ_\pm(\mathbf{n}, \mathbf{n}')/d\Omega'$ are determined by Eqs. (11.46) and (11.47). Without repeating the calculation leading to the unpolarized equation (11.24), we

[3] This is because ζ (and the S_i introduced in the next chapter) are scalars as opposed to pseudo-scalars.

can read off from Eqs. (11.46) and (11.47) the correctly normalized quantities:

$$\frac{d\Theta}{d\Omega'} = \frac{3\sigma_{\rm T}}{32\pi}\left[2(\cos^2\theta+1)\Theta' + (\cos^2\theta-1)Q'_+ + (\cos^2\theta-1)Q'_-\right], \tag{11.58}$$

$$\frac{dQ_\pm}{d\Omega'} = \frac{3\sigma_{\rm T}}{32\pi}\left[2(\cos^2\theta-1)\Theta' + (\cos\theta\pm1)^2Q'_+ + (\cos\theta\mp1)^2Q'_-\right], \tag{11.59}$$

where $\Theta' \equiv \Theta(\mathbf{n}')$, $Q'_\pm \equiv Q_\pm(\mathbf{n}')$ and θ is the angle between $\mathbf{n}$ and $\mathbf{n}'$. Also, $Q_\pm(\mathbf{n})$ and $Q_\pm(\mathbf{n}')$ are defined in the bases of Figure 11.1.

To obtain the result with $\mathbf{n}$ and $\mathbf{n}'$ in generic directions, we perform a rotation using Table A.1 and Eq. (A.24). The phase factors in Eq. (A.24) are cancelled by phase factors picked up by $Q_\pm$, so that we get

$$\frac{d\Theta}{d\Omega'} = \sigma_{\rm T}\times\sum_m \tag{11.60}$$

$$\left[\left(\frac{\delta_{m0}}{4\pi}+\frac{1}{10}Y_{2m}\overline{Y}'_{2m}\right)\Theta' - \frac{1}{10}\sqrt{\frac{3}{2}}\left(Y_{2m}\overline{Y}^{+\prime}_{2m}Q'_+ + Y_{2m}\overline{Y}^{-\prime}_{2m}Q'_-\right)\right],$$

$$\frac{dQ_\pm}{d\Omega'} = \frac{\sigma_{\rm T}}{10}\sum_m\left(-\sqrt{6}Y^\pm_{2m}\overline{Y}'_{2m}\Theta' + 3Y^\pm_{2m}\overline{Y}^{+\prime}_{2m}Q'_+ + 3Y^\pm_{2m}\overline{Y}^{-\prime}_{2m}Q'_-\right). \tag{11.61}$$

Here, unprimed quantities are functions of $\mathbf{n}$, primed quantities are functions of $\mathbf{n}'$, and an overline denotes complex conjugation.

It is clear that the $|m| = 0$, 1 and 2 correspond respectively to the scalar, vector, and tensor modes. We are here interested in the scalar mode $m = 0$. Projecting out the multipoles, we find that polarization has no effect for $\ell = 0$ and 1. For $\ell \geq 2$ we find

$$\boxed{C[\Theta_\ell] = \dot\tau\left(-\Theta_\ell + \delta_{\ell 2}P\right)}, \tag{11.62}$$

$$\boxed{C[E_\ell] = -\dot\tau\left(E_\ell + \sqrt{6}\,\delta_{\ell 2}P\right)}, \tag{11.63}$$

where

$$\boxed{10P \equiv \Theta_2 - \sqrt{6}E_2}. \tag{11.64}$$

11.7 Initial conditions and the transfer functions

Finally we have a closed system of equations for the perturbations.

- The continuity and Euler equations for the cold dark matter (CDM), (8.44).
- The continuity and modified Euler equation for the baryons, (8.45) and (11.30).
- The continuity equation (11.15), Euler equation (11.16) and Boltzmann hierarchy (11.19) for the neutrinos.

- The continuity equation (11.14) and Euler equation (11.29) for the photons, along with the Boltzmann hierarchy for the photon temperature perturbation (11.55) and polarization (11.54).
- Equations (8.40), (8.41) and (8.43) which give δ, V and Π.
- The constraint equations (8.35) and (8.36), giving Φ in terms of δ and V, and $\Psi - \Phi$ in terms of Π.

They can be solved when an initial condition is specified.

It is important to distinguish between the mathematical limit $\eta \to 0$, and the physical situation for the early Universe. The equations to be solved assume that the Universe has just the four components $i = B$, c, γ and ν, with the neutrinos decoupled. To apply the $\eta \to 0$ limit we have to assume that this state of affairs holds indefinitely far back, but we know that it fails before neutrino decoupling at $T \gtrsim 1\,\text{MeV}$. But starting just below $T = 1\,\text{MeV}$ is enough in practice, because all cosmological scales are far outside the horizon then. We therefore consider the initial condition to be specified at $\eta = 0$, taking the equations to be valid right back to that epoch.

Without at first imposing any initial condition, one can identify independent solutions of the system of equations, called modes. One can choose the modes so that each perturbation goes like some power of conformal time. In other words, one can choose them so that each perturbation is either finite or infinite at $\eta = 0$. Only 'regular' modes, in which all quantities are finite, are usually considered, in keeping with the idea that the early Universe should be almost unperturbed. (A complication in the case of neutrino isocurvature perturbations will be considered later.)

There are four regular modes [1], which may be chosen as the adiabatic mode specified by the time-independent value of ζ, and three isocurvature modes specified by the time-independent values of the isocurvature perturbations S_B, S_c and S_ν defined in the next chapter. In the mathematical limit $\eta \to 0$, all perturbations for a given regular mode either go to zero or are related to the fundamental quantity ζ or S_i. The fluid velocities and the anisotropic stress go to zero, so that the local evolution is that of an unperturbed universe. The initial condition for the regular modes is therefore consistent with the separate universe assumption.

We focus on the adiabatic mode, defined by the initial conditions $\zeta \neq 0$ and $S_i = 0$. Given these, regularity at $\eta \to 0$ requires Θ_ℓ and $\Theta_{\nu\ell}$ to vanish for $\ell > 0$. To see how the non-zero quantities are determined by the initial condition, we first generalize Eq. (8.48) to include the neutrino perturbation. The non-decaying solution of Eq. (8.39) is

$$-\zeta = \Phi + \frac{1}{2}\Psi. \tag{11.65}$$

Using Eqs. (11.21), (11.16), (8.38) and (8.43) we find

$$\boxed{\Psi_{\mathbf{k}} = -\frac{2}{3}\frac{1}{1+\frac{4}{15}R_\nu}\zeta_{\mathbf{k}}} \qquad \boxed{\Phi_{\mathbf{k}} = -\frac{2}{3}\frac{1+\frac{2}{5}R_\nu}{1+\frac{4}{15}R_\nu}\zeta_{\mathbf{k}}}, \tag{11.66}$$

where $R_\nu = \rho_\nu/(\rho_\nu + \rho_\gamma) = 0.68$. Equation (8.38) then gives the total density contrast δ and the adiabatic condition S_i gives the individual δ_i.

The presently observed multipoles are

$$E_{\ell m} = \frac{4\pi}{(2\pi)^3} i^\ell \int E_\ell(\mathbf{k}) Y^{+*}_{\ell m}(\hat{\mathbf{k}}) d^3\mathbf{k}, \tag{11.67}$$

with E_ℓ evaluated at the present epoch. Considering just the adiabatic mode we define a transfer function

$$\boxed{E_\ell(\mathbf{k}) = T_{E\ell}\zeta_{\mathbf{k}}}. \tag{11.68}$$

An equivalent definition is

$$E_{\ell m} = \frac{4\pi}{(2\pi)^{3/2}} \int_0^\infty T_{E\ell}(k)\zeta_{\ell m}(k)\,\frac{dk}{k}. \tag{11.69}$$

Using Eq. (6.63) we then get

$$\boxed{C_\ell^{TE} = 4\pi \int_0^\infty T_{E\ell}(k) T_\ell(k) \mathcal{P}_\zeta(k)\,\frac{dk}{k}}, \tag{11.70}$$

$$\boxed{C_\ell^{EE} = 4\pi \int_0^\infty T^2_{E\ell}(k) \mathcal{P}_\zeta(k)\,\frac{dk}{k}}, \tag{11.71}$$

where T_ℓ is the temperature transfer function appearing in Eq. (10.25).

11.8 Line-of-sight integral

The formalism that we have given is mathematically complete. As it stands though, it is unsuitable both for computation and for analytic estimates. We now explain how a small addition to the formalism remedies both of these defects. For further detail see Refs. [2, 3].

We begin with Eq. (11.7), the Boltzmann equation for $\Theta(\eta, \mathbf{k}, \mathbf{n})$. Taking the collision term from our calculation of the multipole collision terms, it is

$$\dot{\Theta}(\eta, \mathbf{k}, \mathbf{n}) = -ik\mu(\Theta + \Psi) + \dot{\Phi} + \dot{\tau}\left[-\Theta + \Theta_0 - iV_B\mu - 5PP_2(\mu)\right]. \tag{11.72}$$

From this expression follows the line-of-sight integral for the presently observed anisotropy:

$$\Theta(\eta_0, \mathbf{k}, \mathbf{n}) = \int_0^{\eta_0} e^{-\tau} S d\eta - \Psi(\eta_0, \mathbf{k}), \tag{11.73}$$

$$S(\eta, \mathbf{k}, \mathbf{n}) \equiv \dot{\Phi} + \dot{\Psi} + \dot{\tau}\left[\Psi + \Theta_0 - iV_B\mu - 5PP_2(\mu)\right], \tag{11.74}$$

$$\tau(\eta_0, \eta) = \int_\eta^{\eta_0} \dot{\tau}(\eta') d\eta', \tag{11.75}$$

where $\mathbf{x} = (\eta - \eta_0)\mathbf{n}$ and as always $\dot{\tau} = a n_e \sigma_T$. Equation (11.73) is valid because, if we allow η_0 to float, its derivative with respect to η_0 reproduces the Boltzmann equation. The last term of Eq. (11.73) has no effect on the anisotropy.

Using Eq. (A.27) we can project out multipoles from the line-of-sight integral to find (for $\ell \geq 2$)

$$\begin{aligned}\Theta_\ell(\eta_0, \mathbf{k}) &= \int_0^{\eta_0} d\eta\, g(\eta) \left[(\Theta_0 + \Psi)\, j_\ell + V_B j'_\ell + \frac{P}{2}\left(3j''_\ell + j_\ell\right)\right] \\ &+ \int_0^{\eta_0} d\eta\, e^{-\tau(\eta)} \left(\dot{\Psi} + \dot{\Phi}\right) j_\ell.\end{aligned} \tag{11.76}$$

Here $g = \dot{\tau} e^{-\tau}$ is the visibility function, and P is given by Eq. (11.64). The argument of the spherical Bessel function is $k(\eta_0 - \eta)$. The arguments of the other functions (Θ_0, V_B, P, Φ and Ψ) are $(\mathbf{k}, \eta)$.

Repeating the calculation with $Q_\pm$ instead of $\Theta_\pm$ one finds

$$E_\ell(\mathbf{k}) = -\frac{3}{2}\sqrt{\frac{(\ell+2)!}{(\ell-2)!}} \int_0^{\eta_0} d\eta\, g(\eta) P \frac{j_\ell}{(kx)^2}. \tag{11.77}$$

The virtue of these equations from the computational viewpoint is that they involve only the lowest three multipoles. Those multipoles can be calculated from the Boltzmann hierarchy truncated at a fairly low order. In contrast, the direct use of the Boltzmann hierarchy involves all observable multipoles, and truncation at a very large order.

They are also of great theoretical interest. If we make the sudden-decoupling approximation $g = \delta(\eta - \eta_{\rm ls})$, and keep only the monopole and dipole in accordance with the tight-coupling approximation, Eq. (11.72) reduces to Eq. (10.29). Keeping the finite width of g, and multiplying the tight-coupling approximation by the Silk damping factor, we arrive at Eq. (10.57).

Much less trivially, we can also extend the sudden-decoupling approximation to include polarization. To do that we need the $\ell = 2$ hierarchy equations. In the tight-coupling approximation the time derivatives are negligible and so are the

$\ell > 3$ contributions to the right-hand side. The $\ell = 2$ equations therefore read

$$0 = \frac{3}{5}k\Theta_1 + \dot{\tau}(-\Theta_2 + P) \tag{11.78}$$

$$0 = \dot{\tau}(E_2 + \sqrt{6}P), \tag{11.79}$$

with $10P = \Theta_2 - \sqrt{6}E_2$. These are equivalent to

$$\frac{4}{\sqrt{6}}E_2 = -\Theta_2 = -\frac{8}{15}\frac{k}{\dot{\tau}}\Theta_1. \tag{11.80}$$

Making the sudden-decoupling approximation for $g(\eta)$ in the line-of-sight integral, one arrives at the sudden-decoupling approximation for E_ℓ. By comparing it with the sudden-decoupling approximation for Θ_ℓ, it allows one to understand the order of magnitude of $E_\ell(\mathbf{k})/\Theta_\ell(\mathbf{k})$, and hence of C_ℓ^{EE}/C_ℓ.

Reionization gives a contribution to Eq. (11.77), which makes the argument of j_ℓ arbitrarily small. The effect of reionization is therefore significant at low ℓ, and the observed magnitude of E_ℓ is sensitive to the reionization epoch. An analytic approximation is provided by giving $g(\eta)$ a delta-function contribution centred on $\eta_{\rm ion}$. Through Eq. (11.77) it should be possible to determine $\Theta_2(\eta_{\rm ion}, \mathbf{k})$ from observation. Then, allowing for the modest evolution to the present, we have a determination of $\Theta_2(\eta_0, \mathbf{k})$ that may be more accurate than the one deduced from direct observation of the present temperature multipoles a_{2m}.

Exercises

11.1 Show that Eq. (11.27) for the temperature multipole collision term follows from Eqs. (11.25) and (11.26).

11.2 Verify Eq. (11.33), giving the intensity measured by a detector of plane-polarized radiation in terms of the Stokes parameters.

11.3 Verify Eq. (11.36), giving the effect of a rotation on the Stokes parameters.

11.4 Verify Eq. (11.42) for the components of the electric field of a Thomson-scattered wave, starting from the expression for the field $\mathbf{E}$ itself given three lines earlier.

11.5 Verify Eq. (11.66) giving the initial values of the gravitational potentials.

11.6 Derive $C_\ell^{\rm EE}/C_\ell$ by the method suggested after Eq. (11.80).

11.7 Repeat the previous exercise with g approximated as a delta function centred on $z_{\rm ion}$ and taking $\Theta_2(\eta_{\rm ion}, \mathbf{k})$ to be the observed quantity. In this way, obtain an estimate of the reionization contribution to E_ℓ as a fraction of the total, and compare your result with Figure 11.2.

References

[1] M. Bucher, K. Moodley and N. Turok. The general primordial cosmic perturbation. *Phys. Rev. D*, **62** (2000) 083508.

[2] U. Seljak and M. Zaldarriaga. A line of sight approach to cosmic microwave background anisotropies. *Astrophys. J.,* **469** (1996) 437.

[3] W. Hu and M. J. White. CMB anisotropies: total angular momentum method. *Phys. Rev. D*, **56** (1997) 596.

12

Isocurvature and tensor modes

We have so far focussed on the scalar mode, with the adiabatic initial condition. Now we study a range of further possibilities. At the time of writing observation is consistent with the hypothesis that these possibilities are not realised in Nature, but one has to work out their consequences before one can place limits on their magnitude.

We begin with the scalar mode with isocurvature initial conditions, and then move to the tensor mode. Next we study the vector mode, which might be generated for instance by cosmic strings or a primordial magnetic field. Although at the time of writing none of these has been observed, they are of interest because even their absence places significant limits on scenarios of the very early Universe. Finally, we look at the effect of spatial curvature, corresponding to $\Omega_0 \neq 1$.

12.1 Isocurvature modes

Possible departures from the adiabatic condition (5.29) are conveniently specified by three **isocurvature density perturbations**. Expressed in terms of the energy densities the isocurvature perturbations are

$$S_{\rm c} = \delta_{\rm c} - \frac{3}{4}\delta_{\rm r}, \tag{12.1}$$

$$S_B = \delta_B - \frac{3}{4}\delta_{\rm r}, \tag{12.2}$$

$$S_\nu = \frac{3}{4}\delta_\nu - \frac{3}{4}\delta_{\rm r}, \tag{12.3}$$

where $\delta_i \equiv \delta\rho_i/\rho_i$ are the density contrasts and $\rho_{\rm r} = \rho_\nu + \rho_\gamma$ is the radiation energy density.

In terms of number densities these definitions are equivalent to

$$S_i = \frac{\delta(n_i/n_{\rm r})}{n_i/n_{\rm r}}, \tag{12.4}$$

(i = B, c and ν). This shows that the S_i are gauge-invariant, and that they are conserved before horizon entry when $n_i \propto \tilde{a}^{-3}(\mathbf{x}, t)$ at each position.

We are going to see that there are three uncoupled isocurvature modes, with the following initial conditions taken to hold in the mathematical limit $t \to 0$.

(i) Cold dark matter (CDM) isocurvature with $\zeta = S_B = S_\nu = 0$ and $S_c \neq 0$.
(ii) Baryon isocurvature with $\zeta = S_c = S_\nu = 0$ and $S_B \neq 0$.
(iii) Neutrino isocurvature with $\zeta = S_B = S_c = 0$ and $S_\nu \neq 0$.

In most of the literature, the isocurvature perturbations are defined using δ_γ instead of δ_r. This makes no difference for the CDM and baryon isocurvature modes. But the neutrino isocurvature mode that we have defined would be a linear combination of all three modes were we to use δ_γ instead of δ_r. As we see in Chapter 26, our neutrino mode is the one likely to be generated because the total energy density in the early Universe is likely to be what matters.

Although the CDM and baryon isocurvature modes are very different from a theoretical viewpoint, they can hardly be distinguished observationally. To see this consider the combination of the CDM and baryon isocurvature modes with $\delta\rho_B = -\delta\rho_c$ initially. It specifies zero initial perturbation in both the radiation density and the matter density. This state of affairs would persist were it not for the random motion of the baryons, but that is negligible by definition except on scales below the baryon Jeans mass. By observing the 21 cm anisotropy or in some other way, it might in the future be possible to probe such scales. At present though, observation cannot strongly constrain S_B and S_c separately, but only the combination

$$S_m \equiv \frac{\delta\rho_m}{\rho_m} - \frac{3}{4}\frac{\delta\rho_\gamma}{\rho_\gamma}, \tag{12.5}$$

which may be written

$$S_m = \frac{\rho_c S_c + \rho_B S_B}{\rho_c + \rho_B}. \tag{12.6}$$

Including the curvature perturbation, we have three modes defined by the primordial quantities $\zeta(\mathbf{x})$, $S_m(\mathbf{x})$ and $S_\nu(\mathbf{x})$. The matter isocurvature fraction may be defined by $\alpha_m = \mathcal{P}_{S_m}/\mathcal{P}_\zeta$ and similarly for α_ν. Assuming that $\mathcal{P}_{S_i}$ is practically scale independent like $\mathcal{P}_\zeta$, the *r.m.s.* values will satisfy $S_i \sim \sqrt{\alpha_i}\zeta$ (using a minimal box size as explained in Section 6.3.2).

Observation places an upper bound on these fractions, which depends on what is assumed about the correlation between the modes. The simple possibilities are no correlation, and $S_i(\mathbf{x}) = \sqrt{\alpha_i}\zeta(\mathbf{x})$ corresponding to full correlation. Current upper bounds on $\alpha_{\rm m}$ at 95% confidence level level are something like

$$\begin{aligned}\sqrt{\alpha_{\rm m}} &< 0.3, && \text{(no correlation)},\\ \sqrt{\alpha_{\rm m}} &< 0.06, && \text{(full correlation)}.\end{aligned} \tag{12.7}$$

The bounds on α_ν are similar.

We are unlikely ever to require a second-order treatment of the evolution of the isocurvature mode after horizon entry. If needed, it would start with the definition $S_i = \delta(n_i/n_{\rm r})/(n_i/n_{\rm r})$, which measures the deviation from the exact adiabatic relation (5.4).

12.2 Matter isocurvature mode

12.2.1 Initial condition

As was the case for the adiabatic mode, we need the initial condition holding in the mathematical early-time limit $\eta = 0$. As with all of the regular modes, the fluid velocities and anisotropic stress vanish. Using Eq. (11.65) with Eqs. (11.21), (11.16), (8.38) and (8.43), one finds $\Phi = \Psi = 0$, so that *spacetime is initially unperturbed in the matter isocurvature mode.* Consistently with this statement, Eq. (8.38) gives $\delta = \delta_{\rm r} = 0$. The initial matter density contrast is $\delta_{\rm m} = S_{\rm m}$.

We will be interested in the growth of Φ and $\delta_{\rm r}$ during radiation domination (still before horizon entry). We will ignore the neutrino perturbation, so that $\Phi = \Psi$. Then Eq. (5.28) gives $4\zeta = (a/a_{\rm eq})S_{\rm m}$. Putting this into Eq. (8.39) gives

$$8\Phi = 8\Psi = -S_{\rm m}\frac{a}{a_{\rm eq}}. \tag{12.8}$$

Including the neutrino perturbation just multiplies Φ and Ψ by different numerical factors of order one. The initial radiation density can be obtained from Eqs. (8.57) and (12.8) as

$$\frac{1}{4}\delta_{\rm r} = -\frac{S_{\rm m}}{2}\frac{a}{a_{\rm eq}}. \tag{12.9}$$

We will also be interested in the situation before horizon entry but during matter domination, which arises in the case of very large scales. Using Eq. (5.28) we find

$$S_{\rm m} = 3\zeta = -5\Phi. \tag{12.10}$$

12.2.2 Matter transfer function

In view of Eq. (12.10) we define an isocurvature matter transfer function by

$$\Phi(\mathbf{k}) = -\frac{1}{5}T_{\rm iso}(k)S_{\rm m}(\mathbf{k}). \tag{12.11}$$

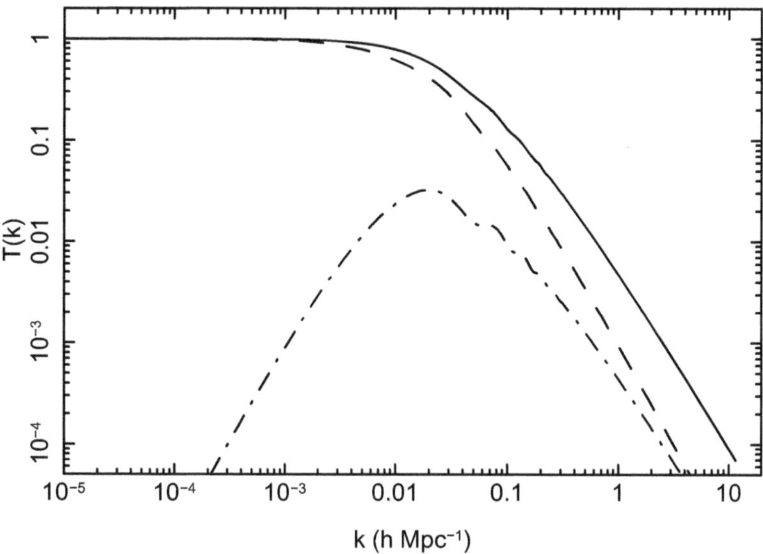

Fig. 12.1. The matter transfer functions. Top down: adiabatic, matter isocurvature and neutrino isocurvature (data courtesy C. Gordon and A. Lewis).

To estimate it on scales $k \gg k_{\rm eq}$, we may simply notice that there is negligible evolution of $\delta_{\rm m}$ during radiation domination, because the perturbation lives in practically unperturbed spacetime. As a result the transfer function is given approximately by Eq. (8.56) without the logarithm. This estimate agrees with the transfer function calculated from the Boltzmann hierarchy, exhibited in Figure 12.1.

Including both the adiabatic and isocurvature modes we have

$$\Phi(\mathbf{k}) = -\frac{3}{5}T(k)\zeta(\mathbf{k}) - \frac{1}{5}T_{\rm iso}S_{\rm m}(\mathbf{k}). \qquad (12.12)$$

To compare with observation we need the spectrum of Φ. If ζ and $S_{\rm m}$ are uncorrelated, then

$$\mathcal{P}_\Phi(k) = \frac{9}{25}T^2(k)\mathcal{P}_\zeta(k) + \frac{1}{25}T^2_{\rm iso}(k)\mathcal{P}_{\rm m}(k), \qquad (12.13)$$

where $\mathcal{P}_{\rm m}$ is the spectrum of $S_{\rm m}$. As we will see, however, early-Universe scenarios can correlate $S_{\rm m}(\mathbf{k})$ and $\zeta(\mathbf{k})$. In that case we need to write

$$\frac{1}{(2\pi)^3}\langle\zeta(\mathbf{k})S_{\rm m}(\mathbf{k}')\rangle \equiv \frac{2\pi^2}{k^3}\delta^3(\mathbf{k}+\mathbf{k}')\mathcal{P}_{\zeta{\rm m}}(k), \qquad (12.14)$$

so that

$$\mathcal{P}_\Phi(k) = \frac{9}{25}T^2(k)\mathcal{P}_\zeta(k) + \frac{6}{25}T(k)T_{\rm iso}(k)\mathcal{P}_{\zeta{\rm m}}(k) + \frac{1}{25}T^2_{\rm iso}(k)\mathcal{P}_{\rm m}(k). \qquad (12.15)$$

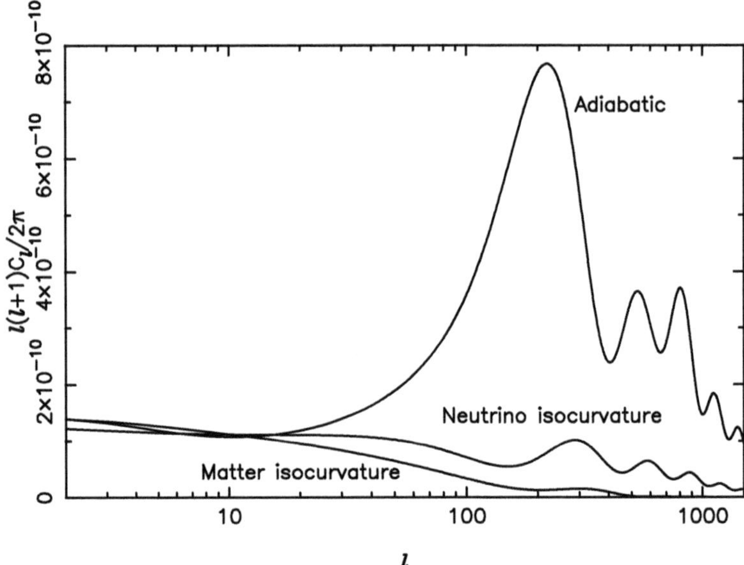

Fig. 12.2. The top curve is for the pure adiabatic mode, as in Figure 10.3. The next curve is for the neutrino isocurvature mode, and the lowest for the matter isocurvature mode. All are normalized to agree with observation at $\ell \simeq 10$, where the Sachs–Wolfe effect dominates (data courtesy C. Gordon).

In the completely correlated case $S_{\rm m}(\mathbf{k}) = X\zeta(\mathbf{k})$ where X is a number. Then

$$\mathcal{P}_\Phi(k) = \left(\frac{3}{5}T(k) + \frac{1}{5}XT_{\rm iso}(k)\right)^2 \mathcal{P}_\zeta(k). \tag{12.16}$$

12.2.3 Acoustic oscillation and CMB anisotropy

Matching to the initial condition (12.9), Eqs. (8.70) and (8.71) become

$$\boxed{\frac{1}{4}\delta_\gamma(\mathbf{k}) \simeq -(1+R)\Phi(\mathbf{k}) - \left[(k_{\rm eq}/k)e^{-k^2/k_{\rm D}^2}\sin(kr_{\rm s})\right] S_{\rm m}(\mathbf{k})}, \tag{12.17}$$

$$\boxed{\frac{1}{4}\dot\delta_\gamma(\mathbf{k}) = -\frac{k}{3}V_\gamma(\mathbf{k}) \simeq k_{\rm eq}c_{\rm s}e^{-k^2/k_{\rm D}^2}\cos(kr_{\rm s})S_{\rm m}(\mathbf{k})}, \tag{12.18}$$

with $\Phi(\mathbf{k}) = T_{\rm iso}(k)S_{\rm m}(\mathbf{k})$.

In strong contrast with the adiabatic case, the amplitude of the acoustic oscillation falls off sharply with k. The peaks of the cosmic microwave background (CMB) anisotropy have the same fall-off, as seen in Figure 12.2. The effect of the acoustic oscillation on the matter transfer function is quite invisible, as seen (or rather not seen) in Figure 12.1.

Finally, consider the Sachs–Wolfe regime. Using Eq. (12.10), Eq. (8.38) gives $\delta_{\rm m} = (2/5)S_{\rm m}$, and then the definition of $S_{\rm m}$ together with the condition $S_\nu = 0$ gives

$$\frac{1}{4}\delta_\gamma = \Phi. \tag{12.19}$$

The temperature anisotropy in the Sachs–Wolfe regime is therefore

$$\boxed{\Theta(\mathbf{e}) = \frac{1}{4}\delta_\gamma + \Phi = 2\Phi = -\frac{2}{5}S_{\rm m}}\,, \tag{12.20}$$

where $S_{\rm m}$ is evaluated at the point of last scattering. It is the same as in the adiabatic case, with the replacement $\zeta \to 2S_{\rm m}$.

To calculate the spectrum of the CMB anisotropy, we have to add the adiabatic and isocurvature modes. The spectrum is then calculated, taking account of the possible correlation between ζ and $S_{\rm m}$ just as we did in Eq. (12.15).

12.2.4 Super-large scale contribution

Analogously with the case of the curvature perturbation, a super-large scale contribution to the spectrum of the isocurvature perturbation would enhance the mean-square quadrupole and dipole. We will just consider the latter. It corresponds to a difference between $V_{\rm r}$ and $V_{\rm m}$; in other words, to relative motion between the CMB and the matter. A universe possessing such a feature is sometimes called a **tilted universe**, because the slice orthogonal to our worldline is tilted relative to that of the CMB worldline. Note that this has nothing to do with the word 'tilt' when used to imply a non-scale-invariant spectrum.

An observational bound on the degree of tilt is provided by the velocity v of the galaxies within around $100h^{-1}\,{\rm Mpc}$ of our position, relative to the CMB rest frame. It seems clear that this velocity is significantly smaller than that of the Sun, but how much smaller remains somewhat controversial, and so, to be conservative, we assume that at least it is no bigger. With this crude estimate we can ignore cosmic variance, yielding $C_{\rm 1VL}^{1/2} \lesssim v/c = 1.2\times 10^{-3}$.

As with the Grishchuk–Zel'dovich effect, the spectrum of $S_{\rm m}$ would have to rise very steeply on scales bigger than the horizon to give an observable effect.[1] Let us represent the very large scale contribution by a delta function $\mathcal{P}_S = \langle S_{\rm VL}^2\rangle\delta(\ln k - \ln k_{\rm VL})$. Because we deal with a scale far outside the horizon, we have $\Theta(\mathbf{e}x_{\rm ls}) = -S(\mathbf{e}x_{\rm ls})/3$, and using Eqs. (6.58) and (6.63), we find that

$$C_{\rm 1VL}^{1/2} = \sqrt{\frac{2}{3}\frac{2\pi}{3}}\frac{k_{\rm VL}}{H_0}\left\langle S_{\rm VL}^2\right\rangle^{1/2}. \tag{12.21}$$

[1] The same is true of tilt caused by S_ν, which we are not going to consider.

Because S is defined as a fractional perturbation, the biggest value that makes sense is $\langle S^2_{\rm VL}\rangle \sim 1$. If this bound is saturated, the observational constraint on $C_{1\rm VL}$ requires $k_{\rm VL}^{-1} \gtrsim 10^3 H_0$.

12.3 Neutrino isocurvature mode

12.3.1 *Initial condition*

The neutrino isocurvature mode has the initial condition $S_\nu \neq 0$ and $\zeta = S_{\rm m} = 0$. In this mode, the initial neutrino distribution function is perturbed, as one sees from the fact that the neutrino density contrast is non-zero on the uniform-density slicing (it is equal to $4S_\nu/3$). As a result, the neutrino anisotropic stress creates a non-zero, time-independent, value for $\Phi - \Psi$. From Eq. (11.65), the initial condition $\zeta = 0$ implies $\Psi = -2\Phi$. Using Eqs. (11.21), (11.16), (8.38) and (8.43), one finds

$$\boxed{\Phi = -\frac{4}{3}\frac{R_\nu}{4R_\nu + 15}S_\nu}\,. \tag{12.22}$$

12.3.2 *Matter transfer function*

Let us define the isocurvature neutrino matter transfer function by

$$\boxed{\Phi(\mathbf{k}) = T_{\nu\rm iso}(k) S_{\nu\mathbf{k}}}\,. \tag{12.23}$$

To estimate it, consider first the very large scales entering the horizon after matter domination. On those scales, the initial value $\zeta = 0$ still holds and Eq. (8.48) gives $T_{\nu\rm iso}(k) = 0$.

Now consider scales which are well inside the horizon at photon decoupling. From Eqs. (8.38) and (12.22) and the condition $S_{\rm m} = 0$, the initial matter density contrast in the neutrino isocurvature mode is

$$\delta_{\rm m} = -\frac{4R_\nu}{4R_\nu + 15}S_\nu. \tag{12.24}$$

This may be compared with the result for the adiabatic mode, $\delta_{\rm m} = (4/3)\zeta$. The subsequent evolution of $\delta_{\rm m}$ will be similar for the two modes. Recalling the factor $-3/5$ in the definition (8.52) of the adiabatic transfer function $T(k)$, we therefore expect

$$\boxed{\frac{T_{\nu\rm iso}(k)}{T(k)} \simeq \frac{5R_\nu}{4R_\nu + 15} \simeq 0.15}\,. \tag{12.25}$$

This estimate of $T(k)$ agrees well with the exact result shown in Figure 12.1.

12.3.3 Acoustic oscillation and CMB anisotropy

From Eq. (8.38), the initial density contrast is $\delta = 4\Phi$, which from the definition of S_ν gives the initial photon density contrast as $\delta_\gamma = -X S_\nu$ with

$$X = \frac{16}{3}\frac{R_\nu}{4R_\nu + 15} + \frac{4}{3}R_\nu. \tag{12.26}$$

The acoustic oscillation at decoupling is therefore

$$\boxed{\frac{1}{4}\delta_\gamma(\mathbf{k}) \simeq -(1+R)\Phi(\mathbf{k}) - \frac{X}{4}e^{-k^2/k_{\rm D}^2}\cos(kr_{\rm s})S_\nu(\mathbf{k})},$$

$$\boxed{\frac{1}{4}\dot\delta_\gamma(\mathbf{k}) = -\frac{k}{3}V(\mathbf{k}) \simeq -\frac{X}{4}kc_{\rm s}e^{-k^2/k_{\rm D}^2}\sin(kr_{\rm s})S_\nu(\mathbf{k})}, \tag{12.27}$$

with $\Phi(\mathbf{k}) = T_{\nu\,\rm iso}(k)S_\nu(\mathbf{k})$. Since $T_{\nu\,\rm iso}$ falls off so sharply, the peaks of C_ℓ are quite accurately at the positions $\ell \simeq n\ell_1$ described in Section 10.7. This is seen in Figure 12.2.

Consider finally the Sachs–Wolfe regime. As $\Phi = 0$, the temperature anisotropy is $\delta_\gamma/4$, evaluated at last scattering. But the further conditions $\zeta = S_{\rm m} = 0$ mean that the radiation density contrast vanishes, leading to

$$\boxed{\Theta(\mathbf{e}) = \frac{1}{4}\delta_\gamma = -\frac{1}{3}\frac{R_\nu}{1-R_\nu}S_\nu}, \tag{12.28}$$

with S_ν evaluated at last scattering.

12.4 The primordial lepton number perturbation

In contrast with the primordial matter isocurvature perturbation, there are severe difficulties in generating the neutrino isocurvature mode. They stem from the fact that the neutrinos are in thermal equilibrium until the run-up to nucleosynthesis begins at $T \sim 1\,\text{MeV}$. As a result, a neutrino isocurvature perturbation can only be present if there is a very large lepton number density, with a significant isocurvature perturbation $S_L \equiv \delta(n_L/n_{\rm r})/(n_L/n_{\rm r})$.

To calculate S_ν in terms of S_L, we need to calculate the effect of a neutrino chemical potential, defined just before neutrino decoupling, upon the neutrino density after neutrino decoupling. To a good approximation, the distribution function of the neutrino maintains its thermal equilibrium form, and to leading order in the chemical potential we have [1]

$$\frac{\rho_\nu}{\rho_\gamma} = \frac{21}{8}\left(\frac{T_\nu}{T_\gamma}\right)^4\left[1 + \frac{30}{7}\left(\frac{\mu_\nu}{\pi T_\nu}\right)^2\right]. \tag{12.29}$$

If n_L vanishes, so does μ_ν and there is no neutrino isocurvature perturbation. Using Eq. (4.13) this gives

$$\boxed{S_\nu = \frac{45}{7}(1 - R_\nu)\left(\frac{\mu_\nu}{\pi T_\nu}\right)^2 S_L = 0.46\left(\frac{n_L}{n_\gamma}\right)^2 S_L}\,. \tag{12.30}$$

Since S_L will be at most of order 1, we need n_L/n_γ not too far below its observational bound if the isocurvature perturbation is to be significant. In other words, the lepton number density has to be many orders of magnitude bigger than the baryon number density n_B. Now comes the problem: we see in Chapter 21 that thermal equilibrium in the early Universe is likely to give $n_B/n_L \sim 1$, requiring $n_L \sim 10^{-9} n_\gamma$.

In principle there is an escape route from the relation (12.30), coming from the fact that the perturbations on each scale only become directly observable as horizon entry for that scale is approached. Horizon entry occurs well after neutrino decoupling for the larger and best-explored cosmological scales. In principle one could imagine that somehow there is significant neutrino creation *after* the neutrino decoupling yet *before* horizon entry for the larger cosmological scales. In practice, though, significant modification of the standard cosmology from nucleosynthesis onwards is very hard to achieve. From a theoretical viewpoint the energy is too low to expect anything much to happen, and there are also strong observational constraints. To date there is no proposal for creating a neutrino isocurvature perturbation in this way.

The same remarks apply more strongly to a different type of neutrino isocurvature mode, known as the neutrino velocity isocurvature mode [2]. The initial condition for this is a non-zero value for $V_\nu - V_\gamma$, corresponding to relative motion between the photons and neutrinos.[2] Such a perturbation cannot exist before decoupling since the frequent collisions prevent it. It would have to be created between the primordial epoch and horizon entry. But the creation of significant bulk motion before horizon entry is impossible in a gaseous universe (because there can be no significant particle movement), and is impossible quite generally if the separate universe picture is valid. No proposal for creating a neutrino velocity isocurvature mode exists at present, nor does one seem likely.

[2] The mode corresponds to decaying gravitational potentials, but the resulting singularity at early times can be avoided by using a different gauge.

12.5 Tensor mode

12.5.1 Primordial tensor perturbation

A primordial tensor perturbation is inevitably created during inflation at some level. After horizon entry it oscillates as a gravitational wave of primordial origin, and one day it may become possible to directly observe this wave. But for the time being the best hope of detecting the tensor mode is through the CMB anisotropy. Some classes of inflation models predict a high enough amplitude for the tensors that they will be detectable in this way.

In the tensor mode, the line element is

$$\boxed{ds^2 = a^2(\eta)\left[-d\eta^2 + (\delta_{ij} + 2h_{ij})\,dx^i dx^j\right]}\,, \tag{12.31}$$

where h_{ij} is traceless and transverse. The Einstein field equation gives

$$\boxed{\ddot{h}_{ij} + 2aH\dot{h}_{ij} + k^2 h_{ij} = 8\pi G\Sigma^{\mathrm{T}}_{ij}}\,, \tag{12.32}$$

where Σ^{T}_{ij} is the traceless and transverse part of the anisotropic stress. The metric perturbation has the decomposition (8.6) so that it is specified by two scalars $h_{+,\times}(\mathbf{k},\eta)$. Similarly, Σ^{T}_{ij} is specified by two scalars $\Sigma_{+,\times}(\mathbf{k},\eta)$.

Well before horizon entry the anisotropic stress is supposed to vanish. Then Eq. (12.31) says that $h_{+,\times}(\mathbf{k},\eta)$ have time-independent values. These define the primordial tensor perturbation and we shall denote them by $h_{+,\times}(\mathbf{k})$ without an argument.

Within the inflationary there is no correlation between tensor and scalar perturbations. Also, the stochastic properties of the tensor perturbation are generally assumed to be invariant under the parity transformation. Assuming invariance under the parity transformation, $\langle h_+(\mathbf{k})h_\times(\mathbf{k}')\rangle$ vanishes and we can define the spectrum $\mathcal{P}_h$ of the primordial tensor perturbation by

$$\boxed{16\langle h_+(\mathbf{k}))h_+(\mathbf{k}')\rangle = 16\langle h_\times(\mathbf{k})h_\times(\mathbf{k}')\rangle = (2\pi)^3\frac{2\pi^2}{k^3}\mathcal{P}_h(k)\delta^3(\mathbf{k}-\mathbf{k}')}\,. \tag{12.33}$$

The tensor fraction is defined as

$$\boxed{r \equiv \frac{\mathcal{P}_h}{\mathcal{P}_\zeta}}\,. \tag{12.34}$$

At the time of writing there is no detection, only an upper bound $r \lesssim 0.1$. Barring a detection, upcoming observations will push this limit down to $r \lesssim 10^{-2}$ to 10^{-3}, but it will be practically impossible to do better.

In Section 24.7 we see how the non-gaussianity of the tensor modes (correlation between modes of different $\mathbf{k}$) is calculated within the inflationary cosmology, as well as the correlation between tensor and scalar modes. At least within the context of general relativity these correlations are very small, and will surely be impossible to observe.

12.5.2 Tensor Sachs–Wolfe effect

As we are in the tensor mode, there is no perturbation in the energy densities of the components of cosmic fluid, nor in their velocities. Adopting the sudden-decoupling approximation, the temperature anisotropy comes entirely from the Sachs–Wolfe effect, which may be ascribed to the successive redshifts seen by a sequence of observers. Well after matter domination the stress perturbation is completely negligible. On scales entering the horizon well after matter domination the solution of Eq. (12.32) is

$$h_{+,\times}(\eta) = \left[3\sqrt{\frac{\pi}{2}} \frac{J_{3/2}(k\eta)}{(k\eta)^{3/2}} \right] h_{+,\times}. \tag{12.35}$$

The fractional perturbation in the distance between adjacent observers, with separation $\delta\mathbf{r}$, is $h_{ij}e_i e_j/2$, where $\mathbf{e} = \delta\mathbf{r}/|\delta\mathbf{r}|$. The perturbation in the velocity gradient is therefore $\partial h_{ij}/\partial t$, and the gravitational Sachs–Wolfe effect is

$$\boxed{\frac{\delta T(\mathbf{e})}{T} = -\frac{1}{2} \int_{\eta_{\rm ls}}^{\eta_0} e_i e_j \frac{\partial h_{ij}(x,\eta)}{\partial \eta} d\eta}, \tag{12.36}$$

where the photon trajectory is $x(\eta) = \eta_0 - \eta$.

The dominant contribution is in the low multipoles $\ell \ll 100$ that correspond to scales well outside the horizon at decoupling. For higher multipoles, corresponding to smaller scales, the amplitude of the gravitational waves at last scattering has been reduced from its primordial value by the redshift.

For $\ell \ll 100$, the time dependence of $h_{+,\times}$ is given by Eq. (12.35), and we can take $\eta_{\rm ls} = 0$ under the pretence that matter domination extends into the infinite past. Taking $\mathcal{P}_h$ to be scale-independent, the Sachs–Wolfe effect alone gives [3]

$$\ell(\ell+1)C_\ell = \frac{\pi}{36}\left(1 + \frac{48\pi^2}{385}\right)\mathcal{P}_h c_\ell, \tag{12.37}$$

where $c_2 = 1.118$, $c_3 = 0.878$, and $c_4 = 0.819$ with $c_\infty = 1$. This approximation is good also if $\mathcal{P}_h(k)$ has moderate scale dependence, provided that it is evaluated at the scale $k \simeq \ell H_0/2$, which dominates the ℓth multipole.

The full calculation of the tensor contribution, described in the next subsection, uses the Boltzmann equation. Figure 12.3 gives the result of that calculation and compares it with the adiabatic mode. The spectra $\mathcal{P}_\zeta$ and $\mathcal{P}_h$ are those predicted by

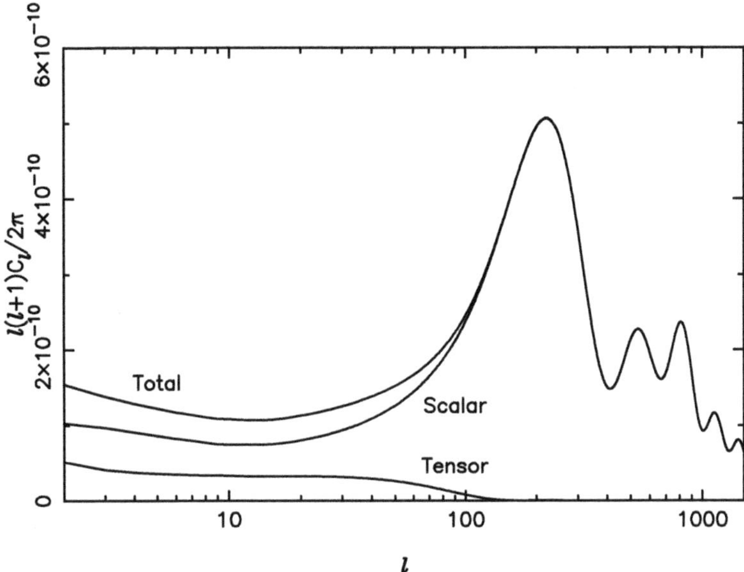

Fig. 12.3. The scalar and tensor temperature anisotropies for chaotic inflation with a quadratic potential, corresponding to tensor fraction $r = 0.16$.

a particular inflation model (ϕ^2 chaotic inflation), which gives a practically scale-invariant $\mathcal{P}_h$ corresponding to $r = 0.16$. The spectrum $\mathcal{P}_\zeta$ is normalized to fit the data, and has the spectral index $n = 0.96$ predicted by the model.

12.5.3 Tensor Boltzmann equation and polarization

Boltzmann equation

In the tensor mode we define $\Theta_\ell(\mathbf{k})$ by Eq. (10.19), except that $Y_{\ell 0}(\mathbf{n})$ is replaced by $Y_{\ell 2}(\mathbf{n})$. Similarly, E_ℓ and B_ℓ are defined by Eq. (11.51), except that $Y^{\pm}_{\ell 0}$ is replaced by $Y^{\pm}_{\ell 2}$.[3] The initial condition we take $h \equiv -(h_+ + ih_\times)/\sqrt{6}$ to be non-zero, and $h_+ - ih_\times$ to be zero. (As we assume parity invariance of the stochastic properties we could have reversed the roles of $h_+ \pm ih_\times$.)

To work out the Boltzmann hierarchy we have to project out these multipoles from Eqs. (11.60) and (11.61) to find the collision term. We will use the notation

[3] We are using the conventions of Ref. [4] except that their Θ_ℓ, E_ℓ and B_ℓ are $2\ell + 1$ times ours.

of (A.25), which allows us to write the hierarchy in the following form:

$$(2\ell+1)\dot{\Theta}_\ell = k\left({}_0\kappa_\ell^m\Theta_{\ell-1} - {}_0\kappa_{\ell+1}^m\Theta_{\ell+1}\right) - \dot{\tau}\Theta_\ell + S_\ell, \tag{12.38}$$

$$\begin{aligned}(2\ell+1)\dot{E}_\ell &= k\left[{}_2\kappa_\ell^m E_{\ell-1} - \frac{2m(2\ell+1)}{\ell(\ell+1)}B_\ell - {}_2\kappa_{\ell+1}^m E_{\ell+1}\right] \\ &- (2\ell+1)\dot{\tau}E_\ell - \sqrt{6}\,\dot{\tau}P\delta_{\ell 2},\end{aligned} \tag{12.39}$$

$$(2\ell+1)\dot{B}_\ell = k\left[{}_2\kappa_\ell^m B_{\ell-1} - \frac{2m(2\ell+1)}{\ell(\ell+1)}E_\ell - {}_2\kappa_{\ell+1}^m B_{\ell+1}\right] - \dot{\tau}B_\ell. \tag{12.40}$$

As before, $10P \equiv \Theta_2 - \sqrt{6}E_2$. The last term of the first expression is

$$S_\ell \equiv \left(\dot{\tau}P - \dot{h}\right)\delta_{\ell 2}. \tag{12.41}$$

The terms proportional to $\dot{\tau}$ are collision terms, and the term proportional to $\dot{h}$ represents the Sachs–Wolfe effect.

Written in this form, Eqs. (12.38)–(12.40) are valid in all three modes ($m = 0$, 1 and 2 corresponding to scalar, vector, and tensor) with different S_ℓ. The same equations apply to the neutrinos, without collision terms.

The evolution of h is determined by Eq. (12.32), with the anisotropic stress coming from the photons and neutrinos. It may be written

$$\ddot{h}_{ij} + 2aH\dot{h}_{ij} + k^2 h_{ij} = (aH)^2\left[(1-R_\nu)\Pi_\gamma + R_\nu\Pi_\nu\right]. \tag{12.42}$$

The photon anisotropic stress is given by $\Pi_\gamma = 8\theta_2$ and the same expression holds for Π_ν with θ_2 referring to the neutrinos.

In the tensor mode we define a temperature transfer function by

$$\boxed{\Theta_\ell(\mathbf{k}) = T_\ell(k)h(\mathbf{k})}, \tag{12.43}$$

and the polarization transfer functions $T_{E\ell}(k)$ and $T_{B\ell}(k)$ are defined in the same way. It can be shown, using the same unitarity argument as for the scalar mode, that the gaussianity of $h(\mathbf{k})$ means that each the CMB multipole has an independent gaussian distribution. The correlators C_ℓ, C_ℓ^{TE} and C_ℓ^{EE} are given by Eqs. (10.25), (11.70) and (11.71) and there is a similar expression C_ℓ^{BB}. Figure 12.4 shows the correlators of the polarization.

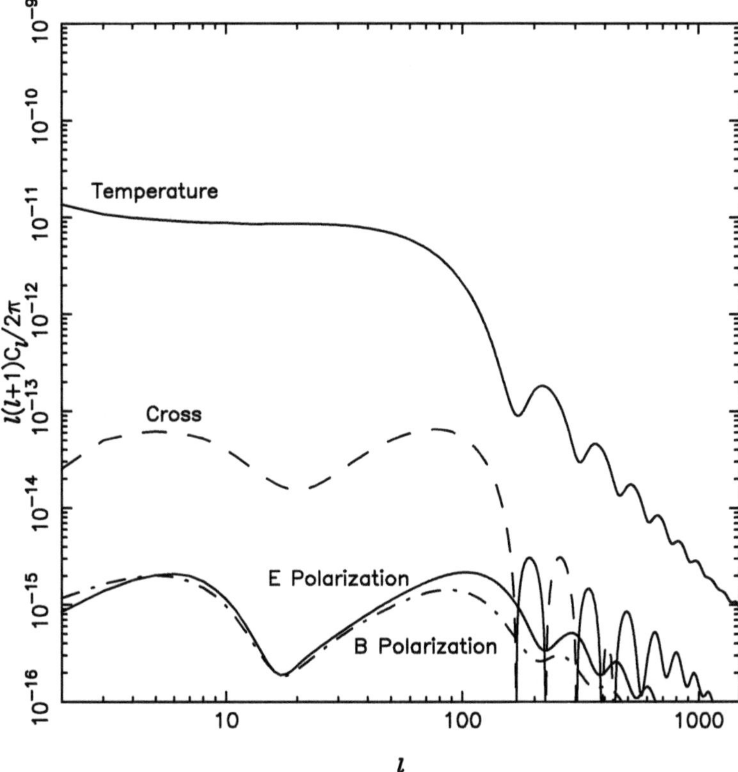

Fig. 12.4. The CMB temperature and polarization produced by a tensor perturbation with tensor fraction $r = 0.16$.

Line-of-sight integrals

The line-of-sight integrals for the tensor mode are

$$\boxed{\Theta_\ell = \int_0^{\eta_0} d\eta \left[-g(\eta)P - e^{-\tau}\dot{H} \right] j_\ell^{22}}, \tag{12.44}$$

$$\boxed{E_\ell = \sqrt{6} \int_0^{\eta_0} d\eta g(\eta) P \epsilon_\ell^2}, \tag{12.45}$$

$$\boxed{B_\ell = \sqrt{6} \int_0^{\eta_0} d\eta g(\eta) P \beta_\ell^2}. \tag{12.46}$$

The functions at the end of each expression are defined in Appendix A, and their argument is $k(\eta_0 - \eta)$.

Making the sudden-decoupling approximation $g = \delta(\eta - \eta_{\rm ls})$, we can understand the main features seen in Figure 12.4. At high ℓ, quite a good approximation is

$$\left|\frac{\beta_\ell^2}{\epsilon_\ell^2}\right|^2 = \frac{8}{13}, \qquad (12.47)$$

and then C_ℓ^{BB}/C_ℓ^{EE} is in the same ratio, in rough agreement with Figure 12.4.

As with the scalar case, we can use the line-of-sight integral to obtain a sudden-decoupling approximation for the polarization. Repeating the procedure there we find

$$\frac{4}{\sqrt{6}} E_2 = -\Theta_2 = \frac{20}{3}\frac{\dot{h}}{\dot{\tau}}. \qquad (12.48)$$

Also, very roughly, $|\epsilon_{\ell 2}|^2 \sim |j_\ell^{22}|^2$. This allows one to estimate C_ℓ^{EE}/C_ℓ at high ℓ, in rough agreement with Figure 12.4. The effect of reionization can also be estimated, analogously with the scalar case.

We end this discussion of the CMB anisotropy by mentioning the interaction of the gravitational wave with the neutrinos. In contrast with the tightly-coupled photons, the momentum distribution of the neutrinos has a quadrupole component at all times, which drains energy away from the gravitational waves. The effect may be significant for the polarization though not for the temperature anisotropy.

12.6 Seeds and the vector mode

We have been assuming that the cosmic fluid is a gas, but it may have other components. Magnetic fields certainly exist at present, on scales up to the size of galaxies. If they exist on larger scales and at earlier times, they will affect the evolution of cosmological perturbations, giving a possibly observable effect. Also there may exist cosmic strings, or other topological defects like domain walls and textures, as described in Section 21.4. These also may affect the evolution of the perturbations with an observable consequence.

The inclusion of magnetic fields and topological defects into the analysis is relatively easy, because their evolution isn't significantly affected by the usual cosmological perturbations (i.e. those of the gas). Objects with that property are sometimes called seeds. Once the evolution of the seed is specified, its effect on ordinary cosmological perturbations can be calculated and compared with observation. In addition to the scalar and tensor modes, there will be a vector mode.

The vector mode of relativistic perturbations resembles the vector mode of Newtonian perturbations. The flow of the cosmic fluid has vorticity, but no shear and no perturbation in the expansion rate. There is no perturbation in the energy density or pressure, only in the anisotropic stress. The crucial property of the vector mode is that it decays in a purely gaseous universe, requiring the seeds to support it.

For the vector mode we can choose the gauge where the only non-zero metric perturbations are h_{i0}. The non-zero Fourier components with the z axis pointing along $\mathbf{k}$ can be taken to be $h_{x0} = -ih_{y0} \equiv W/\sqrt{2}$.[4] We also need the baryon fluid velocity, whose non-zero Fourier components are $v_x = -iv_y \equiv V_B/\sqrt{2}$. As with the tensor mode, the main observational effect of the vector mode will be in the CMB anisotropy.[5] The Boltzmann hierarchy is given by Eqs. (12.38)–(12.40) with

$$S_1 = \dot{\tau} V_B + \dot{W} \qquad S_2 = \dot{\tau} P. \tag{12.49}$$

The line-of-sight integral for Θ_ℓ is (for $\ell \geq 2$)

$$\Theta_\ell = \int_0^{\eta_0} d\eta \left\{ g(\eta) \left[(V_B - W) j_\ell^{11} + P j_\ell^{21} \right] + e^{-\tau} \frac{1}{\sqrt{3}} k W j_\ell^{21} \right\}, \tag{12.50}$$

and those for E_ℓ and B_ℓ are given by Eqs. (12.45) and (12.46) with ϵ^1 and β^1 instead of ϵ^2 and β^2.

Making the sudden-decoupling approximation $g = \delta(\eta - \eta_{\rm ls})$ we deduce the main features of the CMB multipoles in the vector mode. At high ℓ

$$\left| \frac{\beta_\ell^1}{\epsilon_\ell^1} \right|^2 \simeq 6, \tag{12.51}$$

giving C_ℓ^{BB}/C_ℓ^{EE} in the same ratio (recall that the ratio was only $\simeq 8/13$ for the tensor mode). Also,

$$\frac{4}{\sqrt{6}} E_2 = -\Theta_2 = -\frac{4\sqrt{3}}{9} \frac{k}{\dot{\tau}} \Theta_1, \tag{12.52}$$

as in the scalar case except for a numerical factor. This allows one to estimate the relative magnitude of the polarization and temperature anisotropy.

12.7 Spatial curvature

We have so far taken the background to be spatially flat, corresponding to $\Omega_0 = 1$. As we saw in Section 4.5 this assumption is justified because observation gives a tight bound on $|\Omega_0 - 1| \lesssim 10^{-2}$. To obtain that bound though, one has to know how to obtain predictions for $\Omega_0 \neq 1$. In this section we explain briefly how that is done.

In a closed or open universe the line element is given by Eq. (3.65) or (3.66). According to these expressions, the effect of spatial curvature becomes big at co-moving distances

$$x \sim |K|^{-1/2} = |1 - \Omega_0|^{-1/2} H_0^{-1}. \tag{12.53}$$

[4] Vector mode quantities depend on the slicing but not the threading. Our W is the V of Ref. [4].
[5] The vorticity of the fluid velocity field should be detectable in principle.

If $|\Omega_0 - 1|$ were not small, this would grossly distort the geometry of the nearby Universe (redshift $z \ll 1$) which we know from galaxy surveys is not the case. In considering the CMB anisotropy, we need therefore consider only the regime $|\Omega_0 - 1| \lesssim 1$.

In the presence of spatial curvature, the Fourier decomposition doesn't make sense, and one uses instead the spherical decomposition (6.58) with the radial functions $j_\ell(kx)$ functions modified so that $Z_{k\ell m}$ is still an eigenfunction of the Laplacian with eigenvalue $-k^2$.[6] For $\Omega_0 > 1$, corresponding to a closed universe, the k^2 takes on the discrete values $k^2/K = n^2 - 1 - |m|$ with $n = 3, 4, \ldots$. and $|m| = 0, 1, 2$ for respectively the scalar, vector and tensor modes. For $\Omega_0 < 1$, any value $k^2 > 0$ is allowed.[7]

With the modified radial functions in place, the Boltzmann hierarchy and the line-of-sight integrals go through with fairly minor changes [6]. By far the most important effect of spatial curvature is on the CMB anisotropy, because it alone provides detailed information about the Universe on distance scales of order H_0^{-1}. The effect of spatial curvature on the scales explored in detail by galaxy surveys is by comparison negligible. That doesn't mean, though, that measurements of the CMB alone give a good constraint on Ω_0. Such is not the case, because the effect of a given change in Ω_0 on the CMB could be practically cancelled by a change in other cosmological parameters. To get a good bound on Ω_0, one must consider both the CMB anisotropy Ω_0 and data from galaxy surveys. The bound $|1 - \Omega_0| \lesssim 10^{-2}$ comes from fitting these data, assuming the adiabatic initial condition and the ΛCDM model modified by the addition of the extra parameter $|1 - \Omega_0|$. Allowing additional parameters, in particular isocurvature and tensor perturbations make the bound weaker, but the effect is not dramatic unless many parameters are allowed.

If $|1 - \Omega_0|$ is really of order 10^{-2}, the curvature scale is only of order $10H_0^{-1}$. With such a low scale, the discussion in Section 10.6.2 of the Grishchuk–Zel'dovich effect has to be redone [5], though there is no change in the basic conclusion that the absence of an observed quadrupole enhancement rules out a sharp increase in $\mathcal{P}_\zeta(k)$ as k^{-1} goes outside the horizon. However, we will see in Part IV that $|1 - \Omega_0|$ is actually expected to be many orders of magnitude below 1 within the inflationary cosmology.

[6] The Laplacian is the spatial part of the d'Alembertian (3.22).

[7] The eigenfunctions with $k^2 > 1$ form a complete set, but all $k^2 > 0$ have to be included to give the most general random field [5].

Exercises

12.1 Verify Eq. (12.21), giving the CMB dipole generated by a very large scale matter isocurvature perturbation.

12.2 Estimate C_ℓ^{EE}/C_ℓ in the tensor mode, using the method mentioned after Eq. (12.48).

12.3 Estimate C_ℓ^{EE}/C_ℓ in the vector mode, using the method mentioned after Eq. (12.52).

References

[1] D. H. Lyth, C. Ungarelli and D. Wands. The primordial density perturbation in the curvaton scenario. *Phys. Rev.,* D **67** (2003) 023503.

[2] M. Bucher, K. Moodley and N. Turok. The general primordial cosmic perturbation. *Phys. Rev.,* D **62** (2000) 083508.

[3] A. A. Starobinsky. Cosmic background anisotropy induced by isotropic flat-spectrum gravitational-wave perturbations. *Sov. Astron. Lett.*, **11** (1985) 133.

[4] W. Hu and M. J. White. CMB anisotropies: total angular momentum method. *Phys. Rev.,* D **56** (1997) 596.

[5] D. H. Lyth and A. Woszczyna. Large scale perturbations in the open universe. *Phys. Rev. D* **52** (1995) 3338.

[6] W. Hu, U. Seljak, M. J. White and M. Zaldarriaga. A complete treatment of CMB anisotropies in a FRW universe. *Phys. Rev. D* **57** (1998) 3290.

Part III

Field theory

13

Scalar fields and gravity

In Part III we deal with some essential aspects of field theory. Except in the last two sections of the present chapter, the entire discussion is in flat spacetime. After a brief introduction to field theory in general, in this chapter we deal with the classical theory of scalar fields.

13.1 Field theory

13.1.1 Basic concepts

A quantum field theory generally yields various species of elementary particle, each corresponding to one of the fields. In particular the photon corresponds to the electromagnetic field. Fields are classified as bosonic or fermionic according to the spin of the particle. One generally considers only fields corresponding to particles with spin-0 (scalar fields) and spin-$1/2$ particles (spinor fields), plus a special type of field corresponding to particles with spin-1 (vector field) known as a gauge field.[1] The particles corresponding to gauge fields are called gauge bosons. One doesn't generally consider fields of higher spin, because these are difficult to accommodate within the theory and the corresponding elementary particles are not observed. The only exceptions are the spin-2 graviton that is supposed to correspond to a quantized weak gravitational wave, and the spin-$3/2$ gravitino that will be its partner in a supergravity theory.

At the quantum level, each field corresponds to an operator. Bosonic fields can be regarded as classical in a suitable regime. The fields live in curved spacetime described by the metric tensor $g_{\mu\nu}$. The metric tensor is in general regarded as a classical field, there being no consistent quantization for $g_{\mu\nu}$. (An exception is the weak gravitational wave that is presumed to correspond to the graviton.)

The term 'quantum field theory' generally indicates a theory which is formulated

[1] The terms scalar, vector and spinor refer to the Lorentz transformation of the fields; scalar fields are scalars, vector fields are 4-vectors, and spinor fields have the transformation (15.48) given later.

in flat spacetime, the equivalence principle being used to generalize the theory to curved spacetime. As we write the Standard Model of particle physics describes the interactions of all known elementary particles, after some as yet unknown extension to accommodate neutrino mass. A satisfactory description of the early Universe requires further extension of the Standard Model.

Scalar fields can play a number of roles. Their values may determine the masses and couplings of elementary particles. In particular, the present values of one or more Higgs fields are supposed to determine the masses of all presently known elementary particles except the neutrino. The mass of each particle is proportional to the strength of its interaction with the Higgs field. In the Standard Model the couplings are fixed quantities, but extensions of the Standard Model suggested by string theory contain additional scalar fields called moduli, whose values determine the couplings.

The dependence of masses and couplings on scalar fields leads to the concept of spontaneous symmetry breaking. The idea is that a field theory may possess a high degree of symmetry for special values of the scalar fields, which is broken when these fields acquire different values. Zero values for the Standard Model Higgs field would unify the electromagnetic and weak interactions.

Scalar fields also appear in the context of supersymmetry. Supersymmetry pairs each spin-$1/2$ field with scalar partners (one for each spin state) and also pairs each spin-1 (gauge) field with a spin-$1/2$ field. It is motivated by the desire to stabilize quantum field theory against quantum corrections, and is possible though not mandatory according to string theory.

All of this is very relevant for cosmology, because according to present ideas the values of some scalar fields in the early Universe are likely to change with time. During inflation, scalar fields are usually supposed to be the only significant component of the cosmic fluid. At that stage, the value of at least one scalar field is supposed to be very different from its present value. After inflation, scalar fields may again play a crucial role as they settle down to their present values.

13.1.2 *Effective field theory*

A quantum field theory is supposed to be an effective one, valid only for energy scales below some ultra-violet cutoff $\Lambda_{\rm UV}$. Equivalently, with units $c = \hbar = 1$, it is valid only for distance scales bigger than $\Lambda_{\rm UV}^{-1}$. Roughly speaking, the effective field theory will describe particle collisions with centre of mass energy up to $\Lambda_{\rm UV}$, and will describe the early Universe back to the epoch with energy density $\Lambda_{\rm UV}^4$.

A given effective field theory may yield as an approximation a different effective field theory, with a lower cutoff. Heavy elementary particles appearing in the former (high-energy) theory may not appear in the latter (low-energy) theory. On the other hand, bound states of the high-energy theory may become 'elementary particles' in the low-energy theory. More generally, a function of fields in the high-energy theory may appear simply as a field in the low-energy theory.

There is presumably a maximum cutoff, above which no field theory makes sense. At higher energies and smaller distances, a more complete theory such as string theory presumably applies.

With Einstein gravity, the maximum cutoff cannot exceed the Planck scale $M_{\rm Pl}$. This is because quantum field theory and general relativity come into conflict at that scale. For example, if we use flat spacetime field theory to calculate the vacuum fluctuation of a massless scalar field within a region of size R, we find that it carries *rms* energy $E \sim 1/R$. (This is required by the dimensions since R is the only relevant quantity, and is seen explicitly from the calculation that we describe in Section 15.4.) If we assume that the theory is valid down to $R \lesssim M_{\rm Pl}^{-1}$, the vacuum fluctuation gives $E \gtrsim M_{\rm Pl}^2 R$. But then R is below the Schwarzschild radius, meaning that spacetime has strong curvature corresponding to the existence of black holes. There doesn't seem to be any way of handling this 'spacetime foam' in the context of field theory and general relativity.

As we see in Section 16.4, the measured values of the gauge couplings suggest the existence of a Grand Unified (field) Theory or GUT. This would require the maximum cutoff for field theory to be bigger than 10^{16} GeV. The possibility of a much smaller maximum cutoff is nevertheless widely considered.

13.1.3 String theory and the landscape

According to string theory, the fundamental objects are one-dimensional strings living in nine space dimensions. There are also objects called p branes, with p space dimensions $0 < p \leq 9$. (The term 'brane' derives from 'membrane'.) The most important type are called D-p branes, or just D branes if the dimension is not specified. The electromagnetic, weak and strong forces that we experience might be confined to a particular D-3 brane, or to a three-dimensional intersection of D-p branes, with gravity able to penetrate to the region outside, known as the bulk. An important role may be played by D strings, which are D branes with just one of our space dimensions. A higher dimensional spacetime containing branes is usually called a braneworld, and is a typical setting for string theory.

As the energy scale is reduced below some 'string scale', string theory will yield 3-dimensional field theory as an approximation. (Recall that the energy scale might be identified with the centre of mass energy for collisions, or with $\rho^{1/4}$ in the early Universe.) The maximum cutoff for the field theory will be set by the string scale.

It is not expected that string theory will yield a unique field theory. Rather, it is expected that string theory will yield a huge number of different possible field theories corresponding to different vacuum states. Even field theory by itself may yield a huge number of possible vacuum states, corresponding to minima of the scalar field potential that we come to later. The set of possible vacua is known as the **landscape**.

Our Universe corresponds to just one of the vacua. At first sight, one might hope that this vacuum is selected by some initial condition, or by being the lowest energy level. But that may not work, because at the quantum level each vacuum may tunnel into all others given enough time.

The possible vacua (constituting the landscape) are a particular example of what has been called a **multiverse**, meaning an ensemble of possible universes more or less (maybe a lot less) like ours. The other universes may exist in reality, as in the case we are considering at the moment. Alternatively the other universes might exist only as possible quantum states, as may be the case for the other possible realizations of the ensemble of perturbed universes that we consider in Section 24.2.

When considering a multiverse, one should remember that many universes may be inhospitable to life, offering the possibility that the observable Universe is the way it is because we are here to observe it. Versions of this idea are referred to as the **anthropic principle**. We shall see how anthropic considerations may be relevant for the cosmological constant (Section 23.5), the origin of cold dark matter (CDM, Section 23.1), inflation (Section 20.1), and the origin of primordial perturbations (Section 25.1).

Although it may be attractive, an anthropic explanation for any particular observable raises the spectre that the same explanation might apply to lots of other observables, leaving little left for theoreticians to do in the future. Arguably, even parameters like particle masses and coupling constants may take on practically any value as one moves around the landscape, rendering superfluous any attempt to explain them from first principles. There is at present no consensus on this issue.

Notwithstanding these conceptual difficulties, string theory and in particular the braneworld are widely considered, because a given setup may lead to fairly definite predictions. In particular, it may determine the form of the field theory holding below the string scale, and it may also generate a departure from Einstein gravity.

13.2 Action and Lagrangian

As far as we know, the fundamental laws of physics can be derived from an **action**

$$\boxed{S = \int_{-\infty}^{\infty} L \, dt} \,. \qquad (13.1)$$

The Lagrangian L depends on the 'coordinates' q_n (also called 'degrees of freedom') needed to specify the system under consideration, and their time derivatives $\dot{q}_n$. A realistic theory will usually involve only first derivatives. The Lagrangian has dimensions of energy, corresponding to a dimensionless action, and this fixes the dimensions of the degrees of freedom. (Remember that we are setting $\hbar = c = 1$.)

The classical evolution of the degrees of freedom is given by the action principle. This states that $\delta S = 0$, where δS is the change in S resulting from a small change in the time dependence of the 'coordinates'. The change is arbitrary, except for a boundary condition that depends on the system.

Field theory has an infinite number of degrees of freedom. For the rest of this section we warm up by considering first one degree of freedom, and then a finite number. We pay particular attention to the case of a harmonic oscillator, which in the field theory case will correspond to a Fourier component of a field which has no interactions.

13.2.1 Single degree of freedom

We take the Lagrangian to be a function only of q and $\dot{q}$, but allow explicit t-dependence; $L(q, \dot{q}, t)$. The evolution of q is given by the action principle $\delta S = 0$ with

$$\delta S = \int_{-\infty}^{\infty} dt \left(\frac{\partial L}{\partial q} \delta q + \frac{\partial L}{\partial \dot{q}} \delta \dot{q} \right) . \tag{13.2}$$

Integrating the second term by parts gives

$$\delta S = \int_{-\infty}^{\infty} dt \left[\frac{\partial L}{\partial q} - \frac{\partial}{\partial t} \left(\frac{\partial L}{\partial \dot{q}} \right) \right] \delta q + \int_{-\infty}^{\infty} dt \frac{\partial}{\partial t} \left(\frac{\partial L}{\partial \dot{q}} \delta q \right) . \tag{13.3}$$

The second term is just a number which is required to vanish by the action principle. To ensure that we can demand that $\delta q(t)$ vanishes sufficiently rapidly at infinity. Alternatively we can take the integral to run over only a finite time interval say $-T < t < T$, with q and $\dot{q}$ vanishing or periodic at the boundaries $t = \pm T$, with $T \to \infty$ at the end of the calculation. Dropping the second term, the action principle $\delta S = 0$ leads to

$$\boxed{\frac{\partial L}{\partial q} - \frac{\partial}{\partial t} \left(\frac{\partial L}{\partial \dot{q}} \right) = 0} . \tag{13.4}$$

This type of equation, generalized to more degrees of freedom, seems to have been chosen by Nature for a wide variety of systems. It is a second-order differential equation that can be solved given an initial value for q and $\dot{q}$.

Instead of the Lagrangian L one can work with the Hamiltonian H. The relation between them is given by

$$\boxed{H(q,p,t) \equiv p\dot{q}(q,p,t) - L(q,\dot{q}(q,p,t),t)}, \qquad \boxed{p = \partial L/\partial \dot{q}} \tag{13.5}$$

The variable p is called the **canonical momentum**. The Lagrangian equation of motion (13.4) is equivalent to

$$\boxed{\dot{q} = \frac{\partial H}{\partial p}} \qquad \boxed{\dot{p} = -\frac{\partial H}{\partial q}}. \tag{13.6}$$

According to these equations, the time dependence of any function $A(q,p,t)$ is given by

$$\frac{dA}{dt} = \frac{\partial H}{\partial p}\frac{\partial A}{\partial q} - \frac{\partial A}{\partial p}\frac{\partial H}{\partial q} + \frac{\partial A}{\partial t}. \tag{13.7}$$

Equation (13.7) shows that H is time independent (conserved) if it has no explicit time dependence. It is then, by definition, the energy of the system. No explicit time dependence for H is equivalent to L having no explicit time dependence, which in turn is equivalent to the invariance of the action under the coordinate changes (time translations) $q(t) \to q(t+T)$ where T is any constant. Therefore *energy is conserved if and only if the action is invariant under time translations.* This holds for any number of degrees of freedom.

For a non-relativistic particle with unit mass, moving in one dimension under the influence of a potential, the Lagrangian, Hamiltonian and canonical momentum are

$$L = \frac{1}{2}\dot{q}^2 - V(q), \qquad H = \frac{1}{2}p^2 + V(q), \qquad p = \dot{q}. \tag{13.8}$$

These give the Newtonian equation of motion

$$\ddot{q} + \frac{dV(q)}{dq} = 0. \tag{13.9}$$

The Hamiltonian is conserved, being the sum of the kinetic and potential energies.

For a harmonic oscillator we have

$$L = \frac{1}{2}\dot{q}^2 - \frac{1}{2}\omega^2 q^2, \qquad H = \frac{1}{2}p^2 + \frac{1}{2}\omega^2 q^2, \qquad p = \dot{q}. \tag{13.10}$$

The equation of motion is

$$\ddot{q} + \omega^2 q = 0. \tag{13.11}$$

The solutions of Eq. (13.11) correspond to harmonic oscillation with angular frequency ω, and we may call any system with this Lagrangian a harmonic oscillator. If q is the position of a particle with unit mass, the potential is $\omega^2 q^2/2$ corresponding to force $-\omega^2 q$.

With a view to quantization, it is useful to write

$$q = \sqrt{\frac{1}{2\omega}}\left(ae^{-i\omega t} + a^* e^{i\omega t}\right). \tag{13.12}$$

Then

$$H = \omega |a|^2. \tag{13.13}$$

13.2.2 N degrees of freedom

Consider next N degrees of freedom, with L a function of the coordinates $q_n(t)$ and their first derivatives. Now Eq. (13.4) becomes a system of N coupled equations;

$$\frac{\partial L}{\partial q_n} - \frac{\partial}{\partial t}\left(\frac{\partial L}{\partial \dot{q}_n}\right) = 0. \tag{13.14}$$

The Hamiltonian is

$$H(q_n, p_n, t) \equiv \sum p_n \dot{q}_n - L, \tag{13.15}$$

with $p_n \equiv \partial L/\partial \dot{q}_n$. The Hamiltonian equations of motion are

$$\dot{q}_n = \frac{\partial H}{\partial p_n} \qquad \dot{p}_n = -\frac{\partial H}{\partial q_n}. \tag{13.16}$$

Also, Eq. (13.7) becomes

$$\frac{dA}{dt} = \sum_n \left(\frac{\partial H}{\partial p_n}\frac{\partial A}{\partial q_n} - \frac{\partial A}{\partial p_n}\frac{\partial H}{\partial q_n} + \frac{\partial A}{\partial t}\right). \tag{13.17}$$

If L has no explicit time dependence the action is invariant under time translations. Then H is time independent by virtue of the equation of motion, and is by definition the energy of the system.

If $L = \sum L_n(q_n, \dot{q}_n)$, then each q_n evolves independently of the others so that we deal with N uncoupled systems. In particular, if

$$L = \sum_n \frac{1}{2}\dot{q}_n^2 - \frac{1}{2}\omega_n^2 q_n^2, \qquad p_n = \dot{q}_n, \tag{13.18}$$

$$H = \sum_n \frac{1}{2}p_n^2 + \frac{1}{2}\omega_n^2 q_n^2 = \sum_n \omega_n |a_n|^2, \tag{13.19}$$

$$q_n = \sum_n \sqrt{\frac{1}{2\omega_n}}\left(a_n e^{-i\omega_n t} + a_n^* e^{i\omega_n t}\right) \tag{13.20}$$

then we deal with N independent harmonic oscillators. Allowing N to be infinite, we will see in Section 15.4 that this Hamiltonian describes the uncoupled Fourier components of a free scalar field.

In writing the Hamiltonian equations (13.15)–(13.17), we assume that the equations $p_n = \partial L/\partial \dot{q}_n$ can be solved to give $\dot{q}_n(q_1, p_1, q_2, \dots, t)$ as functions of independent variables p_n. If that is not the case, there are constraint equations relating the p_n. The system of harmonic oscillators is obviously unconstrained, and since this will be our only detailed application of the Hamiltonian formalism we will ignore the issue of constraints.

13.2.3 Field theory action

In Newtonian mechanics, and in ordinary quantum mechanics, we are dealing with N particles and the number of degrees of freedom is finite. They correspond to the position coordinates of each particle, plus relevant internal degrees of freedom such as the two angles needed to specify the orientation of a diatomic molecule. In classical or quantum field theory, we deal with fields that are functions of position. The number of degrees of freedom is now infinite because we need an infinite set of numbers to specify the field. These are the value of the field at each point in space, labelled by position $\mathbf{x}$ which is *not* a degree of freedom.

For the moment we are concerned with flat spacetime. To respect the relativity principle the action in flat spacetime should be a scalar (Lorentz invariant). To achieve this, the Lagrangian for the fields must be of the form $L = \int \mathcal{L}\, d^3x$, where the **Lagrangian density** $\mathcal{L}$ is Lorentz invariant and has dimensions [energy]^4. The action is then

$$\boxed{S = \int d^4x\, \mathcal{L}} \,. \tag{13.21}$$

To respect invariance under spacetime translations, $\mathcal{L}$ must not depend explicitly on the coordinates. Its independence on t makes L independent of t, which as we saw leads to energy conservation. As we shall see, its independence on $\mathbf{x}$ leads to momentum conservation.

Lagrangian densities differing by a 4-divergence lead to the same action if the fields fall off sufficiently rapidly at infinity; they are physically equivalent at the classical level and, at least perturbatively, at the quantum level also. Field values are physically equivalent if they are related by what is called a gauge transformation. Where it matters, we will take it for granted that a single Lagrangian density has been chosen from the equivalence class.

The dependence of the Lagrangian density on spacetime derivatives of the fields is restricted by physical considerations. In addition to Lorentz invariance, one would like the theory to have a particle interpretation after quantization. In the classical limit, one would like equations that can be solved with physically reasonable boundary conditions. To satisfy these requirements, one usually imposes the following additional restrictions on the Lagrangian density: (i) it depends only on the first spacetime derivatives of the fields, as opposed to second and higher derivatives, and (ii) it is quadratic in the derivatives of bosonic fields, and linear in the derivatives of fermionic fields.

13.3 Scalar field in flat spacetime

13.3.1 Lagrangian

For a single scalar field ϕ, the simplest Lagrangian density consistent with the requirements mentioned at the end of the previous section is

$$\boxed{\mathcal{L} = -\frac{1}{2}\partial^\mu\phi\partial_\mu\phi - V(\phi)} \,. \tag{13.22}$$

In this expression, $V(\phi)$ is some function, called the scalar field **potential**, and the other term is called the kinetic term. Note that V has dimension $[\text{energy}]^4$, while ϕ has dimension $[\text{energy}]^1$. The names 'potential' and 'kinetic term' are chosen because of the analogy with the Lagrangian of a single particle with unit mass, moving in one dimension. For a homogeneous field $\phi(t)$ this analogy is exact.

The most general Lagrangian density consistent with the requirements is

$$\mathcal{L} = -\frac{1}{2}G(\phi)\partial^\mu\phi\,\partial_\mu\phi - V(\phi). \tag{13.23}$$

Assuming that G is positive, we can always define a new field $\tilde{\phi}(\phi)$ which will recover Eq. (13.22). Negative G would give Eq. (13.22) with the opposite sign for the kinetic term. The physical interpretation of that case is problematic, and isn't usually considered.

13.3.2 Field equation

The evolution of ϕ is given by $\delta S = 0$, where

$$\delta S = \int d^4x \left[\frac{\partial\mathcal{L}}{\partial\phi}\delta\phi + \frac{\partial\mathcal{L}}{\partial(\partial_\mu\phi)}\delta(\partial_\mu\phi)\right]. \tag{13.24}$$

Integrating the second term by parts, in each of the four variables x^μ, gives

$$\delta S = \int d^4x \left\{\frac{\partial\mathcal{L}}{\partial\phi} - \frac{\partial}{\partial x^\mu}\left[\frac{\partial\mathcal{L}}{\partial(\partial_\mu\phi)}\right]\right\}\delta\phi + \int d^4x \frac{\partial}{\partial x^\mu}\left[\frac{\partial\mathcal{L}}{\partial(\partial_\mu\phi)}\delta\phi\right]. \tag{13.25}$$

The last term is just a number which is required by the action principle to vanish. To ensure that it vanishes we can demand that ϕ vanishes sufficiently rapidly at infinity. Alternatively it can be killed by working in a finite box in spacetime with periodic boundary conditions.

Through the first term of Eq. (13.25) the action principle gives

$$\boxed{\frac{\partial \mathcal{L}}{\partial \phi} - \frac{\partial}{\partial x^\mu}\left(\frac{\partial \mathcal{L}}{\partial(\partial_\mu \phi)}\right) = 0}. \tag{13.26}$$

This is called the **field equation**. Taking $\mathcal{L}$ from Eq. (13.22) gives

$$\ddot{\phi} - \nabla^2\phi + V'(\phi) = 0, \tag{13.27}$$

where the prime denotes $d/d\phi$. It can be written

$$\boxed{-\Box\phi + V'(\phi) = 0}, \tag{13.28}$$

where $\Box$ is the d'Alembertian defined by Eq. (2.25).

For a homogeneous field

$$\boxed{\ddot{\phi} + V'(\phi) = 0}. \tag{13.29}$$

In accordance with the mechanical analogy, this is the equation for a unit-mass non-relativistic particle with position ϕ and potential $V(\phi)$.

13.3.3 Hamiltonian

To construct the Hamiltonian, we should regard the entire set of field values $\phi(\mathbf{x}, t)$ as the degrees of freedom, labelled by parameter $\mathbf{x}$. To use Eq. (13.15) for the Hamiltonian though, we need discrete labels. One way of doing this would be to replace $\mathbf{x}$ by a lattice of points in space. Going to the continuum limit, the conjugate momentum and Hamiltonian are

$$\Pi \equiv \frac{\partial \mathcal{L}}{\partial \dot{\phi}}, \qquad H = \int d^3x \Pi\dot{\phi} - L. \tag{13.30}$$

This is equivalent to introducing a Hamiltonian density

$$\boxed{\mathcal{H} \equiv \Pi\dot{\phi} - \mathcal{L}}, \tag{13.31}$$

in terms of which $H = \int d^3x \mathcal{H}$.

Discretizing t as well as $\mathbf{x}$ one arrives at lattice field theory, which we shall not pursue. Another way of introducing discrete labels is to write the field as a Fourier series in a finite box, and use the discrete set of Fourier components as the degrees of freedom. We see how that works for a free field in Section 15.4.

13.4 Energy–momentum tensor

We shall see in Section 13.8.1 that the energy–momentum tensor of scalar (and vector) fields is given in terms of their Lagrangian density by Eq. (13.61). In the case of a single scalar field, Eq. (13.61) becomes

$$\boxed{T^\mu_\nu = \partial^\mu\phi\partial_\nu\phi - \delta^\mu_\nu\left[\frac{1}{2}\partial^\alpha\phi\partial_\alpha\phi + V(\phi)\right]}. \tag{13.32}$$

The momentum density is

$$\boxed{T^0_i = \dot{\phi}\partial_i\phi}. \tag{13.33}$$

A homogeneous field has zero momentum density and its stress is isotropic. Its energy density and pressure are

$$\boxed{\rho = \frac{1}{2}\dot{\phi}^2 + V(\phi)}, \qquad \boxed{P = \frac{1}{2}\dot{\phi}^2 - V(\phi)}. \tag{13.34}$$

In the mechanical analogy, ρ is the energy of the unit-mass particle, the first term being its kinetic energy.

More generally, these expressions hold at each point in the local rest frame of the field. According to the definition given after Eq. (2.39), this is the frame in which the momentum density vanishes. In the local rest frame, ρ and P are given by the above expressions.

A scalar field is a perfect fluid in the terminology of relativity, because its stress is isotropic in the local rest frame. The adiabatic expansion of this perfect fluid is described by the field equation. In contrast with the case of matter or radiation there is no equation of state relating P and ρ, because the same energy density can correspond to different values of the pressure if the energy density is distributed differently between the potential and kinetic terms.

The potential $V(\phi)$ is supposed to have a minimum at which $V = 0$. With ϕ fixed at the minimum the energy density and pressure are zero. The value of ϕ at the minimum therefore corresponds to the vacuum, and is called the **vacuum expectation value** or vev.[2] The word 'expectation' is included because the field has a quantum fluctuation about the classical value. The vev is written as $\langle\phi\rangle$.

13.5 Nearly free scalar field

We are usually concerned with solutions of the field equations that correspond to interacting plane waves. After quantization they correspond to the existence of interacting particles. Of course, the interaction should not be too strong, or the

[2] We ignore here the cosmological constant, which will be considered in Section 23.5.

waves/particles lose their identity. Let us see how things work out for a single scalar field.

Sufficiently near the vev, we can expect the potential to be quadratic:

$$\boxed{V = \frac{1}{2}m^2\phi^2}. \tag{13.35}$$

Then Eq. (13.27) becomes linear,

$$\boxed{\ddot{\phi} - \nabla^2\phi + m^2\phi = 0}. \tag{13.36}$$

This is the **Klein–Gordon equation**, whose solution is a sum of plane waves. We see in Section 15.4 that, after quantization, each plane wave corresponds to particles with mass m and momentum equal to the wave-vector. The particles don't interact, and neither do the plane-wave solutions. We are dealing with a non-interacting, or **free**, field.

Let us first suppose that ϕ is spatially homogeneous. Then Eq. (13.29) gives

$$\boxed{\ddot{\phi}(t) + m^2\phi(t) = 0}. \tag{13.37}$$

The solution corresponds to harmonic oscillation with angular frequency m. The corresponding particles have zero momentum.

Taking $V(\phi)$ to be an even function, let us include the next term in the power series expansion:

$$V = \frac{1}{2}m^2\phi^2 + \frac{1}{4}\lambda\phi^4. \tag{13.38}$$

This adds a term $\lambda\phi^3$ to the left-hand side of the field equation (13.36). We can still make the plane-wave expansion, but the plane waves interact with each other and, after quantization, so do the particles.

The potential may have higher-order terms as well. Still taking the potential to be even, we can write

$$\boxed{V(\phi) = \frac{1}{2}m^2\phi^2 + \frac{1}{4}\lambda\phi^4 + \lambda_6\frac{\phi^6}{\Lambda^2} + \lambda_8\frac{\phi^8}{\Lambda^4} + \ldots}. \tag{13.39}$$

It is usually supposed that the dimensionless coefficients λ_n will be very roughly of order 1, if we take Λ to be the cutoff of the effective field theory. In that case, we can usually drop the higher-order terms in the regime $\phi \ll \Lambda$.

13.6 Several fields

13.6.1 Canonical normalization

For N real fields the Lagrangian density is usually taken to be

$$\mathcal{L} = -\frac{1}{2}\sum_{n=1}^{N} \partial^\mu \phi_n \partial_\mu \phi_n - V(\phi_1, \ldots, \phi_N), \tag{13.40}$$

where V is now a function of all the fields. Fields with this Lagrangian density are said to be **canonically normalized**.

The field equations are

$$-\Box \phi_n + \frac{\partial V}{\partial \phi_n} = 0. \tag{13.41}$$

For homogeneous fields they read

$$\ddot{\phi}_n + \frac{\partial V}{\partial \phi_n} = 0. \tag{13.42}$$

The energy–momentum tensor is

$$T^\mu_\nu = \sum_n \partial^\mu \phi_n \partial_\nu \phi_n - \delta^\mu_\nu \left[\frac{1}{2} \sum_n \partial^\alpha \phi_n \partial_\alpha \phi_n + V(\phi_1, \ldots, \phi_N) \right]. \tag{13.43}$$

For homogeneous fields,

$$\rho = \frac{1}{2}\sum \dot{\phi}_n^2 + V, \qquad P = \frac{1}{2}\sum \dot{\phi}_n^2 - V. \tag{13.44}$$

Equations (13.42) and (13.44) are analogous to those for a collection of non-relativistic particles with unit mass, with positions ϕ_n moving in one dimension with a potential V.

If the potential is of the form

$$V = V_1(\phi_1) + V_2(\phi_2) + \ldots, \tag{13.45}$$

we deal with a set of uncoupled fields. Each field has its own field equation and energy–momentum tensor, as if there were no other fields. In particular, if

$$V = \frac{1}{2}\sum m_n^2 \phi_n^2, \tag{13.46}$$

we deal with a set of free fields.

As in the single-field case, there is supposed to be a minimum at which $V = 0$, attained when every field is at its vev. In scenarios of the early Universe, one or more of the scalar fields is supposed to be displaced from its vev, causing V to be positive.

It is often useful to combine two canonically normalized real fields ϕ_1 and ϕ_2 into a single complex field, defined by convention as

$$\phi = \frac{1}{\sqrt{2}} (\phi_1 + i\phi_2) \,. \tag{13.47}$$

For a single complex field the Lagrangian density is

$$\boxed{\mathcal{L} = -\partial^\mu \phi^* \partial_\mu \phi - V(\phi, \phi^*)} \,, \tag{13.48}$$

and the field equation is

$$\boxed{-\Box\phi + \frac{\partial V}{\partial \phi^*} = 0} \,. \tag{13.49}$$

13.6.2 Field space

The kinetic term is invariant under the following transformations

$$\phi_n \to \phi_n + A_n \qquad \phi_n \to \sum_m R_{nm}\phi_m . \tag{13.50}$$

where A_n are constants and R_{nm} is an orthogonal matrix so that $\sum \phi_n^2$ is unchanged. It is therefore useful to think of the field values as Cartesian coordinates in a Euclidean 'field space'. The kinetic term is unchanged if we shift the origin in field space, or make a rotation about the origin.

If we rotate the free-field potential (13.46) we arrive at

$$V = \frac{1}{2} \sum_{nm} m_{nm} \phi_n \phi_m . \tag{13.51}$$

The symmetric object m_{ij} is called the mass matrix. Conversely, starting with Eq. (13.51) we can arrive at Eq. (13.46) by a suitable rotation in field space. This corresponds to the identification of normal modes of oscillation for a mechanical system.

13.6.3 Curved field space

The most general Lorentz-invariant Lagrangian density satisfying the requirements at the end of Section 13.2 is of the form

$$\mathcal{L} = -\sum_{m,n} G_{nm} \partial^\mu \phi_m \partial_\mu \phi_n - V. \tag{13.52}$$

The real symmetric object G_{nm} is a function of the fields, called the kinetic function. It can be regarded as a metric G_{nm} in field space, defining the infinitesimal distance

$$d\phi^2 = \sum_{nm} G_{nm} d\phi_n d\phi_m. \tag{13.53}$$

To avoid the problems discussed after Eq. (13.23) we assume that $G_{nm} = \delta_{nm}$ can be chosen at any point in field space. This makes the field space a Riemannian manifold, as defined in Section 3.4.

Now Eq. (13.44) is replaced by

$$\rho = \frac{1}{2} \sum G_{nm} \dot\phi_n \dot\phi_m + V, \qquad P = \frac{1}{2} \sum G_{nm} \dot\phi_n \dot\phi_m - V. \tag{13.54}$$

Canonical normalization corresponds to $G_{nm} = \delta_{nm}$ throughout field space, making the field space flat.

If only one field has significant time dependence, say $\phi \equiv \phi_1$, then for any $G_{11}(\phi)$ we can find a new field $\tilde\phi(\phi)$ which is canonically normalized. More generally, according to the local flatness theorem, canonical normalization can always be imposed to high accuracy, in some region surrounding any point in field space. The size of this region is typically of order the ultra-violet cutoff of the effective field theory.

13.7 Field theory in curved spacetime

Considering first flat spacetime, let us see how to write the scalar field action with generic coordinates. Transforming the spacetime volume element, the action will be of the form

$$\boxed{S = \int d^4x \sqrt{-g}\, \mathcal{L}}, \tag{13.55}$$

where g is the determinant of $g_{\mu\nu}$, and $\mathcal{L}$ is given by Eq. (13.40) with $\partial^\mu \equiv g^{\mu\nu}\partial_\nu$. The scalar field equation is Eq. (13.28) with the d'Alembertian (3.22).

According to the equivalence principle, the action (13.55) and the corresponding field equation will apply also in curved spacetime. For the Robertson–Walker metric (3.63) the scalar field equation is

$$\ddot\phi - a^{-2}\nabla^2\phi + 3H\dot\phi + V'(\phi) = 0, \tag{13.56}$$

where ∇^2 is evaluated using the comoving coordinates.

For a homogeneous field

$$\boxed{\ddot\phi + 3H\dot\phi + V'(\phi) = 0}. \tag{13.57}$$

The term $3H\dot\phi$ corresponds in the mechanical analogy to friction, and its effect is called Hubble drag or damping. If Hubble drag is negligible the expansion of the

Universe has no effect on the dynamics of the scalar field. For several homogeneous scalar fields we have

$$\ddot{\phi}_n + 3H\dot{\phi}_n + \frac{\partial V}{\partial \phi_n} = 0. \tag{13.58}$$

This derivation of the curved spacetime field theory through the equivalence principle goes through in exactly the same way with vector fields included. For spinor fields, the equivalence principle still works, but in a different way because spinor fields don't carry spacetime indices [1].

13.8 Gravity from the action principle

13.8.1 Einstein gravity

The Einstein equation in vacuum (i.e. with $T_{\mu\nu} = 0$) corresponds to the **Einstein–Hilbert action**

$$\boxed{S = \frac{1}{2}\int d^4x\sqrt{-g}M_{\rm Pl}^2 R}, \tag{13.59}$$

The field equation is obtained from the action principle, by varying $g_{\mu\nu}$.

Replacing the vacuum by 'matter' consisting of scalar fields, the Einstein equation corresponds to

$$\boxed{S = \int d^4x\sqrt{-g}\left(\frac{1}{2}M_{\rm Pl}^2 R + \mathcal{L}_{\rm mat}\right)}, \tag{13.60}$$

where $\mathcal{L}_{\rm mat}$ is the Lagrangian density obtained from flat spacetime field theory using the equivalence principle, as described in the previous section. Varying $g_{\mu\nu}$, the action principle gives the Einstein field equation, with the energy–momentum tensor

$$T_{\mu\nu} = -2\frac{\partial \mathcal{L}_{\rm mat}}{\partial g^{\mu\nu}} + g_{\mu\nu}\mathcal{L}_{\rm mat}. \tag{13.61}$$

Including vector fields, this derivation of the Einstein field equation goes through in exactly the same way, and Eq. (13.61) continues to apply. Including spinor fields, we have to proceed differently, but the Einstein equation can still be derived from the action principle, resulting in a formula for $T_{\mu\nu}$ that includes the spinor field contributions [1].

13.8.2 Beyond Einstein gravity

One doesn't expect the action (13.60) to be exactly correct. Rather, analogously with the case of ordinary field theory, one expects the action describing gravity to contain an infinite number of terms involving the curvature tensor, suppressed by powers of the ultra-violet cutoff which is usually taken to be $M_{\rm Pl}$.

For terms involving only the curvature tensor, $\mathcal{L}$ might be any scalar function of the curvature tensor. The simplest possibility is to add terms of the form $\lambda_n M_{\rm Pl}^{2(1-n)} R^{n+1}$, with n a positive integer and λ_n a dimensionless constant.[3] In this way one arrives at what is called $f(R)$ gravity;

$$S = \int d^4x \sqrt{-g}\,(f(R) + \mathcal{L}_{\rm mat}) \,. \tag{13.62}$$

As the additional terms are multiplied by inverse powers of $M_{\rm Pl}$, their effect will disappear at sufficiently large distances. Observation places upper bounds on the values of the dimensionless constants λ_n. Of course the observational bounds correspond to very large values, since values of order 1 would make the additional terms ineffective at all distances bigger than $M_{\rm Pl}^{-1}$. The true values of λ_n are presumably far below the observational bounds, perhaps of order 1.

Similar considerations apply to any modification of gravity involving only the curvature tensor. The situation becomes much more complicated for modifications of gravity that include scalar fields, known as scalar–tensor theories. These are considered in Chapter 19.

Exercises

13.1 Verify Eq. (13.10), giving the Hamiltonian of a harmonic oscillator.

13.2 Verify Eq. (13.28), the field equation for a scalar field, by inserting the Lagrangian density (13.22) into Eq. (13.26).

13.3 Put the scalar field Lagrangian density (13.22), with $\partial^\mu = g^{\mu\nu}\partial_\nu$, into the general expression (13.61) for the energy–momentum tensor, to verify Eq. (13.32) for the energy–momentum tensor of a scalar field.

13.4 Specializing to the case of a homogeneous scalar field, demonstrate that the energy–momentum tensor (13.32) reduces to Eq. (13.34) for the energy density and pressure.

13.5 Read off the metric components corresponding to the Robertson–Walker line element (3.63), and put them into Eq. (3.22) to derive an expression for the d'Alembertian $\Box$ in comoving space coordinates. Insert your expression into the general scalar field equation (13.28), to obtain the homogeneous scalar field equation (13.57).

[3] We cannot tolerate $n < -1$ because then $\mathcal{L}$ would blow up in the flat spacetime limit. The case $n = -1$ corresponds to a cosmological constant, which nowadays is regarded as a contribution to $\mathcal{L}_{\rm mat}$ rather than as a modification of Einstein gravity.

Reference

[1] S. Weinberg. *Gravitation and Cosmology, Principles and Applications for General Relativity* (New York: John Wiley and Sons, 1972).

14

Internal symmetry

The field theory action is constructed to be invariant under Lorentz transformations and spacetime translations. These don't mix different fields. Now we consider **internal symmetries**, corresponding to transformations which mix different fields but don't involve spacetime.

Internal symmetries come in two kinds, called **global** symmetry and **gauge** symmetry. A gauge symmetry must be exact, but a global symmetry may only be approximate.

What is the motivation for considering a symmetry? One might first choose an action, guided by observation and/or aesthetics, and then look to see what is its symmetry group. That was the case when Quantum Electrodynamics was first formulated. Nowadays, the more usual procedure is to first choose a symmetry group (guided again by observation and/or aesthetics) and then to write down the most general action consistent with the symmetry group, which turns out to be quite strongly constrained. Often, the action arrived at in this way turns out to have additional global symmetries which were not imposed from the start, called accidental symmetries.

In this chapter we look at some examples of internal symmetry, as it applies to scalar fields at the classical level. Although we won't generally write down formulas involving spin-$1/2$ fields, it will be important to remember their existence.

14.1 Symmetry groups

To say that the action is invariant under some transformation is to say that it is the same before and after the transformation. It follows that the action is invariant also under the inverse of the transformation. Also, if the action is invariant under the transformations $A \to B$ and $B \to C$, it is invariant under the combined transformation $A \to C$. In mathematical language, this means that a symmetry deals with a **group** of transformations.

Our discussion will be elementary, without invoking group theory in any serious way. We just need the following basic features. A group is a set of elements, with a multiplication rule $a = b.c$ satisfying $(a.b).c = a(b.c)$, and containing an identity I such that $Ia = a$ for every element with every element having an inverse such that $aa^{-1} = I$. No other properties are needed for the set to qualify as a group, and groups with the same multiplication table are regarded as the same. If multiplication commutes the group is said to be **Abelian**.

Of special interest are groups whose elements are matrices, with matrix multiplication identified as group multiplication. Different sets of matrices can have the same multiplication table (i.e. be the same group) and each such set is said to form a **representation** of the group. The group is usually named after a particular representation. The $U(1)$ group corresponds to complex numbers of unit modulus and its elements are labelled by one parameter. The $U(N)$ group corresponds to the $N \times N$ unitary matrices. The $SU(N)$ group corresponds to the $N \times N$ unitary matrices with determinant $+1$. The elements of $SU(2)$ are labelled by 3 parameters and those of $SU(3)$ by 8 parameters.

A group may contain subgroups. If every element of a group C can be written as the product of elements drawn from subgroups A and B we write $C = A \otimes B$ (also $C/A = B$), and say that C is the direct product of groups A and B. For instance, $U(N) = SU(N) \otimes U(1)$.

14.2 Abelian global symmetries

14.2.1 Z_N symmetry

Consider one real scalar field and the transformation

$$\phi \to -\phi. \tag{14.1}$$

Acting twice this transformation gives $\phi \to -\phi \to \phi$. We are dealing with the Z_2 group which has just two distinct elements. Symmetry requires that $V(\phi)$ is an even function. If $\phi = 0$ is the vev, $V(\phi)$ will be of the form (13.39).

The point $\phi = 0$ in field space isn't moved by the symmetry group. Such a point is called a **fixed point** of the symmetry. Symmetry groups usually have a fixed point. As in this example, the first derivatives of the scalar field potential usually vanish at a fixed point.

Instead of Eq. (13.38) we may have

$$V(\phi) = \frac{1}{2}m^2\phi^2 + \frac{1}{4}\lambda\phi^4 + \epsilon\phi^3, \tag{14.2}$$

with ϵ some small coupling which breaks the Z_2 symmetry. This is called **explicit**

symmetry breaking, in contrast with spontaneous symmetry breaking that we encounter later.

We will also have occasion to consider symmetry under the following group of transformations, acting on a complex field:

$$\phi \to e^{i2n\pi/N}\phi, \tag{14.3}$$

with $n = 1, 2, \ldots, N-1$ and N a positive integer. This is the Z_N group, with N distinct elements.

14.2.2 Global $U(1)$ symmetry

Now consider a single complex field ϕ, and suppose that V depends only on $|\phi|$. For example the potential might be the analogue of Eq. (13.38):

$$\boxed{V(\phi) = m^2|\phi|^2 + \frac{1}{4}\lambda|\phi|^4}\,. \tag{14.4}$$

Then each of the canonically normalized real fields has mass-squared m^2.

Since V depends only on $|\phi|$, it is invariant under the $U(1)$ group of transformations

$$\boxed{\phi(x) \to e^{i\lambda}\phi(x)}\,. \tag{14.5}$$

Assuming further that λ is independent of spacetime position, the kinetic term of the Lagrangian density is also invariant, and we have a symmetry. It is called a global symmetry because the value of λ is the same throughout spacetime.

As a consequence of the $U(1)$ symmetry, there is a conserved current[1]

$$\boxed{j^\mu \equiv -i\left(\phi^*\partial^\mu\phi - (\partial^\mu\phi^*)\phi\right)}\,. \tag{14.6}$$

The charge density is

$$j^0 = i\left(\phi^*\dot{\phi} - \dot{\phi}^*\phi\right) = -2|\phi|^2\dot{\theta}, \tag{14.7}$$

where θ is the phase of ϕ.

If there are several complex fields, we can define a $U(1)$ transformation by writing

$$\boxed{\phi_n \to e^{i\lambda q_n}\phi_n}\,. \tag{14.8}$$

This $U(1)$ will be a symmetry if V has a suitable form, and then the constants q_n are called charges. The charges are defined only up to an overall constant, allowing us to choose $q = 1$ in the single-field case.

[1] Recall that a conserved current is an object j^μ satisfying the continuity equation $\partial_\mu j^\mu = 0$.

There can be two or more $U(1)$ symmetries, corresponding to different choices for the charges. For instance with complex fields ϕ_1 and ϕ_2 one might have

$$V = m_1^2|\phi_1|^2 + m_2^2|\phi_2|^2 + M\left(\phi_1^2\phi_2 + \text{c.c.}\right). \tag{14.9}$$

Without the last term this has the two symmetries, $U_A(1)$ with $q_1 = 1$ and $q_2 = 0$ and $U_B(1)$ with $q_1 = 0$ and $q_2 = 1$, and the full symmetry is $U_A(1) \otimes U_B(1)$. Including the last term it has the single symmetry $U_C(1)$ with $q_1 = 1$ and $q_2 = -2$. As in this example, one uses a subscript to distinguish different symmetries involving the same symmetry group.

Each of the $U(1)$ symmetries comes with a conserved current, given by

$$\boxed{j^\mu \equiv -i\sum_n q_n\left[\phi_n^*\partial^\mu\phi_n - (\partial^\mu\phi_n^*)\phi_n\right]}. \tag{14.10}$$

The charge of each field determines its contribution to the current. The charge density is

$$j^0 = i\sum_n q_n\left(\phi_n^*\dot{\phi}_n - \dot{\phi}_n^*\phi_n\right) = -2\sum_n q_n|\phi_n|^2\dot{\theta}_n. \tag{14.11}$$

As with any symmetry, a $U(1)$ can act on fields of non-zero spin, which contribute to the conserved current.

Partly on the basis of string theory, it is often supposed that there are no exact global symmetries. A small breaking of a global $U(1)$ might come from a non-renormalizable term with a dimensionless coupling λ of order 1, so that Eq. (14.4) becomes

$$V(\phi) = m^2|\phi|^2 + \frac{1}{4}\lambda|\phi|^4 + \lambda_d\left(\frac{\phi^{d+4}}{M_{\text{Pl}}^d} + \text{c.c.}\right). \tag{14.12}$$

In any case, the symmetry-breaking part of the potential is periodic in the phase θ, and typically is sinusoidal. In the presence of the symmetry breaking, the current isn't exactly conserved, i.e. $\partial_\mu j^\mu$ is non-zero.

14.3 Non-Abelian continuous global symmetries

14.3.1 Global $SU(2)$ symmetry

Now we consider non-Abelian symmetries, confining ourselves to continuous symmetries. The simplest and most important such symmetry is $SU(2)$ symmetry.

Consider a pair $\Phi \equiv (\phi_1, \phi_2)$ of complex fields (called a doublet), and define

$$|\Phi|^2 \equiv \sum_n |\phi_n|^2. \tag{14.13}$$

If V depends only on $|\Phi|^2$ the Lagrangian density will be invariant under the transformations

$$\phi_n \to U_{nm}\phi_m, \tag{14.14}$$

where U is a unitary matrix and a sum over j is understood. Taking U to be independent of spacetime position, the kinetic term is invariant too.

Any unitary matrix can be written as the product of an $SU(N)$ matrix and a phase factor. The Lagrangian density is therefore invariant under an $SU(2)\otimes U(1)$ group of transformations. We focus on the $SU(2)$ because the $U(1)$ will give nothing new.

Every $SU(N)$ matrix is of the form $U = e^{iF}$ with F Hermitian and traceless. An $SU(2)$ matrix can therefore by written

$$U = \exp(i\lambda_n \hat{T}_n) \equiv \exp(i\boldsymbol{\lambda}\cdot\hat{\mathbf{T}}), \tag{14.15}$$

(summation over a understood) with $\hat{T}_n = \sigma_n/2$ and σ_n the Pauli spin matrices

$$\sigma_1 = \begin{pmatrix} 0 & 1 \\ 1 & 0 \end{pmatrix}, \qquad \sigma_2 = \begin{pmatrix} 0 & -i \\ i & 0 \end{pmatrix}, \qquad \sigma_3 = \begin{pmatrix} 1 & 0 \\ 0 & -1 \end{pmatrix}. \tag{14.16}$$

The normalization of the matrices $\hat{T}_n$ is chosen so that their commutator satisfies

$$[\hat{T}_n, \hat{T}_m] = i\epsilon_{nmk}\hat{T}_k. \tag{14.17}$$

It is often useful to take λ to be infinitesimal, to arrive at

$$U = I + i\lambda_n\hat{T}_n. \tag{14.18}$$

A venerable example of $SU(2)$ is isospin symmetry, formulated originally in the context of nuclear physics and nowadays seen as an accidental symmetry of the Standard Model (Section 16.5.2). The nucleon fields (p, n) form a doublet, and we need to consider higher-dimensional representations of $SU(2)$ such as the triplet of pion fields. The necessary mathematics is exactly the same as for angular momentum, in particular spin. The $N\times N$ matrices U providing a representation of $SU(2)$ are of the form (14.15) with the three $N\times N$ matrices $\hat{T}_n$ satisfying Eq. (14.17). The matrices U are unitary and the $\hat{T}_n$ are Hermitian. The latter are called the **generators** of the $SU(2)$ group.

If Φ is an $N\times 1$ column vector with elements ϕ_n, then $\sum|\phi_n|^2$ is invariant under the $SU(2)$ transformation. If V is any unitary matrix, the **similarity transformation** $\Phi\to V\Phi$ with $U\to VUV^{-1}$ gives an equivalent representation. Assuming that U cannot be made block diagonal by a similarity transformation (i.e. that we deal with an irreducible representation of $SU(2)$), Φ is called a **multiplet**. Then we can choose $\hat{T}_3$ to be diagonal with elements given by

$$\hat{T}_3 = \ \mathrm{diag}\,(J, J-1, \ldots, -J), \qquad 2J+1 = N. \tag{14.19}$$

Consider now the $U(1)$ subgroup of $SU(2)$, obtained by taking $\boldsymbol{\lambda} = (0, 0, \lambda)$. Acting on a multiplet Φ, the $U(1)$ transformation is $\phi_n \to \exp(i\lambda T_3)\phi_n$, where T_3 is the ath eigenvalue of $\hat{T}_3$ given by Eq. (14.19). Corresponding to this $U(1)$ is a conserved current, given by Eq. (14.10) with $q_n = T_3$. The sum of the charges T_3 vanishes for each multiplet.

For the case $N = 2$ that we already considered, the charges are $T_3 = \pm 1/2$. We shall also need the case $N = 3$, where Φ is a triplet. Then we can choose Φ so that $\lambda_n \hat{T}_n$ is the matrix for 3-dimensional rotations about the n direction through an angle λ_n. Let us denote the elements of Φ with this choice by W_n. We can regard the W_n as components of a 3-dimensional 'vector' $\mathbf{W}$. The components ϕ_n satisfying Eq. (14.19) are given by $(W_1 + iW_2)/\sqrt{2}$, W_3 and $(W_1 - iW_2)/\sqrt{2}$ and the corresponding charges are $T_3 = 1, 0$ and -1.

The W_n are usefully collected into a 2×2 matrix by writing

$$W \equiv \mathbf{W} \cdot \mathbf{T} \equiv W_n \hat{T}_n, \tag{14.20}$$

with $\hat{T}_n = \sigma_n/2$. Then their $SU(2)$ transformation can be written

$$W \to U(\boldsymbol{\lambda}) W U^{-1}(\boldsymbol{\lambda}), \tag{14.21}$$

where $U(\boldsymbol{\lambda})$ is defined by Eq. (14.15).

14.3.2 Lie groups

All of this has a simple and very powerful generalization. For any continuous internal symmetry, as well as for the Lorentz transformation, the symmetry group is a **Lie group**. A Lie group is one whose matrix representation can be written (at least in some neighbourhood of the identity) in the form (14.15): $U = \exp(\sum_a i\lambda_a \hat{G}_a)$. The generators $\hat{G}_a$ have a commutation relation [1]

$$[\hat{G}_a, \hat{G}_b] = iC^c_{\ ab} \hat{G}_c, \tag{14.22}$$

and the coefficients $C^a_{\ bc}$ are called the **structure constants** of the group. It is often convenient to take the λ_a to be infinitesimally small, so that

$$U = I + i \sum_a \lambda_a \hat{G}_a. \tag{14.23}$$

If the structure constants vanish we are dealing with an Abelian Lie group, corresponding to a $U(1)$ or a direct product of $U(1)$'s. If the structure constants are ϵ_{abc} we are dealing with the $SU(2)$ group, or with the rotation group $O(3)$ (the set of orthogonal 3×3 matrices). As we know from the theory of spin, representations of $SU(2)$ are representations of the rotation group up to a $\pm$ sign (double-valued representations).

We will also be interested in $SU(3)$,which has eight generators. It can be shown that two of the generators can be chosen to be diagonal, which means that there are two conserved currents of the form (14.10) and two conserved charges. For each charge, the sum of the charges within a multiplet is zero.

A single field (real or complex) is called a **singlet**. If it is complex, it can transform under a $U(1)$ symmetry but not under any non-Abelian symmetry (taking the transformation in each case to act on the phase of the field). In the context of gauge symmetry, that we come to later, the term 'gauge singlet' is often used in a stronger sense, to mean that a field that doesn't transform under *any* gauge group, not even under a $U(1)$.

14.4 Noether's theorem and conserved quantities

The conserved currents that we encountered for $U(1)$ and $SU(2)$ are examples of **Noether's theorem**. We discuss the theorem and its consequences in this section. Our discussion is valid for any number of fields of any spin, and to avoid introducing new notation we allow the symbol ϕ_n to denote any component of any field.

We suppose that the action is invariant under infinitesimal transformations of the form

$$\phi_n(x) \to \phi_n(x) + i\lambda \mathcal{F}_n(x), \tag{14.24}$$

where λ is independent of x (a global symmetry). The variation δS of the action vanishes under this transformation, but it doesn't vanish if we allow λ to depend on x. It is easy to show [1] that the variation is of the form

$$\delta S = -\int d^4x j^\mu(x) \partial_\mu \lambda(x), \tag{14.25}$$

and that j^μ *is conserved by virtue of the field equations satisfied by* ϕ_n. This is Noether's theorem, and j^μ is called a Noether current. Clearly, there is a Noether current for each independent infinitesimal symmetry transformation.

Assuming that the Lagrangian density (not just the action) is invariant, as is the case for internal symmetries, it can be shown [1] that

$$j^\mu = -i \sum_n \frac{\partial \mathcal{L}}{\partial(\partial_\mu \phi_n)} \mathcal{F}_n. \tag{14.26}$$

From Eq. (14.26) we derive a formula for the conserved charge F:

$$F = -i \int d^3x \sum_n \frac{\partial \mathcal{L}}{\partial \dot{\phi}_n} \mathcal{F}_n. \tag{14.27}$$

It can be shown [1] that this quantity is still conserved even if it is only the Lagrangian, as opposed to the Lagrangian density, that is invariant.

Table 14.1. *Mechanical conserved quantities*

Invariance	*Conserved quantity*
Time translation	Energy E
Translation along x^i	Momentum component p_i
Rotation about x^i	Angular momentum component J_i
Boost along x^i	K_i given by Eq. (2.55)

Noether's theorem applies to any global symmetry, which means that it applies to spacetime translations and Lorentz transformations. We discuss these briefly, before returning to internal symmetries. A full account is given in, for instance, Ref. [1].

For an infinitesimal translation $x^\mu \to x^\mu + X^\mu$ along the μ direction ($\mu = 0, 1, 2, 3$), the effect on the fields is $\phi_n \to \phi_n + X^\mu \partial_\mu \phi_n$. Fixing μ we can take $\lambda = X^\mu$ and $\mathcal{F}_n = -i\partial_\mu \phi_n$ in Eq. (14.24). For each μ there is a conserved charge, obtained by integrating the time component of the corresponding Noether current, which we identify with a component p_μ of the 4-momentum. The 3-momenta are given by Eq. (14.27) because the Lagrangian is invariant under spatial translations, and it can be shown that the energy is equal to the Hamiltonian obtained by integrating Eq. (13.31) (with $\sum \Pi_n \phi_n$ instead of $\Pi\phi$).

One can find an expression for the four Noether currents, in terms of the fields. The corresponding object $T_{\mu\nu}$ isn't symmetric, but there is a prescription for symmetrizing it and after that is done $T_{\mu\nu}$ is identified with the energy–momentum tensor. In specific cases (in particular for canonically normalized scalar fields) one can easily show that $T_{\mu\nu}$ defined in this way is the same as the $T_{\mu\nu}$ defined as the source of Einstein gravity using the action principle, as described in Section 13.8.1. The same thing is presumably true in general, though there doesn't seem to be a published proof.

Consider next the Lorentz transformation. It is specified by six parameters. The six conserved 'charges' obtained from the Noether currents turn out to be the angular momentum components J_i (corresponding to invariance under rotation about the i-axis) and the quantities K_i, derived from the energy–momentum tensor as described in Section 2.5.3 (corresponding to Lorentz boosts along the i direction). The currents $M^{\lambda\mu\nu}$ corresponding to these 'charges' are different from the Noether currents, and neither set of currents is of much interest. The connection between spacetime symmetries and conserved quantities is summarized in Table 14.1.

We should point out here that it is only the conserved 'charge' $\int d^3x j^0$, derived from a conserved current for an isolated system, that is guaranteed to have physical meaning. In Section 2.6.2 we saw how to construct the current itself if the fluid

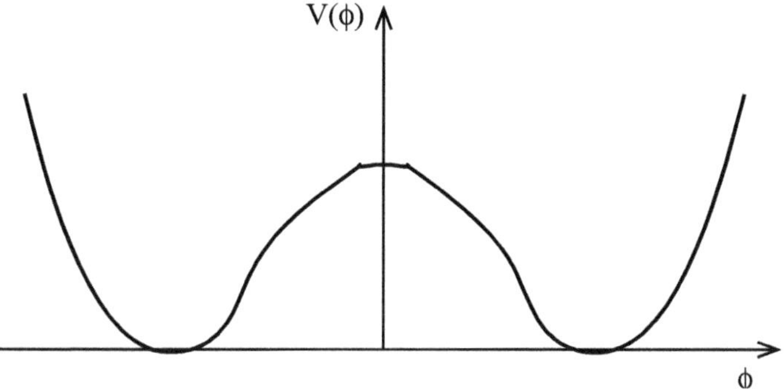

Fig. 14.1. The double-well potential.

under consideration is a gas, but that is a special case. In general, a conserved current has a clear physical meaning only if it is the source of an interaction. That is the case for the energy–momentum tensor, assuming that Einstein's field equation is correct. It is also the case for the electromagnetic current, and we shall see shortly how analogues of the electromagnetic current exist for any gauge symmetry.

14.5 Spontaneously broken global symmetry

14.5.1 Global discrete symmetry

In the cases considered so far, the vacuum corresponds to a fixed point of the symmetry. Now consider the potential

$$\boxed{V(\phi) = \frac{1}{4}\lambda\left(\phi^2 - f^2\right)^2 \equiv V_0 - \frac{1}{2}m^2\phi^2 + \frac{1}{4}\lambda\phi^4}, \tag{14.28}$$

where $m^2 = \lambda f^2$ and $V_0 = \lambda f^4/4$. If the sign in front of m^2 were positive we would be dealing with a self-interacting field with mass m. With the negative sign, m is often called a tachyonic mass. One often writes $+m^2\phi^2$, with the understanding that m^2 is negative.

The potential (14.28) still has the Z_2 symmetry (invariance under $\phi \to -\phi$), but there are now two minima as shown in Figure 14.1, located at $\phi = \pm f$. Either of the minima can serve as the vacuum. Taking for example the one with positive ϕ, we can define a new field $\tilde{\phi} = \phi - f$. Then we have near the minimum

$$V = \frac{1}{2}\tilde{m}^2\tilde{\phi}^2 + A\tilde{\phi}^3 + B\tilde{\phi}^4 + \dots, \tag{14.29}$$

where $\tilde{m} = \sqrt{2}\lambda m$ and we are not interested in the precise values of A and B. The minima represent possible vacua, but an observer confined to just one of the

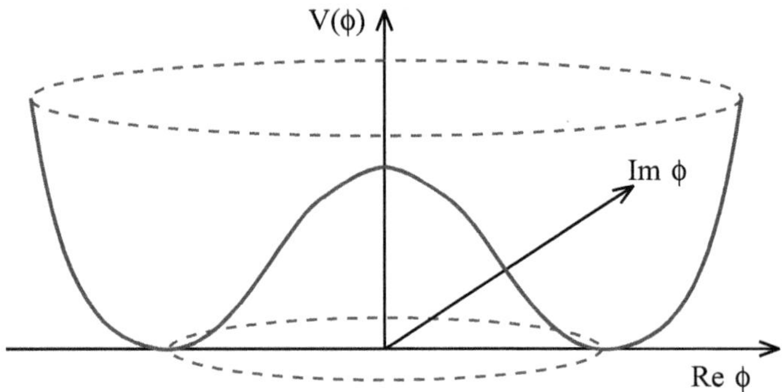

Fig. 14.2. The Mexican-hat potential.

vacua will see no trace of the original discrete symmetry. We say that it has been **spontaneously broken**.

Consider now a second field χ, which interacts with ϕ but which has no mass term:

$$V(\phi) = V_0 - \frac{1}{2}m^2\phi^2 + \frac{1}{4}\lambda\phi^4 + \frac{1}{4}\lambda'\phi^2\chi^2. \tag{14.30}$$

The observer in the vacuum $\langle\phi\rangle = f$ will see the potential

$$V = \frac{1}{2}\tilde{m}^2\tilde{\phi}^2 + \frac{1}{2}m_\chi^2\chi^2 + A\tilde{\phi}^3 + B\tilde{\phi}^4 + \ldots, \tag{14.31}$$

with $2m_\chi^2 = \lambda'\langle\phi\rangle^2$. The spontaneous breaking has generated a mass term for χ.

14.5.2 Global $U(1)$ and $SU(2)$ symmetry

A continuous symmetry can be spontaneously broken in a similar manner. The archetypal example is a complex field ϕ with the $U(1)$ symmetry Eq. (14.5) and the potential

$$\boxed{V(\phi) = \frac{1}{4}\lambda\left(|\phi|^2 - f^2\right)^2 = V_0 - m^2|\phi|^2 + \frac{1}{4}\lambda|\phi|^4}. \tag{14.32}$$

Considered as a function defined in the complex ϕ plane, V has a 'Mexican-hat' shape, shown in Figure 14.2. The vacuum now consists of the circle at the bottom of the rim, with radius f. Points on the circle are parameterized by the phase θ of ϕ, and the distance around the rim in field space is parameterized by the field[2] $\sigma \equiv \sqrt{2}f\theta$.

The field σ which parameterizes the vacuum (strictly speaking, the massless

[2] To be precise, σ is the distance in field space (Eq. (13.53)). In most situations the curvature of the rim can be ignored and then σ can be regarded as a canonically normalized field.

particle species corresponding to it) is called a **Nambu–Goldstone boson**. Acting on the vacuum, the $U(1)$ symmetry takes the form of a **shift symmetry**

$$\sigma \to \sigma + \text{const.} \tag{14.33}$$

If the symmetry is explicitly broken, the Nambu–Goldstone boson field acquires a potential which gives the particle mass. It is then called a **pseudo-Nambu–Goldstone boson** (PNGB). The symmetry-breaking potential is typically sinusoidal:

$$\boxed{V(\theta) = \Lambda^4 \left(1 - \cos\theta\right)}\,, \tag{14.34}$$

where Λ characterizes the strength of the explicit symmetry breaking. We chose $\theta = 0$ to correspond to the vacuum. The mass of the PNGB, read off from $V = \frac{1}{2}m^2\sigma^2 + \ldots$, is

$$\boxed{m = \frac{\Lambda^2}{\sqrt{2}\,f}}\,. \tag{14.35}$$

A $U(1)$ shift symmetry can act on a single real field, $\sigma \to \sigma +$ const, without reference to any other field. The field σ is still called a Nambu–Goldstone boson or, after explicitly breaking the symmetry, a PNGB.

As with a discrete symmetry, the spontaneous breaking of a global $U(1)$ symmetry will give mass to all of the fields coupling to the symmetry-breaking field, even if they have no mass term in the Lagrangian density .

The spontaneous breaking of $SU(2)$ gives three Nambu–Goldstone bosons, and more generally the spontaneous breaking of any non-Abelian global symmetry will give a Nambu–Goldstone boson for each parameter of the group. Explicit breaking is expected so that one will have PNGBs.

As we noticed in Section 14.2.2, explicit breaking of a global symmetry is expected on general grounds. If the symmetry is continuous and is explicitly broken (as here) there are more definite reasons to expect explicit breaking; no massless spin-0 particle is observed and neither is the modification to gravity that would be generated by the interaction of the corresponding field with ordinary matter. The interaction would need to be implausibly weak to avoid such modification.

14.5.3 Dynamical symmetry breaking

In the examples we gave, the Nambu–Goldstone boson field is elementary, i.e. it appears in the Lagrangian density. But spontaneous symmetry breaking works just as well if it is a product of elementary fields. It is then called a **condensate**. The condensate field has to be a scalar, but there is no need for the constituent fields to be scalars. In practice they are usually taken to be spin-$1/2$ fields. Then there is no need for the theory to contain elementary scalar fields.

When ϕ is a condensate, the Nambu–Goldstone boson should be regarded as a bound state, formed from those fields making up the condensate. From the Lagrangian density of the theory we can calculate the interactions of the Nambu–Goldstone boson, and its mass if it is a PNGB. The calculation is far easier than if we were dealing with a normal bound state. For example the mass and interactions of the pion, which is a PNGB corresponding to a quark–antiquark condensate, are much easier to calculate than the masses and interactions of the proton and neutron which are not PNGBs. (The global symmetry of which the pion is a PNGB is described in Section 16.5.2.)

All of this having being said, scenarios for the very early Universe (in particular, models of inflation) which invoke spontaneous symmetry breaking generally take the scalar field to be elementary. For this reason we shall not explore the condensate possibility.

14.6 $U(1)$ gauge symmetry

We have so far considered global symmetries, where the group parameters are independent of the spacetime position. Now we consider gauge transformations, where the group parameters can be functions of spacetime. We begin with $U(1)$ gauge symmetry. Allowing λ in Eq. (14.8) to depend on spacetime position, and replacing q_n by gq_n, we arrive at

$$\phi_n \to e^{i\lambda(x)gq_n}\phi_n. \tag{14.36}$$

The constant g is called the gauge coupling, and the numbers q_n are called the charges.

Only the products gq_n are physically significant, but it is important to pull out the overall factor g. This is because, as we are about to see, the individual magnitudes of the gq_n's are physically significant, and not just their ratios as would be the case for a global symmetry. By pulling out g we allow ourselves to choose the q_n as small integers, or the ratio of small integers, just as we would for a global symmetry. Pulling out g becomes more than a matter of convenience when one considers the dependence of quantities on the renormalization scale, discussed in Section 15.8.3. It turns out that only the overall coupling g runs, and not the individual q_n. We shall see in some detail how these things work when we come to the Standard Model.

With λ dependent on spacetime position, the kinetic term $-\partial_\mu\phi_n^*\partial^\mu\phi_n$ isn't invariant under the transformation (14.36). To achieve invariance one needs to include a field A_μ, called a gauge field, whose transformation is

$$A_\mu(x) \to A_\mu(x) + \partial_\mu \lambda(x). \tag{14.37}$$

One also needs to define a **covariant derivative** $\mathcal{D}_\mu$, which acting on a field ϕ_n is defined by

$$\boxed{\mathcal{D}_\mu \equiv \partial_\mu - igq_n A_\mu(x)}. \tag{14.38}$$

Then $\mathcal{D}_\mu \phi_n$ transforms in the same way as ϕ_n.

The Lagrangian density is invariant if it has the following form:

$$\boxed{\mathcal{L} = -\frac{1}{4} F_{\mu\nu} F^{\mu\nu} - \sum_n (\mathcal{D}^\mu \phi_n)^* \mathcal{D}_\mu \phi_n - V}, \tag{14.39}$$

$$\boxed{F_{\mu\nu} \equiv \partial_\mu A_\nu - \partial_\nu A_\mu}, \tag{14.40}$$

with V invariant under the transformation (14.36) (which is equivalent to its invariance under the global transformation (14.8)).

The first term of the Lagrangian density is included so that A_μ is a dynamical field, which after quantization will correspond to a species of spin-1 particle.[3] The quantity $F_{\mu\nu}$ is called the field strength, and is gauge invariant.

By taking $\lambda(x)$ to be independent of position, the gauge symmetry gives a global symmetry as a special case. This global symmetry has no effect on A_μ, and Noether's theorem applies with the current given by Eq. (14.25). Note that δS doesn't vanish in this equation, because S is supposed to be evaluated with $\epsilon \to \epsilon(x)$ but no compensating change in A_μ.

From the action principle one finds the field equations

$$\mathcal{D}_\mu \mathcal{D}^\mu \phi_n + \frac{\partial V}{\partial \phi_*} = 0, \tag{14.41}$$

$$\partial_\nu F^{\mu\nu} = j^\mu. \tag{14.42}$$

In Eq. (14.42), j^μ is the conserved current of the $U(1)$ gauge symmetry, given by Eq. (14.10) with the replacements $\partial_\mu \to \mathcal{D}_\mu$ and $q_n \to gq_n$. From the definition of $F_{\mu\nu}$,

$$\partial_\mu F_{\nu\alpha} + \partial_\alpha F_{\mu\nu} + \partial_\nu F_{\alpha\mu} = 0. \tag{14.43}$$

Inserting the Lagrangian density into Eq. (13.61) gives the energy–momentum tensor, which may be written

$$T_{\mu\nu} = T^{\text{scalar}}_{\mu\nu} + T^{\text{vector}}_{\mu\nu}. \tag{14.44}$$

[3] By a dynamical field, we mean one whose time derivative is in the Lagrangian density, so that the action principle gives its time dependence. Otherwise, the action principle just gives a constraint equation, relating the field to other fields at each instant of time.

The first term is the contribution of the scalar fields, given by Eq. (13.32) with the replacement $\partial_\mu \to \mathcal{D}_\mu$. The second term is the contribution of $F_{\mu\nu}$, given by

$$\boxed{T_{\mu\nu}^{\text{vector}} = \frac{1}{4} g_{\mu\nu} F_{\rho\sigma} F^{\rho\sigma} - F_{\mu\rho} F_\nu^\rho}. \tag{14.45}$$

If we identify q_n with the electric charge carried by ϕ_n, we arrive at what is called scalar electrodynamics, where $F_{\mu\nu}$ is the electromagnetic field strength. With a different identification of q_n, we would arrive at a field strength that isn't the electromagnetic one. Further discussion of this issue within the context of the Standard Model is given in Chapter 16.

The covariant derivative $\mathcal{D}_\mu$ has much in common with the curved spacetime object of the same name. At a given spacetime point, the transformation (14.37) lets us choose $A_\mu = 0$ so that $\mathcal{D}_\mu = \partial_\mu$. Then the scalar field equation, the conserved current and the energy–momentum tensor have the same form as if there were no electromagnetic field. This is analogous to choosing a locally inertial frame in curved spacetime. If we can choose $A_\mu = 0$ throughout some region of spacetime, then $\mathcal{D}_\mu = \partial_\mu$ everywhere throughout the region and the field strength tensor $F_{\mu\nu}$ vanishes. This is the analogue of flat spacetime, with $F_{\mu\nu}$ the analogue of spacetime curvature.

Once we allow $\lambda(x)$ to depend on spacetime position, there can be no restriction on the magnitude of its gradient $\partial_\mu \lambda$. As a result, it is impossible to define an explicitly broken version of a gauge symmetry. Were we to add a symmetry breaking term to the Lagrangian density, its physical effect could be arbitrarily large no matter how small we made its coefficient. In other words, a gauge symmetry cannot be explicitly broken.

14.7 $SU(2)$ gauge symmetry

Allowing the three functions λ_n in Eq. (14.15) to depend on spacetime position, we arrive at the $SU(2)$ gauge transformations $\Phi \to U\Phi$. To make the kinetic term invariant under these transformation, we need a triplet of gauge fields W_μ^n, which are conveniently collected into a 2×2 matrix as in Eq. (14.20):

$$W_\mu(x) \equiv \mathbf{W}_\mu(x) \cdot \mathbf{T} \equiv W_\mu^n \hat{T}_n. \tag{14.46}$$

Its $SU(2)$ gauge transformation is defined as

$$W_\mu \to -\frac{i}{g}(\partial_\mu U)U^{-1} + UW_\mu U^{-1}, \tag{14.47}$$

The covariant derivative (acting on Φ or W_μ) is defined as

$$\mathcal{D}_\mu \equiv \partial_\mu - igW_\mu. \tag{14.48}$$

Then $\mathcal{D}_\mu\Phi$ transforms in the same way as Φ. We also need a field strength matrix

$$W_{\mu\nu} \equiv \mathcal{D}_\mu W_\nu - \mathcal{D}_\nu W_\mu = \partial_\mu W_\nu - \partial_\nu W_\mu - ig[W_\mu, W_\nu], \tag{14.49}$$

where $[W_\mu, W_\nu] \equiv W_\mu W_\nu - W_\nu W_\mu$ is the commutator of the matrices. Its transformation is the same as for a global symmetry:

$$W_{\mu\nu} \to U W_{\mu\nu} U^{-1}. \tag{14.50}$$

The following Lagrangian density is invariant under the $SU(2)$ gauge transformation:

$$\boxed{\mathcal{L} = -|\mathcal{D}_\mu\Phi|^2 + V(|\Phi|^2) - \frac{1}{2}\operatorname{Tr} W_{\mu\nu}W^{\mu\nu}}, \tag{14.51}$$

where

$$\boxed{|\mathcal{D}_\mu\Phi|^2 \equiv \sum_n (\mathcal{D}_\mu\phi_n)^*\mathcal{D}^\mu\phi_n}. \tag{14.52}$$

The commutator in the definition of $W_{\mu\nu}$ means that the gauge fields are interacting. The gauge coupling g specifies the strength of this interaction, and also of the interaction of the gauge fields with Φ.

Making the λ_n independent of spacetime position we recover global $SU(2)$, with the conserved charge T_3. The components ϕ_n of the doublet Φ have $T_3 = \pm 1/2$. For the triplet W^n_μ, we have $T_3 = \pm 1$ for $(W^1_\mu \pm iW^2_\mu)/\sqrt{2}$, and $T_3 = 0$ for W^3_μ (the same as for the triplet W_n of Section 14.3.1).

Equations (14.47)–(14.52) apply to any $SU(N)$ gauge symmetry, with W_μ a Hermitian traceless $N \times N$ matrix, and $\Phi = (\phi_1, \ldots, \phi_N)$. With different choices for W_μ they apply to a wide class of continuous groups, called simple Lie groups, which include the groups $SU(N)$ and $U(1)$. Later we will include spin-$1/2$ fields. In all cases, invariance under a simple Lie group is achieved with a single gauge coupling g.

14.8 Spontaneously broken gauge symmetry

In this section we see what happens if a gauge symmetry is spontaneously broken. A field whose vev spontaneously breaks a gauge symmetry is called a **Higgs field**, and the particles corresponding to its normal modes of oscillation about the vacuum are called Higgs bosons. All fields which couple to a Higgs field acquire mass, even if there is no mass term in the Lagrangian density. This is called the Higgs mechanism. Gauge bosons can acquire mass only through the Higgs mechanism, because the gauge symmetry forbids a mass term in the Lagrangian density.

14.8.1 Spontaneously broken $U(1)$ gauge symmetry

We assume just one Higgs field ϕ with unit charge. We use Eq. (14.32) to spontaneously break the $U(1)$ symmetry, so that the Lagrangian density is

$$\boxed{\mathcal{L} = -|\mathcal{D}_\mu\phi|^2 - \frac{1}{2}\lambda^2\left(|\phi|^2 - \frac{1}{2}v^2\right)^2 - \frac{1}{4}F_{\mu\nu}^2}, \tag{14.53}$$

where $|\mathcal{D}_\mu\phi|^2 \equiv (\mathcal{D}^\mu\phi)^*(\mathcal{D}_\mu\phi)$ and $F_{\mu\nu}^2 \equiv F_{\mu\nu}F^{\mu\nu}$. By virtue of the local $U(1)$ symmetry we can choose ϕ to be everywhere real. This is called the unitary gauge. In the unitary gauge $\langle\phi\rangle = v/\sqrt{2}$ with v real.

Writing $\phi = (v + \chi(x))/\sqrt{2}$ the Lagrangian density becomes

$$\mathcal{L} = -\frac{1}{2}(\partial_\mu\chi)^2 - \frac{1}{8}\lambda^2\chi^2\,(2v+\chi)^2 - \frac{1}{2}g^2A_\mu^2(v+\chi)^2 - \frac{1}{4}F_{\mu\nu}F^{\mu\nu}, \tag{14.54}$$

where $A_\mu^2 \equiv A^\mu A_\mu$ and $(\partial_\mu\chi)^2 \equiv (\partial^\mu\chi)(\partial_\mu\chi)$. There is a term $-\lambda^2v^2\chi^2/2$ which gives the Higgs boson a mass λv. There is also a term $-g^2v^2A_\mu^2/2$ which (after quantization) gives the gauge boson a mass gv. The other terms specify the self interaction of the Higgs field, and its interaction with the gauge boson.

As in the global symmetry case, the field parameterizing the vacuum can be either elementary or a condensate. (In the simple example we studied it is elementary corresponding to the phase of H.) In contrast with the global case, this field doesn't correspond to a particle (i.e. there is no Nambu–Goldstone boson) because it can be set equal to zero by the gauge transformation. That is a general feature of gauge symmetries.

14.8.2 Spontaneously broken $SU(2)$ gauge symmetry

To spontaneously break $SU(2)$ symmetry we need at least two Higgs fields, forming a Higgs doublet. We write it as $H \equiv (H_1, H_2)$. Taking the potential to be quartic, we write

$$\boxed{\mathcal{L} = -|\mathcal{D}_\mu H|^2 - \frac{1}{2}\lambda^2\left(|H|^2 - \frac{1}{2}v^2\right)^2 - \frac{1}{2}\operatorname{Tr}W_{\mu\nu}^2}, \tag{14.55}$$

where $|\mathcal{D}_\mu H|^2 \equiv \sum|\mathcal{D}_\mu H_n|^2$ and $|H|^2 \equiv \sum|H_n|^2$ and $W_{\mu\nu}^2 \equiv W_{\mu\nu}W^{\mu\nu}$. By virtue of the local gauge symmetry we can choose $H_1 = 0$ and H_2 real. This is called the unitary gauge. Writing $H_2 = (v+\chi)/\sqrt{2}$ the Lagrangian density becomes

$$\mathcal{L} = -\frac{1}{2}(\partial_\mu\chi)^2 - \frac{1}{2}g^2\sum_n(W_\mu^n)^2(v+\chi)^2 - \frac{1}{8}\lambda^2\chi^2(2v+\chi)^2 - \frac{1}{2}\operatorname{Tr}W_{\mu\nu}^2, \tag{14.56}$$

where $(W^n_\mu)^2 \equiv W^n_\mu W^{n\mu}$.

We see that the Higgs boson has mass λv and that each of the gauge bosons has mass gv.

14.9 Discrete gauge symmetry

Suppose now that we have a $U(1)$ gauge symmetry acting on complex fields ϕ_1 and ϕ_2 so that $\phi_n(x) \to e^{iq_i\alpha(x)}\phi_n(x)$. Suppose that the charges are $q_1 = 1$ and $q_2 = 2$. In the special case that $\alpha = \pi$ throughout spacetime, this gives $\phi_2 \to +\phi_2$ but $\phi_1 \to -\phi_1$. Finally, suppose that $\langle\phi_2\rangle = 0$ while $\langle|\phi_1\rangle| \neq 0$. Then, well below the scale set by the height V_0 of the potential (its value at $\phi_1 = \phi_2 = 0$), there is a 'low energy' effective field theory in which $|\phi_1|$ is fixed at its vev, so that $\phi_1 = \langle|\phi_1|\rangle e^{i\theta}$ with only θ corresponding to a field. In the low-energy theory, the only trace of the gauge symmetry is the discrete symmetry $\theta \to \theta + \pi$. This is the simplest example of a discrete gauge symmetry [3].

More complicated examples are easily constructed. The importance of discrete gauge symmetries lies in the fact that they cannot be explicitly broken. As we remarked earlier, one expects on general grounds that a discrete symmetry will be explicitly broken. Therefore, whenever the explicit breaking of discrete symmetry appearing in a field theory is required to be completely negligible, one usually supposes that it is a gauge symmetry. Some authors justify the assumption of a discrete gauge symmetry by exhibiting an underlying continuous gauge symmetry, while others just declare that the discrete symmetry is a gauge symmetry. As far as the application of the symmetry is concerned, it makes no difference which route is adopted. Cases that we shall encounter, in which a discrete gauge symmetry is desirable, are the R-parity of the MSSM (Section 17.5.1) and Z_N symmetries with high N invoked in connection with the axion (Section 16.6) as well as with certain models of inflation (Sections 28.7 and 28.8).

Exercises

14.1 Show that the elements I and A of the Z_2 group satisfy $A^2 = I$.

14.2 Show that the current (14.6) is conserved if and only if V is independent of the phase of ϕ.

14.3 Writing $\phi = (\phi_1 + i\phi_2)/\sqrt{2}$, show that the $U(1)$ transformation (14.5) corresponds for infinitesimal λ to $\phi_1 \to \phi_1 - \lambda\phi_2$ and $\phi_2 \to \phi_2 + \lambda\phi_1$. Hence calculate $\mathcal{F}_1$ and $\mathcal{F}_2$ in Eq. (14.24), and verify that Eq. (14.26) coincides with Eq. (14.6).

14.4 Verify that the Lagrangian density (14.39) is invariant under the local $U(1)$ transformation.

14.5 Verify Eq. (14.54), giving this Lagrangian density after spontaneous symmetry breaking.

14.6 Verify that the Lagrangian density (14.51) is invariant under the local $SU(2)$ transformation.

14.7 Verify Eq. (14.56), giving this Lagrangian density after spontaneous symmetry breaking.

References

[1] S. Weinberg. *The Quantum Theory of Fields, Volume I* (Cambridge: Cambridge University Press, 1995).

[2] S. Weinberg. *Gravitation and Cosmology, Principles and Applications for General Relativity* (New York: John Wiley and Sons, 1972).

[3] L. M. Krauss and F. Wilczek, Discrete gauge symmetry in continuum theories. *Phys. Rev. Lett.*, **62** (1989) 1221.

15

Quantum field theory

In this chapter we consider quantum field theory. We begin with some basic material, including canonical quantization and its application to the free scalar field. Then we sketch quite briefly how higher spins and interactions are handled. Finally, we introduce, with the aid of a particular example, the important concept of 'running' parameters.

15.1 Schrödinger and Heisenberg pictures

A quantum theory invokes a Hilbert space, which is an infinite-dimensional vector space of a particular kind. Following the usual practice we pretend that Hilbert space can be treated like a finite-dimensional vector space. Then each linear operator correspond to a matrix which acts on the components of vectors. In quantum theory we allow the components to be complex, and we deal with Hermitian operators and unitary operators (corresponding respectively to Hermitian and unitary matrices).

At a given time, each physical state corresponds to a state vector $|X\rangle$ which is normalized ($\langle X|X\rangle = 1$) and whose phase is arbitrary. Each observable corresponds to a Hermitian operator $\hat{A}$, whose eigenvectors can be taken to be orthonormal ($\langle n|m\rangle = \delta_{nm}$). If the observable is measured when the state vector is $|X\rangle$, the probability of finding a value A_n is

$$\text{probability} = |\langle n|X\rangle|^2, \tag{15.1}$$

where $|n\rangle$ is the eigenvector of $\hat{A}$ with eigenvalue A_n.[1] The expectation value is $\langle X|\hat{A}|X\rangle$. Immediately after a value A_n has been found, the state vector is $|n\rangle$.

The time dependence of the system is specified by a Hamiltonian operator, denoted by $\hat{H}(t, \hat{q}_1, \hat{p}_1, \hat{q}_2, \hat{p}_2, \ldots)$. The operators $\hat{q}_n$ correspond to degrees of freedom, and the operators $\hat{p}_n$ to their canonical conjugates.

[1] To keep things simple, we here take the eigenvalues to be discrete and non-degenerate.

In what is known as the **Schrödinger picture**, the degrees of freedom are time independent while the state vector satisfies the Schrödinger equation

$$\frac{d}{dt}|t\rangle = -i\hat{H}|t\rangle. \tag{15.2}$$

If $\hat{H}$ is time independent the corresponding observable is by definition the energy. This is supposed to be the case for any isolated system. An observable is then conserved if its operator is time independent and commutes with $\hat{H}$, and in particular energy is conserved.

The Schrödinger equation is equivalent to

$$\boxed{|t\rangle = \hat{U}(t)|0\rangle}, \qquad \boxed{\dot{\hat{U}} = -i\hat{H}\hat{U}}, \tag{15.3}$$

with $\hat{U}$ unitary. Now let us make the replacements

$$|t\rangle \to |0\rangle \quad = \quad \hat{U}^{-1}(t)\,|t\rangle, \tag{15.4}$$

$$\hat{A} \quad \to \quad \hat{U}^{-1}\hat{A}\hat{U}, \tag{15.5}$$

This is a new way of assigning state vectors and operators to physical states and observables, which is completely equivalent to the original one but which makes the state vectors time independent. The new setup is called the **Heisenberg picture**, and is more convenient for quantum field theory.

From Eq. (15.5), we learn that in the Heisenberg picture an operator $\hat{A}(q_n, p_n, t)$ satisfies

$$\boxed{\frac{d\hat{A}}{dt} = i[\hat{H}, \hat{A}] + \frac{\partial \hat{A}}{\partial t}}, \tag{15.6}$$

where $[\hat{H}, \hat{A}]$ is the commutator. The last term gives the time dependence of $\hat{A}$ in the Schrödinger picture. Setting $\hat{A} = \hat{H}$, we learn that $\hat{H}$ is time independent in the Heisenberg picture if and only if it is time independent in the Schrödinger picture. From now on *we work exclusively in the Heisenberg picture.*

In the Heisenberg picture, the Hamiltonian operator can be derived from a Lagrangian operator $L(t, \hat{q}_1, \dot{\hat{q}}_1, \hat{q}_2, \dot{\hat{q}}_2, \ldots)$, with the relation between the two the same as in the classical case. That in turn can be derived from an action $S = \int dt L$. (We are not going to put hats on the Lagrangian, Lagrangian density or action operators.) The degrees of freedom satisfy the same equations as their classical counterparts:

$$\dot{\hat{q}}_n = \frac{\partial \hat{H}}{\partial \hat{p}_n} \qquad \dot{\hat{p}}_n = -\frac{\partial \hat{H}}{\partial \hat{q}_n}. \tag{15.7}$$

It follows that $d\hat{A}/dt$ is also given by the classical equation:

$$\frac{d\hat{A}}{dt} = \sum_n \left(\frac{\partial \hat{H}}{\partial \hat{p}_n} \frac{\partial \hat{A}}{\partial \hat{q}_n} - \frac{\partial \hat{A}}{\partial \hat{p}_n} \frac{\partial \hat{H}}{\partial \hat{q}_n} + \frac{\partial \hat{A}}{\partial t} \right). \tag{15.8}$$

For Eq. (15.8) to be compatible with Eq. (15.6) we need

$$\frac{\partial \hat{H}}{\partial \hat{q}_n} = -i[\hat{H}, \hat{p}_n], \qquad \frac{\partial \hat{H}}{\partial \hat{p}_n} = i[\hat{H}, \hat{q}_n]. \tag{15.9}$$

The $\hat{q}_n$ and $\hat{p}_n$ cannot all commute, or the right-hand sides would vanish. Therefore, if we define a function $\hat{A}$ as (say) a power series in the q_n and p_n, the order in which they appear has in general to be specified. For this reason one has in general to be careful about the definition of the derivatives $\partial/\partial\hat{q}_n$ and $\partial/\partial\hat{p}_n$ acting on an operator. We will not pursue this question, since we perform calculations only for the system of harmonic oscillators for which there is no ambiguity.

In many situations, the quantum theory is obtained from the classical theory simply by promoting the classical degrees of freedom q_n (and their conjugate momenta p_n) to operators. That is the case for classical mechanics, and for bosonic fields in a quantum field theory. Then one may expect that the quantum theory has a classical limit, corresponding to states $|X\rangle$ for which the q_n have sharply defined values during the time period under consideration.

The fermionic fields in a quantum field theory are quite different. The $\hat{q}_n$ that correspond to them aren't Hermitian and there is no limit in which a fermion field can be regarded as classical. Note though that a product of two fermion fields is bosonic (from the rule for adding angular momenta). It can form a condensate, as discussed in Section 14.5.3, which can be regarded as classical.

In this book, almost all of our discussion will focus on the simplest bosonic fields, namely scalar fields, which play the dominant role in most scenarios of the early Universe. For fermionic fields, we confine ourselves to the brief exposition of spin-$1/2$ fields in Section 15.6, which is needed to write down the Lagrangian density of the Standard Model and of supersymmetric theories.

15.2 Symmetry and conserved currents

We have seen how the Lagrangian formalism can describe classical fields. In that context we introduced the concept of a symmetry, whereby the action and hence the physical predictions are invariant under some group of transformations. We identified two kinds of symmetry group; spacetime symmetries, where the group consists of translations and Lorentz transformations, and internal symmetries where the group transforms the fields at a each point in spacetime. We noted that any continuous global symmetry group gives a set of conserved currents, one for each parameter of the group, and hence a set of conserved charges. For the latter, we wrote down a formula that is valid whenever the Lagrangian (and not just the action) is invariant.

The entire discussion works just as well if we regard the degrees of freedom as operators. We see in this section that there is, however, a simpler way of arriving at the conserved quantity associated with a continuous symmetry. It actually applies to any quantum theory, not just quantum field theory, though the latter is our interest here. We see also that in the quantum regime, discrete symmetries also give conserved quantities.

15.2.1 Symmetries in quantum theory

Consider a group of transformations acting on the states:

$$|X'\rangle = \hat{V}|X\rangle, \tag{15.10}$$

where the operators $\hat{V}$ are a representation of the group.[2] If it is a symmetry group, it must not change the overlaps $|\langle X|Y\rangle|$. According to a theorem of Wigner, it is always possible to choose the phases of the state vectors so that either (i) $\hat{V}$ is unitary so that $\langle n'|m'\rangle = \langle n|m\rangle$, or (ii) $\hat{V}$ is anti-unitary which means[3] $\langle n'|m'\rangle = \langle m|n\rangle$ and

$$\hat{V}\left(a|X\rangle + b|Y\rangle\right) = a^*\hat{V}|X\rangle + b^*\hat{V}|Y\rangle. \tag{15.11}$$

A careful proof of this theorem is given in Ref. [1]. For a continuous group, the unitary option must hold because it holds for the identity. The unitary option turns out to hold for all transformations of physical interest except those that include time reversal, and we assume it from now on.

The symmetry group must act on the operators representing observables too:

$$\hat{A}' = \hat{V}\hat{A}\hat{V}^{-1}. \tag{15.12}$$

Only then will the probability Eq. (15.1), of obtaining a given value from the measurement of a given observable, be the same before and after the transformation.

15.2.2 Conserved quantities

We are assuming that $\hat{H}$ is time independent, so that energy is conserved. This corresponds to invariance under the group of coordinate changes (time translations)

[2] For the following analysis to hold, we have to allow the possibility that the representation is multi-valued. The only case of interest is the Lorentz group where we allow double-valued representations corresponding to half-integer spin.

[3] In contrast with a unitary or Hermitian operator, an anti-unitary operator doesn't correspond to a matrix acting on the components of vectors in the Hilbert space.

$t \to t + T$ with T any constant, and to the operator $\hat{V} = e^{-i\hat{H}T}$.[4] For other symmetry groups, $\hat{V}$ typically has nothing to do with time.[5] Then $\hat{V}$ will commute with $\hat{H}$ and, as we now see, there will be more conserved quantities.

Continuous symmetries

The elements of a continuous symmetry group are labelled by one or more continuous parameters. Varying just one parameter, the unitarity of $\hat{V}$ allows us to write

$$\hat{V} = \exp(i\lambda\hat{F}), \tag{15.13}$$

with $\hat{F}$ Hermitian. Taking λ to be infinitesimal, we can write

$$\hat{V}(\lambda) = I + i\lambda\hat{F}, \tag{15.14}$$

where I is the unit operator. Since $\hat{V}$ is unitary, $\hat{F}$ is Hermitian. Also, since we are supposing that $\hat{V}$ commutes with $\hat{H}$, then $\hat{F}$ does too. We see that for each parameter of the group, there is a conserved quantity $\hat{F}$ given by Eq. (15.14). The effect of the infinitesimal transformation on the operators, given by Eq. (15.12), is

$$\hat{A}(\lambda) = \hat{A}(0) + i\lambda[\hat{F}, \hat{A}(0)]. \tag{15.15}$$

These are general results. Coming to field theory, our finding of a conserved quantity for each parameter of the group is in agreement with Noether's theorem for global continuous symmetries. When the Lagrangian is invariant, as it is for internal symmetries, as well as for spatial translations and rotations, it can be shown [1] that the F of Eq. (14.27), promoted to an operator, is precisely the $\hat{F}$ of equation (15.14).

Considering a coordinate change $z \to z + Z$ (translation along the z-axis) by an infinitesimal amount Z, we write

$$\boxed{\hat{V}(Z) = I - i\hat{p}_z Z}, \tag{15.16}$$

and identify $\hat{p}_z$ with the operator corresponding to the z-component of momentum. Considering instead an infinitesimal rotation about the z-axis through angle θ, we write

$$\hat{V} = I + i\hat{J}_z\theta, \tag{15.17}$$

and identify $\hat{J}_z$ as the operator for the z-component of angular momentum. Notice that the definition of momentum parallels the definition of energy.

As is well known, these definitions of energy, momentum and angular momentum make sense in the context of non-relativistic quantum mechanics, and they

[4] This doesn't mean that we have to be working in the Schrödinger picture, since old and new coordinates are supposed to refer to the same point in spacetime.

[5] The exceptions are Lorentz boosts of the Lorentz group, and supersymmetry.

make sense in field theory too. They agree with the definitions arrived at through Noether's theorem in Section 14.4, because (i) the momentum and angular momentum operators are given by Eq. (14.27) and (ii) the Hamiltonian operator is taken to be the object derived from the Lagrangian (given by Eq. (13.30) with $\sum \Pi_n \phi_n$ instead of $\Pi\phi$).

Discrete symmetry

For a discrete symmetry the case of most interest is Z_2 symmetry. Then $\hat{V}^{-1} = \hat{V}$ which means that $\hat{V}$ itself is Hermitian. It can be taken as the conserved quantity, with eigenvalues ± 1.

An example is the parity transformation, considered in Section 16.5.4, which interchanges right- and left-handedness.[6] Another is the R-parity of Section 17.5.1, which is an internal symmetry.

Additive and multiplicative conserved quantities

Consider now what happens if we have two independent systems, each carrying some amount of a given conserved quantity. Two independent systems correspond to independent Hilbert spaces $|X_i\rangle$, from which we form the direct product space $|X_1\rangle|X_2\rangle$. Then $\hat{V} = \hat{V}_1\hat{V}_2$, with $\hat{V}_i$ acting only on $|X_i\rangle$. If the conserved quantity corresponds to a continuous symmetry, it is therefore additive, the eigenvalues satisfying $c = c_1 + c_2$.

For a Z_2 symmetry, the conserved quantity is instead multiplicative, the eigenvalues satisfying $u = u_1 u_2$. As we see in Section 23.2, the multiplicative nature of R-parity could be responsible for the stability of cold dark matter (CDM).

15.3 Harmonic oscillator

The only quantum theory that we shall need to work out in detail is that of a set of harmonic oscillators. As shall see, it describes the Fourier components of a free scalar field. We begin with a single harmonic oscillator. Promoting q to an operator, Eq. (13.10) becomes

$$L = \frac{1}{2}\dot{\hat{q}}^2 - \frac{1}{2}\omega^2\hat{q}^2, \qquad \hat{H} = \frac{1}{2}\hat{p}^2 + \frac{1}{2}\omega^2\hat{q}^2, \qquad \hat{p} = \dot{\hat{q}}. \tag{15.18}$$

To satisfy the consistency condition (15.9) we need the **canonical commutation relation**

$$[\hat{q}, \hat{p}] = i. \tag{15.19}$$

[6] In this case one should in principle allow $\pm i$ as well, corresponding to the representation of the Z_2 being double-valued, but that option isn't usually needed.

We define an operator $\hat{a}$, which corresponds to the quantity a that appeared in Eq. (13.12), via

$$\boxed{\hat{q}(t) = \frac{1}{\sqrt{2\omega}} \left[e^{-i\omega t} \hat{a} + e^{i\omega t} \hat{a}^\dagger \right]}. \tag{15.20}$$

From Eq. (15.19) it satisfies

$$\boxed{[\hat{a}, \hat{a}^\dagger] = 1}. \tag{15.21}$$

Calculating $\hat{p} = \dot{\hat{q}}$, the Hamiltonian in Eq. (13.10) becomes

$$\hat{H} = \tfrac{1}{2} \left(\hat{a}^\dagger \hat{a} + \hat{a} \hat{a}^\dagger \right) \omega = \left(\hat{a}^\dagger \hat{a} + \tfrac{1}{2} \right) \omega. \tag{15.22}$$

This can be written

$$\boxed{\hat{H} = \left(\hat{n} + \tfrac{1}{2} \right) \omega}, \qquad \hat{n} \equiv \hat{a}^\dagger \hat{a}. \tag{15.23}$$

We want to know the eigenvalues of $\hat{H}$, or equivalently of $\hat{n}$, which give the energy levels of the harmonic oscillator. In the Schrödinger picture these are obtained by solving the Schrödinger equation, leading to the well-known result that the energy levels are $(n + \frac{1}{2})$ where n is a non-negative integer:

$$\hat{H} |n\rangle = \left(n + \tfrac{1}{2} \right) \omega \, |n\rangle, \qquad \hat{n} |n\rangle = n |n\rangle. \tag{15.24}$$

Since $\hat{H}$ is Hermitian, the state vectors are orthogonal and they are chosen to be orthonormal, $\langle n | m \rangle = \delta_{nm}$. They provide a basis for the Hilbert space of the quantum theory.

The same result can be obtained in the Heisenberg picture as follows. First we define the ground state by

$$\hat{a} |0\rangle = 0 |0\rangle. \tag{15.25}$$

Then we define $|1\rangle \equiv \hat{a}^\dagger |0\rangle$. Continuing the process we build the entire basis: $\hat{a}^\dagger |n\rangle = \sqrt{n+1} \, |n+1\rangle$.

For a set of oscillators, the Hamiltonian is

$$\boxed{\hat{H} = \sum_i \left(\hat{n}_i + \tfrac{1}{2} \right) \omega_i}, \qquad \hat{n}_i \equiv \hat{a}_i^\dagger \hat{a}_i, \tag{15.26}$$

$$[\hat{a}_i, \hat{a}_j^\dagger] = \delta_{ij}, \qquad [\hat{a}_i, \hat{a}_j] = 0. \tag{15.27}$$

15.4 Quantized free scalar field

The Lagrangian density for a real scalar field operator, given by Eqs. (13.22) and (13.35), is

$$\mathcal{L} = -\frac{1}{2}\partial^\mu\hat{\phi}\partial_\mu\hat{\phi} - \frac{1}{2}m^2\hat{\phi}^2. \tag{15.28}$$

The corresponding Hamiltonian is given by Eq. (13.30).

15.4.1 Fourier series

Since the solution of the field equation is a sum of plane waves, we know that the Hamiltonian will be a sum of harmonic oscillator Hamiltonians. To calculate these Hamiltonians in terms of the Fourier components we write[7]

$$\begin{aligned}\hat{\phi}(\mathbf{x},t) &= L^{-3}\sum_{\mathbf{k}}\left[\phi_k(t)\hat{a}_{\mathbf{k}}e^{i\mathbf{k}\cdot\mathbf{x}} + \phi_k^*(t)\hat{a}_{\mathbf{k}}^\dagger e^{-i\mathbf{k}\cdot\mathbf{x}}\right] \\ &= L^{-3}\sum_{\mathbf{k}}\left[\phi_k(t)a_{\mathbf{k}} + \phi_k^*(t)a_{-\mathbf{k}}^\dagger\right]e^{i\mathbf{k}\cdot\mathbf{x}}.\end{aligned} \tag{15.29}$$

The **mode function** $\phi_k(t)$ depends only on k. It satisfies the oscillator equation

$$\boxed{\ddot{\phi}_k + E_k^2\phi_k = 0}, \qquad \boxed{E_k \equiv \sqrt{k^2+m^2}}, \tag{15.30}$$

and we choose the solution

$$\boxed{\phi_k(t) = \sqrt{\frac{1}{2E_k}}\,e^{-iE_kt}}. \tag{15.31}$$

Inserting Eq. (15.29) into Eq. (13.30), one finds

$$\boxed{\hat{H} = \sum_{\mathbf{k}}\left(\hat{n}_{\mathbf{k}} + \tfrac{1}{2}\right)E_k} \qquad \boxed{\hat{n}_{\mathbf{k}} \equiv L^{-3}\hat{a}_{\mathbf{k}}^\dagger\hat{a}_{\mathbf{k}}}. \tag{15.32}$$

This has the form Eq. (15.23) with $L^{-3/2}\hat{a}_{\mathbf{k}}$ instead of $\hat{a}_i$. The commutation relation of the former is therefore

$$\boxed{L^{-3}[\hat{a}_{\mathbf{k}}, \hat{a}_{\mathbf{k}'}^\dagger] = \delta_{\mathbf{k}\mathbf{k}'}}. \tag{15.33}$$

The operators $\hat{n}_{\mathbf{k}}$ commute for different $\mathbf{k}$. As a result, we can find orthonormal states $|n_{\mathbf{k}_1}, n_{\mathbf{k}_2}, \ldots\rangle$ that are eigenvectors of every $\hat{n}_{\mathbf{k}}$ with some eigenvalue $n_{\mathbf{k}}$. Starting with the $|0, 0, \ldots\rangle$, we can build the states with non-zero $n_{\mathbf{k}}$ by acting with $\hat{a}_{\mathbf{k}}^\dagger$ just as we did for the harmonic oscillator. The states $|n_{\mathbf{k}_1}, n_{\mathbf{k}_2}, \ldots\rangle$ provide a basis for the Hilbert space of the quantum theory, which is called the Fock space.

[7] Remember that the discussion is at this stage in flat spacetime, so that $\mathbf{x}$ is the physical distance and $\mathbf{k}$ is the physical wave-vector (momentum).

The expression (15.32) shows that an increase of $n_{\mathbf{k}}$ by 1 increases the energy by an amount E_k. The following argument shows that it also increases the momentum by an amount $\mathbf{k}$.

Under an infinitesimal translation $z \to z + Z$, $\hat{\phi}(\mathbf{x})$ changes by an amount $Z\partial_z\hat{\phi}(\mathbf{x})$. According to the definition (15.16) of the momentum operator $\hat{p}_z$, that change has to be generated by Eq. (15.15) with $\lambda = -Z$, $\hat{F} = \hat{p}_z$, $\hat{A}(0) = \hat{\phi}(z)$ and $\hat{A}(\lambda) = \hat{\phi}(z+Z)$. Since $\hat{\phi}$ is given by Eq. (15.29), this requires $[\hat{p}_z, \hat{a}_{\mathbf{k}}^\dagger] = k_z\hat{a}_{\mathbf{k}}^\dagger$. Acting with $\hat{p}_z$ on a state with momentum component p_z, we then have

$$\hat{p}_z\hat{a}_{\mathbf{k}}^\dagger|p_z\rangle = \left(\hat{a}_{\mathbf{k}}^\dagger\hat{p}_z + k_z\hat{a}_{\mathbf{k}}^\dagger\right)|p_z\rangle = (p_z + k_z)\hat{a}_{\mathbf{k}}^\dagger|p_z\rangle. \tag{15.34}$$

Assuming that the state $|0, 0, \ldots\rangle$ has zero momentum, the momentum of a state with definite $n_{\mathbf{k}}$ is

$$\boxed{\hat{\mathbf{p}} = \sum_{\mathbf{k}} \hat{n}_{\mathbf{k}}\mathbf{k}}\,. \tag{15.35}$$

In view of these results, we interpret the state $|0, 0, \ldots\rangle$ as the vacuum (containing no particles) and identify $n_{\mathbf{k}}$ as the number of particles with momentum $\mathbf{k}$ and energy $E_k = \sqrt{m^2 + k^2}$ (making the particle mass m). The operator $\hat{a}_{\mathbf{k}}^\dagger$ is called a creation operator because it creates a particle, while $\hat{a}_{\mathbf{k}}$ is called an annihilation operator because it destroys one.

To arrive at Eq. (15.35) we used the definition (15.16) of momentum. We could instead have used the equivalent definition (13.33) (with ϕ promoted to an operator). Inserting the Fourier expansion of ϕ and integrating over all space we would again arrive at Eq. (15.35). Conversely, we could show that $\hat{a}_{\mathbf{k}}^\dagger$ creates a particle with energy E_k, by considering the time dependence of $\hat{\phi}$ and remembering that $\hat{H}$ is the generator of time translations.

In this account, we started with the classical Hamiltonian. After promoting the fields to operators we found that the commutation relation (15.33) is required for consistency of the quantum theory, and then we found that $\hat{a}_{\mathbf{k}}^\dagger$ creates particles. That is the usual approach, often called **canonical quantization**. An alternative approach, adopted for instance in Ref. [2], is to take the commutator (15.33) as the starting point, invoked to make $\hat{a}_{\mathbf{k}}^\dagger$ the particle creation operator, and to then show that the resulting quantum field theory is consistent.

15.4.2 Fourier integral

Going to the limit of infinite box size using Eqs. (6.8)–(6.10) we have

$$\hat{\phi}(\mathbf{x},t) = \frac{1}{(2\pi)^3}\int \left[\phi_k(t)\hat{a}(\mathbf{k}) + \phi_k^*(t)\hat{a}^\dagger(-\mathbf{k})\right] e^{i\mathbf{k}\cdot\mathbf{x}} d^3k, \quad (15.36)$$

$$\hat{H} = \frac{1}{(2\pi)^3}\int E_k \left(\hat{a}^\dagger(\mathbf{k})\hat{a}(\mathbf{k}) + \frac{1}{2}L^3\right) d^3k, \quad (15.37)$$

and the commutation relations

$$(2\pi)^{-3}[\hat{a}(\mathbf{k}), \hat{a}^\dagger(\mathbf{k}')] = \delta^3(\mathbf{k}-\mathbf{k}'), \qquad [\hat{a}(\mathbf{k}), \hat{a}(\mathbf{k}')] = 0. \quad (15.38)$$

The factor $1/\sqrt{E_k}$ in ϕ_k produces the Lorentz invariant combination $d^3k/\sqrt{E_k}$ which makes $\phi(\mathbf{x},t)$ a scalar. We could have motivated Eq. (15.36) in this way, and then motivated the commutation relation (15.38) by requiring that $a^\dagger(\mathbf{k})$ creates particles with momentum $\mathbf{k}$, without reference to the canonical commutation relation.

Dividing the last term of Eq. (15.37) by L^3 gives the vacuum energy density:

$$\langle\rho\rangle = \frac{1}{2}\times\frac{4\pi}{(2\pi)^3}\int_0^{\Lambda_{\rm UV}} k^2 E_k dk \simeq \frac{\Lambda_{\rm UV}^4}{16\pi^2}, \quad (15.39)$$

where the final approximation sets $m=0$. Using Eq. (13.32), a similar calculation for the pressure gives $\langle P\rangle = \frac{1}{3}\langle\rho\rangle$. As they stand, these vacuum contributions cannot be absorbed into the cosmological constant, which has $P=-\rho$. This situation arises because the cutoff $\Lambda_{\rm UV}$ on 3-momentum breaks Lorentz invariance. It can be shown that these values of $\langle\rho\rangle$ and $\langle p\rangle$ hold also if the field describes a particle with non-zero spin, in a definite spin state.

In most applications of quantum field theory gravity is ignored, and then $\langle\rho\rangle$ and $\langle p\rangle$ have no effect. In the context of cosmology, one should do the calculation of $\langle\rho\rangle$ and $\langle p\rangle$ in a Lorentz-invariant manner; then $\langle\rho\rangle = -\langle p\rangle$ can be absorbed into the cosmological constant, though there is no consensus at present about their expected magnitude [3]. In turn, the cosmological constant can be regarded as the vacuum expectation value of the potential V.

15.4.3 Quantized free complex scalar field

Now we suppose that there is invariance under a $U(1)$ transformation, acting on the phases of one or more complex fields ϕ_n through Eq. (14.8) which defines the charges q_n. We also suppose that the symmetry isn't spontaneously broken, so that the real and imaginary parts of ϕ_n are the things to be quantized. We are going to show that a particle of the nth species carries charge q_n, while its antiparticle carries charge $-q_n$.

Writing $\phi = (\phi_1 + i\phi_2)/\sqrt{2}$, and applying Eqs. (15.36) and (15.38) to each ϕ_n, we arrive at

$$\hat{\phi}(\mathbf{x}, t) = \frac{1}{(2\pi)^3} \int d^3k \left[\phi_k(t)\hat{a}(\mathbf{k}) + \phi_k^*(t)\hat{b}^\dagger(-\mathbf{k}) \right] e^{i\mathbf{k}\cdot\mathbf{x}}, \tag{15.40}$$

with

$$[\hat{a}(\mathbf{k}), \hat{a}^\dagger(\mathbf{k}')] = [\hat{b}(\mathbf{k}), \hat{b}^\dagger(\mathbf{k}')] = (2\pi)^3 \delta^3(\mathbf{k} - \mathbf{k}'), \tag{15.41}$$

and the other commutators vanishing. The procedure that we used for a real field shows that $\hat{a}^\dagger(\mathbf{k})$ and $\hat{b}^\dagger(\mathbf{k})$ both create a particle with energy E_k and momentum $\mathbf{k}$. To find the charge carried by these particles, we define the charge operator as the $\hat{F}$ in Eq. (15.14), with $\hat{V}(\lambda)$ the unitary operator corresponding to the transformation (14.8). To reproduce Eq. (14.8) for a given field ϕ_n, we need the creation operators to satisfy

$$[\hat{a}^\dagger(\mathbf{k}), \hat{F}] = q_n \hat{F}, \qquad [\hat{b}^\dagger(\mathbf{k}), \hat{F}] = -q_n \hat{F}. \tag{15.42}$$

It follows that $\hat{a}^\dagger$ creates a particle species with charge q_n while $\hat{b}^\dagger$ creates one with charge $-q_n$. These species have the same mass, and correspond to a particle–antiparticle pair. By convention, we identify the species with charge q_n as the particle, and the species with charge $-q_n$ as the antiparticle.

Instead of using the charge defined through (15.14), we could have used the equivalent expression $\int d^3x j^0$, with j^0 given by Eq. (14.11) and the ϕ_n promoted to operators. As with the analogous procedure for momentum, that would involve more work.

15.5 Vector field

The Lagrangian of a massless vector field A_μ is given by the first term of Eq. (14.39). We want to write the field as a superposition of plane waves, and to do that we must make a choice among the equivalent A_μ related by the gauge transformation Eq. (14.37). This is called choosing a gauge.

We begin by requiring

$$\partial_\mu A^\mu = 0. \tag{15.43}$$

A gauge satisfying this condition is called a Lorenz gauge.[8] In a Lorenz gauge, the Maxwell equation (14.42) becomes $\Box A^\mu = j^\mu$. We are interested in a free field corresponding to $\Box A^\mu = 0$, which has the desired plane-wave solutions. For a plane wave travelling along the z direction, we can choose the gauge so that

[8] This name refers to an early advocate of the gauge. It is often called a Lorentz gauge, presumably through some error that propagated through the literature.

$A_z = 0$, which fixes it up to a rotation about the z direction. Promoting A_μ to an operator, we can then write

$$\hat{A}_\mu(\mathbf{x},t) = \int \frac{d^3k}{(2\pi)^3}\,\phi_k(t)\left[e^R_\mu(\mathbf{k})\hat{a}_R(\mathbf{k}) + e^L_\mu(\mathbf{k})\hat{a}_L(\mathbf{k})\right]e^{i\mathbf{k}\cdot\mathbf{x}} + \text{h.c.}, \quad (15.44)$$

where in a coordinate system with z along $\mathbf{k}$, the polarization vectors are defined by[9]

$$e^R_\mu = -\frac{1}{\sqrt{2}}(0,1,i,0), \qquad e^L_\mu = \frac{1}{\sqrt{2}}(0,1,-i,0). \quad (15.45)$$

The annihilation and creation operators satisfy[10]

$$[a_\lambda(\mathbf{k}), a^\dagger_{\lambda'}(\mathbf{k}')] = (2\pi)^3\delta^3(\mathbf{k}-\mathbf{k}')\delta_{\lambda\lambda'}, \quad (15.46)$$

where λ takes on the values L and R. From the same argument as for a scalar field, each of the creation operators creates a particle with momentum $\mathbf{k}$ and energy $E_k = k$. Also, by repeating the argument leading to Eq. (15.34) for angular momentum, we learn that $\hat{a}^\dagger_R(\mathbf{k})$ increases the spin component along $\mathbf{k}$ by 1 while $\hat{a}^\dagger_L(\mathbf{k})$ decreases it by 1.[11]

The spin component along $\mathbf{k}$ is called helicity. A Lorentz boost changes the momentum, but (since our discussion applies in any inertial frame) it doesn't change the helicity. The same is true for any massless particle, by virtue of the commutation relations of the generators of the Lorentz transformation [1].

The Lagrangian of a massive spin-1 field has an additional term $-m^2 A_\mu A^\mu/2$. To quantize such a field we again assume $\partial_\mu A^\mu = 0$.[12] As there is no gauge invariance no other condition can be imposed, and we must write

$$\hat{A}_\mu(\mathbf{x},t) = \int \frac{d^3k}{(2\pi)^3}\phi_k(t)\left[e^R_\mu(\mathbf{k})\hat{a}_R(\mathbf{k}) + e^L_\mu(\mathbf{k})\hat{a}_L(\mathbf{k}) + e^0_\mu\hat{a}_0(\mathbf{k})\right]e^{i\mathbf{k}\cdot\mathbf{x}} + \text{h.c.} \quad (15.47)$$

The mode function now oscillates with angular frequency $E_k = \sqrt{k^2+m^2}$, identifying m as the particle mass. The extra polarization vector is given by $e^0_\mu = (-k/m, 0, 0, E_k/m)$, again taking the z-axis to point along $\mathbf{k}$. The definition of e^0_μ gives A_μ the correct transformation under a boost along the $\mathbf{k}$ direction. Since e^0_μ is invariant under rotations about $\mathbf{k}$, the particles created by $\hat{a}^\dagger_0$ have zero helicity, while those created by $\hat{a}^\dagger_{R,L}$ still have helicity ± 1.

[9] The minus sign in the first expression accords with the convention of Ref. [1].

[10] These are required by Eq. (14.45) if $a^\dagger(\mathbf{k})$ is to create particles with momentum $\mathbf{k}$, and can also be derived from canonical quantization.

[11] Instead of this simple argument, we could calculate the momentum density operator from Eq. (14.45) and then calculate the angular momentum operator from Eq. (2.54). That procedure, and its analogue for spin-1/2 fields, is adopted in many textbooks.

[12] Allowing $\partial_\mu A^\mu$ to be non-zero introduces an extra degree of freedom, which is equivalent to introducing a real scalar field and setting $\partial_\mu A^\mu = 0$ [1].

A Lorentz boost in a generic direction mixes the polarization vectors and hence the helicity states.

We conclude that quantization of a 4-vector field, with the condition $\partial_\mu A^\mu = 0$, corresponds to particles with spin 1. If the field is massless the helicity must be ± 1, otherwise all three helicities ± 1 and 0 are allowed.

15.6 Spin-1/2 field

Spin-1/2 particles are described by spinor field operators.[13] These are double-valued representations of the Lorentz group, from which one constructs Lorentz-invariant Lagrangians. To make sense of the theory, different components of the spinor fields have to anti-commute.

For a field theory involving massive particles different from their antiparticle, whose interactions respect P and C invariance, the most convenient description is provided by the four-component Dirac spinors. Examples of such theories are Quantum Electrodynamics and Quantum Chromodynamics, which we encounter in the next chapter. In general though, two-component Weyl spinors are in some respects more convenient and we shall use them exclusively when writing down Lagrangian densities with only an occasional mention of Dirac spinors.

15.6.1 Lorentz transformation and free-field Lagrangians

We use Greek indices to label the components of the Weyl spinors. Weyl spinors are of two kinds, called left handed and right handed. For a left-handed Weyl spinor ψ_α, the infinitesimal Lorentz transformation consisting of a Lorentz boost with velocity $\mathbf{v}$, plus a rotation about the vector $\boldsymbol{\theta}$ through an angle $|\boldsymbol{\theta}|$, has the following effect on the spinor evaluated at a fixed spacetime point:

$$\psi_\alpha \longrightarrow \sum_\beta (I - i\boldsymbol{\theta} \cdot \boldsymbol{\sigma}/2 - \mathbf{v} \cdot \boldsymbol{\sigma}/2)_{\alpha\beta}\, \psi_\beta, \tag{15.48}$$

where $\boldsymbol{\sigma} = (\sigma_1, \sigma_2, \sigma_3)$ is the set of Pauli matrices. For a right-handed Weyl spinor, the sign in front of $\boldsymbol{\theta}$ is reversed. Only one kind is needed and we choose the left-handed spinors.

To define Lorentz-invariant products we follow the sign conventions of Ref. [1] with slightly different notation. The Lorentz-invariant product of Weyl spinors ψ and χ is

$$\psi\chi \equiv \psi_\alpha \epsilon^{\alpha\beta} \chi_\beta = \chi\psi, \tag{15.49}$$

[13] One also considers ordinary functions that have the same transformation properties as the spinor fields, which are themselves called spinors. (These are needed to write a spinor field in terms of creation and annihilation operators, and also when using the Dirac equation to formulate the relativistic quantum mechanics of electrons and muons.) In this book we deal only with spinor fields, which from now on we call simply spinors.

where $\epsilon^{\alpha\beta}$ is antisymmetric with $\epsilon^{12} = 1$. Its Hermitian conjugate is

$$\psi^* \chi^* \equiv \chi^\dagger_\beta \epsilon^{\alpha\beta} \psi^\dagger_\alpha = \chi^* \psi^*. \tag{15.50}$$

The Lorentz-invariant product of a spinor and its derivative is Hermitian, and defined by

$$i\psi^* \not\partial \psi \equiv i\psi^\dagger_\alpha (\bar\sigma^\mu)^{\alpha\beta} \partial_\mu \psi_\beta. \tag{15.51}$$

Here $\bar\sigma^\mu$ transforms as a 4-vector, and in a given Lorentz frame one can write $(\bar\sigma^\mu)^{\alpha\beta} = \epsilon^{\alpha\delta}\epsilon^{\beta\gamma}(\bar\sigma^\mu)_{\delta\gamma}$ with $(\bar\sigma^\mu)_{\delta\gamma} = -(I, \sigma_1, \sigma_2, \sigma_3)_{\delta\gamma}$ where I is the unit matrix.[14]

A free massive spin-1/2 particle which is its own antiparticle can be described by a single left-handed Weyl spinor, and the free-field Lagrangian density is

$$\boxed{\mathcal{L} = -i\psi^* \not\partial \psi - \frac{1}{2}\left(m\psi\psi + \text{c.c.}\right)}. \tag{15.52}$$

Here c.c. means that stars are put on the preceding term, though we are not talking about complex conjugation for ψ. By a choice of the spinor the parameter m can be chosen to be real and positive, and is then the mass of the particle. That choice is usually made though sometimes a complex m is considered; then $|m|$ is the mass. The mass term is called a **Majorana** mass term and the particle is said to be a Majorana particle.

A free massive spin-1/2 particle different from its antiparticle, such as an electron, can be described by a pair of left-handed Weyl spinors. We denote the spinors of say the electron by e and $\bar e$, and similarly for the other quarks and leptons.[15] The Lagrangian density for the free field is

$$\boxed{\mathcal{L} = -ie^* \not\partial e - i\bar e^* \not\partial \bar e - m\left(e\bar e + \text{c.c.}\right)}. \tag{15.53}$$

The last term determines the mass m of the particle and is called a Dirac mass term.

A free massless particle is described by a single left-handed Weyl spinor, with the Lagrangian

$$\boxed{\mathcal{L} = -i\psi^* \not\partial \psi}. \tag{15.54}$$

This applies to the neutrinos within the context of the Standard Model as described in Chapter 16. There we denote the spinors simply by the particle names; ν_e, ν_μ and ν_τ.

To give these Lagrangian densities the required energy dimension 4, the spinor fields must have dimension 3/2.

[14] The shorthand slash notation defined here is inspired by the standard usage for Dirac spinors but isn't generally used for Weyl spinors. The σ_i of Ref. [1] are -1 times ours.

[15] In contrast with the Dirac notation, the bar on $\bar e$ does not mean that it is related to e. It is an independent field.

15.6.2 Spin-1/2 particles

To arrive at the particle concept, each Fourier component of a spinor is written in terms of creation and annihilation operators. This is usually done with the four-component Dirac spinors, each of which can be regarded as a pair of Weyl spinors. We will just give the end result for the Weyl spinors.

Each Fourier component $\psi(\mathbf{k}, t)$ of a Weyl spinor comes with two creation operators, corresponding to the fact that it has two components. By considering the behaviour of the spinor field $\psi(x)$ under spacetime displacements, one finds that the particles have momentum $\mathbf{k}$ and energy $E_k = \sqrt{k^2 + m^2}$, just as for spin 0 and 1. By considering the behaviour under rotations, one finds that in a given inertial frame it is possible to choose the creation operators so that one of them creates a particle with helicity $+1/2$ (right-handed spin) while the other creates a particle with helicity $-1/2$ (left-handed spin). As is the case for any spin, a Lorentz boost doesn't change the helicity for particles whose mass is zero or negligible ($\ll k$), but otherwise it mixes them. If ψ has a Majorana mass term, no other spinor is involved and we deal with a species which is its own antiparticle.

Suppose now that there is invariance under a $U(1)$ transformation $\psi \to e^{iq_\psi \lambda}\psi$. Repeating the argument we used for a complex scalar field, it is found that one of the creation operators creates particles with 'charge' q_ψ while the other creates particles with 'charge' $-q_\psi$. If the mass is non-zero, invariance under the $U(1)$ transformation forbids the Majorana mass term (15.52). We must use instead the Dirac mass term (15.53), with the barred fields carrying opposite 'charges'. Then we have four creation operators altogether, creating a particle with two spin states and an antiparticle with two spin states. If instead the mass is zero, there can be invariance under a $U(1)$ transformation $\psi \to e^{iq_\psi \lambda}\psi$ that acts on just a single Weyl spinor.

From now on we focus on the important case of zero or negligible mass,[16] negligible meaning $m \ll k$. Then helicity is Lorentz invariant, and it becomes a label R or L that we can attach to the particle's name. Right-handed and left-handed electrons are denoted by respectively e_R and e_L, and similarly for the other particles.

Using the transformation of the spinor under rotations and the $U(1)$ transformation $\psi \to e^{iq_\psi \lambda}\psi$, one can show that left-handed particles carry 'charge' q_ψ while right-handed particles carry 'charge' $-q_\psi$.[17] We will adopt the convention that unbarred fields ν, e correspond to left-handed particles ν_L, e_L and so on while barred fields $\bar{e}$ correspond to right-handed particles e_R and so on. (In each case the field corresponds also to antiparticles with the opposite helicity.) We have no

[16] We see in Section 22.1 that all quarks and leptons are expected to be massless in the very early Universe.

[17] This connection between helicity and charge comes from the fact that we are using left-handed Weyl spinors, and would be reversed if we used right-handed Weyl spinors.

need of barred neutrino fields, because right-handed neutrinos are not observed (and neither are left-handed antineutrinos).

According to this assignment of fields to particles, invariance under a $U(1)$ transformation $e \to e^{iq_L}e$ corresponds to left-handed electrons e_L having charge q_L, while invariance under a transformation $\bar{e} \to \bar{e}^{-iq_R}\bar{e}$ corresponds to right-handed electrons e_R having charge q_R. (We take the electron as an example; it could be any particle.) If there is invariance under both transformations, with $q_L = q_R \equiv q$, we can simply say that the electron has 'charge' q. That is the case for electric charge, baryon number and lepton number, but in Section 16.1 we encounter charges for which it is not the case.

As explained for instance in Ref. [1], the four-component Dirac spinor Ψ describing say an electron can be written $\Psi = (e_L, e_R)$ where $e_L = e$ and $e_R = \bar{e}^\dagger$. (We use of the same symbol for the field and the corresponding relativistic particle.) Because of the Hermitian conjugation, e_R is a right-handed Weyl spinor, and a $U(1)$ transformation $\bar{e} \to e^{-iq_R\lambda}\bar{e}$ has the effect $e_R \to e^{-iq_R\lambda}e_R$. Then the transformation of the previous paragraph with $q_L = q_R \equiv q$ becomes simply $\Psi \to e^{iq\lambda}\Psi$.

15.7 Free scalar field with time-dependent mass

In this section we study the quantum theory of a field with the potential

$$\boxed{V(\phi) = \frac{1}{2}m_\phi^2(t)\phi^2}, \tag{15.55}$$

corresponding to a time-dependent mass. Such a potential cannot exist in the fundamental Lagrangian, but it can exist in an effective Lagrangian describing the behaviour of ϕ in the presence of some time-dependent classical field. The latter might be another scalar field or a gauge field. In Section 21.2 we encounter this setup in the cosmological context, with the time-varying field a scalar field. A similar setup describes the production of an electron–positron pair by a time-dependent electromagnetic field.

The field operator is given in terms of the annihilation and creation operators by Eq. (15.36). The mode function $\phi_k(t)$ is now the solution of

$$\boxed{\ddot{\phi}_k(t) + E_k^2(t)\phi_k(t) = 0}, \qquad \boxed{E_k^2(t) \equiv k^2 + m^2(t)}. \tag{15.56}$$

In the mechanical analogy we are dealing with a unit-mass particle which oscillates with a time-dependent frequency.

We are interested in the case where $E_k(t)$ becomes time independent in the early-time limit $t \to -\infty$. We assume that the usual quantum field theory is applicable in this limit. Then the mode function is the solution of Eq. (15.56) with the initial condition (15.31). The solution of Eq. (15.56) has a time-independent Wronskian, whose normalization is fixed by the initial condition:

$$\dot{\phi}_k \phi_k^* - \dot{\phi}_k^* \phi_k = -i. \tag{15.57}$$

The evolution is trivial if E_k varies slowly on its own timescale;

$$|\dot{E}_k| \ll E_k^2. \tag{15.58}$$

This is called adiabatic variation.[18] In this case, the mode function is given by

$$\phi_k(t) = \sqrt{\frac{1}{2E_k(t)}} \exp\left(-i \int^t E_k(t) dt\right). \tag{15.59}$$

The lower limit on the integral can be chosen arbitrarily. Over a time interval $E_k^{-1} \ll \Delta t \ll E_k/|\dot{E}_k|$, the mode function has the initial form except for a constant phase that can be absorbed into $\hat{a}_{\mathbf{k}}$. Then we recover the particle interpretation. The particle number doesn't change with time, remaining zero if it is zero initially.

We are interested in the case where adiabaticity fails for a while, after which it is restored. The late-time mode function will then be a superposition of positive and negative frequencies:

$$\boxed{\phi_k(t) = \sqrt{\frac{1}{2E_k}} \left(\alpha_k e^{-iE_k t} + \beta_k e^{iE_k t}\right)}, \qquad (t \to \infty). \tag{15.60}$$

We define a new annihilation operator by

$$\widetilde{\hat{a}}_{\mathbf{k}} = \alpha_k \hat{a}_{\mathbf{k}} + \beta_k \hat{a}_{\mathbf{k}}^\dagger. \tag{15.61}$$

This is called a **Bogolyubov transformation**. In terms of the new annihilation operator, $\hat{\phi}$ has the original form (15.29);

$$\hat{\phi}(\mathbf{x}, t) = L^{-3} \sum_{\mathbf{k}} \sqrt{\frac{1}{2E_k}} \left[e^{-iE_k t} \widetilde{\hat{a}}_{\mathbf{k}} + e^{iE_k t} \widetilde{\hat{a}}_{-\mathbf{k}}^\dagger\right] e^{i\mathbf{k}\cdot\mathbf{x}}. \tag{15.62}$$

The operator $\widetilde{\hat{\mathbf{n}}}_{\mathbf{k}} \equiv \widetilde{\hat{a}}_{\mathbf{k}}^\dagger \widetilde{\hat{a}}_{\mathbf{k}}/L^3$ corresponds to the expectation value of the late-time occupation number. If the initial state is the vacuum, we find from the commutator (15.33) that the expectation value of the late-time occupation number is $|\beta_k|^2$.

The time independence of the Wronskian gives

$$|\alpha_k|^2 - |\beta_k|^2 = 1. \tag{15.63}$$

If the occupation number is very large, then $|\alpha_k| = |\beta_k|$ to a very good approximation.

[18] The meaning of 'adiabatic' here has no scientific connection with the thermodynamic meaning.

A spin-$1/2$ particle can be created in the same way. Because fermion creation and annihilation operators anti-commute, the late-time occupation number cannot exceed one.

15.8 Quantized interactions

In this section we consider interacting quantum field theory, still in the context of flat spacetime. The interactions add a term H_I to the Hamiltonian H_0 that described the free fields. The effect of H_I is usually handled perturbatively, just as in ordinary quantum mechanics.

The perturbative treatment can be a good approximation only if the couplings responsible for H_I are sufficiently small. For instance, the coupling in Eq. (13.38) has to satisfy $|\lambda| \ll 1$. If instead the couplings are big, we typically deal with a strongly coupled field theory. Then it is no longer the case that each field corresponds to a particle species; instead, the field might correspond to bound states of the would-be particles. We shall not consider strongly coupled field theory.

The calculation of a given quantity in perturbation theory can be represented by Feynman diagrams. There are 'tree-level' diagrams containing no loops, and diagrams containing one or more loops. At a given order of perturbation theory the number of diagrams is finite.

15.8.1 Renormalization

Part of the effect of H_I can be taken into account by altering (renormalizing) the values of the couplings and masses appearing in the Lagrangian. The original values of these parameters are called bare values, and the values after renormalization are called dressed values. The latter are the physical values, to be compared directly with observation. Focussing on a scalar mass, we have

$$m^2_{\rm physical} = m^2_{\rm bare} + m^2_{\rm rad}, \tag{15.64}$$

where the second term is the effect of the interaction, often called the **radiative correction**.

For scalar fields, the radiative correction is problematic because it is generically of order the ultra-violet cutoff. In order to use the field theory, we want the cutoff to be much bigger than relevant energy scales and in particular much bigger than $m_{\rm physical}$. A generic field theory will therefore require an accurate cancellation for scalars:

$$\boxed{m^2_{\rm bare} \simeq -m^2_{\rm rad}}\,. \tag{15.65}$$

A similar cancellation will be required for other parameters, like the coupling λ in Eq. (13.38).

If the interactions are what is called **renormalizable** that is the end of the story. The action contains a finite number of parameters, and when the parameters are renormalized all predictions of the theory remain finite as the cutoff goes to infinity. If instead the interactions are non-renormalizable, new infinities are encountered at every order of perturbation theory, which can be removed only by adding new interaction terms. The non-renormalizable theory therefore progressively loses predictive power as one increases the order.

Given a set of fields, there are only a finite number of renormalizable interaction terms. Renormalizable terms are those whose coefficients have non-negative energy dimension. The coefficients of non-renormalizable terms in contrast have negative energy dimension, and are usually supposed to be at most of order 1 in units of the ultra-violet cutoff.[19] In that case, a non-renormalizable theory will *usually* be adequate when relevant energy scales are well below the cutoff. We met an example in Eq. (13.39). Another is provided by the kinetic function G_{nm} of Eq. (13.52). In its power-series expansion

$$G_{nm}(\phi) = G_{nm0} + \sum_{d=1}^{\infty} G_{nmd} \left(\frac{\phi}{\Lambda_{\rm UV}} \right)^d + \ldots, \tag{15.66}$$

all terms are non-renormalizable, except the first which without loss of generality can be taken as δ_{nm}. Thus, renormalizability means that we can choose the scalar field kinetic term to be minimal, and the same is true of the kinetic terms of spin-$1/2$ fields and gauge fields.

There are at least two caveats to the statement that a renormalizable theory will usually be adequate for energy scales well below the ultra-violet cutoff. First, in order to motivate the form of the renormalizable theory one may wish to derive it as an approximation to a non-renormalizable theory. In particular, one may wish to understand in this way aspects of the Standard Model or its supersymmetric extension. Second, and most importantly for our purpose, one may need to consider non-renormalizable terms in the potential if the renormalizable terms are for some reason very small. We will see how non-renormalizable terms can be important for models of inflation.

Finally, we need to mention that the renormalization may have an important side-effect: in the presence of spinor fields, it can happen that the renormalization procedure breaks a symmetry of the action. One then says that the theory has an **anomaly**. In the case of a global symmetry one can accept an anomaly. In the case of a gauge theory one must forbid an anomaly, and this imposes powerful restrictions on the form of the action.

[19] See Ref. [2] for an authoritative discussion of this and many other issues of field theory.

15.8.2 *Keeping radiative corrections under control*

We saw that the fine-tuned cancellation (15.65) is needed to keep the scalar masses under control in a generic field theory. This is generally regarded as undesirable. The most popular mechanism for avoiding a cancellation is supersymmetry, described in Chapter 17. Unbroken supersymmetry completely removes some of the radiative correction to scalar masses, and more generally keeps all radiative corrections under control.

If a field is a PNGB, the radiative corrections are under control even without supersymmetry. To see why, recall that a PNGB field ϕ is one for which there is an approximate shift symmetry, meaning that the Lagrangian is approximately invariant under the $U(1)$ transformation $\phi \to \phi + \text{constant}$. If the shift symmetry were exact, this would require that ϕ had only derivative couplings (those involving spacetime gradients of ϕ). Such couplings are suppressed by inverse powers of the cutoff scale (non-renormalizable). Going to the approximate symmetry, we have ordinary (non-derivative) couplings whose magnitude is under control in the sense that it is determined by the amount of symmetry breaking.

The PNGB alternative to supersymmetry is difficult to implement for a field such as the Standard Model Higgs field, which has big couplings. A scheme that can work for such a field is called the Little Higgs mechanism, described in the (different) context of inflation in Section 28.11. It removes the one-loop contribution to the Higgs mass, allowing the ultra-violet cutoff to be a couple of orders of magnitude bigger than the Higgs mass without fine tuning.

The only other known method of controlling the radiative correction for a Higgs field is to invoke an extra space dimension x^5, and to make the Higgs field the component A^5 of a gauge field. We describe this in the context of inflation in Section 28.13.2.

15.8.3 *Coleman–Weinberg potential and renormalization group equations*

Now we focus on the radiative corrections in a renormalizable field theory. Part of these corrections can be handled by altering the form of the scalar field potential, giving what is called the **effective potential**. At tree level, the radiative corrections just renormalize the masses and couplings. At higher order, the effective potential acquires an extra piece which varies logarithmically.

In this section we consider the one-loop contribution. In the direction of a given field ϕ, with all other fields at their vev, it is given by

$$\boxed{V_{\text{loop}}(\phi) = \sum_i \frac{\pm g_i}{64\pi^2} M_i^4(\phi) \ln \frac{M_i^2(\phi)}{Q^2}} . \tag{15.67}$$

This is called the Coleman–Weinberg potential. The sum goes over all particle species, with the plus/minus sign for bosons/fermions, with g_i the number of spin states. The quantity $M_i^2(\phi)$ is the effective mass-squared of the species, obtained from the Lagrangian by fixing ϕ at a given value. For a scalar field ϕ_i we have $M_i^2 = \partial^2 V/\partial\phi_i^2$, which is valid for ϕ itself as well as other scalar fields.

The quantity Q is called the **renormalization scale**. It has to be introduced to make sense of the renormalization procedure. If the loop correction were calculated to all orders, the potential would be independent of Q. In a given situation, Q should be set equal to a typical energy scale so as to minimize the size of the loop correction and its accompanying error.

As an example, consider the potential

$$\boxed{V(\phi,\chi) = V_0 + \frac{1}{2}m^2\phi^2 + \frac{1}{2}m_\chi^2\chi^2 + \frac{1}{4}\lambda\phi^4 + \frac{1}{4}g\chi^2\phi^2}. \qquad (15.68)$$

For $\sqrt{g}\phi \gg m_\chi$ it gives $M_\chi^2(\phi) = g\phi^2/2$, which gives a contribution to the Coleman–Weinberg potential[20]

$$V_{\text{loop}} = \frac{1}{32\pi^2}\frac{g^2\phi^4}{4}\ln\frac{\phi}{Q}. \qquad (15.69)$$

Adding this contribution to the potential (15.68) with $\chi = 0$ gives

$$\boxed{V = V_0 + \frac{1}{2}m^2\phi^2 + \frac{1}{4}\left[\lambda + \frac{g^2}{32\pi^2}\ln\frac{\phi}{Q}\right]\phi^4}. \qquad (15.70)$$

To make the potential independent of Q at the one-loop level, one should make λ depend on Q:

$$\frac{d\lambda(Q)}{d\ln Q} = \frac{g^2}{32\pi^2}. \qquad (15.71)$$

Such dependence is called **running**.

In general there will be other interactions, and all masses and couplings must run to make physical quantities independent of Q. The coupled differential equations that determine the running are called **renormalization group equations** (RGEs). Having specified the renormalized parameters (masses and couplings) at one value of the renormalization scale Q, the RGEs determine them at all other scales.

As mentioned earlier, one should choose the renormalization scale to match the energy scales of the problem at hand. In computing the effective potential, the field ϕ may be the most relevant energy scale. Then one might set $Q = \phi$ and ignore the loop correction. One then arrives at the **renormalization group improved**

[20] We have absorbed a factor $2/g$ into Q^2.

potential

$$\boxed{V(\phi) = \frac{1}{2}m^2(Q)\phi^2 + \frac{1}{4}\lambda(Q)\phi^4}, \tag{15.72}$$

with $Q = \phi$.

Exercises

15.1 Taking $\hat{H}$ to be the harmonic oscillator Hamiltonian (15.23), evaluate the quantum evolution equation (15.6) taking $\hat{A} = \hat{q}$, and show that it coincides with Eq. (15.8) (using Eq. (15.7)).

15.2 Use $\langle 0|0\rangle = 1$ and $[\hat{a}, \hat{a}^\dagger] = 1$ to show that the harmonic oscillator states $\hat{a}^\dagger|n\rangle = \sqrt{n+1}\,|n+1\rangle$ are orthonormal.

15.3 By what factor does Eq. (15.39) overestimate the cosmological constant, as compared to its observed value, if $\langle V\rangle = 0$ and the ultra-violet cutoff of the field theory is the Planck scale?

15.4 Derive the complex scalar field commutation relations (15.41).

15.5 Setting $m(t)$ to zero for negative t, and to a constant m for positive t, solve the mode function equation (15.56) with the initial condition (15.31) by matching the value and derivatives at $t = 0$. Hence calculate β_k and the expectation value of the late-time occupation number $|\beta_k|^2$.

15.6 Verify Eq. (15.69), for the contribution to the Coleman–Weinberg potential coming from $M_\chi(\phi)$.

References

[1] S. P. Martin. A supersymmetry primer. arXiv:hep-ph/9709356.

[2] S. Weinberg. *The Quantum Theory of Fields, Volume I* (Cambridge: Cambridge University Press, 1995).

[3] L. Hollenstein, M. Jaccard, M. Maggiore and E. Mitsou. Zero-point quantum fluctuations in cosmology. arXiv:1111.5575.

16

The Standard Model

In this chapter we consider the Standard Model, and see how it may be extended to include neutrino mass. We also describe Peccei–Quinn symmetry and the axion. The treatment is sketchy, dealing only with things that will be needed when we come to consider scenarios of the early Universe.

The Standard Model describes the interactions of the quarks and leptons, which are divided into three generations (also called families) according to the following scheme:

$$\begin{array}{ccccccc} u & c & t & \quad & e & \mu & \tau \\ d & s & b & \quad & \nu_e & \nu_\mu & \nu_\tau \end{array} .$$

The quarks are named up, down, strange, charm, bottom and top, while the leptons are the electron, muon, and tau along with their corresponding neutrinos. The 'charges' of the first generation are listed in Table 16.1, and those of the second and third generation are the same. The quark and lepton masses are given in Table 16.2. The Standard Model treats the neutrinos as massless.

The gauge symmetry group of the Standard Model is $SU_C(3)\otimes SU_L(2)\otimes U_Y(1)$, the subscripts identifying the set of fields which transform under the group. The factor $SU_C(3)$ describes the strong interaction, possessed only by the quarks and the eight gluons that are the gauge bosons of $SU_C(3)$. The factor $SU_L(2)\otimes U_Y(1)$ describes the electroweak interaction, possessed by leptons, quarks and the gauge bosons γ, Z and $W^\pm$. As shown in Table 16.3, the Z and $W^\pm$ have non-zero masses because $SU_L(2)\otimes U_Y(1)$ is spontaneously broken, leaving only the electromagnetic subgroup $U_{\rm em}(1)$ unbroken.

The formulation of the Standard Model was complete by the early 1970s, and since then observations (mostly at particle accelerators) have confirmed it with ever-increasing accuracy. The only direct indication that the Standard Model is incomplete comes from the existence of neutrino mass.

To implement the spontaneous breaking, the Standard Model invokes two ele-

Table 16.1. *Lepton and quark 'charges'*

	ν_e	e	u	d
Q	0	-1	$2/3$	$-1/3$
B	0	0	$1/3$	$1/3$
L	1	1	0	0

Table 16.2. *Lepton and quark masses in GeV*

e	μ	τ	u	c	t	d	s	b
0.000511	0.106	1.78	0.002	1.3	173	0.004	0.09	4.24

Table 16.3. *Gauge boson electric charges and masses in GeV*

	γ	g	Z	$W^{\pm}$
Q	0	0	0	± 1
mass	0	0	91.19	80.42

mentary Higgs fields which correspond to one electrically neutral Higgs particle. As we write there is no direct detection of this particle, but radiative corrections to transition amplitudes are sensitive to its mass. For the Standard Model to agree with observation one needs the Higgs mass to be roughly of order $100\,\mathrm{GeV}$, corresponding to $\lambda \sim 1$ in Eq. (14.32). If there is a Higgs particle with this mass it will be found at the Large Hadron Collider (LHC).

The excellent agreement of the Standard Model with observation strongly constrains the form of any modification that one might propose, if it is relevant at energies below $10\,\mathrm{TeV}$ or so. In particular, it is difficult to make the Higgs field a condensate in such a 'low-energy' modification. The most widely considered low-energy modification compatible with the data involves supersymmetry, as described in the next chapter. The LHC will soon discriminate between low-energy modifications, or perhaps find that none are needed.

16.1 Electroweak Lagrangian

We will initially describe the electroweak interaction, which corresponds to the $SU_L(2)\otimes U_Y(1)$ gauge group, under the pretence that the strong interaction doesn't exist.

The electroweak interaction distinguishes between left and right handedness, and the charges associated with it have different values for relativistic left- and right-handed particles. A symmetry of this kind is said to be **chiral.**[1]

We use the Weyl spinors and the notation described in Section 15.6. We pretend for the moment that there is only the first generation of quarks and leptons: u, d, ν_e and e. Their Weyl spinors are denoted by u, $\bar{u}$, d, $\bar{d}$, ν_e, e and $\bar{e}$.

We consider first the $U_Y(1)$ transformation. We denote its gauge field by B_μ, and the corresponding field strength is $B_{\mu\nu} \equiv \partial_\mu B_\nu - \partial_\nu B_\mu$. We denote the $U_Y(1)$ gauge coupling by g'. Under the $U_Y(1)$ transformation, each Weyl spinor is multiplied by a factor $e^{\pm i\lambda(x)g'Y}$, with the plus sign for unbarred fields and the minus sign for barred fields. The 'charge' Y is called hypercharge, and the sign is chosen so that the corresponding relativistic particle species carries hypercharge $+Y$, in accordance with the discussion at the end of Section 15.6.2. The hypercharges of the first generation are shown in Table 16.4, and are the same for the other generations.[2]

For the $SU_L(2)$ we use the notation of Section 14.7. The barred fields (corresponding to right-handed relativistic particles) are $SU_L(2)$ singlets. The Higgs doublet $H = (H_1, H_2)$ transforms as $H_i = \sum_j U_{ij} H_j$ with U an $SU(2)$ matrix. We also need $H_{\rm c}$ defined by $H_{\rm c}^i = \epsilon^{ij} H_j^*$ which has the same transformation. (In this definition, ϵ^{ij} is antisymmetric with $\epsilon_{12} = 1$, and the star denotes complex conjugation if we regard H as classical, or Hermitian conjugation if we regard it as an operator.) The doublets $L \equiv (\nu_e, e)$ and $Q = (u, d)$ have the same transformation. Writing $H = (H_1, H_2)$ and $L = (L_1, L_2)$, the product $HL \equiv \epsilon^{ij} H_i L_j$ is invariant under $SU_L(2)$, and so are the similarly-defined products QH and QH_c.

The covariant derivative acting on a barred field is just

$$D_\mu = \partial_\mu + ig'YB_\mu(x). \tag{16.1}$$

(This is Eq. (14.38), with different notation.) The covariant derivative acting on a doublet must accommodate the $SU_L(2)$ as well. Regarding the doublet as a column matrix,

$$D_\mu = I\partial_\mu - iIg'YB_\mu(x) - igW_\mu^a \hat{T}_a, \tag{16.2}$$

where g is the $SU_L(2)$ gauge coupling and I is the unit matrix. The hypercharges for H, L and Q are $Y = 1/2, -1/2$ and $1/6$, as shown in Table 16.4. Extending the notation of Eq. (15.57), the gauge invariant kinetic term for L is $L^* \not{D} L \equiv L_i^* \not{D} L_i$, and similarly for Q.

The $SU_L(2)$ group of transformations has a subgroup $U_L(1)$. As shown in Table

[1] This term derives from the Greek word for hand. It was originally (and is still) used to describe organic molecules that happen to exist on the Earth with only one handedness.

[2] The normalization of Y is conventional. Some authors normalize it to be a factor ± 2 bigger than ours.

Table 16.4. *Electric charge, T_3 and hypercharge. As explained in the text,* $Y = Q - T_3$.

	ν_L	e_L	u_L	d_L	e_R	u_R	d_R	H_1	H_2	W^+
Q	0	-1	$2/3$	$-1/3$	-1	$2/3$	$-1/3$	1	0	1
T_3	$1/2$	$-1/2$	$1/2$	$-1/2$	0	0	0	$1/2$	$-1/2$	1
Y	$-1/2$	$-1/2$	$1/6$	$1/6$	-1	$2/3$	$-1/3$	$1/2$	$1/2$	0

16.4 it multiplies each field by a phase $e^{i\lambda T_3}$ where $T_3 = \pm 1/2$ for the upper and lower components of a doublet and $T_3 = 0$ for a singlet.

The electromagnetic U_{em} is a subgroup of $SU_L(2) \times U_Y(1)$, different from both $U_Y(1)$ and $U_L(1)$. It multiplies each field by a phase $e^{\pm i\lambda Q}$, with the plus sign for unbarred fields and the minus sign for barred fields. For each field, the values of Q, Y and T_3 are related by

$$\boxed{Q = Y + T_3}\,. \tag{16.3}$$

This relation is shown in Table 16.4.

With these ingredients in place we can write down the electroweak Lagrangian density :

$$\begin{aligned} \mathcal{L}_{\text{ew}} &= -\frac{1}{2} \operatorname{Tr} W_{\mu\nu} \cdot W^{\mu\nu} - \frac{1}{4} B_{\mu\nu} B^{\mu\nu} \\ &- iL^* \not{D} L - iQ^* \not{D} Q - i\bar{e}^* \not{D} \bar{e} - i\bar{d}^* \not{D} \bar{d} - i\bar{u}^* \not{D} \bar{u} \\ &- |D_\mu H|^2 - \frac{1}{2}\lambda^2 \left(|H|^2 - \frac{1}{2} v^2 \right)^2 \\ &- \left(y^e \bar{e} L H + y^d \bar{d} Q H - y^u \bar{u} Q H_c + \text{c.c.} \right) . \end{aligned} \tag{16.4}$$

Given the fields, this is the most general renormalizable Lagrangian density consistent with the gauge symmetries up to a 4-divergence.

The terms in first two lines specify the kinetic terms of the gauge and spin-$1/2$ fields. The first term also specifies the self-interaction of W_μ^a through the use of the covariant derivative. The first term of the third line specifies the kinetic term of the Higgs field, and its gauge interactions. The second term of the third line specifies the potential of the Higgs field. Each term in the last line specifies an interaction of the Higgs field with fermions, called a **Yukawa interaction**. Here 'c.c.' means that the preceding terms are repeated with a star attached to everything. We choose the phases of the fermion fields so that the Yukawa couplings y^e and so on are real.

16.2 Electroweak theory: particles and interactions

The Higgs doublet breaks an $SU(2)$ subgroup of $SU_L(2) \otimes U_Y(1)$, leaving unbroken the electromagnetic subgroup $U_{\rm em}(1)$. Since electric charge is different from hypercharge, $U_{\rm em}(1)$ isn't the same as $U_Y(1)$. To handle this, we make a rotation in the space of the gauge fields B_μ and W^3_μ to arrive at the electromagnetic gauge field A_μ and an orthogonal gauge field Z_μ;

$$A_\mu = B_\mu \cos\theta_{\rm W} + W^3_\mu \sin\theta_{\rm W}, \tag{16.5}$$
$$Z_\mu = -B_\mu \sin\theta_{\rm W} + W^3_\mu \cos\theta_{\rm W}. \tag{16.6}$$

Inverting this and putting it into Eq. (16.2), the relation $Q = T_3 + Y$ gives the angle $\theta_{\rm W}$ and the electromagnetic gauge coupling e:

$$\boxed{\frac{1}{e^2} = \frac{1}{g^2} + \frac{1}{g'^2}}, \qquad \boxed{\sin\theta_{\rm W} = \frac{g'}{g}}. \tag{16.7}$$

As we are dealing with a non-Abelian symmetry group, the gauge bosons have self-interactions. Writing out the kinetic terms, we see that only the fields $W^\pm_\mu \equiv (W^1_\mu \pm iW^2_\mu)/\sqrt{2}$ interact with the electromagnetic field. After quantization they correspond to gauge bosons $W^\pm$ with electric charge $Q = \pm 1$.

To handle the Higgs field we work in the unitary gauge defined in Section 14.8.2, so that $H_2 = 0$ and $H_1 \equiv (v + \chi)/\sqrt{2}$ is real. After quantization, χ corresponds to the electrically neutral Higgs particle.

The gauge boson masses come from the term $|D_\mu H|^2$ of the Lagrangian. Setting H equal to its vev one finds

$$M_Z = \frac{1}{2}\bar{g}v, \qquad M_{W^\pm} \equiv M_W = \frac{1}{2}gv, \tag{16.8}$$

with $\bar{g} \equiv \sqrt{g^2 + g'^2}$. These are tree-level masses, to which must be added radiative corrections.

The scale $M_W \sim M_Z \sim 100\,\text{GeV}$ defines the **electroweak scale**. At energies far below the electroweak scale, electroweak scattering and decay processes (transitions) divide into electromagnetic transitions and weak interaction transitions. The amplitudes for the latter are accurately described by Feynman graphs with a single W or Z exchange, and are far smaller than electromagnetic transition amplitudes justifying the term 'weak'. For weak interactions described by W exchange, each transition amplitude is a numerical factor times g^2/M_W^2, which is a numerical factor times $1/v^2$. The Fermi constant is defined by $\sqrt{2}G_{\rm F} = 1/v^2$, and accurate measurements of the transition amplitudes give $v = 246\,\text{GeV}$.

The gauge couplings run with renormalization scale Q.[3] At $Q = M_Z$ the values

[3] We are using Q to mean three different things, because in each case it is the standard notation. These things are the renormalization scale, electric charge and the quark doublet.

agreeing with observation are

$$\alpha_2 \equiv \frac{g^2}{4\pi} = \frac{1}{29.6}, \qquad \alpha_1 \equiv \frac{g'^2}{4\pi} = \frac{1}{98.4}. \tag{16.9}$$

These correspond to $\sin^2\theta_W = 0.231$. The calculated running agrees with observation in the accessible regime $Q \lesssim 100\,\text{GeV}$.

From Eq. (16.7) the electromagnetic coupling at $Q = M_Z$ is $\alpha = 1/128$. This may be compared with standard value $1/137$, which holds at the scale $Q \ll m_\text{e}$ appropriate for atomic physics and classical electrodynamics.

Putting Eq. (16.9) into Eq. (16.8) gives tree-level masses $M_W = 80.3\,\text{GeV}$ and $M_Z = 91.5\,\text{GeV}$. Radiative corrections bring these numbers into agreement with the observed masses.

The tree-level electron and quark masses are given by the last line of the Lagrangian. Comparing with Eq. (15.53) we have

$$m_e = \frac{1}{\sqrt{2}} y_e v, \qquad m_d = \frac{1}{\sqrt{2}} y_d v, \qquad m_u = \frac{1}{\sqrt{2}} y_u v. \tag{16.10}$$

To reproduce the electron mass we need $y_e \simeq 10^{-6}$, and to reproduce the u and d quark masses we need y_d and y_u of order 10^{-5}. Theory so far provides no reason for these small numbers.

16.3 Electroweak theory with three generations

Now we consider the electroweak interaction including all three generations. Each of the fields u_R, d_R, L and Q acquires a generation index $i = 1, 2, 3$. The terms in the second line of the Lagrangian become $-iL_i^* \not{D} L_i$ and so on (with a sum over i understood). The last line of the Lagrangian, which we denote by $\mathcal{L}_\text{yuk}$, has the form

$$\begin{aligned} \mathcal{L}_\text{yuk} &= -\left(y_{ij}^e \bar{e}_i L_j H + y_{ij}^d \bar{d}_i Q_j H + y_{ij}^u \bar{u}_i Q_j H_\text{c} + \text{c.c.} \right) \\ &\equiv -\left(\bar{e}^\text{T} y^e L H + \bar{d}^\text{T} y^d Q H + \bar{u}^\text{T} y^u Q H_\text{c} + \text{c.c.} \right). \end{aligned} \tag{16.11}$$

In the second line of this expression, we used matrix notation and regarded the fields $\bar{e}_i$ and so on as column vectors, T being the transpose.

The matrices y^e, y^u and y^d can be complex and are not restricted by any symmetry. By an appropriate choice of the fermion fields though, we can make y^e and y^u real and diagonal and can bring y^d into a simple form. (The roles of y^u and y^d could be exchanged, but this is the standard convention.) To do this, one uses the fact that, given any matrix y', one can choose unitary matrices A and B such that $y \equiv A y' B$ is diagonal.

Considering the quark terms, we make the transformations,

$$\begin{aligned} y^u &\to Ay^uB, \qquad y^d \to Cy^dDD^{-1}B \equiv Cy^dDV, \\ \bar{u}^T &\to \bar{u}^TA^{-1}, \qquad \bar{d}^T \to \bar{d}^TC^{-1}, \qquad Q \to B^{-1}Q. \end{aligned} \tag{16.12}$$

This does not affect the rest of the Lagrangian, and we can choose the matrices A, B, C and D so that the new y satisfy $y^d_{ij} = y^d_i\delta_{ij}$ and $(Cy^uD)_{ij} = y^u_i\delta_{ij}$ with y^d_i and y^u_i real. A similar transformation gives $y^e_{ij} = y^e_i\delta_{ij}$, so that

$$\mathcal{L}_{\text{yuk}} = -\left(\sum y^e_i\bar{e}_iL_iH + \sum y^d_i\bar{d}_iV_{ij}Q_jH + \sum y^u_i\bar{u}_iQ_iH_c + \text{c.c.}\right). \tag{16.13}$$

The unitary matrix V has nine parameters, but five of them can be removed by making a transformation

$$V \to XVY, \qquad \bar{d}^T \to \bar{d}^TX^{-1}, \qquad Q \to Y^{-1}Q, \qquad \bar{u}^T \to \bar{u}^TY, \tag{16.14}$$

where X and Y are diagonal unitary matrices whose elements are just phase factors.[4] It can be shown that this transformation allows us to bring V into a form specified by three angles θ_{ij} and a phase δ:

$$V = \begin{pmatrix} c_{12}c_{13} & s_{12}c_{13} & s_{13}e^{-i\delta} \\ -s_{12}c_{23} - c_{12}s_{23}s_{13}e^{i\delta} & c_{12}c_{23} - s_{12}s_{23}s_{13}e^{i\delta} & s_{23}c_{13} \\ s_{12}s_{23} - c_{12}c_{23}s_{13}e^{i\delta} & -c_{12}s_{23} - s_{12}c_{23}s_{13}e^{i\delta} & c_{23}c_{13} \end{pmatrix}, \tag{16.15}$$

where $c_{ij} \equiv \cos(\theta_{ij})$ and $s_{ij} \equiv \sin(\theta_{ij})$. This is called the CKM matrix. The measured values are

$$s_{12} = 0.22, \qquad s_{23} = 0.041, \qquad s_{13} = 0.004, \qquad \delta = 1.0\,\text{radians}. \tag{16.16}$$

The hierarchy $s_{12} \gg s_{23} \gg s_{13}$ almost decouples the third generation, and means that the matrix elements are almost real except for $s_{13}e^{-i\delta}$. The reason for the hierarchy isn't known.

The expression Eq. (16.13) gives the quark and charged lepton masses, whose measured values are given in Table 16.2. The lepton and up-quark masses are read off immediately:

$$\begin{aligned} m_e &= \frac{1}{\sqrt{2}}y^e_1v, \qquad m_\mu = \frac{1}{\sqrt{2}}y^e_2v, \qquad m_\tau = \frac{1}{\sqrt{2}}y^e_3v, \\ m_u &= \frac{1}{\sqrt{2}}y^u_1v, \qquad m_c = \frac{1}{\sqrt{2}}y^u_2v, \qquad m_t = \frac{1}{\sqrt{2}}y^u_3v. \end{aligned} \tag{16.17}$$

The down-quarks d, s and b have at this stage a mass matrix instead of definite

[4] Six parameters are required to specify the phases, but one of them can be chosen as α defined by $X = Y^{-1} = e^{i\alpha}I$. The parameter α has no effect on V and no effect on the Lagrangian. This is the B (baryon number) symmetry of Section 16.5.

masses. This is because the fields $d_i \equiv (d, s, b)$ do not yet correspond to the physical particles. To achieve that, we need to diagonalize the down-quark mass matrix by making a final transformation

$$V \to UVU^{-1}, \qquad \bar{d}^T \to \bar{d}^T U^{-1}, \qquad d \to Ud. \tag{16.18}$$

Then

$$m_d = \frac{1}{\sqrt{2}} y_1^d v, \qquad m_s = \frac{1}{\sqrt{2}} y_2^d v, \qquad m_b = \frac{1}{\sqrt{2}} y_3^d v. \tag{16.19}$$

The final transformation acts only on the down part of Q. If the electroweak interaction were written in terms of the final fields we would lose the manifest invariance under the $SU_L(2)$ gauge transformation. This corresponds to the fact that a quark with definite mass is changed by the electroweak interaction into a time-dependent linear superposition of quarks with definite mass.

16.4 Quantum Chromodynamics (QCD)

In this section we describe the strong interaction, corresponding to the $SU_C(3)$ gauge group, under the pretence that the electroweak interaction doesn't exist except insofar as it generates quark masses. That corresponds to what is called Quantum Chromodynamics (QCD). From a mathematical viewpoint, QCD is a generalization of Quantum Electrodynamics (QED), which describes the (purely electromagnetic) interaction of the photon, electron and μ ignoring all other particles.

16.4.1 Lagrangian and the QCD scale

The QCD interaction is the same for each of the quarks. We therefore introduce a 'flavour' index $f = 1$ to 6 and denote the quark fields by q_f so that $q_1 = u$, $q_2 = d$ and so on.[5] The quark fields also carry a colour label $c = 1$ to 3. Let us regard say u as a column vector with elements u_c, and $\bar{u}$ as a column vector with elements $\bar{u}_c$. Then the $SU(3)$ transformation is $u \to Uu$ and $\bar{u}^T \to \bar{u}^T U^{-1}$, where U is an $SU(3)$ matrix. The combination $\sum_c \bar{u}_c u_c$ is clearly invariant under this transformation.

To allow U to depend on the spacetime position we need the covariant derivative defined by

$$D_\mu q_{fc} = \partial_\mu q_{fc} - i g_s \sum_{c'} (G_\mu)_{cc'} q_{fc'}, \tag{16.20}$$

where G_μ is 3×3 matrix of gluon fields with the transformation (14.47). Also, we

[5] Nowadays, 'flavour' is often taken to mean 'generation' but we stick to the original usage.

need the gluon field strength matrix $G_{\mu\nu}$, defined by Eq. (14.49) with W replaced by G.

With these in place we can write down the QCD Lagrangian:

$$\mathcal{L}_{\rm QCD} = -\frac{1}{2}\,{\rm Tr}\,G_{\mu\nu}G^{\mu\nu} - i\sum_{f=1}^{6}\sum_{c=1}^{3}\left(q^*_{fc}\,\not{D}q_{fc} + \bar{q}^*_{fc}\,\not{D}\bar{q}_{fc}\right)$$
$$-\sum_{f=1}^{6}\sum_{c=1}^{3} m_{fc}\left(\bar{q}_{fc}q_{fc} + {\rm c.c.}\right). \tag{16.21}$$

At $Q = M_Z$ the observed value of the strong coupling is given by

$$\alpha_3 \equiv \frac{g_{\rm s}^2}{4\pi} = \frac{1}{8.42}. \tag{16.22}$$

The calculated Q dependence of $\alpha_3(Q)$ agrees with observation in the accessible regime $100\,{\rm MeV} \lesssim Q \ll 1\,{\rm TeV}$. As shown in Figure 16.1, $\alpha_3(Q)$ increases as the scale decreases.

At $Q \gg 100\,{\rm MeV}$ we have $\alpha_3(Q) \ll 1$. This regime is relevant for collision energies $\gg 100\,{\rm MeV}$, and for a fluid of quarks and gluons whose spacing is $\ll 100\,{\rm MeV}$. In that regime, the quarks and gluons interact only weakly, and the fluid of quarks and gluons is gaseous. The fact that quarks and gluons interact only weakly in the regime $Q \gg 100\,{\rm MeV}$ is called **asymptotic freedom**.

16.4.2 Confinement of colour

The scale $\Lambda_{\rm QCD} \simeq 200\,{\rm MeV}$, at which $\alpha_3(Q) = 1$, is called the QCD scale. Temporarily restoring c and $\hbar$ we have $\Lambda_{\rm QCD}^{-1} \simeq 10^{-15}\,{\rm m}$, which is about the size of a nucleon. In the regime $Q \lesssim \Lambda_{\rm QCD}$, the QCD interaction becomes non-perturbative, and most calculations have to be done using lattice field theory though some analytic methods are available.

The most remarkable finding from these calculations concerns the eight conserved charges, that correspond to the eight parameters of the $SU_C(3)$ group. The charges are collectively called **colour**. Since $SU_C(3)$ is a non-Abelian group, the gluons are self-interacting which means that gluons as well as quarks carry colour. (For an Abelian theory it would just be the quarks.) What is found is that the colour vanishes, within any region $\gg \Lambda_{\rm QCD}^{-1}$. In other words *colour is confined to regions with size* $\lesssim \Lambda_{\rm QCD}^{-1}$.

Before moving on we should address the following possible concern. Since the operators corresponding to the eight colours don't all commute, the can't in general have well-defined values at the same time. How, therefore, can we say that they all vanish? The answer is that they *can* all have the value zero, because that simply

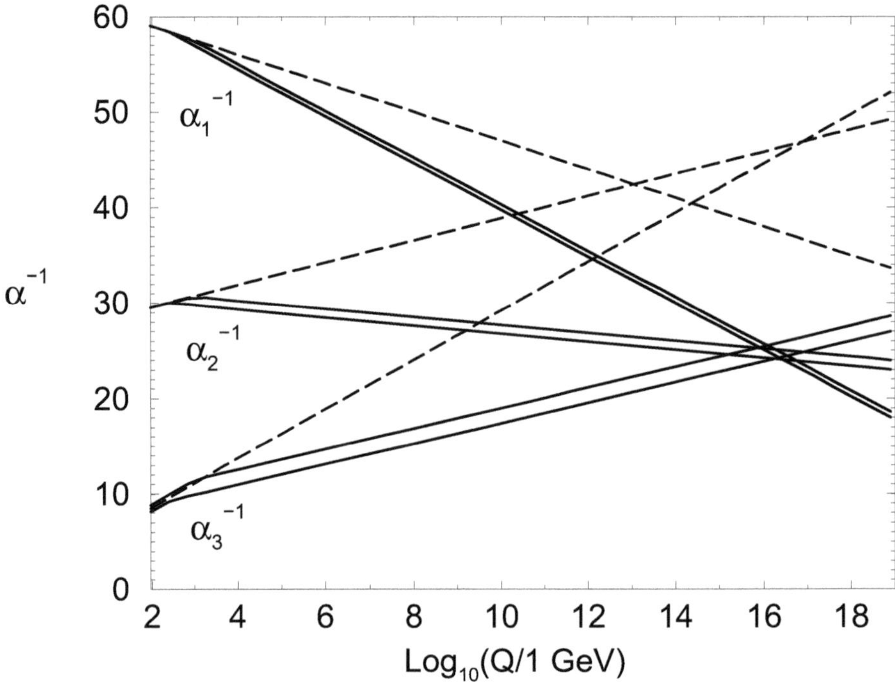

Fig. 16.1. The full lines show the running of the gauge couplings with supersymmetry, the broken lines their running without supersymmetry. The lines are straight in a given range of Q, insofar as no relevant particles have masses in that range. The pairs of lines correspond to choosing the supersymmetric partners to have masses 250 GeV and 1 TeV. In the case of α_3 the pair allow in addition a range $0.113 < \alpha_3(M_Z) < 0.123$ spanning the central observed value. The α_i^{-1} in the range $\sim 100\,\mathrm{MeV} \lesssim Q \lesssim 100\,\mathrm{GeV}$ correspond roughly to the continuation of the dashed lines, since few particle species have mass in this range. Reproduced with permission from Ref. [1].

corresponds to a state which is invariant under the $SU_C(3)$ transformations. The situation is analogous to the familiar one for the components angular momentum. In general they can't all have well-defined values at the same time, but they *can* all vanish because that corresponds to a state which is invariant under rotations.

A very important consequence of colour confinement is that an isolated colour charge cannot exist. If it did there would be a gluon field outside it, falling like $1/r$ just like the electric field, and this would be in contradiction with confinement because the gluon field itself carries colour. From this consequence it immediately follows that isolated quarks and gluons cannot exist, which is in accordance with observation. Instead, quarks and gluons can exist only if their spacing is $\lesssim \Lambda_{\mathrm{QCD}}^{-1}$, about the size of a nucleon.

In fact, nucleons are bound states of quarks and gluons. A proton consists of two u quarks plus a d quark plus an indefinite number of quark–antiquark pairs and gluons. (As we are in the quantum regime, there is no need for the numbers of those objects to have well-defined values.) A neutron consists of two d quarks and a u quark, plus the pairs. The quark content gives the proton and neutron baryon number $B = 1$, the proton charge $Q = 1$ and the neutron charge $Q = 0$, as observed.

More generally, the bound states of quarks and gluons are classified as **hadrons**, consisting of three quarks plus quark pairs and gluons, and **mesons** consisting entirely of quark pairs plus gluons. From the rules for adding angular momentum it follows that hadrons are fermions while mesons are bosons. Also, hadrons carry $B = 1$ while mesons carry $B = 0$. All of the bound states carry colour zero, in accordance with colour confinement.

The proton is the lightest hadron, and is therefore stable insofar as B is conserved, in accordance with observation. The next lightest hadron is the neutron, which decays to a proton plus leptons, but very slowly because it is hardly heavier than the proton. All other hadrons have very short lifetimes and so do the mesons.

To a useful approximation nuclei can be regarded as bound states of nucleons, but at a deeper level they are bounds states of quarks and gluons, with colour charge zero and B and Q positive integers. Counting the quarks in a heavy nucleus, one finds that their average spacing is $\ll \Lambda_{\text{QCD}}^{-1}$. As a result, the interior of a heavy nucleus can be regarded as a gas of quarks and gluons. The properties of very heavy nuclei, observed in collisions of the ions, are in accordance with this picture. As we see in Section 22.2, the cosmic fluid of the very early Universe is also a gas containing free quarks and gluons.

16.4.3 Grand Unified Theory (GUT)

As seen in Figure 16.1, the calculated gauge couplings (dashed lines) almost cross at a scale $Q \sim 10^{15}$ GeV. This suggests the existence of a Grand Unified Theory (GUT), whose gauge symmetry group involves a single gauge coupling. The idea is that the GUT gauge group is spontaneously broken down to the Standard Model gauge group by some GUT Higgs fields. The GUT Higgs potential is supposed to have the usual Mexican-hat form with coupling $\lambda \sim 1$, so that it generates masses for the Higgs and gauge fields of order the GUT scale M_{GUT}. Then the Standard Model couplings will cross when the renormalization scale is $Q \simeq M_{\text{GUT}}$.

If the GUT hypothesis is accepted, one expects that there are no particles with mass in the range TeV to M_{GUT} that couple to the Standard Model gauge group. The reason is that such particles would usually spoil the unification. This **desert hypothesis** greatly simplifies the discussion of what may lie beyond the Standard Model. In particular it simplifies discussion of the very early Universe, back to the time when the energy density is of order M_{GUT}^4. The desert hypothesis doesn't exclude the existence of a **hidden sector**, consisting of fields different from the Standard Model fields, which have different gauge symmetries.

The fact that the calculated couplings don't quite unify means that some extension of the Standard Model is required, if it is to be compatible with the GUT and desert hypotheses. In the next chapter we see that the Minimal Supersymmetric Standard Model (MSSM) gives remarkably accurate unification, as shown by the solid lines in Figure 16.1.

16.5 The complete Lagrangian and its accidental symmetries

16.5.1 Lagrangian

Now we can write down the complete Standard Model Lagrangian. We suppress the colour indices and it is understood that the covariant derivative for quark fields includes the additional term given by Eq. (16.20). As in Eq. (16.11), we regard the three-generation objects e_i etc. as column vectors. Then the Lagrangian is

$$\begin{aligned}
\mathcal{L}_{\rm SM} &= -\frac{1}{2}\,{\rm Tr}\,G_{\mu\nu}G^{\mu\nu} - \frac{1}{2}\,{\rm Tr}\,W_{\mu\nu}\cdot W^{\mu\nu} - \frac{1}{4}B_{\mu\nu}B^{\mu\nu} \\
&- iL^{*T}\,\not{D}L - iQ^{*T}\,\not{D}Q - i\bar{e}^{*T}\,\not{D}\bar{e} - i\bar{d}^{*T}\,\not{D}\bar{d} - i\bar{u}^{*T}\,\not{D}\bar{u} \\
&+ |D_\mu H|^2 - \frac{1}{2}\lambda^2\left(|H|^2 - \frac{1}{2}v^2\right)^2 \\
&- \left(\bar{e}^T y^e LH + \bar{d}^T y^d QH - \bar{u}^T y^u QH_{\rm c} + {\rm c.c.}\right). \qquad (16.23)
\end{aligned}$$

Given the fields, this is the most general renormalizable Lagrangian consistent with the gauge symmetries up to 4-divergences.

As we now see, the Lagrangian possesses some accidental global symmetries, which were not imposed from the beginning.

16.5.2 Baryon number, lepton number and isospin

The continuous global transformations act only on the spinors, not on the Higgs field or gauge fields. We begin with the $U_B(1)$ transformation, which acts only on quark fields with $q_{fc} \to e^{i\lambda B}$ and $\bar{q}_{fc} \to e^{-i\lambda B}\bar{q}_{fc}$. The Lagrangian is invariant under this transformation and the corresponding charge is baryon number B; by convention we choose $B = 1/3$ as the baryon number carried by each quark.

Next we consider the $U_{L_i}(1)$ transformations, which act on the ith generation of leptons with $L_i \to e^{i\lambda L_i} L_i$ and $\bar{e}_i \to e^{-i\lambda L_i}\bar{e}_i$, where L_i in the exponent is the lepton number (a separate one for each generation). The Lagrangian is invariant under these transformations. By convention we choose $L_i = 1$ as the lepton number carried by each lepton.

The B and L_i symmetries are broken by anomalies, leaving only $L_i - L_j$ and $B - L$ as the conserved quantities, where $L = \sum L_i$ is the total lepton number [2]. The transitions breaking B and L conservation are usually called sphaleron transitions, and can occur with significant rate only in the early Universe.

The QCD Lagrangian also has some approximate global symmetries, broken only by the small masses of the u and d quarks. Because the strong interaction dominates the electroweak interaction, these are also approximate global symmetries of the Standard Model Lagrangian, broken by the electromagnetic interaction as well as by the quark masses.[6]

One of these symmetries is a global $SU(2)$ called **isospin symmetry**. Let us define $\psi \equiv (u, d)$ and $\bar{\psi} \equiv (\bar{u}, \bar{d})$ and regard both as column vectors. The isospin $SU(2)$ transformation is then $\psi \to U\psi$ and $\bar{\psi}^T \to \bar{\psi}^T U^{-1}$. (In terms of the doublet $\Psi \equiv (\Psi_u, \Psi_d)$ of Dirac spinors these become simply $\Psi \to U\Psi$.) The QCD Lagrangian is invariant under the isospin $SU(2)$ transformation to the extent that the small u and d quark masses can be set equal to zero. (It would also be invariant if these masses were the same, but that is not the case.) Writing U in the form (14.15), we learn that the u and d quarks carry conserved isospin charges $T_3 = \pm 1/2$. Although we use the same symbol T_3 as for the $SU_L(2)$ charge, there is of course no physical relationship between the two. (Isospin came first, and $SU_L(2)$ used to be called weak isospin.)

Isospin symmetry requires that the hadrons and nuclei form multiplets of approximately the same mass, and gives relations between the interactions of the multiplets. The nucleons (p, n) form an isospin doublet and the pions (π^+, π_0, π_-) form an isospin triplet. (The symbols can be thought of as state vectors, though it is useful to define a pion field too since the pion is a PNGB.) To the extent that the s quark is also light, isospin can be extended to a global $SU(3)$ symmetry describing bigger multiplets of hadrons and quarks.

16.5.3 Global chiral symmetries

Now we come to global chiral symmetries, which distinguish between left and right handedness. Like isospin, they act on the fields of the u and d quarks and are symmetries of QCD insofar as the masses of these quarks can be neglected.

The **chiral** $SU(2)$ **symmetry** transformation is $\psi \to U\psi$ and $\bar{\psi}^T \to \bar{\psi}^T U$. In contrast with isospin, this symmetry is spontaneously broken. The breaking is dynamical, and the PNGB is the pion triplet. The pion fields are condensates, made out of the fields corresponding to the u and d quarks.

Chiral $SU(2)$ extends to a chiral $SU(3)$ symmetry to the extent that the mass of

[6] We have in mind here the low-energy regime where the weak interaction is distinct from the electromagnetic interaction. The global symmetries of QCD are broken by the weak interaction, but at a negligible level.

the s quark can be ignored. Then there are eight PNGBs, namely the three pions, the four kaons and the eta. The invariance of the Standard Model Lagrangian under the chiral symmetries is broken by electromagnetism, and electromagnetism also causes them to be broken by an anomaly.

Finally, QCD has the **chiral** $U(1)$ **symmetry**, which multiplies u, d, $\bar{u}$ and $\bar{d}$ by the same phase factor. (To the extent that the s quark is light, we can allow the chiral $U(1)$ symmetry to multiply also s and $\bar{s}$ by the same factor.) Chiral $U(1)$ symmetry is spontaneously broken by a condensate made out of the light quark fields, but is in addition broken by an anomaly present within QCD. As a result of the anomaly, the would-be PNGB of chiral $U(1)$ has its mass pushed up above the GeV scale, and loses its identity through mixing with other mesons.

16.5.4 P, C and T

The parity transformation P interchanges right and left, the charge conjugation transformation C interchanges particle and antiparticle, and time reversal T reverses the arrow of time. These properties do not uniquely define the effect of the P, C and T transformations on the fields present in a field theory, but there is a standard choice. Here we will just explain the situation verbally.

Any Lorentz-invariant quantum field theory is invariant under the combined transformation PCT. This gives some striking predictions, notably that a stable particle (elementary or composite) and its antiparticle have exactly the same mass, which are verified to high accuracy.

The electroweak Lagrangian grossly violates P and C invariance, because the $SU_L(2)$ gauge interaction interaction treats left- and right-handed (barred and unbarred) fields differently. As is clear from the assignments in Section 15.6.2, the P transformation interchanges left- and right-handed fields, and so does the C transformation.

The combined CP transformation takes a left-handed field into itself, and a right-handed field into itself. This allows the possibility that the electroweak interaction is invariant under CP, which turns out to be the case *except* for the effect of the phase δ in the CKM matrix. That phase gives CP violation, but at a very low level because of the hierarchy $s_{12} \gg s_{23} \gg s_{13}$.

The QCD Lagrangian is invariant under C, P and T, which at first sight means that QCD itself is invariant under these transformations. That would make the Standard Model invariant under CP except for the tiny effect of the CKM phase, which would be in excellent agreement with observation. Unfortunately though, the chiral $U(1)$ anomaly spoils the prediction that QCD has CP invariance, because it adds to the effective QCD Lagrangian a term[7]

$$\mathcal{L}_\theta = -\frac{\theta}{64\pi^2}\epsilon^{\alpha\beta\mu\nu}\mathrm{Tr}\, G_{\alpha\beta}G_{\mu\nu}. \tag{16.24}$$

The T transformation takes x^0 to $-x^0$, which means that it reverses $\epsilon^{\alpha\beta\mu\nu}$ and hence $\mathcal{L}_\theta$. The term $\mathcal{L}_\theta$ therefore violates T invariance. Since any field theory is invariant under CPT, this term also violates CP invariance.

The strongest observational constraint on T invariance comes from an upper bound on the neutron dipole moment, which is required to vanish by T invariance. It can be shown to require $\theta < 10^{-8}$. Within the context of the Standard Model there is no explanation for this small number, because θ is just a parameter like the Yukawa couplings.

16.6 Peccei–Quinn symmetry and the axion

The favoured way of guaranteeing the small value of θ that we just encountered is to supplement the Standard Model, in such a way that θ becomes a dynamical field whose vev is zero. This θ field is the PNGB of a global $U(1)$ symmetry called Peccei–Quinn (PQ) symmetry. The corresponding particle, called the axion, is a cold dark matter (CDM) candidate. In this section we focus on things that are needed in the cosmological context. The underlying particle physics is described for instance in Ref. [2], and detailed model-building in Ref. [3].

To get off the ground, let us suppose that the PQ symmetry is spontaneously broken by a single complex field ϕ, whose potential $V(|\phi|)$ has the usual Mexican-hat form giving a minimum at $|\phi| = f_\mathrm{a}$. Writing $\phi = f_\mathrm{a}e^{i\theta}$ the canonically normalized axion field is

$$a = \sqrt{2}f_\mathrm{a}\theta. \tag{16.25}$$

The PQ symmetry is broken explicitly by the axial $U(1)$ anomaly of QCD. This generates a potential which is well approximated by Eq. (14.34):

$$V(\theta) = \Lambda_\mathrm{a}^4 (1 - \cos\theta). \tag{16.26}$$

The axion mass is $m_\mathrm{a} = \Lambda_\mathrm{a}^2/\sqrt{2}f_\mathrm{a}$, and Λ_a is of order the QCD scale. Making a precise calculation one finds

$$\boxed{\frac{m_\mathrm{a}}{6.2\times 10^{-6}\,\mathrm{eV}} = \frac{10^{12}\,\mathrm{GeV}}{f_\mathrm{a}}}\,. \tag{16.27}$$

[7] It can be shown that the term is a 4-divergence, but because of the anomaly it still has an effect. We are implicitly assuming here that the phases of the quark fields have been chosen to make their masses real. If that choice isn't made, the physically significant quantity becomes θ minus the sum of the phases of the masses, and we can trade θ for any one of those phases.

In more complicated models, PQ symmetry may be spontaneously broken by more than one field. These fields might not be canonically normalized, and the PQ symmetry might not have a fixed point in field space so that there is no Mexican-hat potential. Such things happen if the fields breaking PQ symmetry are the moduli of string theory as reviewed for instance in Ref. [4]. In all cases though, the axion potential is well approximated by Eq. (16.26), and we can define $f_{\rm a}$ as the factor in Eq. (16.25) that relates θ to the canonically normalized axion field. (Recall that any one field can always be canonically normalized.)

The axion field has a coupling to the gluon field strength, whose magnitude is determined by the requirement that it solves the CP problem. In a viable model the axion must have additional couplings and two possibilities are usually considered. In what are called DFSZ models an additional Higgs doublet $\tilde{H}$ is introduced, and a couples to the gauge-invariant product $H\tilde{H}$. (Both of the required Higgs fields occur in the supersymmetric extension of the Standard Model described in Chapter 17.) In what are called KSVZ or hadronic axion models, a couples instead to an additional heavy quark. Taking the axion interactions into account, collider physics and astrophysics constraints give $f_{\rm a} > 10^9\,\text{GeV}$, corresponding to an extraordinary small axion mass of at most $10^{-2}\,\text{eV}$ [5]. Because the mass is so small the axion lifetime is much bigger than the age of the Universe, making the axion a CDM candidate. As we see in Section 23.1, $f_{\rm a} \sim 10^{12}\,\text{GeV}$ is typically required to get the right axion CDM density.

The potential (16.26) is a quantum effect. One might expect that there is additional PQ symmetry breaking, present already at the classical level. Supposing that just one field ϕ breaks the symmetry, the breaking might be as in Eq. (14.12) so that altogether we have

$$V(\theta) = \Lambda_{\rm a}^4 \left[\cos\theta + |\lambda_d| \left(\frac{f_{\rm a}}{M_{\rm Pl}} \right)^{d-4} \frac{f_{\rm a}^2}{m_{\rm a}^2} \cos[d(\theta - \theta_0)] \right] . \tag{16.28}$$

The phase θ_0 is determined by the phase of λ_d and there is no reason for it to be small. So that PQ symmetry can do its job of setting $\langle\theta\rangle = 0$, the classical breaking should be negligible. Taking $|\lambda_d| = 1$ and $f_{\rm a} = 10^{12}\,\text{GeV}$ this requires $d \gtrsim 10$.

There are various ways of understanding why the classical breaking is so small. In a supersymmetric theory one can forbid terms with $d < 10$ by a discrete symmetry (which may be a gauge symmetry so that it cannot be broken). We discuss that mechanism in Section 17.3. Alternatively one can suppose that the QCD axion is a string axion (the imaginary part of a modulus) which could sufficiently suppress λ_d.

16.7 Neutrino mass

The observed phenomenon of neutrino oscillation shows that neutrinos have mass, and that the mass eigenstates are mixtures of the states ν_e, ν_μ and ν_τ appearing in the Standard Model. In this section we describe neutrino mixing, and consider the possible role of additional 'sterile' neutrino species.

Within the Standard Model, the neutrino is massless. There is a left-handed neutrino and a right-handed antineutrino. Including the mass, this makes no sense because a Lorentz boost along the direction of the neutrino motion can reverse the momentum and hence change a left-handed neutrino into a right-handed one. It is usually supposed that the so-called neutrino and antineutrino are really the same particle and that there is no antiparticle. Then the neutrino has the Majorana mass term discussed in Section 15.6. Alternatively, the neutrino might have a distinct antiparticle just like the electron. Then it has a Dirac mass term. An observation of neutrinoless double beta decay $\mathrm{n}+\mathrm{n} \to \mathrm{p}+\mathrm{p}+e+e$ (occurring within a nucleus) would rule out the Dirac possibility.

16.7.1 Neutrino mixing matrix

The states $|\nu_\alpha\rangle$ ($\alpha = e, \mu, \tau$) that appear in the Standard model are not the same as the mass eigenstates $|\nu_i\rangle$ ($i = 1, 2, 3$). Instead, they are linear combinations:

$$|\nu_\alpha\rangle = \sum_i U^*_{\alpha i} |\nu_i\rangle, \tag{16.29}$$

where U is a unitary matrix. (We are using the convention of the Review of Particle Physics [6].) In that case there will be transitions $\alpha \leftrightarrow \beta$, which violate conservation of the individual lepton numbers but conserve total lepton number.

We can write $U = XVY$, where V has the same form as the CKM quark mixing matrix and X and Y are diagonal unitary matrices whose elements are just phase factors. The phases in X can be absorbed into the phases of the states ν_α. In terms of the Standard Model fields, regarded as column vectors, this is achieved by the transformation $L \to X^{-1}L$ and $\bar{e}^T \to \bar{e}^T X$. The Standard Model Lagrangian is invariant under this transformation, corresponding to the conservation of the three lepton numbers.

What about the phases in Y? If the neutrinos have Majorana masses, these phases cannot be absorbed into the phases of the mass eigenstates, because those have been fixed to make the masses real. However, an overall phase of those states is equivalent to an overall phase of the states $|\nu_\alpha\rangle$ (since the factor $Ie^{i\alpha}$ commutes with U). Such a phase has no physical significance because the total lepton number is conserved.[8] We can therefore set one of the phase factors equal to 1. Taking it

[8] Neutrino masses are significant only for low-energy collisions, which conserve L.

to be the 33 element, U is of the form

$$\begin{aligned} U &= \begin{pmatrix} c_{12}c_{13} & s_{12}c_{13} & s_{13}e^{-i\delta} \\ -s_{12}c_{23} - c_{12}s_{23}s_{13}e^{i\delta} & c_{12}c_{23} - s_{12}s_{23}s_{13}e^{i\delta} & s_{23}c_{13} \\ s_{12}s_{23} - c_{12}c_{23}s_{13}e^{i\delta} & -c_{12}s_{23} - s_{12}c_{23}s_{13}e^{i\delta} & c_{23}c_{13} \end{pmatrix} \\ &\times \ \text{diag}\left(e^{i\alpha_1/2}, e^{i\alpha_2/2}, 1\right), \end{aligned} \tag{16.30}$$

where $c_{ij} \equiv \cos(\theta_{ij})$ and $s_{ij} \equiv \sin(\theta_{ij})$.

If the neutrinos instead have Dirac masses, they can be treated within the Standard Model as we see in Section 16.7.3. Then the phases α_1 and α_2 can be eliminated by redefining the fields, just as was done for the quark mixing matrix.

16.7.2 Neutrino oscillation

For a free neutrino with zero momentum, the Hamiltonian is just the neutrino mass and in the Schrödinger picture the evolution of a mass eigenstate is

$$|i, \tau\rangle = e^{-im_i\tau}|i, 0\rangle. \tag{16.31}$$

(We denoted time in the rest frame (proper time) by τ, and $i = 1, 2, 3$ labels the neutrino.) In a frame where the momentum is $p_i \gg m_i$

$$m_i\tau_i = E_i t - p_i L \simeq (E_i - p_i)L \simeq \frac{m_i^2 L}{2E_i}. \tag{16.32}$$

Here L is the distance travelled by the neutrino in time t, and the last equality follows from $E_i = \sqrt{p_i^2 + m_i^2}$.[9]

Consider now a state $|\alpha\rangle$ produced in a process such as $pe \to n\nu_e$. It will have a well-defined momentum and energy, $p \simeq E$. Using Eq. (16.29) and its inverse, the state when the neutrino has travelled a distance L is

$$|\alpha, L\rangle = \sum_{\beta,i} \left[U^*_{\alpha i} U_{\beta i} e^{-i(m_i^2 L/2E)}\right] |\beta, 0\rangle. \tag{16.33}$$

The modulus-squared of the square bracket is the probability for the transition $\alpha \to \beta$, that can be measured by observing transitions like $n\nu_e \to pe$. We see that the phases α_i have no effect; only the angles θ_{ij} and the phase δ can be determined through the neutrino oscillation.

Observations of the oscillation can be made, using neutrinos produced by cosmic rays and neutrinos produced at colliders. Inside the Sun, analogous mixing occurs whose description has to take into account the interaction of the neutrinos with the

[9] We are pretending that the spacetime location and 4-momentum of the neutrino can simultaneously be specified. A more correct treatment using wave packets gives essentially the same result in situations of observational interest.

stellar medium. That mixing too can be detected by observing the flux of neutrinos coming from the Sun. At the time of writing, observation gives $\sin^2(2\theta_{12}) = 0.86 \pm 0.03$, $\sin^2(2\theta_{23}) > 0.92$ and $\sin^2(2\theta_{13}) < 0.19$ with no information about the phase δ.

16.7.3 The seesaw mechanism and sterile neutrinos

The favoured explanation for the small neutrino mass is known as the seesaw mechanism. To describe it, we will pretend that there is only one neutrino species ν_e whose field we denote simply by ν.

The seesaw mechanism invokes another field $\bar{\nu}$, which doesn't have the electroweak or QCD interactions. Regarding $(\nu, \bar{\nu})$ as a column vector x, the most general mass term involving both fields is $\mathcal{L}_{\text{mass}} = -mx$ where m is a real symmetric mass matrix.

Let us denote the elements of the mass matrix as follows:

$$m \equiv \begin{pmatrix} m_\nu/2 & m_{\text{dirac}} \\ m_{\text{dirac}} & M/2 \end{pmatrix}. \tag{16.34}$$

What orders of magnitude do we expect for them? The Dirac mass can be obtained by adding to the Standard Model Lagrangian a Yukawa coupling $-y_\nu \bar{\nu} L H_{\text{c}}$. This will give $m_{\text{dirac}} = y_\nu v/\sqrt{2}$. The mass m_ν is forbidden by the gauge symmetries of the Standard Model plus the requirement of renormalizability. It could be generated only by adding a non-renormalizable term to the Standard Model, which presumably means that it is orders of magnitude smaller than m_{dirac}. Now comes the interesting part. Unlike the others, the mass term $M\bar{\nu}\bar{\nu}$ can be inserted directly into the Lagrangian because $\bar{\nu}$ is invariant under the Standard Model gauge group. There is no reason why M should be at the electroweak scale. Assuming it to be orders of magnitude bigger, we have

$$m = \begin{pmatrix} 0 & y_\nu v/\sqrt{2} \\ y_\nu v/\sqrt{2} & M/2 \end{pmatrix}, \tag{16.35}$$

with $y_\nu v \ll M$. This matrix is diagonalized by a tiny rotation of the basis to give

$$m \simeq \begin{pmatrix} m_\nu/2 & 0 \\ 0 & M/2 \end{pmatrix}, \tag{16.36}$$

with and $m_\nu = 2y_\nu^2 v^2/M$. As the rotation is tiny, the new ν has practically the same electroweak interaction as the original object while the new $\bar{\nu}$ still has negligible interaction. They have Majorana masses m_ν and M, and each is its own antiparticle.

This is the seesaw mechanism. It gives the Standard Model neutrino a tiny mass, at the expense of introducing a heavy neutrino with negligible electroweak interaction which is called a sterile or right-handed neutrino.[10] To get a neutrino mass of order 10^{-1} eV with y_ν roughly of order 1, we need the sterile neutrino to be very heavy, $M \sim 10^{14}$ GeV.

Including all three generations, we need three sterile neutrinos. If the masses of the sterile neutrinos have roughly the same order of magnitude, that magnitude should be roughly 10^{14} GeV. As we notice in Section 22.3.3, the decay of a heavy sterile neutrino might be responsible for the baryon number of the Universe. More complicated seesaw mechanisms allow a much lighter sterile neutrino, which can have a variety of cosmological consequences that we shall not consider.

The seesaw mechanism isn't mandatory. It could be that only $m_{\rm dirac}$ is present. Then the field $\bar\nu$ could be included in the Standard Model in the same way as the field $\bar e$. It could also be that M is of the same order as $m_{\rm dirac}$.

Exercises

16.1 Show that $H^i_{\rm c} \equiv \epsilon^{ij} H^*_j$ has the same infinitesimal $SU(2)$ transformation as H_i.

16.2 Using the method indicated in the text, verify Eq. (16.7) which gives the electromagnetic gauge coupling in terms of the more fundamental electroweak gauge couplings.

16.3 Evaluate the matrix $W_{\mu\nu}$ in terms of $W^\pm_\mu$, Z_μ and A_μ. Hence evaluate the first term of the electroweak Lagrangian (16.4), and show that Z_μ has no electromagnetic interaction.

References

[1] S. P. Martin. A supersymmetry primer. arXiv:hep-ph/9709356.

[2] S. Weinberg. *The Quantum Theory of Fields, Volume II* (Cambridge: Cambridge University Press, 1996).

[3] J. E. Kim. Light pseudoscalars, particle physics and cosmology. *Phys. Rep.*, **150** (1987) 1.

[4] T. Banks, M. Dine and M. Graesser. Supersymmetry, axions and cosmology. *Phys. Rev. D*, **68** (2003) 075011.

[5] G. G. Raffelt. Astrophysical axion bounds. *Lect. Notes Phys.*, **741** (2008) 51.

[6] C. Amsler *et al.* [Particle Data Group]. Review of particle physics. *Phys. Lett. B*, **667** (2008) 1.

[10] The term right-handed isn't really appropriate, since the heavy neutrino isn't part of the Standard Model and there is no reason to think of relativistic right-handed heavy neutrinos as being particles as opposed to antiparticles.

17
Supersymmetry

Internal symmetries are part of the Standard Model, and definitely exist in Nature. The same is true of invariance under Lorentz transformations. Now we come to supersymmetry, which combines an internal symmetry with the Lorentz transformations (global supersymmetry) or with the general coordinate transformation (supergravity). As we write there is no direct evidence for supersymmetry, but it is a widely considered possibility in the context of both collider physics and cosmology.

The general motivation for supersymmetry is that it tends to stabilize the theory against radiative corrections. In particular, it avoids the extremely fine-tuned cancellation between the bare mass and its radiative corrections, seen in Eq. (15.65). A more concrete motivation is that it allows the three gauge couplings of the Standard Model to unify with remarkable accuracy.

We shall not give a full account of supersymmetry. Instead we describe some basic implications of supersymmetry, focussing mostly on the scalar field potential. Many good introductions to supersymmetry are available, such as Ref. [1] upon which our summary is loosely based.

17.1 The supersymmetry transformation

Acting on a given state, an infinitesimal supersymmetry transformation is of the form

$$|X\rangle \to \left(I + \hat{Q}\right)|X\rangle, \tag{17.1}$$

where the Lorentz transformation of $\hat{Q}$ is that of a spinor. The transformation mixes bosons and fermions. A set of fields related by the supersymmetry transformation is called a supermultiplet. We shall deal with the simplest version of supersymmetry, known as $N = 1$ supersymmetry, which alone seems able to provide a viable extension of the Standard Model. Here, each chiral spin-$1/2$ field is paired with either a complex spin-0 field (making a chiral supermultiplet), or else with a spin-1 gauge field (making a gauge supermultiplet). Supersymmetry may be a global symmetry, but it is usually taken to be a local symmetry. Then it includes gravity and is called supergravity. In supergravity the spin-2 graviton has a spin-$3/2$ partner, the gravitino.

Unbroken supersymmetry requires that the particles within a supermultiplet have the same mass. Such supermultiplets are not observed, which means that supersymmetry is broken. For supergravity the breaking has to be spontaneous, leading to a gravitino with non-zero mass. For global supersymmetry the breaking could be spontaneous and/or explicit. Taken literally, spontaneous breaking of global supersymmetry requires a spin-$1/2$ field called the Goldstino, which is massless in the absence of explicit breaking. But when global supersymmetry is invoked it is usually regarded as an approximation to supergravity. Therefore, one expects that there will be a massive gravitino and no Goldstino.

17.2 Renormalizable global supersymmetry

Supergravity is non-renormalizable because it includes gravity whose coupling $M_{\rm Pl}^{-2}$ has negative energy dimension. Global supersymmetry can be renormalizable. In this section we describe some aspects of renormalizable global supersymmetry, with the aid of toy models that illustrate the principles. To keep things simple we include only chiral supermultiplets.

17.2.1 The supersymmetric potential and superpotential

The potential is the sum of two terms, called the F-term and the D-term:

$$\boxed{V = V_F + V_D}\,. \tag{17.2}$$

The D-term involves only those scalar fields that have gauge interactions. If there is just a $U(1)$ with the transformation (14.36), it is of the form

$$\boxed{V_D = \frac{1}{2}g^2\left(\sum_n q_n|\phi_n|^2 + \xi^2\right)^2}\,. \tag{17.3}$$

The constant ξ^2 is called the Fayet–Iliopoulos term, and can have either sign. It need not be present, and when invoked is usually supposed to come from string theory. Then ξ is typically of order the string scale which presumably is far above the electroweak scale $10^2\,\mathrm{GeV}$. For a set of $U(1)$ interactions, the D-term is a sum of terms of the above form. Non-Abelian interactions give contributions of a similar form but without the possibility of a Fayet–Iliopoulos term.

The vacuum expectation value (vev) of the D-term is usually supposed to vanish in the vacuum. Also, the D-term vanishes throughout the history of the Universe in most scenarios. We will generally ignore it.

The F-term is specified by a holomorphic function W of complex fields ϕ_n, called the superpotential. It is of the form

$$\boxed{V_F = \sum_n |W_n|^2}, \tag{17.4}$$

where $W_n \equiv \partial W/\partial\phi_n$. Renormalizability requires that W contains at most cubic terms so that V is at most quartic. The superpotential determines the mass and Yukawa interactions of the spin-$1/2$ partners ψ_n of the fields ϕ_n, through the formula

$$\boxed{\mathcal{L}_{\text{yukawa}} = -\frac{1}{2}\sum_{nm}\left[W_{nm}(\phi_1,\phi_2,\ldots)\psi_n\psi_m + \text{c.c.}\right]}, \tag{17.5}$$

with $W_{nm} \equiv \partial^2 W/\partial\phi_n\partial\phi_m$, and the spinor notation as in Section 15.6.1. To have a realistic theory one needs to include gauge supermultiplets but we shall not write down the formulas.

We consider just one complex field ϕ, and write

$$W = a\phi + \frac{1}{2}m\phi^2 + \frac{1}{6}y\phi^3. \tag{17.6}$$

(We omit a constant term because it has no effect.) This gives

$$V = \left|a + m\phi + \frac{1}{2}y\phi^2\right|^2, \qquad \mathcal{L}_{\text{yuk}} = -\frac{1}{2}\left(m\psi\psi + y\phi\psi\psi\right) + \text{c.c.} \tag{17.7}$$

Now consider a global $U(1)$ symmetry, with the transformation acting on the phase of ϕ. For the potential to be invariant under this transformation, it must contain only one term.[1] Let us consider the three possibilities in turn.

If only a is non-zero there is an exact shift symmetry $\phi \to \phi +$ const where const is an arbitrary *complex* number, which includes the exact $U(1)$ as a subgroup. We have free ϕ and ψ particles with zero mass.

If only y is non-zero, we can arrive at an exact global $U(1)$ symmetry by allowing the transformation to act on both ϕ and its superpartner: $\phi \to e^{iq_\phi\lambda}\phi$ and $\psi \to e^{iq_\psi\lambda}\psi$ with $q_\phi = -2q_\psi$. If instead only m is non-zero, we arrive at an exact global symmetry by setting $q_\psi = 0$. In both of these cases, the symmetry transformation acts differently on members of a supermultiplet, so that it does

[1] The potential has to be invariant under an internal symmetry transformation, but the superpotential can change by a phase factor without affecting the physics.

not commute with the supersymmetry transformation (17.1). Such a symmetry is called an R-symmetry.

From this simple example, we learn that *the holomorphy of W greatly enhances the ability of an internal symmetry to restrict the form of the potential.* That is a general and very welcome feature of supersymmetry.

17.3 Global supersymmetry breaking

If supersymmetry is unbroken, the Lagrangian is invariant under the supersymmetry transformation (17.1) and so is the vacuum. It can be shown that this corresponds to the vanishing of V in the vacuum. An example is provided by Eqs. (17.6)–(17.7) with $a = 0$.

Supersymmetry breaking can be spontaneous or explicit, and we consider these possibilities in turn.

17.3.1 Spontaneous supersymmetry breaking

For spontaneously broken supersymmetry, the Lagrangian is invariant but the vacuum isn't. It can be shown that this corresponds to $V > 0$ in the vacuum. A trivial example is provided by Eqs. (17.6)–(17.7) with $m = y = 0$. In that example, all members of the supermultiplet are massless and they do not interact.

In general spontaneous supersymmetry breaking splits the masses within a supermultiplet. It can be shown that at tree level

$$\boxed{\sum m^2_{\rm scalars} = \sum m^2_{\rm fermions}}\,, \tag{17.8}$$

with a sum over all chiral supermultiplets.

As a simple example, we consider two supermultiplets with

$$W = \frac{1}{2} y\phi\left(\chi^2 - M^2\right), \tag{17.9}$$

$$V = \lambda|\phi|^2|\chi|^2 + \frac{1}{4}\lambda|\chi^2 - M^2|^2, \qquad \lambda = |y|^2, \tag{17.10}$$

$$\mathcal{L}_{\rm yuk} = -\frac{1}{2}\left(y\phi\psi_\chi\psi_\chi + y\chi\psi_\chi\psi_\phi + \text{c.c.}\right), \tag{17.11}$$

where ψ_ϕ and ψ_χ are the superpartners of ϕ and χ.

There is invariance under the global $U(1)$ transformation $\phi \to e^{i\theta}\phi$, $\psi_\chi \to e^{-i\theta/2}\psi_\chi$, $\psi_\phi \to e^{i\theta/2}\psi_\phi$ (an R-symmetry). Demanding this symmetry, we forbid all terms of higher order in ϕ (as well as terms independent of ϕ). Demanding instead invariance under the Z_N subgroup $\phi \to e^{2\pi i/n}\phi$ etc. (with $n = 0, 1, \ldots N-1$) we forbid powers of ϕ up to ϕ^N. To the extent that exact global continuous symmetries are disfavoured, the latter is preferable. By making the Z_N a discrete gauge symmetry one ensures that it is unbroken.

The potential has a supersymmetric minimum with $\phi = 0$ and $\chi = M$, at which $V = 0$. But at $|\phi|^2 > M^2/2$, the potential in the χ direction has a minimum at $\chi = 0$, at which V has the constant value λM^4. In other words, the potential as a function of $|\phi|$ and χ has a flat valley at $\chi = 0$. Each choice of $|\phi|$ and χ within this valley corresponds to a possible vacuum, in which supersymmetry is spontaneously broken.[2] In this vacuum, the superpartner of χ has mass $y\langle|\phi|\rangle$ while that of ϕ is massless. Replacing ϕ by a fixed number, the canonically normalized real and imaginary parts of χ have masses squared

$$m_i^2 = \lambda \left(|\phi|^2 \mp \frac{1}{2} M^2 \right), \tag{17.12}$$

in accordance with Eq. (17.8).

In this example, the non-supersymmetric vacuum comes with a supersymmetric vacuum. The global minimum of the potential is at the supersymmetric vacuum, which means that the non-supersymmetric vacuum is unstable against quantum tunnelling to the supersymmetric one. (Recall that the supersymmetric vacuum has $V = 0$ while non-supersymmetric vacua have $V > 0$.) Such a vacuum corresponds to a local minimum and is called a false vacuum, as opposed to the global minimum which is called a true vacuum. To achieve a stable non-supersymmetric vacuum one needs a more complicated superpotential, such as $W = -M^2\phi_1 + m\phi_2\phi_3 + y\phi_1\phi_3^2$.

A false vacuum is a viable candidate for the vacuum in which we live, provided that the tunneling rate out of the false vacuum is longer than the age of the Universe. Even so, models of spontaneous supersymmetry breaking were until recently designed so that the non-supersymmetric vacuum corresponds to the global minimum of the potential.

17.3.2 Soft supersymmetry breaking

Now we consider explicit supersymmetry breaking, where the vacuum is invariant under the supersymmetry transformation but the Lagrangian is not.

In general, explicit supersymmetry breaking allows us to add arbitrary terms to the Lagrangian. But to preserve the stability of the theory against radiative corrections, it turns out that the coefficients of the terms should have positive energy dimension. This is called soft supersymmetry breaking.

Soft supersymmetry breaking cannot add quartic terms to the potential and it

[2] We are going to see that the valley gets a slope from the loop correction, so that it is not really a viable vacuum. This is a toy model, which we will use to get a realistic model of inflation.

cannot add Yukawa couplings between the fermions and scalars, as in both cases the coefficient would have zero energy dimension. It therefore preserves the relation $\lambda = |y|^2$ between the self-coupling of a scalar field and its Yukawa coupling to the supersymmetric partner.

Soft supersymmetry breaking can add quadratic terms and cubic terms to the potential. Considering just one field we can have

$$V_{\rm soft} = m^2_{\rm soft}|\phi|^2 + \left(B\phi^2 + {\rm c.c.}\right) + \left(A\phi^3 + {\rm c.c.}\right). \tag{17.13}$$

The coefficients, including $m^2_{\rm soft}$, can be positive or negative. Soft supersymmetry breaking can also contribute to gaugino masses. The above-mentioned contributions, and their analogues including more fields, are essentially the only soft supersymmetry breaking terms.[3]

17.3.3 One-loop correction to the potential

In the direction of any field ϕ, the one-loop correction is given by Eq. (15.67). If supersymmetry were unbroken, each spin-1/2 field would have a scalar or gauge field partner with the same mass and couplings, causing the loop correction to vanish. We now study the one-loop correction in the direction of a complex field ϕ, due to a chiral supermultiplet (χ, ψ). We suppose that there is a $U(1)$ symmetry acting on the phase of ϕ, so that the potential depends only on $|\phi|$. We consider in turn soft breaking and spontaneous breaking. In both cases the relation $\lambda = |y|^2$ between the quartic and Yukawa couplings is preserved, which means that the $\phi^4 \ln\phi$ contribution considered in Section 15.8.3 is absent.

For soft breaking, we consider the one-loop correction to the potential in the direction of a complex scalar field ϕ, which is generated by a superpotential $W = y\phi\chi^2/2$. In the absence of supersymmetry breaking, χ and its superpartner ψ both have mass $y|\phi|$. In the presence of soft supersymmetry breaking, we suppose that the canonically normalized real and imaginary parts of χ are mass eigenstates, with soft masses-squared $m^2_{\chi 1}$ and $m^2_{\chi 2}$. Then

$$m_\psi(\phi) = +y|\phi|, \qquad m^2_{\chi i}(\phi) = m^2_{\chi i} + \lambda|\phi|^2, \tag{17.14}$$

with $\lambda = y^2$. Calculating $V_{\rm loop}$ from Eq. (15.67), the leading term proportional to $|\phi|^4 \ln|\phi|$ vanishes by virtue of the relation $\lambda = y^2$. For $y|\phi| \gg m_\chi$ the next term, proportional to $|\phi|^2 \ln|\phi|$, dominates. Going to the canonically normalized

[3] Linear terms in the potential are soft, but are usually forbidden by a symmetry. Contributions to the masses of the superpartners of scalar fields are soft, but they can be absorbed into W and the other soft terms.

real part of ϕ, and absorbing a constant into Q, this gives the potential

$$\boxed{V = \frac{1}{2}\left[m^2 + \frac{y^2}{16\pi^2} m_\chi^2 \ln\frac{\phi}{Q}\right]\phi^2}, \tag{17.15}$$

where $2m_\chi^2 \equiv m_{\chi 1}^2 + m_{\chi 2}^2$ and m is the tree-level soft mass of ϕ. Note that m_χ^2 can be positive or negative.

The contribution of a gauge supermultiplet is of the same form, with y^2 replaced by $-g^2$ where g is a gauge coupling, and with m_χ^2 replaced by the necessarily positive gaugino mass-squared.

The one-loop approximation represented by Eq. (17.15) is expected to be valid over a limited range of ϕ, with Q set equal to a value of ϕ within that range. If a large range of ϕ is under consideration, one should use instead the renormalization group improved expression

$$V = \frac{1}{2} m^2(\phi)\,\phi^2, \tag{17.16}$$

with the running mass calculated from the renormalization group equations (RGEs).

For spontaneous breaking we consider Eq. (17.9) and take $|\phi| \gg |\phi_c|$. Now the term $\propto |\phi|^2 \ln|\phi|$ vanishes as well, by virtue of Eq. (17.8), leaving

$$\boxed{V^{\rm loop} \simeq \frac{y^4 M^4}{64\pi^2} \ln\frac{|\phi|}{Q}}. \tag{17.17}$$

17.4 Supergravity

Supergravity becomes global supersymmetry in the limit $M_{\rm Pl} \to \infty$, where gravity is switched off. The supergravity potential is the sum of a D-term and an F-term, but we ignore the former on the expectation that its vev is zero and that it continues to vanish in the early Universe. To specify the F-term we need, in addition to W, an object called the Kahler potential K. It is not a holomorphic function of the fields, but instead a real function of the fields and their complex conjugates. It turns out that there is invariance under the Kahler transformation defined by $W \to e^X W$ and $K \to K - M_{\rm Pl}^2(X + X^*)$ where X is any holomorphic function of the fields.

The F-term is

$$\boxed{V = V_+ - 3M_{\rm Pl}^2 m_{3/2}^2(\phi_1, \phi_2, \ldots)}, \tag{17.18}$$

with

$$\boxed{3m_{3/2}^2(\phi_1, \phi_2, \ldots) \equiv M_{\rm Pl}^{-4} e^{K/M_{\rm Pl}^2} |W|^2}, \tag{17.19}$$

and

$$\boxed{V_+ \equiv e^{K/M_{\rm Pl}^2} \sum_{m,n} \left[\left(W_n + M_{\rm Pl}^{-2} W K_n \right) K^{mn} \left(W_m + M_{\rm Pl}^{-2} W K_m \right)^* \right]}. \tag{17.20}$$

Here $K_n \equiv \partial K / \partial \phi_n$ and K^{mn} is the inverse of the matrix $K_{mn} \equiv \partial^2 K / \partial \phi_m \partial \phi_n^*$.

The Kahler potential also determines the kinetic terms of the chiral supermultiplets. The kinetic term of the scalar fields is given by

$$\mathcal{L}_{\rm kin} = -\sum_{m,n} K_{mn} \partial_\mu \phi_m \, \partial^\mu \phi_n^*. \tag{17.21}$$

A renormalizable Kahler potential will be quadratic in the fields, and we can write $K = \sum |\phi_n|^2$. (Any additional quadratic terms can be pushed into W by a Kahler transformation.) This is called the minimal Kahler potential and it corresponds to canonical normalization of the fields.

Allowing non-renormalizable potentials and considering just one field direction for simplicity, K and W can be taken to have the form

$$K = |\phi|^2 + \sum_{d>1} \kappa_d M_{\rm Pl}^{2(1-d)} |\phi|^{2d}, \tag{17.22}$$

$$W = a + \frac{1}{2} m \phi^2 + \frac{1}{4} y \phi^3 + \sum_{d>3} y_d M_{\rm Pl}^{3-d} \phi^d. \tag{17.23}$$

With ultra-violet cutoff $M_{\rm Pl}$ one expects $|\kappa_d| \sim 1$ and $|y_d| \sim 1$.

With supergravity, supersymmetry becomes a local (gauge) symmetry, which means that it must be broken spontaneously rather than explicitly. Spontaneous supersymmetry breaking corresponds to $\langle V_+ \rangle > 0$, just as in global supersymmetry (which indeed may provide a good approximation to V_+). In contrast with global supersymmetry though, V contains a negative contribution. This contribution must cancel V_+ in the vacuum, so that the vacuum energy density vanishes.[4] As is implied by the notation, the vev of $m_{3/2}(\phi_1, \ldots)$ is the gravitino mass, denoted by $m_{3/2}$. Denoting the vev of V_+ by $M_{\rm S}^4$, the vanishing of V in the vacuum requires

$$\boxed{\langle V_+ \rangle \equiv M_{\rm S}^4 = 3 M_{\rm Pl}^2 m_{3/2}^2}. \tag{17.24}$$

This condition can be achieved only with suitable forms for W and K, and in general only after fine-tuning. For typical global supersymmetry potentials, like those encountered in the previous section, the right-hand side of Eq. (17.24) is much smaller than the left-hand side because all dimensionful quantities are typically far below the Planck scale. The usual attitude is to go ahead with the global

[4] We ignore here the issue of dark energy, which is discussed in Section 23.5.

supersymmetry calculation, on the assumption that it won't be affected by whatever modification of the Lagrangian ensures Eq. (17.24).

Suppose now that the masses of the chiral supermultiplets with unbroken symmetry are negligible. Then one typically finds that soft supersymmetry breaking gives similar magnitudes for the masses-squared to the scalars, denoted by $|m_0^2|$, and similar masses to the spin-$1/2$ particles, denoted by $m_{1/2}$.

A given M_S is expected to make $|m_0|^2$ *at least* of order $M_S^4/M_{Pl}^2 \simeq m_{3/2}^2$. To understand this, suppose that $M_S^4 \equiv \langle V_+ \rangle$ comes from a single canonically normalized field ϕ_1, and that K^{11} has the following terms:

$$K^{11} \supset M_{Pl}^{-2} \sum_i \lambda_i |\phi_i|^2. \tag{17.25}$$

In a generic theory with ultra-violet cutoff M_{Pl}, one can expect such terms with $|\lambda_i| \sim 1$, giving the advertised result $|m_0|^2 \sim M_S^4/M_{Pl}^2$. Further investigation reveals more possible contributions to $|m_0^2|$, of the same order.

If the minimal contributions to $|m_0^2|$ dominate, one says that supersymmetry breaking is gravity mediated, because the strength of the interaction between ϕ and the other fields is controlled entirely by M_{Pl}. A special scheme, called anomaly mediation, reduces the minimal contribution by a factor of 10^{-2} or so but it is difficult to go further.

On the other hand, it is easy to increase $|m_0^2|$. The ultra-violet cutoff might be $M \ll M_{Pl}$, increasing the expected value of λ_i by a factor $(M_{Pl}/M)^2$. More importantly the field(s) whose vevs give M_S^4 can couple to a field ϕ_i through terms in the superpotential. A maximal coupling of order one would give $|m_0^2| \sim M_S^2$. We conclude that a typical scalar mass-squared is in the range

$$\frac{M_S^4}{M_{Pl}^2} \lesssim |m_0^2| \lesssim M_S^2. \tag{17.26}$$

The spin-$1/2$ masses $m_{1/2}$ may be of order $|m_0|$ or far smaller, depending on the scheme of soft supersymmetry breaking.

17.5 The Minimal Supersymmetric Standard Model

The Minimal Supersymmetric Standard Model (MSSM) invokes renormalizable global supersymmetry. It contains the superpartners of the quarks and leptons, called squarks and sleptons, and those of the gauge bosons, called gauginos. It contains two complex Higgs fields, which as we see later is the smallest possible number allowed by supersymmetry. They correspond to three neutral Higgs particles, plus a charged Higgs particle–antiparticle pair. The superpartners of these particles are called Higgsinos. The neutral gauginos and the neutral Higgsinos mix, and the mass eigenstates are called neutralinos. The superpartners of the Standard Model particles (counting both Higgs) are usually called supersymmetric particles. The MSSM doesn't invoke any other fields.

Unbroken supersymmetry would require that each Standard Model particle has the same mass as its superpartner. This is ruled out by observation, which means that the global supersymmetry possessed by the MSSM must be broken. We first describe the MSSM with the breaking ignored.

17.5.1 Unbroken supersymmetry

The Higgs doublets are called H_u and H_d, and in terms of them the superpotential of the MSSM is

$$W = -y_e \tilde{L} H_d \tilde{\bar{e}} - y_d \tilde{Q} H_d \tilde{\bar{d}} + y_u \tilde{Q} H_u \tilde{\bar{u}} + \mu H_u H_d. \tag{17.27}$$

Here a tilde denotes the scalar partner, and $\tilde{Q} H_u$ is the gauge-invariant combination $\tilde{Q}_i \epsilon^{ij} H_{uj}$ (similarly for the other products of doublets). Using Eq. (17.5), the first three terms give the Yukawa terms of the Standard Model, with H_d instead of H and H_u instead of H_c. We need two Higgs fields because H_c cannot be used, corresponding to the fact that W is holomorphic. Two Higgs are also needed to keep $SU_L(2)$ free of anomalies.

The last term of Eq. (17.27) gives mass-squared μ^2 to each of the Higgs fields. Except for these, the potential V_F calculated from W consists entirely of quartic terms. To this must be added V_D, which to agree with observation must also contain only quartic terms.

In the space of the scalar fields there are **flat directions** where the quartic term is absent. An example is $H_u = (0, \phi)$ and $\tilde{L} = (\phi, 0)$, where the complex field ϕ parameterizes the flat direction. Indeed, there is no contribution $H_u \tilde{L}$ to the F-term, and it can be shown that there is no contribution to the D-term either.

The gauge symmetries allow more cubic terms in the superpotential (17.27), but they would grossly violate B and L conservation in contradiction with observation. To forbid such terms, it is usually supposed that there is invariance under a Z_2 R-symmetry that multiplies each field by $(-1)^{3B+L+2s}$ where s is its spin. The conserved 'parity' is called R-parity, which we will denote by R. Each Standard Model particle has $R = +1$ and each superpartner has $R = -1$.

That isn't quite the end of the story of R-parity. Although it has no motivation from the MSSM, let us suppose that all particles carry the conserved R-parity, even those that do not belong to the MSSM. Now comes the interesting part. *The lightest particle carrying* $R = -1$ *is stable*, because the final state could contain only particles with $R = +1$ and hence would itself have $R = +1$.

The lightest particle with $R = -1$ is called the lightest supersymmetric particle

(LSP), and the lightest MSSM particle with $R = -1$ is called the lightest ordinary supersymmetric particle (LOSP). The LSP might be the LOSP, the gravitino or some other particle. Whatever it is, the LSP must be electrically neutral to avoid conflict with observation, and is then a cold dark matter (CDM) candidate. We discuss these matters in Section 23.2.

17.5.2 Softly broken supersymmetry

Supersymmetry breaking should give masses roughly of order $100\,\mathrm{GeV}$ to the squarks, sleptons and gauginos. They cannot be much smaller or supersymmetry would have been observed, and they should not be much bigger if supersymmetry is to do its job of stabilizing the Higgs potential. As the squark and slepton masses-squared have to be positive, Eq. (17.8) cannot apply, which means that the supersymmetry breaking is soft as opposed to spontaneous. (Negative squark or slepton masses-squared would spontaneously break the electromagnetic gauge symmetry $U_{\mathrm{em}}(1)$, and negative squark masses-squared would spontaneously break $SU_C(3)$ as well.)

Soft supersymmetry breaking may also add cubic terms to the potential, called A-terms, but it cannot add quartic terms. As a result, *flat directions in the space of the scalar fields have no quartic terms even after soft supersymmetry breaking.* As we shall see, the flat directions may be important in the early Universe.

Soft supersymmetry breaking must also contribute to the Higgs masses-squared, to get viable electroweak symmetry breaking. We can use $SU_L(2)$ symmetry to choose $\langle H_u \rangle = (0, v_u)$, and it turns out that the form of the Higgs potential then requires $\langle H_d \rangle = (v_d, 0)$. Both vevs can be chosen as real. As with the Standard Model, these vevs leave $U_{\mathrm{em}}(1)$ unbroken. Defining $v^2 = (v_d^2 + v_u^2)/2$, the tree-level W mass is still given by Eq. (16.8), requiring $v = 246\,\mathrm{GeV}$ at tree level.

As one would expect, $\mu \sim 100\,\mathrm{GeV}$ is needed to successfully break the electroweak symmetry. To avoid inserting this parameter by hand, one can suppose that it is in reality the vev of a field or a product of fields, in such a way that the vev is naturally of order the electroweak scale.

When the running of the gauge couplings is calculated in the MSSM, they unify with remarkable accuracy, as seen in Figure 16.1 on page 278. The unification (GUT) scale is $M_{\mathrm{GUT}} = 2 \times 10^{16}\,\mathrm{GeV}$, which presumably is to be identified with the vev of the GUT Higgs. To achieve this unification, the gaugino masses should not be much above $1\,\mathrm{TeV}$.[5]

Soft supersymmetry breaking adds of order 100 parameters to the MSSM, and the successes of the Standard Model are preserved in only a small part of the pa-

[5] The unification still works if the squark and slepton masses are some orders of magnitude bigger than $1\,\mathrm{TeV}$, a proposal known as Split Supersymmetry. But Split Supersymmetry does not stabilize the Higgs mass.

rameter space. Those successes include the fact that the electroweak interaction acts only within a generation and is the same for each generation, and that there is almost exact CP invariance (if the θ parameter is sufficiently small). To preserve these successes, there presumably are relations between the soft parameters, coming from whatever mechanism is responsible for the soft breaking. The form that such relations need to take is discussed for instance in Ref. [1].

17.5.3 Gravitino mass

The MSSM is supposed to be an approximation to a supergravity theory, involving only a subset of the fields in that theory. This subset is called the MSSM sector. The softly broken supersymmetry of the MSSM sector is supposed to be obtained from the full supergravity potential as an approximation. It is usually supposed that the spontaneous breaking takes place in some 'hidden sector', involving fields which don't have the Standard Model gauge interactions.

As we saw in Section 17.4, the gravitino mass is given by $m_{3/2} = M_S^2/\sqrt{3}M_{Pl}$ where M_S characterizes the strength of spontaneous supersymmetry breaking.[6] The value of M_S that is needed to give scalar masses of order $100\,\text{GeV}$ depends on the strength of the interaction between the hidden sector and the MSSM sector. In a generic supergravity theory, the interaction will be at least of gravitational strength, giving roughly $M_S^2 \sim m_0 M_{Pl} \sim (10^{10}\,\text{GeV})^2$ and $m_{3/2} \sim 100\,\text{GeV}$. This is gravity-mediated supersymmetry breaking. With anomaly mediation one might have $m_{3/2} \sim 10\,\text{TeV}$. This is the biggest gravitino mass that is usually contemplated.

Instead of being of gravitational strength, the interaction between the hidden sector and the MSSM sector may involve one or more gauge couplings, which typically makes M_S smaller by some orders of magnitude. One then says that supersymmetry breaking is **gauge-mediated**. The absolute lower limit is $M_S \sim 1\,\text{TeV}$, which would mean that supersymmetry breaking takes place directly in the MSSM sector, but that is difficult to arrange. It would correspond to $m_{3/2} \sim 10^{-3}\,\text{eV}$.

Gauge mediation is usually formulated in a globally-supersymmetric model, with the understanding that it is an approximation to supergravity. It is under better control than gravity mediation, making it easier to ensure that supersymmetry breaking doesn't spoil the successes of the Standard Model. Although gauge mediation is difficult to arrange if the vacuum corresponds to the absolute minimum of the potential, it becomes relatively easy to arrange if the vacuum corresponds only

[6] We are focussing for simplicity on the scalar masses. The other soft supersymmetry breaking terms have to be compatible with observation too and a viable model for the origin of the soft terms must ensure that.

to a local minimum (a metastable vacuum) [2]. For a useful review of this kind of gauge mediation, see Ref. [3].

17.6 Supersymmetry and the axion

Now we see how the Peccei–Quinn (PQ) mechanism can work in the context of the MSSM. To keep things simple, let us suppose first that Peccei–Quinn symmetry remains spontaneously broken in the limit of unbroken supersymmetry. Ignoring the axion mass, which is negligible in this context, the axion field is then part of a massless chiral supermultiplet. The particles of this supermultiplet are the axion, another spin-0 particle called the saxion and a spin-$1/2$ particle called the axino.

The saxion and axino receive masses from supersymmetry breaking. Although these masses are quite model-dependent, they most typically are of order the gravitino mass. In the case of gravity-mediated supersymmetry breaking, the LSP might then be the LOSP, the gravitino or the axino depending on the details of the theory. In the case of gauge-mediated supersymmetry breaking, the LSP is expected to be either the gravitino or the axino.

Now suppose instead that PQ symmetry breaking is generated by supersymmetry breaking, so that there is no PQ symmetry breaking in the limit of unbroken supersymmetry. Several new features arise in this case, with implications for cosmology and collider physics that are under ongoing investigation at the time of writing [5, 6, 7]. Let us mention the main points.

The tachyonic mass of the scalar field responsible for spontaneous PQ symmetry breaking is now a soft supersymmetry breaking mass. As such, it will be far smaller than f_{a}, making the PQ potential very flat. This may generate a brief era of 'thermal inflation' as described in Section 21.6.

If $N > 1$ complex scalar fields break PQ symmetry a further complication arises, because it ceases to be useful to assign the axion to a particular chiral supermultiplet. Soft supersymmetry breaking now becomes responsible for the masses of all of the $2N$ scalar particles except the axion, and for the masses of all of the N superpartners. We then have in effect $2N - 1$ saxions and N axinos, with only the lightest axino being a possible CDM candidate.

Exercises

17.1 Verify Eq. (17.12), giving the masses-squared generated by spontaneous supersymmetry breaking with the superpotential (17.9), by calculating V as a function of $|\phi|$ and the canonically normalized real and imaginary parts of χ.

17.2 Work out V for the superpotential $-M^2\phi_1+m\phi_2\phi_3+y\phi_1\phi_3^2$ mentioned in

Section 17.3, and show that the global minimum has $V > 0$ corresponding to spontaneously broken supersymmetry.

17.3 Assuming a single field $\phi = \phi_1$, with $K = |\phi|^2 + \kappa|\phi|^4/M_{\rm Pl}^2$, calculate the coefficient λ_1 in Eq. (17.25) for K^{11}.

References

[1] S. P. Martin. A supersymmetry primer. arXiv:hep-ph/9709356.

[2] K. Intriligator, N. Seiberg and D. Shih. Dynamical SUSY breaking in meta-stable vacua. *JHEP*, **0604** (2006) 021.

[3] K. Intriligator and N. Seiberg. Lectures on Supersymmetry Breaking. *Class. Quant. Grav.*, **24** (2007) S741.

[4] E. J. Chun and A. Lukas. Axino mass in supergravity models. *Phys. Lett. B*, **357** (1995) 43.

[5] T. Banks, M. Dine and M. Graesser. Supersymmetry, axions and cosmology. *Phys. Rev. D*, **68** (2003) 075011.

[6] E. J. Chun, H. B. Kim, K. Kohri and D. H. Lyth. Flaxino dark matter and stau decay. *JHEP*, **0803** (2008) 061.

[7] S. Kim, W. I. Park and E. D. Stewart. Thermal inflation, baryogenesis and axions. arXiv:0807.3607 [hep-ph].

Part IV

Inflation and the early Universe

18

Slow-roll inflation

In the final Part of this book, we see how inflation can set the initial conditions for the subsequent Big Bang. Until Chapter 24, all quantities are taken to be homogeneous, which means that we are dealing with the unperturbed Universe.

As seen in Section 3.9, inflation is an initial era during which the expansion rate $\dot{a}$ is accelerating (repulsive gravity). The observable Universe is supposed to be well inside the horizon at the beginning of inflation, and well outside the horizon at the end.

After some preliminary material, we focus in this chapter on the slow-roll inflation paradigm, where the cosmic fluid during inflation consists of a slowly-varying scalar field called the inflaton.

18.1 Inflation defined

Inflation is defined as an era during which the rate of increase of the scale factor accelerates, corresponding to repulsive gravity:

$$\boxed{\text{INFLATION} \quad \Longleftrightarrow \quad \ddot{a} > 0}\,. \tag{18.1}$$

Inflation sometimes is described just as a rapid expansion, though it isn't clear with respect to what the expansion is supposed to be rapid.

There is an equivalent expression of the condition for inflation that gives it a more physical interpretation:

$$\boxed{\text{INFLATION} \quad \Longleftrightarrow \quad \frac{d}{dt}\frac{H^{-1}}{a} < 0}\,. \tag{18.2}$$

Because H^{-1}/a is the comoving Hubble length, the condition for inflation is that the comoving Hubble length, which is the most important characteristic scale of the expanding Universe, is decreasing with time. Viewed in comoving coordinates,

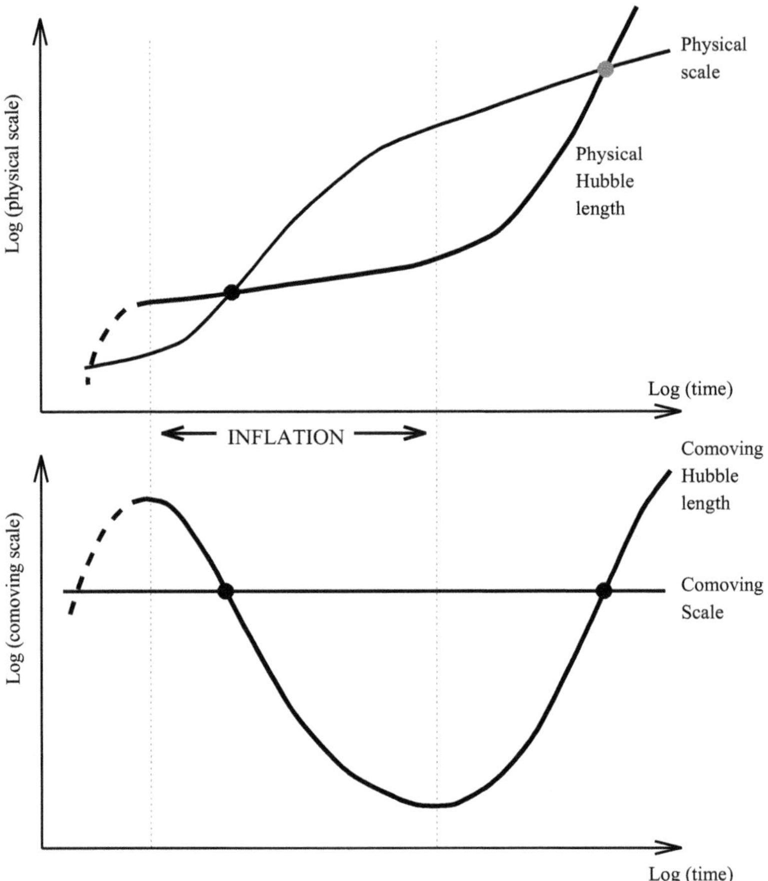

Fig. 18.1. Two views of the size of a comoving region within the observable Universe, relative to the Hubble length (horizon scale). The comoving Hubble length $1/aH$ is decreasing during inflation and increases afterwards at least up to the present. (What happens in the future depends on the nature of the dark energy, as discussed in Section 23.5.) The upper panel shows the physical size of the region, the lower one its comoving size. The vertical axis covers many powers of 10 in scale. The region starts well inside the horizon, then crosses outside some time before the end of inflation, reentering long after inflation is over.

the comoving region that will become the observable Universe actually becomes *smaller* during inflation. This is illustrated in Figure 18.1.

The condition for inflation can also be written

$$\boxed{\text{INFLATION} \quad \Longleftrightarrow \quad -\frac{\dot{H}}{H^2} < 1}\,. \tag{18.3}$$

On the usual assumption that H decreases with time, inflation is an era when

H varies slowly on the Hubble timescale. If this is a strong inequality, $|\dot{H}| \ll H^2$, then H is practically constant over many Hubble times and we have almost-exponential expansion, $a \propto e^{Ht}$. A universe with H exactly constant is called a de Sitter universe.

These equivalent definitions of inflation make no assumption about the theory of gravity. Assuming now Einstein gravity, the condition for inflation can be written as a requirement on the pressure of the cosmic fluid. Directly from the acceleration equation (3.83), with $\Lambda = 0$ or absorbed into ρ and P, we find

$$\boxed{\text{INFLATION} \quad \Longleftrightarrow \quad \rho + 3P < 0}\,. \tag{18.4}$$

Because we always assume ρ to be positive, it is necessary for P to be negative to satisfy this condition, which is independent of the density parameter of the Universe. To have almost-exponential inflation we need $P \simeq -\rho$. We see later how scalar fields can provide the necessary negative pressure.

As we see in Section 24.7, inflation with Einstein gravity generates a primordial tensor perturbation at some level. The present upper bound on such a perturbation requires $\rho^{1/4} < 2\times 10^{16}$ GeV while the observable Universe is leaving the horizon. This is well below the Planck scale.

18.2 Three problems of the pure Big Bang

In the modern view, by far the most important function of inflation is to generate the primordial curvature perturbation described in Chapters 5 and 6. It may generate other primordial perturbations too, including the isocurvature and tensor perturbations studied in Chapter 12.

However, the historical motivation for inflation was rather different, and arose largely on more philosophical grounds concerning the question of whether the initial conditions required for the *unperturbed* Big Bang seem likely or not. The ability of inflation to motivate one or more of the initial conditions was noticed by several authors [1, 2, 3, 4, 5]. Widespread appreciation of the issue followed the paper of Guth [6], which gave inflation its name, and explained how it could solve three problems that exist for a pure Big Bang cosmology.

18.2.1 Flatness problem

In Eq. (3.85) we defined a density parameter Ω which specifies the spatial curvature of the Universe in Hubble units. If the Universe is flat ($\Omega = 1$), then it remains so for all time. Otherwise, the density parameter evolves. During the Big Bang, the rate of expansion $\dot{a}$ is by definition decelerating (attractive gravity) and Ω is moving away from 1. For example, in a nearly flat matter-dominated Universe

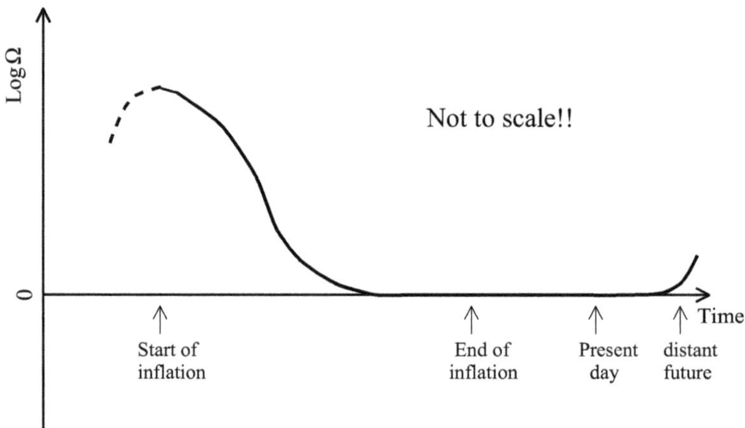

Fig. 18.2. Schematic of the inflationary solution to the flatness problem. Whether there is a definite start time to inflation or not is irrelevant. By definition, during inflation Ω is driven toward 1. At the end of inflation, Ω is supposed to be so close to 1 that it remains very close to 1 up to the present. In the distant future it may once again diverge from one, as shown here, though if the observed dark energy (discussed in Section 23.5) persists it will never do so.

$|1-\Omega| \propto t^{2/3}$, and in a nearly flat radiation-dominated Universe $|1-\Omega| \propto t$. We know observationally that, at the present, Ω is within a few percent of unity, which implies that at much earlier times it must have been extremely close to 1. To obtain our present Universe, at nucleosynthesis for example (when the Universe was around 1 s old) we require $|\Omega(t_{\rm nuc})-1| \lesssim 10^{-16}$. The flatness problem is that such finely tuned initial conditions seem extremely unlikely.

There is an equivalent way of stating the flatness problem. Instead of demanding initial fine tuning of $\Omega - 1$, we can contemplate a generic initial value. Then we find that almost all initial conditions lead either to a closed universe that recollapses almost immediately, or to an open universe in which Ω quickly becomes smaller than is now allowed by observation. For this reason the flatness problem also is phrased sometimes as an age problem — how did our Universe get to be so old?

The flatness problem is solved by inflation, provided that the observable Universe starts well inside the horizon:

$$aH \ll H_0 \quad \text{start of inflation.} \tag{18.5}$$

This situation is shown in Figure 18.2. The solution of the flatness problem rests on the fact that values of Ω at horizon exit and horizon entry are the same. Indeed, since inflation drives Ω toward 1, Ω will be close to 1 at horizon entry for all scales which are well within the horizon at the beginning of inflation.

18.2.2 Horizon problem

The horizon problem is most usefully regarded as referring to the horizon H^{-1}. Going back in time during the Big Bang, the comoving horizon $(aH)^{-1}$ shrinks indefinitely compared with the comoving size H_0^{-1} of the observable Universe. We saw in Section 3.10 that in a Universe dominated by matter and/or radiation, the comoving horizon $(aH)^{-1}$ will of order the effective comoving particle horizon. This is the maximum distance that light can have travelled since the beginning of the Big Bang. On this view the horizon problem is to understand how the observable Universe can be so nearly homogeneous at early times, when it is much bigger than the particle horizon.

Irrespectively of the nature of the cosmic fluid, we argued in Section 5.3.2 that the early Universe at a given epoch can be regarded as a collection of separate universes, each of them with a comoving size of order $(aH)^{-1}$. The separate universes evolve independently, and from this viewpoint the horizon problem is to understand how conditions within them come to be related.

A dramatic illustration of the horizon problem is provided by the cosmic microwave background. The horizon at photon decoupling is only $205\,\mathrm{Mpc}$, while its present value is about $4\,000\,\mathrm{Mpc}$. Microwaves coming from regions separated by more than a degree or so were separated at last scattering by more than the horizon distance. The pure Big Bang model therefore offers no prospect of explaining why the temperature seen in different regions of the sky is so accurately the same; the near-homogeneity must form part of the initial conditions.

Taking the horizon problem to refer to H^{-1}, it is solved by inflation provided that the Universe is inside the horizon at the beginning of inflation. We will see how that allows inflation to do its job of setting the initial condition for the subsequent Big Bang, without reference to the particle horizon. The inflationary solution of the horizon problem is illustrated in Figure 18.3.

18.2.3 Unwanted relics

By 'relics' we mean particles (or topological defects, see Sections 21.4 and 21.5) produced at early times. In a given scenario they are 'unwanted' if their presence is in conflict with observation. Guth was concerned with monopoles created during a GUT phase transition (see Section 21.5).

In contrast with the flatness and horizon problems, the problem of unwanted relics is not inherent to the Big Bang cosmology, but depends on the nature of the fundamental interactions at energy scales beyond the scope of the Standard Model of particle physics.

Consider first relics that exist before inflation. The relics may initially be in thermal equilibrium, or anyhow have significant interaction (with themselves and/or

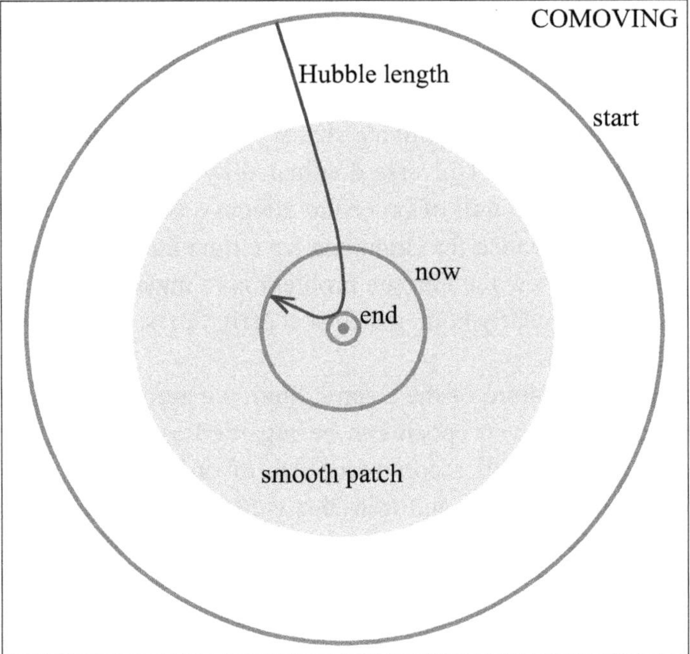

Fig. 18.3. Schematic of the inflationary solution to the horizon problem. The plot is in comoving units. During inflation, the comoving Hubble radius decreases dramatically, allowing our entire observable region (the ring marked 'now' around the central dot indicating our position) to lie within a region (the shaded smooth patch) that was well inside the Hubble radius at the start of inflation. Any initial inhomogeneities (lying outside the smooth patch) finish up on scales vastly larger than our observable Universe.

other particles). But the interaction rate per particle will decrease as the expansion dilutes the number density of particles, and will fall below the Hubble parameter after a few e-folds of inflation so that we deal with practically non-interacting relics.[1] For all known examples of unwanted relics, the amount of inflation occurring after the Universe leaves the horizon is more than enough to remove any conflict with observation.

What about unwanted relics produced *after* inflation? If the relics are produced by radiation in thermal equilibrium, we can get rid of them simply by making the temperature after inflation sufficiently low. But unwanted relics produced by other means may still be a problem. Examples of such relics are mentioned in Sections 21.7 and 23.2.2. Finally, we should mention the possibility that radiation

[1] The Hubble parameter falls more slowly than a^{-1} during inflation, and we expect the rate per particle to fall much faster than a^{-1}. In the simplest case, the rate will fall like the number density and the latter will be proportional to a^{-3}.

during inflation can be continuously produced, corresponding to what is called warm inflation (Section 25.7). In principle that also might give unwanted relics.

18.3 Initial condition for inflation

In the above account we postulated an inflating Robertson–Walker universe, existing well before the observable Universe leaves the horizon. There are various ideas about how that might have happened.

Though not compulsory, it is normally imagined that the emergence of four-dimensional spacetime in the observable Universe is followed promptly by an era of inflation. This is desirable for two reasons. One is to prevent the observable Universe from collapsing if the density parameter Ω is initially bigger than 1 (without being fine-tuned to a value extremely close to 1).

The other reason, which applies also to the case $\Omega < 1$, is that inflation protects an initially homogeneous region from invasion by its inhomogeneous surroundings, This is because the effective event horizon during inflation, which represents the farthest distance that an inhomogeneity can travel before inflation ends, is typically of order the Hubble distance. If the onset of inflation were significantly delayed, one would need an initially homogeneous patch much bigger than the Hubble distance, which sounds unnatural.[2] This is illustrated in Figure 18.3.

There is probably no need to assume that the inflating patch begins with a high degree of homogeneity and isotropy, because inflation at the classical level tends to drive the inhomogeneity and anisotropy to zero. For Einstein gravity there is a precise version of this statement [13, 14], which is similar to the 'no-hair" theorem for black holes. Alternatively, if one assumes that the cosmic fluid during inflation consists of scalar fields and particles, one can argue for the homogeneity of the region corresponding to the observable Universe without reference to the gravity theory, as we see in Section 24.1.

It is not essential to suppose that the initial era of inflation is maintained until the observable Universe leaves the horizon. There could instead be an intervening era of radiation domination, as was envisaged by Guth and was the standard assumption for some time afterwards. Later though, it was realised that an intervening era of radiation domination is an unnecessary complication, and nowadays it is rarely considered. Further discussion of the initial condition for inflation is given in Sections 28.2, 28.5 and 28.6.

[2] An escape from this conclusion would be to invoke a spatially-periodic universe, with a period of order the Hubble distance [12].

18.4 The amount of observable inflation

18.4.1 Observable inflation

We have seen how inflation can solve the flatness, horizon and relic problems of the unperturbed Universe, if the observable Universe starts out well inside the horizon. What about generating primordial perturbations, in particular the primordial curvature perturbation?

As we will see, primordial perturbations are supposed to come from the perturbation in one or more scalar fields, and those perturbations are in turn supposed to be generated on each scale around the time of horizon exit during inflation. As the primordial curvature perturbation is only known to exist on scales $k^{-1} \lesssim H_0^{-1}$, it is again enough for the observable Universe to be start out well inside the horizon.

We will call inflation after the scale $k^{-1} \sim H_0^{-1}$ leaves the horizon **observable inflation**. In words, observable inflation is what takes place after the observable Universe leaves the horizon.[3] Almost all of our discussion will focus on observable inflation, as opposed to the possibly huge amount of expansion that occurred earlier.

As seen in Table 7.2, the smallest cosmological scale is $k^{-1} \sim 10^{-3}\,\mathrm{Mpc}$. It corresponds to the size of the first gravitationally bound objects, and is the smallest scale on which the curvature perturbation can be *directly* observed. On cosmological scales the spectrum of the curvature perturbation is required by observation to be more or less flat. As we shall see, this almost certainly requires inflation to be almost exponential while cosmological scales are leaving the horizon. For the entire range $10^{-3}\,\mathrm{Mpc} \lesssim k^{-1} \lesssim 10^4\,\mathrm{Mpc}$ to leave the horizon, we therefore need around 16 e-folds of inflation (an e-fold corresponding to the expansion of the Universe by a factor e).

The curvature perturbation will also be generated on smaller scales, down to the scale leaving the horizon at the end of inflation. On those scales observation gives only an upper bound on the spectrum, coming from the upper bound on the primordial black hole abundance as described in Section 23.4. It is therefore possible in principle that inflation becomes strongly non-exponential after cosmological scales leave the horizon, though in practice that happens only in specially designed inflation models.

18.4.2 How much observable inflation?

Let us see how much observable inflation is needed. The amount of inflation after a scale k^{-1} leaves the horizon is usually characterized by the number of e-folds,

[3] As seen in Part II, one actually needs scales an order of magnitude or two bigger than H_0^{-1} to fully describe the stochastic properties of the observed perturbations. But those scales leave the horizon only a few e-folds before the observable Universe.

$N(k) = \ln(a_{\rm end}/a_k)$ where the subscripts denote the end of inflation and the epoch of horizon exit.

As we saw earlier, the best characterization of the definition of inflation is that the comoving Hubble length $1/aH$ is decreasing. Therefore, there is a case to be made for quantifying the amount of inflation by a slightly different quantity, that being the ratio of the initial comoving length to the final one given by $\tilde{N}(k) \equiv \ln a_{\rm end}H_{\rm end}/a_k H_k$. Although technically more accurate, during inflation $a(t)$ typically varies much faster than H so that the difference is not very important. We stick to the usual convention and use N to quantify inflation.

During observable inflation we take H to have a constant value H_*. Then the change in N between two scales is given by

$$N(k_2) - N(k_1) = \ln\frac{k_1}{k_2}. \tag{18.6}$$

This gives $N(k)$ if we know it at a given scale.

We choose the pivot scale $k_0 \equiv 0.002\,{\rm Mpc}^{-1}$, used by observers to specify relevant cosmological parameters (Section 6.7). We denote $N(k_0)$ simply as N.

The scale k_0^{-1} is a factor 9 below the scale H_0^{-1}, corresponding to a difference $\Delta N \simeq 2$. To generate the curvature perturbation on the largest cosmological scale we need $N \gtrsim 14$. It follows that N is close to the number of e-folds of observable inflation as defined earlier. As we see in Chapter 28 though, the increasing accuracy of observations means that the precise choice of k_0 may soon be significant.

To determine N, we require an assumption about the complete evolution of the scale factor after inflation. The possibilities for this evolution up to $T \sim 10\,{\rm MeV}$ are described in Chapter 21, with the assumption of Einstein gravity. The (known) evolution after that is described in Chapter 4. Here we give a simplified version of the analysis in Ref. [7].

We begin by assuming the usual post-inflation history. Almost-exponential inflation gives way to matter domination (corresponding to the oscillating scalar field), until radiation domination takes over at a 'reheat temperature' $T_{\rm R}$. This in turn gives way to matter domination after the epoch of matter–radiation equality given by Eq. (4.39). The value of N follows from the relation

$$1 = \frac{a_{\rm exit}H_*}{a_{\rm entry}H_{\rm entry}} = e^{-N}\frac{a_{\rm end}H_*}{a_{\rm reh}H_{\rm reh}}\frac{a_{\rm reh}H_{\rm reh}}{a_{\rm eq}H_{\rm eq}}\frac{a_{\rm eq}H_{\rm eq}}{a_{\rm ent}H_{\rm entry}}. \tag{18.7}$$

In this relation, 'exit' is the epoch when the pivot scale leaves the horizon and 'entry' the epoch when it enters the horizon. Also, 'end' is the end of inflation, 'reh' is the reheat epoch and 'eq' the epoch of matter–radiation equality.

Taking each change in pressure to be instantaneous, one finds

$$\boxed{N = 56 - \frac{2}{3}\ln\frac{10^{16}\,\mathrm{GeV}}{\rho_*^{1/4}} - \frac{1}{3}\ln\frac{10^{9}\,\mathrm{GeV}}{T_{\mathrm{R}}}}\,. \tag{18.8}$$

Here ρ_* is the energy density at the end of inflation. We see in Section 23.2.2 that T_{R} is constrained to be less than 10^9 GeV if there is supersymmetry.

According to most inflation models $\rho_*^{1/4}$ is not many orders of magnitude below 10^{16} GeV, though in principle it need not be far above 1 MeV which would correspond to only $N = 15$. Setting $\rho_*^{1/4}$ and T_{R} to 10^{16} GeV gives $N = 61$. Setting $\rho_*^{1/4} = 10^{16}$ GeV with $T_{\mathrm{R}} = 1$ MeV gives $N = 47$.

The radiation era may be punctuated by one or more eras of matter domination. This would reduce N. A much more drastic possibility for reducing N is to have an era of thermal inflation, as described in Section 21.6. Thermal inflation lasting for N_{late} e-folds will reduce N by N_{late}, and one bout of thermal inflation typically gives $N_{\mathrm{late}} \sim 10$.

Going the other way, N will be increased if $P > \rho/3$ after inflation. Assuming field theory with canonical kinetic terms, the maximum pressure is $P = \rho$. It would correspond to domination by the kinetic term of a homogeneous scalar field, called kination. If kination persists from the end of inflation to $T \sim 1$ MeV it will increase N by $\ln(V_*^{1/4}/1\,\mathrm{MeV})/3$. With $V_*^{1/4} \sim 10^{16}$ GeV this gives $N = 71$. This is the maximum possible value, if the Universe after inflation is to be described by general relativity and field theory with canonical kinetic terms.

18.5 The slow-roll paradigm

During inflation, the energy density and pressure of the cosmological fluid are usually supposed to be dominated by scalar fields. Assuming Einstein gravity and canonically normalized fields, the energy density and pressure are then given by Eq. (13.44), and the inflationary condition $P < -\rho/3$ is achieved if the fields do not vary too rapidly.

In this chapter we suppose that just one field ϕ varies. We call this **single-field inflation** and call ϕ the **inflaton**. After the density parameter has been driven close to 1, the Friedmann equation reads

$$\boxed{3M_{\mathrm{Pl}}^2 H^2 = V(\phi) + \frac{1}{2}\dot{\phi}^2}\,. \tag{18.9}$$

Differentiating it using the field equation (13.57) gives the useful expression

$$\boxed{2M_{\mathrm{Pl}}^2 \dot{H} = -\dot{\phi}^2}\,. \tag{18.10}$$

18.5.1 Slow-roll approximation

During inflation, the **slow-roll approximation** is practically always satisfied.[4] To formulate the slow-roll approximation, we can begin by assuming that inflation is almost-exponential:

$$\boxed{\frac{|\dot{H}|}{H^2} \ll 1}, \tag{18.11}$$

which is equivalent to $3M_{\rm Pl}^2 H^2 \simeq V$. If the latter approximation is valid, it is reasonable to hope that its low derivatives with respect to t will also be valid. Using Eq. (18.10), the first derivative of Eq. (18.11) will read

$$\boxed{3H\dot{\phi} \simeq -V'(\phi)}. \tag{18.12}$$

This is equivalent to neglecting the first term of the exact field equation (13.57):

$$|\ddot{\phi}| \ll 3H|\dot{\phi}|. \tag{18.13}$$

Eq. (18.13) states that $\dot{\phi}$ doesn't change much in one Hubble time. It follows from Eqs. (18.11) and (18.12) that[5]

$$\boxed{\epsilon(\phi) \ll 1} \qquad \text{where} \quad \epsilon \equiv \frac{M_{\rm Pl}^2}{2}\left(\frac{V'}{V}\right)^2. \tag{18.14}$$

Going further, we can hope that the derivative of the approximation (18.12) is also valid:

$$\boxed{\ddot{\phi} \simeq -\frac{\dot{H}}{H}\dot{\phi} - \frac{V''\dot{\phi}}{3H}}. \tag{18.15}$$

Comparing with the exact equation (13.57), we see that Eqs. (18.11), (18.12) and (18.15) imply

$$\boxed{|\eta(\phi)| \ll 1} \qquad \text{where} \quad \eta \equiv M_{\rm Pl}^2\frac{V''}{V} \simeq \frac{V''}{3H^2}. \tag{18.16}$$

The strong inequality (18.11) and the approximations (18.12) and (18.15) constitute the slow-roll approximation. The slow-roll approximation implies the conditions (18.14) and (18.16) on the potential, which we will call **flatness conditions**.

[4] We encounter exceptions in Sections 25.6 (fast-roll inflation), 28.7 (locked inflation) and 19.3 (power-law inflation). They typically give only a few e-folds of inflation.

[5] Instead of assuming Eqs. (18.11) and (18.12) and deriving Eq. (18.14) we could assume Eqs. (18.14) and (18.12) and derive Eq. (18.11). That alternative procedure will be essential when we come to multi-field slow-roll inflation.

18.5.2 Status of the slow-roll approximation

We saw that the slow-roll approximation can be derived by differentiating the condition (18.11) for almost-exponential inflation, which in turn is just a stronger version of the condition for inflation *per se*. One therefore expects that the flatness conditions on the potential are generally *necessary* for almost-exponential inflation, and that inflation will not take place at all if they are badly violated. As we see later, the flatness condition $|\eta| \ll 1$ is needed if we are to generate a viable curvature perturbation during slow-roll inflation.

One may also expect that the flatness conditions are *sufficient* for slow-roll inflation, given any initial value of $\dot{\phi}$ that is not too big. The argument goes by analogy with the motion of a particle with friction, which reaches a terminal velocity. If $|3H\dot{\phi}|$ is initially bigger than $|V'|$ we can expect that the friction term will reduce it to be about the same, while if it is initially smaller we might expect that it grows under the action of V' until it is about the same.

18.5.3 More flatness parameters

Using the slow-roll formula $3H\dot{\phi} \simeq -V'$, we can work out the rates of change of H and ϵ in terms of the potential and its derivatives. Instead of t it is useful to work with $dN \equiv -H\,dt$. We find

$$-\frac{d(\ln H)}{dN} \simeq -\epsilon, \qquad -\frac{d(\ln \epsilon)}{dN} \simeq 4\epsilon - 2\eta. \tag{18.17}$$

By further differentiation we find

$$-\frac{d\eta}{dN} \simeq 2\epsilon\eta - \xi, \qquad -\frac{d\xi}{dN} \simeq 4\epsilon\xi - \eta\xi - \sigma, \tag{18.18}$$

where

$$\xi \equiv M_{\rm Pl}^4 \frac{V'(d^3V/d\phi^3)}{V^2}, \qquad \sigma \equiv M_{\rm Pl}^6 \frac{V'^2(d^4V/d\phi^4)}{V^3}. \tag{18.19}$$

These formulas will be useful when we come to consider the curvature perturbation generated during inflation.

Satisfying the flatness condition $|\eta| \ll 1$ over several e-folds will typically be equivalent to $|\xi| \ll 1$, and satisfying *that* condition over several e-folds will typically be equivalent to $|\sigma| \ll 1$. Requiring that η and ξ are slowly varying on the Hubble timescale will typically be equivalent to a hierarchy

$$|\sigma| \ll |\xi| \ll |\eta|. \tag{18.20}$$

It is clear that these statements can fail to be true if the quantities have oscillating time-dependence, or if there are cancellations.

18.6 Hamilton–Jacobi formulation

18.6.1 The formulation

The Hamilton–Jacobi formulation is a powerful way of rewriting the equations of motion for single-field inflation. It can be derived by considering the scalar field itself to be the time variable, which is possible during any epoch in which the scalar field varies monotonically with time.

For definiteness, throughout we choose the sign of ϕ so that $\dot{\phi} > 0$. Dividing both sides of Eq. (18.10) by $\dot{\phi}$

$$\dot{\phi} = -2\, M_{\rm Pl}^2\, H'(\phi), \tag{18.21}$$

which gives the relation between ϕ and t. This allows us to write the Friedmann equation in the first-order form

$$\boxed{[H'(\phi)]^2 - \frac{3}{2M_{\rm Pl}^2} H^2(\phi) = -\frac{1}{2M_{\rm Pl}^4} V(\phi)}\,. \tag{18.22}$$

Equation (18.22) is the Hamilton–Jacobi equation. It allows us to consider $H(\phi)$, rather than $V(\phi)$, as the fundamental quantity to be specified. Because H, unlike V, is a geometric quantity, inflation is described more naturally in that language.

We can use the Hamilton–Jacobi formalism to write down a slightly different version of the slow-roll approximation than we did earlier, defining slow-roll parameters $\epsilon_{\rm H}$ and $\eta_{\rm H}$ as

$$\epsilon_{\rm H} = 2M_{\rm Pl}^2 \left(\frac{H'(\phi)}{H(\phi)}\right)^2, \qquad \eta_{\rm H} = 2M_{\rm Pl}^2\, \frac{H''(\phi)}{H(\phi)}. \tag{18.23}$$

In the slow-roll limit, $\epsilon_{\rm H} \to \epsilon$ and $\eta_{\rm H} \to \eta - \epsilon$. Some manipulation allows these to be written in various ways, such as

$$\epsilon_{\rm H} = 3\frac{\dot{\phi}^2/2}{V + \dot{\phi}^2/2} = -\frac{\dot{H}}{H^2} = \frac{1}{2}\frac{\dot{\phi}^2}{M_{\rm Pl}^2 H^2}, \tag{18.24}$$

$$\eta_{\rm H} = -\frac{\ddot{\phi}}{H\dot{\phi}} = \epsilon_{\rm H} - \frac{1}{2}\frac{\dot{\epsilon}_{\rm H}}{H\epsilon_{\rm H}}. \tag{18.25}$$

With these new parameters, many results that were approximate in terms of $V(\phi)$ become exact. The definition $\ddot{a} > 0$ of inflation now is given *precisely* by $\epsilon_{\rm H} < 1$, and the number of e-foldings after a given epoch is

$$N \equiv \ln \frac{a(t_{\rm end})}{a(t)} = \int_t^{t_{\rm end}} H\, dt = -\frac{1}{2M_{\rm Pl}^2}\int_\phi^{\phi_{\rm end}} \frac{H}{H'}\, d\phi. \tag{18.26}$$

18.6.2 Attractor theorem

Using the Hamilton–Jacobi approach one can show that all possible inflationary trajectories (solutions of Eq. (13.57) with H given by the Friedman equation) will quickly converge to a common 'attractor' solution, if they are sufficiently close to each other initially. This is the behaviour that we argued for within the slow-roll approximation, but the proof makes no use of that approximation.

For simplicity, we assume that the solutions have $\dot{\phi}$ positive, though the result holds under more general circumstances. Suppose $H_0(\phi)$ is any solution to Eq. (18.22), which can be either inflationary or non-inflationary. Add to this a linear homogeneous perturbation $\delta H(\phi)$; the attractor condition will be satisfied if $\delta H/H_0$ goes quickly to zero as ϕ increases. Substituting $H(\phi) = H_0(\phi) + \delta H(\phi)$ into Eq. (18.22) and linearizing, we find that the perturbation obeys

$$H_0'\,\delta H' \simeq \frac{3}{2M_{\rm Pl}^2} H_0\,\delta H. \qquad (18.27)$$

This equation has the solution

$$\begin{aligned} \delta H(\phi) &= \delta H(\phi_{\rm i}) \exp\left(\frac{3}{2M_{\rm Pl}^2}\int_{\phi_{\rm i}}^{\phi} \frac{H_0(\phi)}{H_0'(\phi)}\,d\phi\right), & (18.28)\\ &= \delta H(\phi_{\rm i}) \exp[-3N(\phi)], & (18.29) \end{aligned}$$

where $\delta H(\phi_{\rm i})$ is the value at some initial point $\phi_{\rm i}$, and $N(\phi)$ is the number of e-foldings of unperturbed expansion from that point. During inflation, $\epsilon_{\rm H} < 1$ and $\delta H/H_0$ falls off faster than e^{-2N}. The attractor solution will therefore be approached quickly for a wide range of initial conditions.

Neither the assumption of linearity nor the assumption that $\dot{\phi}$ doesn't change sign is very restrictive. The linearity condition $|\delta H| \ll H$ is reasonable for inflationary solutions because they all have H in the range $H_{\rm exp}$ to $\sqrt{3/2}\,H_{\rm exp}$, where $3M_{\rm Pl}^2 H_{\rm exp}^2 = V$ corresponds to exponential inflation. A change of sign in $\dot{\phi}$ can matter only if the perturbation takes the field over the top of a maximum in the potential because otherwise it will simply roll up, reverse its direction, and pass back down through the same point, where it can be regarded as a perturbation on the original solution with the same sign of $\dot{\phi}$.

18.7 Inflationary potentials

In this section we look briefly at the types of inflationary potential that have been proposed. Much more detail is given in Chapter 28.

In the slow-roll approximation, the number N of e-folds of inflation after the pivot scale k_0 leaves the horizon at the field value ϕ_0 is

$$N \simeq \frac{1}{M_{\rm Pl}^2} \int_{\phi_{\rm end}}^{\phi_0} \frac{V}{V'} d\phi. \quad (18.30)$$

As we saw in Section 18.4, this relation essentially gives the range of ϕ for observable inflation.

The simplest potentials giving inflation are

$$V \propto \phi^\alpha, \quad (18.31)$$

with α an even integer. These potentials satisfy the flatness conditions in the regime $\phi \gg M_{\rm Pl}$. This kind of inflation was proposed by Linde [8], and is usually called **chaotic inflation**, for a reason we explain in Section 28.2.

Chaotic inflation ends at $\phi \sim M_{\rm Pl}$, after which the inflaton moves towards its vev, and typically oscillates before settling down. At the beginning of Section 13.5 we described such an oscillation in flat spacetime, and it is easy to repeat the discussion in the expanding Universe. The field equation Eq. (13.57) reads

$$\ddot{\phi} + 3H\dot{\phi} + m^2\phi = 0. \quad (18.32)$$

If $m \gg H$ the Hubble drag is small. Then ϕ oscillates with angular frequency m, and with an amplitude which is damped by the Hubble drag. It is useful to consider quantities averaged over one oscillation, denoted by a bar. Ignoring Hubble drag completely, the energy density is $\overline{\rho} = \frac{1}{2}m^2\phi_0^2$, where ϕ_0 is the oscillation amplitude. The oscillator condition $\overline{d^2\phi/dt^2} = m^2\overline{\phi^2}$ gives $\bar{P} = 0$. The oscillating field therefore corresponds to a nearly pressureless fluid. For consistency with the zero-pressure dependence $\bar{\rho} \propto a^{-3}$, the amplitude of the oscillation must fall like $a^{-3/2}$. This may be verified directly by substitution into Eq. (18.32), and using the WKB approximation which takes ϕ_0 to be slowly varying on the timescale of the oscillation.

A different kind of inflation model works near a maximum of the potential, as in Figure 18.4. Setting $\phi = 0$ at the maximum, the potential will generically be of the form

$$V = V_0 - \frac{1}{2}m^2\phi^2 + \ldots \equiv V_0 \left[1 + \frac{1}{2}\eta_0 \frac{\phi^2}{M_{\rm Pl}^2} + \ldots \right], \quad (18.33)$$

where $\eta_0 \equiv -m^2 M_{\rm Pl}^2/V_0$ is the value of η at the maximum. This is known as **hilltop inflation**. The dots in Eq. (18.33) indicate an additional contribution, which generates a minimum at the vev of ϕ. At the minimum, $V = 0$. This potential gives slow-roll inflation near the maximum if $|\eta_0| \ll 1$, and slow roll can continue after other terms become important if they give a sufficiently flat potential. As with chaotic inflation, inflation ends around the time that slow roll fails, after which

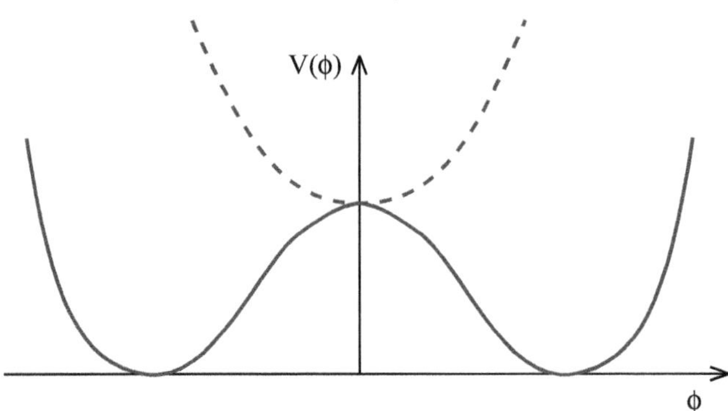

Fig. 18.4. The potential Eq. (18.33) is illustrated by the full line. The potential Eq. (18.34) with $\chi = 0$ is illustrated by the dashed line. The potential Eq. (18.34) *as a function of* χ has the shape of the dashed line for ϕ bigger than a critical value, changing to the shape of the full line when ϕ falls below this value.

ϕ oscillates about its vev. The shape of the potential is illustrated in Figure 18.4 taking V to be an even function.

If $\phi = 0$ is the fixed point of symmetries, then m^2 is what we previously called a tachyonic mass, leading to spontaneous symmetry breaking. That kind of hilltop inflation is called **new inflation**, after the first viable model [9, 10] of inflation using a scalar field.

In the examples of inflation considered so far, the entire inflationary potential comes from the displacement of the inflaton from its vev. This requires the inflationary potential $V(\phi)$ to have a minimum corresponding to the vacuum $V = 0$, and it requires inflation to be ended by a failure of the flatness conditions.

In what are called **hybrid inflation** models, the bulk of the potential is instead generated by the displacement from the vacuum of some 'waterfall field' different from the inflaton. In hybrid inflation models, inflation ends when the waterfall field is destabilized as the inflaton field moves through some critical value. The flatness conditions can still be well satisfied when this happens.

The simplest potential for hybrid inflation is [11]

$$\boxed{V(\phi,\chi) = V(\phi) - \frac{1}{2}m_\chi^2\chi^2 + \frac{1}{4}\lambda\chi^4 + \frac{1}{2}\lambda'\chi^2\phi^2}, \tag{18.34}$$

with

$$\boxed{V(\phi) = V_0 + \frac{1}{2}m^2\phi^2}. \tag{18.35}$$

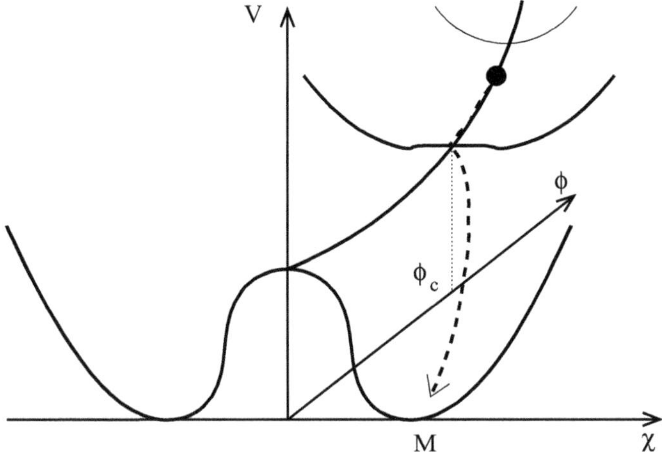

Fig. 18.5. The hybrid inflation potential. The field rolls down the $\chi = 0$ channel from large ϕ, until it encounters the instability point, after which the $\chi = 0$ solution becomes unstable and the fields roll to their true minimum at $\phi = 0$, $\chi = \pm M$.

Setting $V = 0$ at $\langle\phi\rangle \equiv M$, an equivalent expression is

$$V(\phi,\chi) = \frac{1}{4}\lambda\left(\chi^2 - M^2\right)^2 + \frac{1}{2}m^2\phi^2 + \frac{1}{2}\lambda'\chi^2\phi^2, \tag{18.36}$$

$$M^2 = \frac{m_\chi^2}{\lambda} = \left(\frac{4V_0}{\lambda}\right)^{1/2}. \tag{18.37}$$

The couplings are supposed to satisfy

$$0 < \lambda \lesssim 1 \qquad 0 < \lambda' \lesssim 1, \tag{18.38}$$

and the masses

$$\frac{m^2 M_{\rm Pl}^2}{V_0} \equiv \eta_0 \ll 1 \qquad \frac{m_\chi^2 M_{\rm Pl}^2}{V_0} \gg 1. \tag{18.39}$$

This potential is illustrated in Figures 18.5 and 18.4. There is an exact global symmetry $\chi \to -\chi$, which is spontaneously broken in the vacuum, but is restored for $\phi > \phi_c \equiv m_\chi/\sqrt{\lambda'}$. In this regime there is slow-roll inflation with the potential (18.35), corresponding to the dashed line in Figure 18.4. When ϕ falls below ϕ_c, the waterfall field is destabilized. Inflation then ends with the inflaton and waterfall fields descending quickly to their vevs. The end of hybrid inflation may be regarded as a phase transition, as we discuss in more detail in Chapter 21.

Hybrid inflation has several variants. One possibility is to give the potential of the waterfall field a dip in the middle, as shown in Figure 18.6. For instance one

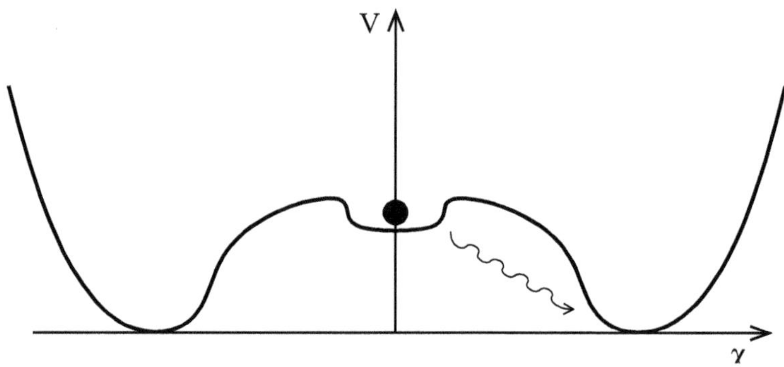

Fig. 18.6. Potential with a metastable vacuum state at the origin. Inflation ends when the field quantum tunnels through to the true minimum.

could replace Eq. (18.34) by

$$V(\phi, \chi) = V(\phi) + \frac{1}{2}m_\chi^2\chi^2 - \frac{1}{4}\lambda\chi^4 + \frac{\chi^6}{\Lambda^2} + \frac{1}{2}\lambda'\chi^2\phi^2. \tag{18.40}$$

At the classical level the waterfall field is now trapped at the origin and inflation will never end. But at the quantum level the waterfall field can tunnel to the vacuum. The tunneling will occur when the inflaton field becomes small enough for the waterfall field to develop a minimum, exactly as in the previous case. The tunnelling initially forms bubbles of the new phase (i.e. of the waterfall field) and the transition is complete only when the bubbles coalesce. We are dealing with what is called a first-order transition, as opposed to a second-order one where there is a smooth transition to the new phase.

At this point one may wonder why the inflaton field is necessary at all. If the tunnelling rate of χ to its vev is sufficiently slow, Eq. (18.40) can support inflation even if $\phi = V(\phi) = 0$. Such a setup is known as old inflation [6], but it doesn't work because the Universe expands too rapidly for the bubbles ever to coalesce, and the region inside a bubble doesn't represent a viable post-inflationary universe.

Exercises

18.1 Demonstrate that the conditions $\ddot{a} > 0$ and $-\dot{H} < H^2$ are equivalent.

18.2 Suppose monopoles form at a temperature of order $3 \times 10^{-4} M_{\rm Pl}$ with a mass of $10^{-3} M_{\rm Pl}$, and that the Universe is then radiation dominated until BBN. Take the present-day limit on monopoles to be $\Omega_{\rm mon} < 10^{-6}$ where $\Omega_{\rm mon}$ indicates the fraction of the total energy density residing in monopoles. (The origin of this bound, known as the Parker bound, is described for instance in Ref. [15].) Calculate an upper bound on the number of monopoles per horizon volume at formation, assuming $g_* \sim 100$ at that time and that monopole annihilation is negligible.

Suppose that monopoles form with a density of order 1 per horizon volume. If exponential inflation occurs after the monopoles have formed, how many e-foldings of inflation are required to satisfy the Parker bound. Compare this with the number of e-foldings required to solve the horizon problem.

18.3 Calculate the redshift at which the pivot scale $k = 0.002\,\mathrm{Mpc}^{-1}$ enters the horizon, and verify Eq. (18.8) under the stated assumptions.

18.4 Show that with exponential inflation and subsequent radiation domination, the number of e-foldings from horizon exit to the end of inflation is the same as the number from the end of inflation to horizon entry.

18.5 For the chaotic inflation potential $V = m^2\phi^2/2$, calculate the value of ϕ when $\epsilon = 1$ and the value of ϕ when $\eta = 1$. Assuming that inflation ends when $\epsilon = 1$, use Eq. (18.30) to calculate the number of e-foldings of inflation that occur for an initial value ϕ_i. Find the slow-roll solution $\phi(t)$. Use your solution for ϕ to calculate $H(t)$ and hence $a(t)$. Demonstrate that the number of e-foldings calculated using the solution for a is the same as that which you calculated using Eq. (18.30). Expand the solution for a at small $t - t_\mathrm{i}$ to demonstrate that the inflation is approximately exponential at the initial stage. Calculate the time constant κ [from $a \sim \exp(\kappa t)$] and demonstrate that it equals the (slow-roll) Hubble parameter during inflation. Calculate the ratio of the effective pressure P_ϕ to the energy density ρ_ϕ sixty e-foldings before the end of inflation.

18.6 For the hybrid inflation potential (18.34), obtain an expression for the number of e-folds $N(\phi_0)$ of inflation after the pivot scale leaves the horizon.

References

[1] R. Brout, F. Englert and E. Gunzig. The creation of the Universe as a quantum phenomenon. *Annals Phys.*, **115** (1978) 78.

[2] A. A. Starobinsky. Quantum effects and regular cosmological models (in Russian). *JETP Lett.*, **30** (1979) 682.

[3] A. A. Starobinsky. Relict gravitation radiation spectrum and initial state of the Universe (in Russian). *Phys Lett. B*, **91** (1980) 99.

[4] D. Kazanas. Dynamics of the Universe and spontaneous symmetry breaking. *Astrophys. J.*, **241** (1980) L59.

[5] K. Sato. On the thermodynamic properties of the quantum vacuum. *Mon. Not. Roy. Astr. Soc.*, **195** (1981) 467.

[6] A. Guth. The inflationary universe: a possible solution to the horizon and flatness problems. *Phys. Rev. D*, **23** (1981) 347.

[7] A. R. Liddle and S. M. Leach. How long before the end of inflation were observable perturbations produced?. *Phys. Rev. D*, **68** (2003) 103503.

[8] A. D. Linde. Chaotic inflation. *Phys. Lett. B*, **129** (1983) 177.

[9] A. D. Linde. A new inflationary universe scenario: a possible solution of the horizon, flatness, homogeneity, isotropy and primordial monopole problems. *Phys. Lett. B*, **108** (1982) 389.

[10] A. Albrecht and P. J. Steinhardt. Cosmology for grand unified theories with radiatively induced symmetry breaking. *Phys. Rev. Lett.*, **48** (1982) 1220.

[11] A. D. Linde. Axions in inflationary cosmology. *Phys. Lett. B*, **259** (1991) 38.

[12] A. Linde. Creation of a compact topologically nontrivial inflationary universe. *JCAP*, **0410** (2004) 004.

[13] R. W. Wald. Asymptotic behavior of homogeneous cosmological models in the presence of a positive cosmological constant. *Phys. Rev. D*, **28** (1983) 2118.

[14] A. A. Starobinsky. Isotropization of arbitrary cosmological expansion given an effective cosmological constant. *JETP Lett.*, **37** (1983) 66.

[15] E. W. Kolb and M. S. Turner. *The Early Universe* (California: Addison-Wesley, 1990).

19

Inflation with modified gravity

The theory of gravity is an integral part of any inflation model. It is easy to write down theories with modified gravity. Motivation for such theories can come from the idea that the Lagrangian will contain all terms not forbidden by a symmetry. It can also come from string theory ideas such as the braneworld. It turns out that the simplest modifications of gravity from these sources lead to inflation models that are essentially equivalent to theories with Einstein gravity. In this book we confine our attention to those modifications, which are dealt with in this chapter.

19.1 Scalar–tensor theories

In scalar–tensor theories, the modification of gravity is caused by one or more scalar fields. We will just consider one field, with the action

$$S = \int d^4x\sqrt{-g}\left[\frac{f(\phi)}{2}R - \frac{1}{2}\partial_\mu\phi\partial^\mu\phi - V(\phi) + \mathcal{L}_{\rm mat}\right]. \qquad (19.1)$$

As in Eq. (13.60), $\mathcal{L}_{\rm mat}$ is obtained from the Lagrangian of flat spacetime field theory using the equivalence principle. It cannot include the field ϕ because the $f(\phi)$ term violates the equivalence principle (and invalidates Einstein gravity).

This action can be put into the Einstein form by what is called a **conformal transformation** of the metric. This is a transformation of the form

$$\boxed{g_{\mu\nu} \to \Omega^2\, g_{\mu\nu}}, \qquad (19.2)$$

with Ω a function of spacetime position. The effect of the transformation on the curvature scalar is [1]

$$R \to \frac{1}{\Omega^2}\left[R + \frac{6}{\Omega}\Box\Omega\right], \qquad (19.3)$$

where $\Box$ is given in terms of the original metric by Eq. (3.22). In our case the appropriate choice [2] is $\Omega^2 = f(\phi)/M_{\rm Pl}^2$.

Before proceeding we should clarify the physical significance of a conformal transformation. It is quite different from a coordinate transformation. The latter doesn't change the curvature of spacetime, it merely transforms the components of the curvature tensor. The conformal transformation instead alters the curvature of spacetime. It is a field redefinition that mixes up the gravitational and matter degrees of freedom. Each choice of field definitions is commonly called a **frame.**[1] The frame in which gravity takes the form of Einstein's theory is known (when it exists) as the **Einstein frame**.

The change from one frame to another doesn't correspond to a change in the physics. True, it changes the spacetime curvature, but it also changes $\mathcal{L}_{\rm mat}$ and the latter provides the definition of the ideal clocks that are to be used in measuring the spacetime curvature. Phenomena that appear to be due to gravity in one frame may appear to have their origin in the matter sector in another. The statement that Einstein gravity is valid at the present epoch amounts to saying that there is an Einstein frame, *and* that in this frame $\mathcal{L}_{\rm mat}$ is given by the Standard Model (or a relevant approximation such as classical electromagnetism). We could use a different frame, but then $\mathcal{L}_{\rm mat}$ would look more complicated.

Returning to the scalar–tensor theories, it is convenient also to redefine ϕ and the potential:

$$\boxed{d\phi \to M_{\rm Pl}\sqrt{\frac{2f+3f'^2}{4f^2}}\,d\phi} \qquad \boxed{V \to \frac{M_{\rm Pl}^4}{f^2}V(\phi)} \,. \tag{19.4}$$

Then the action for R and ϕ has the completely standard form:

$$S = \int d^4x\sqrt{-g}\left(\frac{1}{2}M_{\rm Pl}^2 R - \frac{1}{2}\partial_\mu\phi\partial^\mu\phi - V(\phi) + \tilde{\mathcal{L}}_{\rm mat}\right) . \tag{19.5}$$

Because $g_{\mu\nu}$ has been transformed, the form of $\tilde{\mathcal{L}}_{\rm mat}$ will be quite different from the original.

19.2 Induced gravity and variable Planck mass

In this section, $\mathcal{L}_{\rm mat}$ is supposed to have a negligible effect during inflation. We consider what are called induced gravity and variable Planck mass theories. Among useful references are Refs. [2, 3, 4].

In an induced gravity theory ϕ is supposed to spontaneously break a symmetry, in such a way that f would vanish if the symmetry were unbroken. The Planck scale is given by $M_{\rm Pl}^2 = f(\langle\phi\rangle)$. We consider the simplest action, corresponding

[1] Note that this use of the word frame has nothing to do with its use to denote a coordinate system, as in 'inertial frame'.

to

$$f = \xi\phi^2, \qquad V = \frac{\lambda}{4}\left(\phi^2 - v^2\right)^2, \qquad M_{\rm Pl}^2 = \xi v^2. \tag{19.6}$$

We assume $\xi \gg 1$ because that turns out to be necessary for viable inflation. Then, after making the transformations (19.2) and (19.4) and choosing the origin of ϕ appropriately, the potential at $\phi \gtrsim M_{\rm Pl}$ is well approximated

$$\boxed{V(\phi) = V_0\left(1 - e^{-\sqrt{\frac{2}{3}}\frac{\phi}{M_{\rm Pl}}}\right)} \qquad V_0 \equiv \frac{\lambda M_{\rm Pl}^4}{4\xi^2}. \tag{19.7}$$

Inflation takes place at $\phi \gtrsim M_{\rm Pl}$, after which ϕ oscillates about its vev. From Eq. (19.4), one sees that the original potential during the oscillation is of order $\lambda\phi^4$ making the original field value of order its vev v. But with the original field at its vev we have Einstein gravity in the original frame. We conclude that the distinction between the original frame and the Einstein frame disappears as the oscillation amplitude decays. If ϕ has unsuppressed couplings to the other fields, the decay will be rapid, otherwise the decay will come at first just from Hubble damping.

We see in Section 28.9 that

$$V_0^{1/4} \simeq 6 \times 10^{15}\,\text{GeV} \tag{19.8}$$

is required, if Eq. (19.7) is to generate the observed curvature perturbation. As a result, ϕ can plausibly be identified with a GUT Higgs field. Indeed, setting $v = M_{\rm GUT} = 2 \times 10^{16}$ GeV and using $M_{\rm Pl}^2 = \xi v^2$, we find the reasonable value $\lambda \simeq 10^{-2}$ (corresponding to $\xi = 10^4$).

A variable Planck mass theory is like an induced gravity theory, except that only part of the Planck scale comes from the vev of ϕ. Instead of Eq. (19.6) we have

$$f = M + \xi\phi^2, \qquad V = \frac{\lambda}{4}\left(\phi^2 - v^2\right)^2, \qquad M_{\rm Pl}^2 = M^2 + \xi v^2. \tag{19.9}$$

In the regime $1 \ll \sqrt{\xi} \ll M_{\rm Pl}/v$ this gives the same inflation potential as induced gravity. In contrast with that case though, the choice of the vev $\langle\phi\rangle = v$ can have a negligible effect during inflation. One can still identify ϕ with the GUT Higgs field, but one might instead consider the possibility of identifying it with the Standard Model Higgs field [5].

19.3 Extended inflation

Now we turn to a proposal known as extended inflation [6]. It starts with the Jordan–Brans–Dicke theory of gravity (often called just Brans–Dicke theory). In this theory $V(\phi) = 0$, and it is useful to write Eq. (19.1) in the form

$$S = \int d^4x\sqrt{-g}\left\{\frac{M_{\rm Pl}^2}{2}\left[\Phi R - \frac{\omega}{\Phi}\,\partial_\mu\Phi\partial^\mu\Phi\right] + \mathcal{L}_{\rm matter}\right\}, \qquad (19.10)$$

where

$$\Phi \equiv \frac{f(\phi)}{M_{\rm Pl}^2}, \quad \omega(\Phi) \equiv \frac{f(\phi)}{(df/d\phi)^2}. \qquad (19.11)$$

In the limit of large ω, the kinetic term becomes large and the energetically favoured situation is that Φ be constant in space and time. Then general relativity is recovered after a redefinition of the energy unit. Brans–Dicke theory takes ω to be constant, and present-day tests of general relativity require $\omega > 40\,000$.

Extended inflation includes just one field χ in $\mathcal{L}_{\rm mat}$, whose potential has a metastable minimum at $\chi = 0$ as in Figure 18.6. During inflation χ is trapped at $\chi = 0$, and inflation ends when χ tunnels to the true vacuum. This proceeds by bubble formation just as in the 'old inflation' model, but the modification of gravity can enhance the formation of the bubbles so that they coalesce.

After making the transformations (19.2) and (19.4), we have a hybrid inflation model with Einstein gravity. The potential during inflation is

$$V(\phi) = V_0 \exp\left(-\sqrt{\frac{2}{p}}\frac{\phi}{M_{\rm Pl}}\right), \qquad 2p = \omega + \frac{3}{2}. \qquad (19.12)$$

With this V, the exact field equation (13.57) has a solution

$$\frac{\phi}{M_{\rm Pl}} = \sqrt{2p}\ln\left(\sqrt{\frac{V_0}{p(3p-1)}}\frac{t}{M_{\rm Pl}}\right), \qquad (19.13)$$

and $a = a_0 t^p$.[2] There is inflation for any $p > 1$, called power-law inflation. The slow-roll regime is $p \gg 1$, with $\epsilon = \eta/2 = 1/p$. The field χ serves as a waterfall field, inflation ending when it tunnels to the vacuum via bubble nucleation.

Although it successfully ends inflation, the original extended inflation model was immediately seen to be unviable because the non-observation of the bubbles in the microwave sky requires $\omega < 20$, in severe conflict with the value $\omega > 40\,000$ required by the success of Einstein gravity. One can modify the theory so that the ω in the inflationary potential doesn't endanger the success of Einstein gravity in the present Universe, but it turns out that the value $\omega < 20$ is still ruled out if the curvature perturbation is to be generated from the perturbation of the inflaton field.[3] As a result of all this, viable versions of extended inflation are very different from the original proposal. It should be remembered in this context that all versions

[2] The field equation is non-linear because H is a function of V and $\dot\phi$. It will have an infinity of other solutions, which converge to this one by virtue of the attractor theorem.

[3] Equation (25.5) with $\omega < 20$ requires $n < 0.8$ and $r > 1$.

of extended inflation become simply hybrid inflation in the Einstein frame, albeit with a possibly strange-looking potential.

19.4 R^2 inflation

The R^2 inflation model [7] invokes $f(R)$ gravity (Eq. (13.62)) with $f(R) = R + R^2/6M^2$. On the basis of the discussion in Section 13.8.2, one may expect $M \sim M_{\rm Pl}$.

The field equation for $g_{\mu\nu}$ now involves spacetime derivatives up to fourth order, making the physical interpretation problematic. To avoid that we can introduce a scalar field ϕ, and write down an action which is equivalent to the original one in the sense that it gives the same field equation for $g_{\mu\nu}$. With the new form of the action, this field equation is obtained by combining the coupled field equations for ϕ and $g_{\mu\nu}$, which are of the usual second-order form.

The action for generic $f(R)$ is derived for instance in Ref. [8]. For the R^2 case, in the Einstein frame, it gives inflation with the potential (19.7), with $V_0 = 3M_{\rm Pl}^2 M^2/4$ [3]. The normalization (19.8) then requires $M \sim 10^{-4} M_{\rm Pl}$, as opposed to the value $M \sim M_{\rm Pl}$ that one might expect.

19.5 Modified gravity from the braneworld

When the energy scale is sufficiently low, a braneworld theory provides a four-dimensional effective field theory. In some cases the field theory is of standard form, providing for instance a concrete scenario for soft supersymmetry breaking in the MSSM. We are concerned here with the ability of braneworld cosmology to provide a field theory that will describe inflation. In this case, a field theory derived from braneworld scenario may have a quite unusual form, motivated only by the braneworld. In particular, it may lead to non-Einstein gravity where there is no Einstein frame.

We will briefly consider one such case, derived from a simple scenario known as Randall–Sundrum type II (RS-II). This is usually considered just to be a toy model within which one can explore phenomenology that might persevere in more complex/realistic models. In the RS-II scenario, there is a single brane where we live, and just one extra dimension. In the region outside the brane (the bulk) there is a negative cosmological constant, which strongly warps the spacetime geometry and prevents the fifth dimension from being so evident as to conflict with observations.

In the RS-II scenario, the four-dimensional Einstein equation (3.41) is modified to become

$$R_{\mu\nu} - \frac{1}{2} g_{\mu\nu} R = \frac{1}{M_{\rm Pl}^2} T_{\mu\nu} + \left(\frac{8\pi}{M_5^3}\right)^2 \pi_{\mu\nu} - E_{\mu\nu}. \tag{19.14}$$

The tensor $\pi_{\mu\nu}$ is quadratic in the energy–momentum tensor $T_{\mu\nu}$, but $E_{\mu\nu}$ is a new source of four-dimensional gravity unrelated to $T_{\mu\nu}$. The constant M_5 is the five-dimensional analogue of the four-dimensional Planck scale $M_{\rm Pl}$.

Taking the spacetime on the brane to be a flat Robertson–Walker metric, the Friedmann equation becomes

$$H^2 = \frac{\rho}{3M_{\rm Pl}^2}\left(1+\frac{\rho}{2\lambda}\right)+\frac{\epsilon}{a^4}, \qquad \lambda \equiv \frac{3M_5^6}{32\pi^2 M_{\rm Pl}^2}. \tag{19.15}$$

The coefficient ϵ in the last term comes from $E_{\mu\nu}$, and this term will be inflated away just like radiation. Ignoring it, the only effect of the braneworld is the term quadratic in ρ. In the present Universe one needs $\rho \ll \lambda$ and then the quadratic term is negligible, but in the early Universe it might dominate.

The simplest braneworld inflation scenario takes the matter source to be a slowly rolling scalar field as normal, and provided it is restricted to the brane it will obey the usual equation of motion. However in the high-energy regime the Hubble parameter will be greatly increased for a given value of ϕ, thus increasing the friction term in the wave equation and making slow-roll much easier to achieve.

In contrast with the cases where Einstein gravity can be recovered by a conformal transformation, the braneworld models have in principle the ability to give distinctive predictions for the curvature perturbation. At the time of writing though, it seems that such signatures are absent for this particular model, and for a range of more complicated models. This concludes our study of modified gravity from the braneworld.

Exercises

19.1 Work out the action that is obtained from Eq. (19.1) by using the conformal transformations (19.2) and (19.3). Show that the further transformation (19.3) leads to the action (19.5).

19.2 Referring to the discussion of Section 19.2, verify that $v = M_{\rm GUT}$ and $\lambda \simeq 10^{-2}$ are compatible with the requirement $\xi = 5\times 10^4\sqrt{\lambda}$

19.3 Verify that the function $\phi(t)$ displayed in Eq. (19.13) is an exact solution of the field equation (13.57) for the potential of the previous line, the scale factor having the power law behaviour $a \propto t^p$. Verify the inflation regime is $p > 1$. Would $p = 1$ be enough to solve (i) the flatness and (ii) the horizon problem?

References

[1] R. M. Wald. *General Relativity* (Chicago: University of Chicago Press, 1984).

[2] T. Hirai and K. i. Maeda. Gauge invariant cosmological perturbations in generalized Einstein theories. *Astrophys. J.*, **431** (1994) 6.

[3] D. S. Salopek, J. R. Bond and J. M. Bardeen. Designing density fluctuation spectra in inflation. *Phys. Rev. D*, **40** (1989) 1753.

[4] D. I. Kaiser. Primordial spectral indices from generalized Einstein theories. *Phys. Rev. D*, **52** (1995) 4295.

[5] F. L. Bezrukov and M. Shaposhnikov. The Standard Model Higgs boson as the inflaton. *Phys. Lett. B*, **659** (2008) 703.

[6] D. La and P. J. Steinhardt. Extended inflationary cosmology. *Phys. Rev. Lett.*, **62**, 376 (1989). [Erratum-ibid. **62**, 1066 (1989)].

[7] A. A. Starobinsky. In *Quantum Gravity*, edited by M. A. Markov and P. West (Plenum, New York, 1983).

[8] S. Nojiri and S. D. Odintsov. Introduction to modified gravity and gravitational alternative for dark energy. *Int. J. Geom. Meth. Mod. Phys.*, **4** (2007) 115.

20

Multi-field dynamics

So far, we demanded that only one scalar field varies during inflation. Now we allow two or more fields to vary. In the first section we generalize the slow-roll paradigm, and in the second section we give a more general discussion which applies even if the fields are not responsible for inflation.

20.1 Multi-field slow-roll inflation

Let us suppose that more than one field varies during inflation, corresponding to an inflationary trajectory $\phi_1(t), \phi_2(t), \ldots$. We continue to assume Einstein gravity, with the energy density dominated by the scalar field potential of canonically normalized fields.

We recover the single-field case if the inflationary trajectory is a straight line in field space, because we can then perform a rotation of the field basis so that just a single field ϕ has significant variation. Even if the trajectory is not straight, we recover the single-field case provided that the trajectory lies in a steep-sided valley of the potential. The inflaton field then just corresponds to the distance in field space along the valley bottom.

To get something non-trivial, we suppose that the inflationary trajectory is not straight, and that it is one of a continuous family of possible slow-roll trajectories. The family is obtained by displacing the original trajectory sideways in field space. We then say that there is **multi-field slow-roll inflation**.

To define the multi-field slow-roll approximation, we regard ϕ_n as the components of a vector $\vec{\phi}$ in field space, and the $V_n \equiv \partial V/\partial\phi_n$ as the components of the gradient ∇V. We will use the notation $|\dot{\vec{\phi}}|^2 \equiv \sum_n \dot{\phi}_n^2$ for the norm of a vector. Then Eqs. (18.9) and (18.10) hold with $|\dot{\vec{\phi}}|^2$ instead of $\dot{\phi}^2$, to become

$$\boxed{3M_{\rm Pl}^2 H^2 = V(\phi) + \frac{1}{2}|\dot{\vec{\phi}}|^2}\,, \tag{20.1}$$

and

$$\boxed{2M_{\rm Pl}^2\dot{H} = -|\dot{\vec{\phi}}|^2}. \tag{20.2}$$

The multi-field slow-roll approximation assumes that inflation is almost exponential:

$$\boxed{\frac{|\dot{H}|}{H^2} \ll 1}. \tag{20.3}$$

It also assumes

$$\boxed{3H\dot{\vec{\phi}} \simeq -\nabla V}, \tag{20.4}$$

which is to be understood as the approximation

$$|\ddot{\vec{\phi}}| \ll 3H|\dot{\vec{\phi}}|, \tag{20.5}$$

in the exact equation (13.58). An equivalent approximation is that the vector $\dot{\vec{\phi}}$ doesn't change much in one Hubble time, $|\Delta\dot{\vec{\phi}}| \ll |\dot{\vec{\phi}}|$.

Equations (20.3) and (20.4) imply a flatness condition

$$\boxed{\epsilon \ll 1}, \qquad \epsilon \equiv \frac{1}{2}M_{\rm Pl}^2\frac{|\nabla V|^2}{V^2} \equiv \frac{1}{2}M_{\rm Pl}^2\frac{\sum V_n^2}{V^2}. \tag{20.6}$$

Finally, we demand that the first derivative of Eq. (20.4) provides a valid approximation for $\ddot{\vec{\phi}}$:

$$\ddot{\vec{\phi}} \simeq -\frac{\dot{H}}{H}\dot{\vec{\phi}} - \frac{(\dot{\nabla V})}{3H}, \qquad (\dot{\nabla V})_n = \sum_m V_{nm}\dot{\phi}_m. \tag{20.7}$$

(We defined $V_{nm} \equiv \partial^2 V/\partial\phi_n\partial\phi_m$.) The three equations (20.4), (20.6) and (20.7) define the multi-field slow-roll approximation.

Analogously with the single-field case, the inclusion of the third equation leads to a condition on the second derivatives of the potential. To formulate this condition, it is convenient to define $\eta_{nm} \equiv V_{nm}/3H^2$ and[1]

$$\bar{\eta} \equiv \frac{\sum V_n V_m \eta_{nm}}{\sum V_k V_k}, \qquad \overline{\eta^2} \equiv \frac{\sum V_m V_k \eta_{nm}\eta_{nk}}{\sum V_k V_k}. \tag{20.8}$$

(In a basis where η_{nm} is diagonal, these are weighted means of η_{nn} and η_{nn}^2.) Then, with the exact equation (13.58), Eqs. (20.4) and (20.7) imply

$$\left(\frac{\dot{H}}{3H^2}\right)^2 + \frac{\dot{H}}{3H^2}\frac{\bar{\eta}}{3} + \frac{\overline{\eta^2}}{9} \ll 1. \tag{20.9}$$

[1] With our current assumption of Einstein gravity, the condition (20.3) makes $\eta_{nm} \simeq M_{\rm Pl}^2 V_{nm}/V$. We took the latter as the definition for single-field inflation, but now we adopt the definition using H because it will allow us to go beyond Einstein gravity.

If there are only two or a few fields, this requirement will be more or less equivalent to

$$\boxed{|\eta_{nm}| \ll 1}, \tag{20.10}$$

which we adopt for simplicity.

The flatness conditions (20.6) and (20.10) are consequences of the slow-roll approximation. As in the single-field case, they typically ensure the validity of the slow-roll approximation (20.4) and its first derivative, more or less independently of the initial condition. Adopting the mechanical analogy, this happens because Eq. (20.4) represents a critical velocity which will be approached from below if the initial speed is too small and from above if it is too big.

The slow-roll approximation (20.4) for $\dot{\vec{\phi}}$ means that each multi-field slow-roll trajectory is practically straight during one Hubble time. The straightness of the trajectories has an important implication for the situation during the era of observable inflation, which lasts only 50 or so Hubble times: during that era, the field basis can be chosen so that a given trajectory lies within the space of at most a few of the slowly rolling fields. The others hardly move during observable inflation and are in practice excluded from the discussion. If the trajectory can be chosen to lie almost entirely in one direction, we are in practice back to the single-field case. The term 'multi-field inflation' is used only when the trajectory has appreciable curvature, so that it necessarily lies in the space of at least two fields. One usually considers only two.

A widely discussed example is the two-field chaotic inflation potential,

$$V(\phi_1, \phi_2) = \frac{1}{2}m_1^2\phi_1^2 + \frac{1}{2}m_2^2\phi_2^2 + \ldots, \tag{20.11}$$

As shown in Figure 20.1, the contours of equal potential are ellipses, and the lines of steepest descent are possible inflation trajectories. If the masses are equal the trajectories are straight and we are back to single-field slow-roll inflation.

20.2 Light and heavy fields

20.2.1 Light fields

In the context of Einstein gravity, we define the light fields as those which satisfy the flatness conditions (20.6) and (20.10) at each point in some region of field space around the inflationary trajectory. More generally, we define the light fields just as those which satisfy Eq. (20.10), with the understanding that inflation is almost exponential. In the latter case there is no definite condition on the first derivatives V_n, but they should not be too big or the field will soon find itself in a region of field space where it ceases to be light.

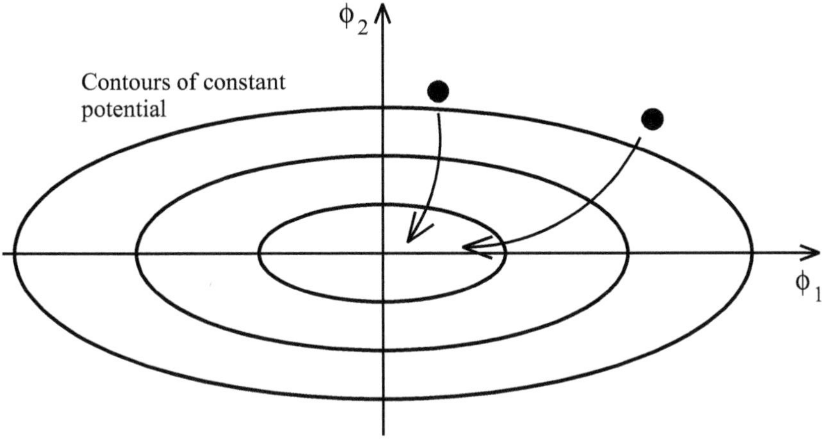

Fig. 20.1. Inflation with $V = \frac{1}{2}m_1^2\phi_1^2 + \frac{1}{2}m_2^2\phi_2^2$ and $m_2 > m_1$. Contours of equal potential are shown, and two lines of steepest descent which represent possible slow-roll trajectories.

The term 'light' is clearly appropriate for a set of fields ϕ_n whose potential is quadratic, because V_{nm} appearing in Eq. (20.10) then specifies the masses-squared (in the basis where V_{nm} is diagonal). It is actually appropriate quite generally, because V_{nm} in any case specifies the masses-squared of the field perturbations which are the objects of most interest (see Chapter 24).

As we see in later chapters, the quantum fluctuation of light fields is promoted during inflation to a classical perturbation, which can be used to generate the primordial curvature and isocurvature perturbations described in Chapters 5 and 6.

Because of their ability to generate primordial perturbations, all light fields are of potential interest. That is the case independently of the inflation mechanism, and no matter how light is the field.

20.2.2 Heavy fields

At the opposite extreme to light fields are **heavy fields**. A heavy field is one whose value ϕ_0 during inflation adjusts to minimize the potential. Near the minimum the potential of a heavy field will have the form

$$V(\phi) = V_0 + \frac{1}{2}m^2(t)\,[\phi - \phi_0(t)]^2 + \dots. \tag{20.12}$$

To be heavy, the field must have mass-squared $m^2 \equiv V''$ much bigger than H^2, in contrast with light fields which have $|m^2| \ll H^2$. We will generally assume that every field is either heavy or light. This means that we ignore the borderline situation of fields with $|m^2| \sim H^2$, as well as more complicated cases where a field is far from the minimum but not slowly rolling.

Usually the value ϕ_0 will be time independent and in the simplest case this value is the vev, but neither of these is mandatory. For instance, the waterfall field χ in the usual hybrid inflation model is fixed at a value different from its vev, while in the 'mutated hybrid' inflation of Section 28.7.2 the waterfall field is not fixed. Similar possibilities exist even if the heavy field has a negligible effect on the inflationary trajectory.

Exercise

20.1 Check that Eq. (20.6) follows from the conditions stated in the text. Do the same for Eq. (20.9).

21

Reheating and phase transitions

In this and the following two chapters, we consider scenarios for the history of the Universe from the end of inflation to neutrino decoupling at $T \sim 1\,\mathrm{MeV}$. Einstein gravity is assumed in all cases. A viable scenario must lead to a radiation-dominated Universe at $T \sim 1\,\mathrm{MeV}$, with properties not too different from those described in Section 4.5. In particular, the abundance of relic particles such as gravitinos or moduli should be below observational bounds. As seen in Section 24.7.3, the detection of a cosmic gravitational wave background may in the far future offer powerful discrimination between different scenarios.

This chapter begins with the reheating process, which establishes thermal equilibrium after inflation and initiates the Hot Big Bang. Then we see how the spontaneous breaking of symmetries may lead to the creation of solitons of various kinds, including in particular cosmic strings and other topological defects. Finally, we discuss the possibility of a short burst of late inflation known as thermal inflation.

21.1 Reheating

21.1.1 Initial reheating

At the end of inflation, the entire energy density of the Universe remains locked in the scalar fields. Everything else has presumably been diluted away by the inflationary period. We have to free this energy density by converting it into other forms, with the ultimate goal of creating the Hot Big Bang radiation that is certainly present when the run-up to nucleosynthesis begins at $T \sim 1\,\mathrm{MeV}$. The conversion process is known as reheating, and may proceed through various stages. Reheating is complete when practically all of the energy is in radiation at thermal equilibrium. The temperature at that stage is called the **reheat temperature**.

After inflation is over the inflaton field(s) moves toward the vacuum, accompanied by the waterfall field(s) if there is hybrid inflation. We pretend for now that there is just the inflaton ϕ, since the presence of other fields doesn't alter the broad picture.

We need to consider the loss of energy by the oscillating field, which was ignored in Section 18.7. If the amplitude of the oscillation is sufficiently small, this loss will correspond to the decay of individual ϕ-particles, so that $a^3\bar{\rho}$ is proportional to $\exp(-\Gamma t)$ where Γ is the particle decay rate. Then there is a modified continuity equation

$$\dot{\bar{\rho}} + (3H + \Gamma)\bar{\rho} = 0. \tag{21.1}$$

The epoch of decay is $t = \Gamma^{-1}$, corresponding to $H \sim \Gamma$.

A mechanism different from single particle decay, known as preheating, may be the dominant effect especially during the early part of the oscillation. For the moment we stay with single particle decay.

One usually assumes that the inflaton decay products are relativistic, and that most of the energy quickly thermalizes. As we will see in Section 22.2, each species is then expected to have the generalized blackbody distribution, corresponding to energy density of order T^4. Since the decay occurs when $H \sim \Gamma$, the reheat temperature is

$$\boxed{T_{\rm reh} \sim \sqrt{M_{\rm Pl}\Gamma}}\,. \tag{21.2}$$

One needs $\Gamma \lesssim m$ or there would not be time for even one oscillation.[1] If the decay occurs through interactions whose strength is measured by some dimensionless coupling g, then $\Gamma \sim (g^2/4\pi)m$.[2] If g is a gauge coupling one expects $g^2/4\pi \sim 10^{-1}$ to 10^{-2}. If it is a Yukawa coupling it might be some orders of magnitude smaller. Finally, if it is a non-renormalizable coupling one expects $g \sim m/\Lambda_{\rm UV}$. A far smaller value is not excluded, but accepting this estimate we arrive at a plausible lower limit $g \sim m/M_{\rm Pl}$, corresponding to a 'gravitational strength' interaction. This corresponds to reheat temperature

$$\frac{T_{\rm reh}}{1\,{\rm GeV}} \sim \left(\frac{m}{10^6\,{\rm GeV}}\right)^{3/2}. \tag{21.3}$$

Reheating must occur before nucleosynthesis gets underway, which requires something like $m \gtrsim 10^4$ GeV if the inflaton has only gravitational-strength interactions.

21.1.2 Second reheating

In the simplest case, the radiation-dominated era initiated by reheating continues without interruption until the present matter-dominated era begins. It could happen

[1] The same requirement applies to the decay rate of any particle, since it corresponds to a quantized oscillation.

[2] The decay rate is proportional to the square of an amplitude as in Eq. (2.69), and the amplitude is proportional to g because it is given by first-order perturbation theory (corresponding to a Feynman diagram with a single vertex).

instead that the radiation-dominated era is interrupted by an era of matter domination. Then there is a second reheating, when this matter decays.

Such a thing will happen if there is a species of non-interacting particle, with non-relativistic random motion so that it is indeed matter as opposed to radiation. Before it decays, the mass density of such a species grows like a relative to the radiation, and can indeed come to dominate.

The number density n of non-interacting particles falls like a^{-3}, and so does the entropy density s of the radiation as long as the radiation is in thermal equilibrium. Then n/s is a constant. But at the second reheating, the entropy of the radiation is increased by some factor γ^{-1} and n/s is reduced (diluted) by a factor γ. To estimate γ we note that $\rho_{\text{before}} = \rho_{\text{r,after}}$, where the subscripts denote times just before and just after reheating and the subscript 'r' denotes the radiation part of the energy density. Since $s \sim \rho_{\text{r}}$ at both epochs this leads to

$$\gamma \sim \frac{\rho_{\text{r,before}}}{\rho_{\text{before}}}. \tag{21.4}$$

Finally, since the ratio ρ_{r}/ρ starts to fall when matter domination begins, and does so like $1/a \propto 1/T$, we have $\gamma \sim T_{\text{R}}/T_1$ where T_1 is the temperature when matter domination begins and T_{R} is the reheat temperature. Since the maximum inflation scale is 10^{16} GeV and the minimum reheat temperature is 1 MeV, γ cannot be less than 10^{-19}. Typically it will be very many orders of magnitude bigger. Still, even a modest dilution factor can have significant consequences, as we will see.

Suppose now that the particle whose decay causes the second reheating corresponds to the oscillation of a homogeneous field σ, whose mass m is less than H at early times. The fields starts to oscillate when $H(t) \sim m$, and its energy density is then $\rho_\sigma \sim m^2\sigma_{\text{osc}}^2$. We assume that the oscillation begins after the first reheating, when the Universe is radiation dominated. The total energy density is $3M_{\text{Pl}}^2 H^2$, and $\rho_\sigma/\rho \sim \sigma_{\text{osc}}^2/M_{\text{Pl}}^2$.[3]

The oscillating field may at first lose energy by preheating. Discounting such energy loss, ρ_σ/ρ increases proportional to the scale factor. Just before decay at $H \sim \Gamma$ one has

$$\boxed{\frac{\Omega_\sigma}{(1-\Omega_\sigma)^{3/2}} \sim \frac{\sigma_{\text{osc}}^2}{M_{\text{Pl}}^2}\sqrt{\frac{m}{\Gamma}}}, \tag{21.5}$$

where $\Omega_\sigma \equiv \rho_\sigma/\rho$. If $\Omega_\sigma/(1-\Omega_\sigma)$ is bigger than 1, there will be matter domination and a second reheating. Oscillation beginning before the first reheating will tend to reduce Ω_σ, while preheating just before the second reheating will tend to decrease it.

[3] We assume that ρ_σ/ρ is significantly less than one. In the opposite case σ would have supported an era of inflation before it started to oscillate.

The Minimal Supersymmetric Standard Model (MSSM) and its extensions include several candidates for an oscillating field that may cause a second reheating. It might be a flat direction of the MSSM [1], a modulus, the scalar partner of a sterile neutrino [2], or the saxion [3].

21.2 Preheating

In the scenario outlined above, the oscillating field which causes reheating is assumed to be equivalent to a collection of particles, each of them decaying as if it were in isolation. In reality, a time-dependent scalar field can decay by quite different mechanisms, which may be far more efficient. Such mechanisms go by the general name of preheating.

Preheating can operate even if there is no oscillation and therefore no particle concept. Such preheating is needed if the inflaton vev is infinite or so large that it is never reached. In that case there is no oscillation, but there will usually be an era of so-called kination, when the energy density is dominated by the kinetic term corresponding to $\rho \simeq P \propto a^{-6}$.

After preheating is over, the cosmic fluid typically consists of a gas which is not in thermal equilibrium. It will contain particle species produced by the preheating. It may or may not contain also the particle species corresponding to the original oscillating field. In any case, the particles present at this stage are supposed eventually to produce thermalized radiation which completes the reheating process. Unless these particles are highly relativistic, they have to undergo single-particle decay before that can happen.

We are going to describe preheating by using the quantum theory of a free field with time-dependent mass, given in Section 15.7. In doing that, we ignore the expansion of the Universe, since the quantum theory applies to flat spacetime. In most cases this is a reasonable approximation during the initial and most important phase of preheating. Indeed, this phase lasts only for a time $\Delta t \ll H^{-1}$; also it produces particles with momenta mostly $k/a \gg H$, which means that we can describe preheating locally by considering a region with size $\Delta x \ll H^{-1}$. In summary, we can work within a region of spacetime whose size is much less than H^{-1}. In such a region the spacetime curvature is negligible, and by virtue of the equivalence principle we can use flat spacetime field theory.

21.2.1 Initial vacuum state

According to the setup described in Section 15.7, we need the initial state to correspond to the vacuum. To be precise, we need the average initial occupation number of the relevant momentum states to be much less than 1, so that practically all of them are in the vacuum. This corresponds to the density of particles being negligible. Enough inflation will certainly ensure that, and the following argument shows that there almost certainly will be enough.

Preheating produces a wide range of k values, as opposed to a range $\Delta k \ll k$. Consider a range $\Delta k \sim k$, centred on some value of k, and denote the average occupation number by $\bar{n}_k$. Then the energy density of the particles is $\gtrsim \bar{n}_k (k/a)^4$ (the inequality being saturated if the mass is negligible). Assuming Einstein gravity, this has to be much less than the energy density at the beginning of almost-exponential inflation;

$$\bar{n}_k \left(\frac{k}{a_{\rm begin}} \right)^4 \ll M_{\rm Pl}^2 H_*^2. \tag{21.6}$$

The scale k^{-1} is certainly smaller than the size of the observable Universe.[4] We conclude that

$$\bar{n}_k \ll e^{-4\Delta N} \frac{M_{\rm Pl}^2}{H_*^2}. \tag{21.7}$$

where ΔN is the number of e-folds that have already occurred when the observable Universe leaves the horizon. The vacuum assumption $\bar{n}_k \ll 1$ is therefore valid provided that $2\Delta N \gtrsim \ln(M_{\rm Pl}/H_*)$, which is a mild requirement.

21.2.2 Ordinary preheating (parametric resonance)

In this section we describe preheating in its simplest form, following Ref. [4]. We set $\langle\phi\rangle = 0$, and assume that by the end of inflation $V(\phi)$ is practically quadratic. To achieve preheating, we assume that ϕ interacts with some field χ so that the potential is

$$V = \frac{1}{2} m_\phi^2 \phi^2 + \frac{1}{2} g^2 \phi^2 \chi^2 + \dots. \tag{21.8}$$

We neglect the mass m_χ of χ, which is supposed to become significant only after preheating is over. Until that happens, the time-dependent mass-squared of χ is

$$m_\chi^2(t) \simeq g^2 \phi^2(t), \tag{21.9}$$

and the mode function $\chi_k(t)$ satisfies

$$\ddot{\chi}_k + E_k^2(t) \chi_k = 0, \tag{21.10}$$

with $E_k^2 = (k/a)^2 + m_\chi^2(t)$.

[4] For the particles produced by preheating, k^{-1} is likely to be very many orders of magnitude smaller. Taking that on board, the following argument for the vacuum assumption could be made much stronger. We leave it as it is partly for safety, but mostly for later use when we come to consider the generation of perturbations on cosmological scales.

As preheating typically takes place in a time $\Delta t \ll H^{-1}$ we can use the formalism of Section 15.7, with χ instead of ϕ. As the change in a is negligible we can identify $\mathbf{k}$ with the physical momentum $\mathbf{k}/a$, by setting $a = 1$. The initial condition for the mode function is Eq. (15.31). We consider scales well inside the horizon at the end of inflation. On those scales, E_k^2 varies adiabatically during inflation and there are no χ particles. After inflation is over, ϕ oscillates with angular frequency m_ϕ and amplitude $\phi_{\rm end}$. This can give non-adiabatic variation of E_k^2.

While ϕ is oscillating we can write $m_\chi^2(t) = g^2\phi_{\rm end}^2 \sin^2(m_\phi^2 t)$, and the mode equation can be written

$$\boxed{\frac{d^2\chi_k}{dz^2} + [A(k) - 2q\sin(2z)]\,\chi_k = 0}\,, \tag{21.11}$$

where

$$q \equiv \frac{g^2\phi_{\rm end}^2}{4m_\phi^2}, \qquad A \equiv \frac{k^2}{m_\phi^2} + 2q, \qquad z \equiv m_\phi t. \tag{21.12}$$

This is the Mathieu equation. In certain regions of the parameter space $\{A, q\}$, its solution can grow rapidly corresponding to what is called parametric resonance, and there will be significant particle production. The strongest growth (broad parametric resonance) can occur if $q \gtrsim 1$.

As it stands this formalism doesn't give a late-time regime where the evolution is again adiabatic. But according to the discussion of Section 15.7, the strong growth of the mode function means that we will end up with a large occupation number, equal to $|\beta_k(t)|^2$, where $\beta_k(t)$ is read off from Eq. (15.60) evaluated at the end of the non-adiabatic era. As a rough estimate, this era can be identified as the epoch when the energy density of the created particles, calculated within the above formalism, becomes comparable with the total energy density of the oscillating ϕ field. At that stage the assumption of a constant oscillation amplitude is obviously unviable, and we may conclude that in reality the tachyonic preheating has drained away a significant amount of energy, which has caused the oscillation amplitude to become too small to create further particles. Investigation along these lines shows that particle production within the broad resonance regime is big enough to rapidly drain away energy from the oscillating field.

To take into account the eventual interaction of the produced particles and the expansion of the Universe, one needs a detailed numerical calculation. In general, such calculations broadly confirm the the simple picture that we have outlined.

Ordinary preheating can apply to any oscillating field. That field might be the inflaton, a waterfall field, or a field causing a second reheating. Also, this type of preheating can work by creating fermions instead of spin-0 particles [5].

21.2.3 Instant preheating

We should mention a variant of ordinary preheating known as **instant preheating** [6]. The production of χ particles by the preheating mechanism in this scenario is usually supposed to cease after the first passage of ϕ through zero. In any case, the essential point is that χ has a coupling to a fermion field, with the right magnitude for it χ to decay into a fermion pair around the time that ϕ has its maximum amplitude. Because the mass $m_\chi(t)$ is large at that time, the fermions can be heavy.

If the energy density of the ϕ oscillation decreases significantly faster than that of the fermions, the latter can dominate and complete the preheating process. In such a case, the preheating process is over after one oscillation of the ϕ field, justifying the term 'instant'.

More usually, the energy density of the ϕ oscillation will decrease like a^{-3} (Hubble drag being insignificant over the timescale of interest). Then the oscillation of ϕ is hardly disturbed by the fermion production, and there is no preheating (if the production of the χ particles occurs only on the first pass of ϕ through zero). Nevertheless, the mechanism of fermion production that we described (usually referred to as instant preheating even when there is no actual preheating) can be important for cosmology as mentioned in Sections 23.3 and 26.6.2.

21.2.4 Tachyonic preheating

Ordinary preheating converts an initially homogeneous oscillating field ϕ into particles of a different species. Tachyonic preheating [7], by contrast, converts it into particles of the same species.

The mode function ϕ_k satisfies Eq. (15.56) with $m^2(t) = V''(\phi)$. Tachyonic preheating occurs when $m^2(t)$ is negative and increasing in magnitude, because a given E_k^2 can then pass through zero, which is a non-adiabatic variation. For negative E_k^2, the oscillation ceases and $|\delta\phi_k|$ increases exponentially.

In practice, ϕ starts out near a maximum of the potential $V(\phi)$. As ϕ moves down the potential towards the vev, tachyonic preheating shuts off when the point of inflection is reached, where m^2 goes through zero and becomes positive. If ϕ oscillates, and returns to the region around the maximum so that m^2 again becomes negative, preheating switches on again, and so on until the oscillation amplitude becomes too small. At that stage we have ϕ particles, whose occupation number is $|\beta_k|^2$, where β_k is read off from the late-time solution (15.60) of the mode function. The occupation number is typically very big, because the mode function is driven to large values while $m^2(t)$ is negative.

This analysis fails when the perturbation $\delta\phi$ becomes comparable to the unperturbed value $\phi(t)$. That may happen very quickly, before even one oscillation is complete. After it happens, those modes with occupation number significantly bigger than 1 can be treated as classical interacting plane waves, whose evolution can be followed numerically, while the other modes can be ignored. Each plane wave decays as a collection of single particles, possibly enhanced by a large final-state occupation number.

Tachyonic preheating can happen to any field which descends from a maximum and then oscillates. After hybrid inflation it can happen to the waterfall field. After hilltop inflation, tachyonic preheating can also happen to the inflaton field, but the mechanism there is not quite the same as the one we described. The initial condition in that case corresponds to the classical perturbation $\delta\phi(\mathbf{x}) \sim H/2\pi$, that is generated from the vacuum fluctuation during inflation, as described in Chapter 24. But the subsequent amplification of this perturbation follows the course that we described. In both of these cases, ordinary preheating may be operating at the same time, but the separation of the two kinds is at least a useful way of thinking.

21.3 Phase transitions and solitons

21.3.1 Spontaneous symmetry breaking

Symmetries that are spontaneously broken in the present Universe may be unbroken in the early Universe because of field interactions. This is referred to as **symmetry restoration**. The simplest example is Eq. (18.34), which we considered in the context of hybrid inflation. The potential has a global symmetry $\chi \to -\chi$, and during inflation the large value of ϕ holds χ at the origin which means that the symmetry is unbroken. When ϕ becomes too small, inflation ends and χ moves away from the origin towards its non-zero vacuum value, which spontaneously breaks the symmetry.

This example is rather special because it involves inflation. More usually, one supposes that the symmetry is restored through finite-temperature effects. According to quantum field theory any scalar field ϕ in thermal equilibrium receives a contribution of order $g^2T^2\phi^2$ to its effective potential, where g is a measure of the coupling strength of ϕ to the particles in the thermal bath.[5] Including this contribution Eq. (14.28) becomes

$$\boxed{V(\phi) = V_0 + \left(g^2T^2 - \frac{1}{2}m^2\right)\phi^2 + \frac{1}{4}\lambda\phi^4}. \tag{21.13}$$

At temperature $T > m/\sqrt{2}g$ the symmetry is restored, but at lower temperature the symmetry is spontaneously broken.

The spontaneous breaking of a symmetry represents a **phase transition**, because the nature of the cosmic fluid changes. In the hybrid inflation example the change

[5] Derivations of this thermal contribution can be found in Refs. [8, 9, 10].

is drastic, because it ends inflation. For a thermal phase transition the change is more subtle, coming from changes in the mass, charge and other properties of the particles making up the cosmic fluid.

If the Mexican-hat potential has a dip in the middle, the point of unbroken symmetry is stable at the classical level. Then the phase transition is of first order and bubbles of the new phase are formed which subsequently coalesce. Usually though there is no dip, and at least in the thermal case the new phase is formed more or less continuously, the process being known as spinodal decomposition.

Within the Standard Model and its extensions, two symmetry-breaking phase transitions may occur in the early Universe; the **electroweak transition** and the **chiral transition**.

Electroweak symmetry is restored above a temperature of a few hundred GeV. The precise temperature, as well as the nature of the transition, depends on the Higgs mass in the Standard Model, and on other parameters as well in extensions of the Standard Model like the MSSM. Constraints from collider physics require the electroweak transition to be of second order assuming the Standard Model or the MSSM, but it might be of first order in a modest extension of the MSSM.

The global chiral $SU(2)$ symmetry described in Section 16.5 is spontaneously broken when the temperature falls below $\Lambda_{\rm QCD} \sim 100\,{\rm MeV}$, and the QCD interaction becomes strong enough to form the quark condensate. This phase transition occurs at about the same time as the quark–hadron transition, before which the quarks exist as free particles. The latter transition occurs when the typical spacing $n^{-1/3} \sim T^{-1}$ becomes of order the confinement distance $\Lambda_{\rm QCD}^{-1} \sim (100\,{\rm MeV})^{-1}$. Although it is a phase transition, the quark–hadron transition doesn't correspond to the breaking of any symmetry.

Lattice field theory calculations show that the quark–hadron transition is of first order. During this transition, the sound speed becomes very small. As a result, the radiation density contrast grows on scales within the horizon. Depending on the nature of the cold dark matter (CDM), this can lead to the early formation of light CDM halos, with various observable effects. The QCD phase transition will also cause a small-scale isocurvature baryon perturbation, which may affect Big Bang Nucleosynthesis (BBN). These issues are reviewed in Ref. [11].

21.3.2 Solitons

At a symmetry-breaking phase transition, the field(s) responsible for the breaking will not be absolutely homogeneous. If the phase transition is thermal the field will have a thermal fluctuation. If instead it happens at the end of inflation as a waterfall field moves away from the origin, the waterfall field will become light near the end of inflation and will acquire a perturbation as described in Chapter

24. When the phase transition begins, the field will therefore be at the origin only in fairly exceptional regions of space. As we will now explain, it can happen that the field in such regions becomes trapped while the rest of space goes into spontaneous symmetry breaking. The trapped regions correspond to concentrations of energy called **topological defects**, and this mechanism for forming them is called the **Kibble mechanism**.

Topological defects are particular examples of **solitons**. As a preliminary definition in flat spacetime, a soliton may be defined as a solution of the field equations whose energy–momentum tensor is time independent and falls off rapidly outside some a limited region. That region may extend indefinitely in one direction (a string) or in two directions (a domain wall) or in none (a pointlike object going by various names). In any case the soliton has a more or less well-defined 'size', (the width of the string or domain wall if it isn't pointlike). The size is usually small compared with the Hubble scale, so that the structure of the soliton can indeed be described using flat spacetime.

If there is a soliton solution of the field equations that satisfies the preliminary definition, Lorentz invariance requires that there be solutions corresponding to solitons moving with constant velocity. Going to the level of approximations, we can further relax the preliminary definition. A curved wall or string makes sense so long as the string (wall) is approximately straight (flat) over a region much bigger than its width. It makes sense to talk about a collection of moving solitons if the distance between them is big compared with their size. It also makes sense to talk about solitons living in curved spacetime, once their structure has been described using a locally inertial frame. Finally, one can allow slowly decaying solitons.

Let us describe under what circumstances solitons may exist. In most cases, the field values within a soliton at rest correspond to a static (time-independent) solution of the field equation Eq. (13.41). Supposing for the moment that the fields have no gauge interaction, their energy density is

$$\rho(\mathbf{x}) = \frac{1}{2}\sum_n \left(\dot{\phi}_n^2 + |\nabla\phi_n|^2\right) + V. \tag{21.14}$$

The lowest-energy solution of the field equations corresponds to the vacuum, where each field is fixed at its vev. For free fields (quadratic potential) there is no other time-independent solution. Indeed, all other solutions just correspond to plane waves which after quantizations become free particles. This remains true if we allow only a small departure from the quadratic potential, so that we have interacting particles. To get solitons we have to allow a large departure, to find solutions of the field equations that do not describe interacting particles.

21.4 Main types of soliton

We now describe the main types of topological defect, mentioning also Q-balls, which are not topological defects. Topological defects and their cosmology are described in detail in Ref. [12].

21.4.1 Domain walls

We consider first an idealized domain wall, taken to lie in a plane in flat spacetime. We suppose that just one real field ϕ exists, whose potential (14.28) spontaneously breaks the Z_2 symmetry $\phi \to -\phi$. There are two possible vevs, $\langle\phi\rangle = \pm f$. Suppose now that ϕ depends only on x, with

$$\begin{aligned} \phi(x,t) &\to f, \qquad x \to +\infty, \\ \phi(x,t) &\to -f, \qquad x \to -\infty. \end{aligned}$$

It would take an infinite amount of energy to change the asymptotic values, which is not going to happen. Therefore ϕ will settle down to the configuration which minimizes the energy while preserving the asymptotic values. We then have a domain wall, which can be taken to lie in the plane $x = 0$.

Given the potential $V(\phi)$ we can find the profile $\phi(x)$. If V is of the form (14.28) with $\lambda \sim 1$, f is the only relevant parameter. Then the thickness of the wall is $d \sim 1/f$ and its energy per unit area is $\sigma \sim f^3$.

A Lorentz boost along the wall doesn't alter the profile $\phi(x)$, which means that it doesn't alter the components of the energy–momentum tensor. Neglecting the thickness of the wall, the non-zero components are therefore

$$T^0_0 = -T^y_y = -T^z_z = \sigma\delta(x). \tag{21.15}$$

This idealized treatment is valid for a piece of wall that is small enough to be practically flat and to be well within the horizon, yet big compared with the wall thickness. Such conditions are usually satisfied. Domain walls can form a closed surface (a bubble of domain wall).

If the $U(1)$ symmetry is restored at early times, domain walls will form through the Kibble mechanism, along the surfaces at which ϕ happens to vanish. The inhomogeneity of ϕ is uncorrelated on scales significantly bigger than the horizon, which means that there will initially be at least of order one domain wall stretching across each horizon volume. Bubbles of domain wall can collapse and disappear once they come within the horizon, but for the type of domain wall that we are considering at the moment the collapse can begin only after horizon entry. As a result there will always be of order one domain wall stretching across each Hubble volume, forming a network.

With a different type of domain wall, the network of walls can completely annihilate. One way to achieve this is to explicitly break the Z_2 symmetry by adding an odd term (say $\epsilon\phi^3$) to Eq. (14.28). Then the pressure on the two sides of the wall will be unequal, driving walls separated by the true vacuum together so that they can annihilate. A different way in which a network of domain walls might disappear is described in Section 23.1.4.

Domain walls are strongly constrained by observation because they are liable to distort the spacetime geometry. Just requiring that a single domain wall stretching across the observable Universe has less energy than the rest of the Universe requires that its energy per unit area is less than $(1\,\mathrm{MeV})^3$.

21.4.2 Cosmic strings

Reviews of cosmic strings are given in Refs. [12, 13]. Let us suppose that there is just one complex field ϕ, with potential (14.32) so that there is a $U(1)$ symmetry. The possible vacuum values correspond to the circle $|\phi| = f$, and there is a topological defect called a cosmic string, whose energy is concentrated along (say) the line $z = 0$. At the centre of the string, $\phi = 0$. Defining polar coordinates by $r^2 = x^2 + y^2$ and $\tan\theta = y/x$, the string corresponds to a configuration

$$\phi = g(r)e^{in\theta}, \tag{21.16}$$

where n is some integer. The radial function $g(r) = |\phi|$ minimizes the energy. When r is bigger than the thickness d of the string, we are outside the string and $|\phi| = f$ to high accuracy. Going once around the string, ϕ goes n times around the circle $|\phi| = f$ so that all points in the vacuum are encountered. As with the domain wall, this configuration is stable because it would take an infinite amount of energy for ϕ to achieve a homogeneous vacuum value corresponding to $|\phi| = f$. The string may carry an electromagnetic current (superconducting cosmic string) but usually it doesn't.

If the $U(1)$ is a global symmetry, we have what is called a global string. Outside the global string there is a canonically normalized field $\sigma = \sqrt{2}fn\theta$. The gradient of this field generates an energy density $\rho_{\rm string}(r) = f^2n^2/r^2$. The energy within a cylinder of radius r around an isolated global string increases like $\ln r$, but in practice there will be a network of strings with the string spacing cutting off the logarithm. If $U(1)$ is a gauge symmetry we have what is called a gauge or local string. Then the would-be scalar field σ doesn't exist (because it can be removed by a choice of gauge) and the energy density outside the string vanishes.

A cosmic string may involve more than one field, and a non-Abelian symmetry. As with a domain wall, the profiles of the fields within a cosmic string are invariant under Lorentz boosts along the string and so is the energy–momentum tensor. The energy per unit length μ is therefore equal to the tension of the string.

This idealized treatment is valid for a piece of string small enough to be practically straight and to be well within the horizon, yet long compared with its thickness. Such conditions are usually satisfied.

A curved string can be of indefinite extent, or be a loop. Far away from a small loop of string with length r, there is the Newtonian gravitational field that would be generated by a point source with mass $r\mu$. A long straight string on the other hand distorts the spacetime metric, so that the circumference of a circle at a distance r from the string is only $(2\pi - \mu/M_{\rm Pl}^2)r$. Effects such as these make the string detectable in a variety of ways.

When two strings cross they typically reconnect, though there is a small probability that they instead pass through each other. Just after reconnection each new string has a cusp, which emits gravitational waves as it smooths out.

If the $U(1)$ symmetry is restored at early times, cosmic strings will form by the Kibble mechanism when it is spontaneously broken, along the lines in space where ϕ happens to vanish. In the case of gauge strings another mechanism called **flux trapping** is also available. As the name implies, the strings in this case form in regions of space where the space–space components of the gauge field strength (analogous to a magnetic field flux) are non-zero. Such regions will be present before the phase transition through thermal or quantum fluctuations, but after the phase transition they can survive only within cosmic strings. At least one string per horizon volume is expected initially.

The subsequent evolution of a network of cosmic strings can be evolved by numerical simulation, if the reconnection probability and other relevant parameters are known. Using the simulation one can examine possible observational signatures of the strings. At present there is no detection. Focussing on local strings, the absence of an observed signature in the cosmic microwave background (CMB) requires $\mu \lesssim (10^{16}\,{\rm GeV})^2$. Pulsar timing measurements constrain the intensity of the gravitational waves emitted by cusps, to yield a similar bound. Future gravitational wave detectors will decrease this bound to around $(10^{12}\,{\rm GeV})^2$, or give a detection.

We should mention strings that are not strictly solitons, because they have zero thickness at the classical level. These are the fundamental strings of string theory (F strings), and D branes which look like strings from the 3-dimensional viewpoint (D strings). When they intersect their reconnection probability is low. As a result they are more numerous, and the CMB constraint is lower than for ordinary cosmic strings.

21.4.3 More solitons

Consider the case of three real fields ϕ_n, and define $|\phi|^2 \equiv \sum \phi_n^2$. If the potential depends only on $|\phi|^2$ there is symmetry under rotations in field space, and if it is of the form (14.28) the symmetry is spontaneously broken. Repeating the previous analysis one finds that there is a stable spherically symmetric field configuration, whose energy is mostly concentrated within a sphere. This is a monopole, which may be either local or global. A magnetic monopole is a local monopole possessing a magnetic moment.

We have been considering stable topological defects. These are formed at a symmetry-breaking phase transition, corresponding to regions of space where the symmetry is permanently unbroken. There can also be unstable topological defects, corresponding to regions where the symmetry is restored for a long time, but not for ever. The most interesting defects of this type are known as textures. They require the spontaneous breaking of a global non-Abelian symmetry. Textures can have an observable effect on the CMB anisotropy, which places an observational upper bound on the energy scale of the relevant phase transition. In that respect they are like cosmic strings.

Finally, we mention a soliton which is not a topological defect, known as a Q-ball (see Ref. [13, 14] for reviews). The simplest example involves a single complex field ϕ, whose potential $V(|\phi|)$ has a minimum at $\phi = 0$ corresponding to an unbroken $U(1)$ symmetry and $V = 0$. A Q-ball may exist if $V/|\phi|^2$ has a minimum at $|\phi| > 0$. It may be a global or a gauge symmetry, and the Q-ball may be stable or unstable. In contrast with all of the other solitons that we considered, the field ϕ is not time independent. Instead, its phase changes linearly with time. In the mechanical analogy, the rotation of the scalar field is holding it up against the inward force of the potential gradient. The Q-ball is so-called because the rotating field means that the Q-ball carries the conserved 'charge' associated with the $U(1)$ symmetry.

21.5 Topological defects and the GUT transition

From these simple examples, one might think that spontaneous symmetry breaking always leads to topological defects. That is not true when one goes to more complicated cases. Instead, the field in the entire region surrounding the point of unbroken symmetry may be able to descend quickly to the same vacuum.

The electroweak transition does not create topological defects, and neither does the chiral symmetry transition. To encounter the creation of topological defects we have to go beyond the Standard Model. One possibility is a GUT transition.

As originally envisaged, restoration of the GUT symmetry would be thermal. This would require a temperature roughly of order $M_{\rm GUT} \sim 10^{16}$ GeV. Above this temperature, the GUT symmetry would be completely restored; in other words, all of the GUT Higgs fields would be zero before the transition. Using the fact that the GUT corresponds to what is called a simple group, it can be shown that the GUT transition will always produce monopoles. The transition will often produce strings too, as well as unstable topological defects. To establish the eventual abundance of the monopoles, taking the latter into account, is not straightforward, because it may depend on the GUT group and even on the GUT interactions. On the other hand, a calculation of the monopole abundance ignoring any other defects does give a monopole abundance far bigger than is allowed by observation. One of the original motivations for inflation was to solve this problem, by avoiding a temperature bigger than M_{GUT} after inflation. That is more or less inevitable if there is Einstein gravity during inflation, because the observational bound on the primordial tensor perturbation requires an inflation scale below 10^{16} GeV.

Instead of restoring GUT symmetry thermally, one can do it in a hybrid inflation model by choosing the waterfall field(s) to be GUT Higgs fields. In that case, the GUT symmetry need not be completely restored during inflation. In other words one or more of the GUT Higgs fields could be at its vev during inflation, or at least be away from the origin. Then one might produce cosmic strings without monopoles, or else no topological defects at all. In the former case, the string tension is not expected to be far below the GUT scale, which means that the strings should be observable in the foreseeable future.

21.6 Thermal inflation

For the electroweak and chiral phase transitions, the zero temperature potential has the usual quartic form (14.32). Let us denote $|\phi|^2$ simply by ϕ^2 so that we deal with the potential (14.28).[6] Then the finite temperature potential has the form (21.13). One usually makes the assumption that a similar potential holds for the GUT transition, which is perhaps supported by the accurate unification of the couplings occurring within the MSSM at a single scale. With that assumption, the phase transition happens as soon as the energy density becomes comparable to the height V of the potential. As a result, the energy density is never dominated by V.

Without tying ourselves to a particular phase transition, let us consider an alternative possibility, that the zero-temperature potential is much flatter than (14.28). Then the finite-temperature term will hold the field at the top of the potential for much longer, so that it dominates the energy density before the phase transition occurs. This gives a short bout of inflation just before the phase transition, called

[6] Recall that several fields might be involved, in which case $|\phi|^2$ will be shorthand for some combination that is invariant under the relevant symmetry.

thermal inflation [15, 16, 17, 18]. As we will see in the next chapter, thermal inflation may play a crucial role in relation to baryogenesis and the creation of CDM.

In this section we will illustrate the idea, by supposing that the quartic term in Eq. (14.32) is replaced by a single higher-order (non-renormalizable) term. Then the finite temperature potential is

$$\boxed{V(\phi,T) = V_0 + \left(g^2T^2 - \frac{1}{2}m^2\right)\phi^2 + \lambda_d \frac{\phi^d}{M_{\rm Pl}^{d-4}}}. \tag{21.17}$$

We suppose that the tachyonic mass $m \sim 100\,\text{GeV}$ comes from soft supersymmetry breaking. We assume $\lambda_d \sim 1$, which makes $\langle\phi\rangle \sim 10^{10}\,\text{GeV}$, $V_0^{1/4} \sim \sqrt{m\langle\phi\rangle} \sim 10^6\,\text{GeV}$ and the mass m_ϕ in the vacuum of order m. As in the Peccei–Quinn (PQ) case more than one field might be involved, and in the case of gauge-mediated supersymmetry breaking the form of the potential might actually be logarithmic while preserving the above estimates [19].

With this potential, the thermal energy density $\rho_{\rm r} \sim T^4$ falls below V_0 when $T \sim V_0^{1/4}$, but the phase transition happens only later when $T \sim m$. Thermal inflation takes place between these two events. Because $a \propto 1/T$, there are about $\ln(V_0^{1/4}/m) \sim 10$ e-folds of thermal inflation. With so few e-folds, thermal inflation can hardly replace ordinary inflation, but it can efficiently remove dangerous relics. The entropy dilution factor (21.4) is given by

$$\gamma \sim \left(\frac{m}{V_0^{1/4}}\right)^3 \frac{T_{\rm R}}{V_0^{1/4}}, \tag{21.18}$$

where the first factor accounts for the expansion during thermal inflation and the second factor accounts for the subsequent expansion until ϕ decays and reheats the Universe to a temperature $T_{\rm R}$.

Let us estimate $T_{\rm R}$. The oscillation of ϕ about its vev begins when thermal inflation ends. Because that vev is large the decay rate will be small. To see this, consider a Yukawa coupling $\lambda\phi\psi\psi$, to a fermion into which ϕ might decay. For the decay to occur we need $m > 2m_\psi$. The inequality will usually be well satisfied, making m the only relevant dimensionless parameter and giving a decay rate $\Gamma \sim \lambda^2 m$. But the large vev of ϕ gives a contribution $m_\psi(\langle\phi\rangle) = \lambda\langle\phi\rangle/2$ to the mass m_ψ of ψ.[7] For the decay to occur we therefore need $\lambda\langle\phi\rangle \lesssim m$, corresponding to a decay rate $\Gamma \lesssim m^3/\langle\phi\rangle^2$. A similar estimate emerges if one looks at other decay possibilities. The corresponding reheat temperature is at most of order $M_{\rm Pl}^{1/2} m^{3/2}/\langle\phi\rangle$, which is usually taken as an actual estimate. This gives

[7] This is from Eq. (15.52); the analysis is the same for a decay into a particle–antiparticle pair with a Dirac mass.

$T_R \sim 10^2$ GeV and the entropy dilution factor (21.18) becomes

$$\gamma \sim 10^{-12} \times 10^{-4} \sim 10^{-16}. \qquad (21.19)$$

As the field ϕ has mass $m \gg H$ during thermal inflation, it doesn't acquire a perturbation. Any fields with $m \lesssim H$ will acquire a perturbation, and could generate primordial perturbations on scales leaving the horizon during thermal inflation. In the discussion of Section 18.4 we implicitly assumed that such a thing does not happen for cosmological scales. That is almost certainly the case, because thermal inflation is typically brief and typically occurs at a low energy scale. In principle though, part or even all of the curvature perturbation might be generated during thermal inflation [20]. If all of the curvature perturbation were generated, earlier inflation would be needed only to set the initial condition for thermal equilibrium.

We end with a couple of remarks about the cosmological conditions required for thermal inflation [18]. Essentially two conditions are required. One is the existence of thermalized radiation. That is quite a mild requirement, because as seen in Section 22.1 thermalized radiation exists well before the actual reheating epoch.

The other requirement is less trivial; the field ϕ should have interactions of normal strength before the phase transition, so that the finite temperature potential will be generated. One might think that this will be automatic if it has gauge couplings to other fields, or Yukawa couplings not too many orders of magnitude below 1, but in fact that alone is not enough. It is necessary also that ϕ vanishes during ordinary inflation, or is anyhow much less than the final value $\langle\phi\rangle \sim 10^{10}$ GeV. This is because a large value would give large masses to the particles with which ϕ interacts, inhibiting thermal equilibrium. Whether this condition is satisfied depends on the initial condition for ϕ and on its interaction with other relevant fields including the inflaton.

21.7 Moduli problem

In this section we consider a type of scalar field known as a modulus, and see why it may pose a serious cosmological problem.

For the present purpose a modulus may be defined as a scalar field with a potential of the form

$$\boxed{V = V_0 f(\phi/M_{\rm Pl})}, \qquad (21.20)$$

This is supposed to hold at least in the range $0 < \phi \lesssim M_{\rm Pl}$, with the function $f(x)$ and its low derivatives of order unity at a generic point. With ϕ at the vev this corresponds to mass $m^2 \sim V_0/M_{\rm Pl}^2$. If the potential has a maximum, it will typically be located at a distance of order $M_{\rm Pl}$ from the vev with the tachyonic mass-squared V'' typically of order $-m^2$.

Fields with this property are expected (though not inevitable) in a field theory derived from string theory. Moduli are commonly supposed to have a lifetime similar to that of the gravitino. Alternatively though, a modulus may have interactions of ordinary strength, in particular gauge interactions. The fixed point of the symmetry group is then called a point of enhanced symmetry. A point of enhanced symmetry might correspond to either the vev or to a maximum of the potential. It is even possible for both of these to be points of enhanced symmetry, involving different symmetry groups.

Usually the field theory is taken to be supersymmetric. Then the moduli are complex, with the real and imaginary parts expected to have the form (21.20). In another terminology, it is the real part of each complex field that is a modulus, while the imaginary part is called a string axion. String axions can be regarded as pseudo-Nambu–Goldstone bosons (PNGBs) and may have a much flatter potential than (the real parts of) moduli. The superpartners of moduli are called modulini.

In the context of supersymmetry, the mass of a modulus is usually supposed to be of order $m \sim m_{3/2}$, corresponding to $V_0 \sim M_{\rm S}^4$. Then, if the lifetime is similar to that of the gravitino the observational constraints on its abundance n/s are similar. Moduli are, however, typically far more abundant than gravitinos, making the moduli problem more serious. Let us see how this works.

The main creation mechanism for moduli is the displacement of the modulus field from its vev in the very early Universe. Supersymmetry breaking during inflation will typically give an additional contribution to the modulus potential of order $H_*^2 \left(\phi - \phi_0\right)^2$ with ϕ_0 typically of order $M_{\rm Pl}$. The modulus potential during inflation will then be

$$V(\phi) \simeq M_{\rm S}^4 f(\phi/M_{\rm Pl}) + H_*^2(\phi - \phi_0)^2, \tag{21.21}$$

Unless the vev is a point of enhanced symmetry there is no reason why ϕ_0 should coincide with the vev, and instead one expects $|\phi_0 - \langle\phi\rangle|$ to be of order $M_{\rm Pl}$. As we see in Section 28.4, the inflation scale $\sqrt{M_{\rm Pl} H_*}$ is expected to be bigger than $M_{\rm S}$ so that the second term of Eq. (21.21) dominates, and ϕ during inflation is displaced from its vev by an amount $\phi_0 \sim M_{\rm Pl}$.

After inflation, ϕ will start to oscillate about its vev when $H(t) \sim m_{3/2}$. At this stage the energy density of the modulus is of order $\sim M_{\rm Pl}^2 m_{3/2}^2$, about the same as the total energy density. There now follows an era of matter domination, ending only when the modulus decays at about the same time as the gravitino. If $m_{3/2} \gtrsim 10\,{\rm TeV}$ the modulus decays well before BBN, and is harmless unless it creates unwanted relics. But in the more usual case $m_{3/2} \lesssim 10\,{\rm TeV}$, matter domination persists to give a universe bearing no resemblance to ours.

Let us see by how much the abundance of the modulus must be reduced, to get rid of this moduli problem. When the modulus starts to oscillate,

$$\boxed{\frac{n}{s} \sim \left(\frac{M_{\rm Pl}}{m_{3/2}}\right)^{1/2} \sim 10^9 \left(\frac{1\,{\rm GeV}}{m_{3/2}}\right)^{1/2}}. \tag{21.22}$$

There now begins an era of matter domination. If $m_{3/2} \lesssim 10\,{\rm MeV}$, the modulus is a CDM candidate and Eq. (4.37) shows that we need a dilution factor of order $10^{-18}(1\,{\rm GeV}/m_{3/2})^{1/2}$. If instead $m_{3/2} \gtrsim 100\,{\rm MeV}$ the modulus abundance is constrained by BBN, the distortion of the CMB, or the diffuse γ-ray background. It turns out that this requires a similar (actually somewhat bigger) dilution factor. Such dilution cannot be supplied by a second reheating, but it might be supplied by thermal inflation.

A different solution to the moduli problem would be to suppose that the vev corresponds to a point of enhanced symmetry. Then the modulus need not be displaced from its vev in the early Universe, and even then its decay rate will be greatly enhanced so that it need present no danger to BBN (provided that the decay creates no unwanted relics).

Exercises

21.1 Show that Eq. (21.1) holds if $a^3\bar\rho$ is proportional to $e^{-\Gamma t}$.

21.2 Verify Eq. (21.2) assuming that the age of the Universe at a given epoch is of order H^{-1}.

21.3 Show that an oscillating scalar field with a quartic potential gives average pressure $P = \rho/3$.

21.4 Show that $\rho \propto a^{-6}$ during an era of kination.

21.5 Show that the equalities (21.15) are preserved by a Lorentz boost along the direction of the domain wall.

21.6 Show that a domain wall stretching across the observable Universe has energy equal to that of the cosmic fluid, if its energy per unit area is roughly of order $(1\,{\rm MeV})^3$.

21.7 For the global string corresponding to Eq. (21.16), set $g = f$ to show that the energy density outside the string is $n^2 f^2/r^2$ as stated in the text.

21.8 For the thermal inflation potential (21.17), evaluate $\langle\phi\rangle$ and m_ϕ in terms of λ_d and m.

21.9 Verify the estimate (21.18) of the entropy dilution factor generated by thermal inflation.

21.10 Verify Eq. (21.22), giving n/s for a modulus when it starts to oscillate.

References

[1] K. Enqvist and A. Mazumdar. Cosmological consequences of MSSM flat directions. *Phys. Rept.*, **380** (2003) 99.

[2] K. Hamaguchi, H. Murayama and T. Yanagida. Leptogenesis from sneutrino-dominated early universe. *Phys. Rev. D*, **65** (2002) 043512.

[3] M. Kawasaki, K. Nakayama and M. Senami. Cosmological implications of supersymmetric axion models. *JCAP*, **0803** (2008) 009.

[4] L. Kofman, A. D. Linde and A. A. Starobinsky, Reheating after inflation. *Phys. Rev. Lett.*, **73** (1994) 3195.

[5] P. B. Greene and L. Kofman. Preheating of fermions. *Phys. Lett. B*, **448** (1999) 6.

[6] G. N. Felder, L. Kofman and A. D. Linde. Instant preheating. *Phys. Rev. D* **59** (1999) 123523.

[7] G. N. Felder, J. Garcia-Bellido, P. B. Greene, L. Kofman, A. D. Linde and I. Tkachev. Dynamics of symmetry breaking and tachyonic preheating. *Phys. Rev. Lett.*, **87** (2001) 011601.

[8] D. H. Lyth and T. Moroi. The masses of weakly-coupled scalar fields in the early universe. *JHEP*, **0405** (2004) 004.

[9] V. Mukhanov. *Physical Foundations of Cosmology* (Cambridge: Cambridge University Press, 2006).

[10] M. Dine. *Supersymmetry and String Theory* (Cambridge: Cambridge University Press, 2007).

[11] D. J. Schwarz. The first second of the universe. *Annalen Phys.*, **12** (2003) 220.

[12] A. Vilenkin and P. Shellard. *Cosmic Strings and Other Topological Defects* (Cambridge: Cambridge University Press, 1994).

[13] M. B. Hindmarsh and T. W. B. Kibble, Cosmic strings, *Rept. Prog. Phys.*, **58** (1995) 477. A. C. Davis and T. W. B. Kibble, Fundamental cosmic strings. *Contemp. Phys*, **46** (2005) 313.

[14] A. Kusenko, Properties and signatures of supersymmetric Q-balls. arXiv:hep-ph/0612159.

[15] P. Binetruy and M. K. Gaillard. Candidates for the inflaton field in superstring models. *Phys. Rev. D*, **34** (1986) 3069.

[16] G. Lazarides and Q. Shafi. Anomalous discrete symmetries and the domain wall problem. *Nucl. Phys. B*, **392** (1993) 61.

[17] D. H. Lyth and E. D. Stewart. Cosmology with a TeV mass GUT Higgs. *Phys. Rev. Lett.*, **75** (1995) 201.

[18] D. H. Lyth and E. D. Stewart. Thermal inflation and the moduli problem. *Phys. Rev. D*, **53** (1996) 1784.

[19] A. de Gouvea, T. Moroi and H. Murayama. Cosmology of supersymmetric models with low-energy gauge mediation. *Phys. Rev. D*, **56** (1997) 1281.

[20] K. Dimopoulos and D. H. Lyth. Models of inflation liberated by the curvaton hypothesis. *Phys. Rev. D*, **69** (2004) 123509.

22

Thermal equilibrium and the origin of baryon number

In this chapter we study thermal equilibrium in the early Universe. Then we look at possible mechanisms for the creation of baryon number (baryogenesis). We pay particular attention to baryogenesis mechanisms that directly involve a scalar field, because they offer the best chance of a primordial isocurvature perturbation.

22.1 Thermal equilibrium before the electroweak phase transition

In this section we show that electroweak symmetry is likely to be restored in the early Universe, with every particle of the Standard Model in thermal equilibrium.

Electroweak symmetry is restored if the temperature is bigger than a critical temperature $T_{\rm EW}$. The critical temperature is of order a few hundred GeV, the precise value depending on the parameters of the Standard Model, or an extension like the Minimal Supersymmetric Standard Model (MSSM).

A particle species is in thermal equilibrium if the rate per particle for all relevant interactions exceeds the Hubble rate H. The dominant interactions are (i) decays and two-body scattering and (ii) the sphaleron transitions that violate B and L conservation. As we are interested in the case that electroweak symmetry is restored, particle masses vanish except for the masses of two Higgs particles which are roughly of order $100\,{\rm GeV}$ and hence less than the temperature. As a result, T is the only relevant dimensionful parameter.

The rate per particle for the decays and two-body scatterings can be calculated from Feynman diagrams, and will be of order T times the square of a coupling (for decays), or the square of a product of two couplings (for typical two-body scatterings).[1] The couplings may be gauge or Yukawa couplings, but in the case

[1] The relevant decays proceed by first-order perturbation theory (corresponding to Feynman diagrams with a single vertex), so that their amplitude is proportional to a coupling. The relevant two-body scattering processes proceed by second-order perturbation theory (corresponding to Feynman diagrams with one particle exchanged), so that their amplitude is proportional to a product of two couplings.

of two-body scatterings it is enough for most purposes to consider only those processes that involve at least one gauge coupling. Gauge couplings are roughly of order 1. The smallest Yukawa coupling is that of the electron, $y_{\rm e} \sim 10^{-6}$. The rate per particle for decays and two-body scatterings is therefore expected to be at least of order 10^{-12} or so.

The sphaleron reaction rate per particle is bigger, of order $10^{-6}T$. (The prefactor is $(\alpha_{\rm W})^5$ times a numerical factor of order 1.) All relevant processes are therefore in thermal equilibrium provided that

$$\boxed{H \lesssim 10^{-12} T \qquad \text{(thermal equilibrium condition)}}\,. \tag{22.1}$$

We want to see if this condition is compatible with $T > T_{\rm EW}$.

Suppose first that reheating occurs before the electroweak transition, $T_{\rm R} > T_{\rm EW}$.[2] Then $3M_{\rm Pl}^2 H^2 \sim T^4$ for $T < T_{\rm R}$ and Eq. (22.1) becomes

$$\boxed{T \lesssim 10^{-12} M_{\rm Pl} \sim 10^6\ \text{GeV}}\,. \tag{22.2}$$

We see that $T > T_{\rm EW}$ is allowed.

Now suppose instead that reheating occurs at $T_{\rm R} < T_{\rm EW}$. Although the radiation dominates only after reheating, it exists before reheating. Therefore, it $T_{\rm R}$ is not too far below $T_{\rm EW}$, the radiation can still thermalize at a temperature $T > T_{\rm EW}$ [1]. Let us see in what regime of parameter space that is possible.

We will call the particle whose decay leads to reheating ϕ. Its decay contributes to the radiation according to the relation

$$d(a^3\rho_{\rm r}) = -P_{\rm r} d(a^3) - d(a^3\rho_\phi) = -\frac{\rho_{\rm r}}{3} d(a^3) - (a^3\rho_\phi)\Gamma dt. \tag{22.3}$$

This gives

$$\dot\rho_{\rm r} + 4H\rho_{\rm r} = -\Gamma\rho_\phi. \tag{22.4}$$

We are interested in the era *after* the epoch $T = T_{\rm dom}$ when ϕ begins to dominate the energy and *before* the reheat epoch at temperature $T = T_{\rm R}$. During this era $\rho_\phi \propto a^{-3} \propto t^{-2}$. Dropping a decaying mode the solution of Eq. (22.4) is then $\rho_{\rm r}(t) = (3/5)\Gamma t \rho_\phi(t)$. Also, $T \simeq \rho_{\rm r}^{1/4}$. It follows that in the regime $T_{\rm R} < T < T_{\rm dom}$,

$$\boxed{\rho^{1/4} \simeq \rho_\phi^{1/4} \simeq \frac{T^2}{T_{\rm R}} = 10^5\ \text{GeV} \left(\frac{T}{300\ \text{GeV}}\right)^2 \left(\frac{1\ \text{GeV}}{T_{\rm R}}\right)}\,. \tag{22.5}$$

The interaction rate of each component of the radiation is related to the temperature in the same way as before, and the condition for thermal equilibrium is still

[2] Here $T_{\rm R}$ is the final reheat temperature if there is more than one reheating.

that the rate be bigger than H. As seen earlier, this requires $H \lesssim 10^{-12}T$. Using Eq. (22.5) one finds there can be thermal equilibrium at $T > T_{\rm EW}$ provided that $T_{\rm R} \gtrsim 1\,{\rm GeV}$, and provided that ϕ dominates by the epoch

$$\boxed{\rho^{1/4} \sim 10^5\,{\rm GeV}\,\frac{1\,{\rm GeV}}{T_{\rm R}}}\,. \tag{22.6}$$

In summary, we find that electroweak symmetry is likely to be restored, with all Standard Model particles in good thermal equilibrium. Equilibrium is likely to fail only if the (final) reheat temperature is less than 1 GeV or so, and then only for the electron and first-generation quarks which have very small Yukawa couplings. Possible consequences of the latter circumstance are mentioned in Ref. [1] but we shall not consider them.

22.2 Thermal equilibrium with non-zero $B - L$

We noticed in Section 16.5.2 that the sphaleron processes violate baryon and lepton number conservation, but conserve $B - L$ and the differences $L_\alpha - L_\beta$ where α and β run over the three generations.[3] In this section we argue that if the density of $B - L$ is non-vanishing, thermal equilibrium determines the separate B and L densities. To be precise,

$$n_B = \frac{24 + 4N_H}{66 + 13N_H}\,n_{B-L}, \tag{22.7}$$

where N_H is the number of Higgs doublets. For the Standard Model, $N_H = 1$ and $n_B = (28/79)n_{B-L}$.

We begin with a general analysis of thermal equilibrium in the presence of non-zero charge densities, where 'charge' means any conserved quantity. Let q_{ia} be the charge of type i carried by species a. (For instance, if i is lepton number L, then $q_{ia} = 1$ for each lepton and 0 for every other species.) We assume that the chemical potentials are of the form

$$\mu_a = \sum_i q_{ia}\mu_i, \tag{22.8}$$

where μ_i are some numbers. This guarantees that the chemical potentials are conserved at every interaction, and we expect that any additional contribution is set to zero because enough reactions are in thermal equilibrium.

[3] At $T \sim 1\,{\rm MeV}$ neutrino mixing requires that the neutrino chemical potentials are the same for each generation, but that will not be the case before the electroweak transition because the neutrino masses are negligible compared with the temperature. In fact, according to the seesaw mechanism they vanish because the Higgs fields have zero vacuum expectation value (vev).

Following Ref. [2] we assume $|\mu_i| \ll T$, which should be checked for consistency at the end of the calculation. Then Eq. (4.12) is valid giving

$$n_a - \bar{n}_a = \frac{T^2}{6}\,\tilde{g}_a \sum_i q_{ia}\mu_i. \tag{22.9}$$

Using $n_i = \sum_a q_{ia}(n_a - \bar{n}_a)$ this gives

$$n_i = \frac{T^2}{6}\sum_j M_{ij}\mu_j, \qquad M_{ij} \equiv \sum_a \tilde{g}_a q_{ia} q_{ja}. \tag{22.10}$$

Inverting this relation gives[4]

$$\mu_i = \sum_j \left(M^{-1}\right)_{ij} n_j \frac{6}{T^2}. \tag{22.11}$$

We want to apply this result to the early Universe before the electroweak phase transition. We noticed in Section 4.3 that the density of electric charge Q has to vanish, or at least be extremely small, because it is the source of the electromagnetic field. Otherwise there would be a $1/r$ electric force between elements of the cosmic fluid, which would outweigh gravity. By the same token the densities of hypercharge, of the three $SU_L(2)$ charges and of the eight $SU_C(3)$ charges must vanish, or be extremely small, since they too are sources of interactions. (Electric charge is the sum of hypercharge and *one* of the $SU_L(2)$ charges, namely T_3, but we are saying here that all three of the latter must be vanishing or nearly so.) Zero values for all three $SU_L(2)$ charges and for all eight $SU_C(3)$ charges are indeed possible, as discussed in Section 16.4.2 for the latter. Actually, we saw in that section that the colour charge densities are required by theory to vanish exactly (after smoothing on a scale $\Lambda_{\rm QCD}^{-1} \sim 10^{-15}\,$m) but the reason for the vanishing is irrelevant for the present purpose; we just need to know that it happens.

The only other conserved charges existing before the electroweak transition are $B - L$ and the differences $L_\alpha - L_\beta$. This is because the isospin and chiral symmetries of QCD (present at low energies as described in Section 16.5) are completely ruined by the electroweak interaction. We are allowing at least $B - L$ to have non-vanishing density, so as to arrive at Eq. (22.7).

What about the species present? There are leptons, quarks and gauge bosons, all with zero mass, and there is a massive Higgs particle for each of the complex fields H_1 and H_2 (with a distinct antiparticle). For the quarks and leptons, we need to consider separately left- and right-handed particles. Except for colour, the charges carried by each particle are discussed in Chapter 16, and displayed in Tables 16.1,

[4] The inverse exists because M is a positive-definite matrix, meaning that $\xi^T M \xi > 0$ unless the column matrix ξ vanishes.

16.3 and 16.4 (pages 270 and 272). Notice that each particle carries a definite amount of T_3. When that is non-zero the amounts of T_1 and T_2 carried are not well defined, but their expectation values vanish which presumably justifies their neglect. Coming to colour, we noticed at the end of Section 14.3.2 that two of the eight $SU(3)$ charges can simultaneously have well-defined non-zero values. The other six are then not well defined, but again their vanishing expectation values presumably justifies their neglect.

We need to calculate the matrix M_{ij} and invert it. That looks like a big job, but it isn't. This is because the total of T_3 summed over an $SU_L(2)$ multiplet vanishes, as does the total of each of the two colour charges summed over the three colours of each quark. As a result, the matrix M_{ij} is block diagonal, with a 2×2 block having i and j running over $B - L$ and Y. The same will be true of the inverse matrix M^{-1}.

The baryon number density is therefore given by

$$n_B = \sum_a B_a(n_a - \bar{n}_a) \tag{22.12}$$

$$= \sum_a \tilde{g}_a B_a \left[(B-L)_a M^{-1}_{(B-L)(B-L)} + Y_a M^{-1}_{Y(B-L)}\right] n_{B-L} \tag{22.13}$$

$$= \left(4M^{-1}_{(B-L)(B-L)} - 2M^{-1}_{Y(B-L)}\right) n_{B-L}. \tag{22.14}$$

The elements of the relevant 2×2 submatrix of M are $M_{B-L,B-L} = 13$, $M_{B-L,Y} = -8$ and $M_{YY} = 10 + N_H$ where N_H is the number of Higgs doublets. Inverting this submatrix and substituting the result into Eq. (22.14) gives the advertised result (22.7).

22.3 Baryogenesis mechanisms

22.3.1 Sakharov conditions

It is generally assumed that the baryon number density of the Universe vanishes at some early time. The subsequent creation of baryon number (baryogenesis) requires the following conditions.

(i) The failure of baryon number conservation, or in other words the breaking of the $U_B(1)$ symmetry.

(ii) The breaking of the symmetries C (interchange of matter and antimatter) and CP (C combined with a change from left- to right-handed coordinates).

(iii) The absence of thermal equilibrium with zero chemical potentials, or else the breaking of CPT symmetry (CP combined with time reversal).

These are known as the Sakharov conditions, after the author of the first baryogenesis paper [3]. The first condition is obviously needed, and so is the second because otherwise any creation of positive B will be cancelled by the creation of an equal amount of negative B. The third is needed because CPT invariance requires that particle–antiparticle pairs have equal mass, so that with zero chemical potential there are equal numbers of particles and antiparticles. Often the third condition is simply stated as the absence of thermal equilibrium, and indeed we have seen that in the absence of any conserved charge the chemical potentials may be expected to vanish if enough reactions are in thermal equilibrium.

The breaking of the four symmetries $U_B(1)$, C, CP and (if the thermal equilibrium condition holds) CPT has to occur in the field theory that applies in the early Universe, but it need not be present in the field theory that applies in the vacuum. Rather, the breaking could be 'spontaneous' in the sense that it is caused by the presence of some scalar field, or other smooth quantity, that is present in the early Universe but not in the vacuum. That presumably would have to be the case for the breaking of CPT invariance, since quantum field theory predicts CPT invariance in the vacuum.

Now we discuss in turn four scenarios for baryogenesis. The first three assume CPT invariance.

22.3.2 Electroweak baryogenesis

We have seen that electroweak symmetry is probably restored by thermal equilibrium in the early Universe. We have also seen that while the symmetry is restored, the density of baryon number B is a multiple of $B - L$, where L is the lepton number. If $B - L$ vanishes, baryogenesis has to occur at or after the electroweak transition.

Let us consider baryogenesis at the electroweak transition. Even within the Standard Model, CP invariance is violated through the phase in the CKM matrix, and baryon number conservation is violated through the sphaleron processes. Of the three Sakharov conditions, that leaves only the requirement of non-equilibrium. It will be satisfied if the electroweak transition is of first order, because bubbles of the new phase then form and non-equilibrium processes can take place at the bubble walls.

Given a Lagrangian specifying the relevant interactions, numerical calculation can identify the part of parameter space in which successful electroweak baryogenesis is possible. For both the Standard Model and the MSSM, this part of parameter space is ruled out by collider observations. Successful electroweak baryogenesis may still be possible in a modest extension of the MSSM.

All along, we have been assuming that electroweak symmetry breaking is a thermal effect. As an alternative, it has been suggested [4] that electroweak symmetry breaking occurs at the end of hybrid inflation, the waterfall field being the Higgs field(s) of the Standard Model (or MSSM). With the scale so low, the loop correction to the inflaton potential caused by its coupling to the waterfall field is too big to allow ordinary hybrid inflation [5]. The scenario could however work with inverted hybrid inflation at the cost of extreme fine tuning [6].

What about baryogenesis after the electroweak transition? In principle the mechanisms of the following subsection could generate baryon number after the electroweak transition, but that is difficult to arrange at such a late stage. For the rest of this section, we therefore suppose that $B - L$ already exists just before the electroweak transition, with the right value to make n_B/s agree with observation. That means that we are looking for a mechanism for generating $B - L$ before the electroweak transition. To create $B - L$ we can create either B or L (or both). Creating L is called leptogenesis. The Sakharov conditions apply to leptogenesis, except that it is lepton number rather than baryon number conservation which has to be violated.

22.3.3 Out of equilibrium decay

One possibility is for the decay of some particle to produce an excess of baryon or lepton number. The decay must occur out of equilibrium, and subsequent interactions must not destroy (wash out) baryon number. It can be shown that for this to work, the decay processes must not belong to the Standard Model or the MSSM [2].

The decaying particle might start out in thermal equilibrium, or it might never be in thermal equilibrium. One candidate for the decay might be a GUT particle, but the most commonly considered candidate is a heavy sterile neutrino, involved in the seesaw mechanism that generates the observed neutrino mass. This creates lepton number, and has been the subject of extensive investigation.

22.3.4 Affleck–Dine baryogenesis

Now we consider what is called Affleck–Dine baryogenesis. It was introduced in Ref. [7] and its study may usefully begin with Refs. [8, 9].

The Affleck–Dine mechanism uses a complex scalar field $\phi = |\phi|e^{i\theta}$ which carries $B - L$. The vev of such a field is zero because neither B nor L is spontaneously broken. For small field values the potential is practically independent of θ, corresponding to $B - L$ conservation.

To create $B - L$ it is supposed that the field is significantly displaced from its vacuum at very early times, and that the potential then depends on θ. It is supposed that $\dot\theta$ is negligible at first because of the Hubble drag. When H falls sufficiently,

ϕ moves towards its vev. Initially the motion has an angular component, which through Eq. (14.11) creates $B - L$.

Affleck–Dine baryogenesis could in principle take place during inflation, either with a field whose potential is too flat to affect inflation or even with the inflaton. This case is not usually considered though. Instead, the field ϕ responsible for Affleck–Dine baryogenesis is supposed to be one which parameterizes a flat direction of the MSSM. The simplest example of a suitable field is the one defined in the unitary gauge by $H_u = (0, \phi)$ and $\tilde{L} = (\phi, 0)$. In this direction, the gauge-invariant field $\Phi^2 \equiv \tilde{L} H_u$ is non-zero, and it carries $B - L = -1$. We work in the unitary gauge so that $\Phi^2 = \phi^2$.[5]

To calculate the baryon number density, one needs to have the effective potential $V(\phi, \phi^*)$ during and after inflation. During inflation it depends on the inflation model and on the interaction of the flat direction with the inflaton. Later it depends on the nature of supersymmetry breaking in the MSSM. A form covering a range of possibilities is [8]

$$\begin{aligned} V &= \left(m_\phi^2(|\phi|) + c_1 H^2\right) |\phi|^2 + |\lambda_d|^2 \frac{|\phi|^{2d-2}}{M_{\rm Pl}^{2d-6}} \\ &+ \lambda_d \left[(c_2 H + A) \frac{\phi^d}{M_{\rm Pl}^{d-4}} + \text{c.c.} \right]. \end{aligned} \tag{22.15}$$

The $|\phi|$ dependence (running) of the soft mass m_ϕ^2 can typically be ignored if supersymmetry breaking is gravity-mediated, but not if it is gauge-mediated. In the latter case $m_\phi^2(|\phi|)|\phi^2|$ will typically decrease at large $|\phi|$. The dimensionless parameters c_1 and c_2 represent the additional supersymmetry breaking that is typically present in the early Universe, as described in Section 28.4, and they are expected to be at most of order 1 in magnitude. If $|c_1|$ is of order 1 we need $c_1 < 0$, so that ϕ is displaced from its vev during inflation (assuming $H < m_\phi$ during inflation).

The crucial point allowing baryogenesis is that the second line generates a potential in the angular direction. The movement of ϕ typically begins when $H \sim m_\phi \sim 100\,\text{GeV}$, after which the potential achieves its final form (no H-dependent part), where $m_\phi \equiv m_\phi(0)$ is the true mass of order $100\,\text{GeV}$. Baryogenesis ends when Hubble drag has driven ϕ to sufficiently small values that the potential is quadratic. Finally, ϕ decays to Standard Model particles which carry the $B - L$. The decay may occur either before or after reheating.

Either ϕ or the inflaton field may be the first to decay and cause reheating. For both of these oscillating fields, there may initially be preheating.

The outcome of all this is a value of $(n_B - n_L)/s$ during thermal equilibrium, given a value for n_B/s. For successful baryogenesis the latter may be of order

[5] Remember that the doublet $\tilde{L}$ is the superpartner of the doublet $L = (\nu_e, e_L)$, and that $\tilde{L}H \equiv \epsilon_{ij} \tilde{L}_i H_j$.

10^{-9} as observed, or much bigger. In the latter case, the observed value can be recovered if there is subsequent entropy production. Affleck–Dine baryogenesis is so efficient that it can survive even the huge entropy production from thermal inflation [10].[6] On the other hand, one can also have Affleck–Dine baryogenesis after thermal inflation [11].

In all of this, we assumed that the Affleck–Dine field decays by oscillating. A completely different possibility is for this field to form Q-balls, carrying L and/or B. The gradual decay of the Q-balls can create baryon number, and this can happen at a late epoch. If that epoch is after the freeze-out of neutralino CDM, described in Section 23.2.1, then neutralinos coming from the Q-ball decay can be the main component of the CDM [12]. It might also happen that the Q-ball is stable or has a lifetime longer than that of the age of the Universe, in which case it is a CDM candidate in its own right [13]. (Of course, a different mechanism is then required for baryogenesis.)

22.3.5 Spontaneous baryogenesis

Now we turn to what is called spontaneous baryogenesis, which occurs during thermal equilibrium through the spontaneous breaking of CPT invariance. It was proposed in Ref. [14], and is reviewed in Ref. [15].

As usually envisaged, spontaneous baryogenesis invokes the following term in the Lagrangian

$$\mathcal{L}_{\text{spon}} = \frac{1}{M} \partial_\mu \phi j^\mu_B . \tag{22.16}$$

(The following discussion is the same for spontaneous leptogenesis.) It is supposed that in the early Universe, ϕ is homogeneous with $\dot{\phi} \neq 0$. Then $\mathcal{L}_{\text{spon}} = \dot{\phi}(t) n_B/M$. Going to the Fourier decomposition, this adds to the Hamiltonian of each free field a term

$$\sum_{\mathbf{k}} \Delta \hat{E}_{\mathbf{k}} \equiv \frac{B\dot{\phi}}{M} \sum_{\mathbf{k}} \left(\hat{n}_{\mathbf{k}} - \hat{m}_{\mathbf{k}} \right) . \tag{22.17}$$

Here B is the baryon number carried by the field, while $\hat{n}_{\mathbf{k}}$ and $\hat{m}_{\mathbf{k}}$ are the occupation number operators for the particle and antiparticle. This is equivalent to changing the chemical potentials of the particle and antiparticle by $\Delta\mu_\pm \equiv \pm B\dot{\phi}/M$. If

[6] If the entropy dilution factor from thermal inflation is much greater than the observed quantity $n_B/s \sim 10^{-10}$, Affleck–Dine baryogenesis must create $n_B/s \gg 1$. In that case, the decay of the flat direction carrying the n_B must occur only after thermal inflation is over. This is because each particle carries only $B \sim 1$, so that the particles in thermal equilibrium carry $|n_B| \lesssim s$.

the original chemical potentials vanish, this gives

$$\frac{n_B}{s} = -\frac{15}{4\pi^2 g_* T}\frac{\dot{\phi}}{M}\sum_a \tilde{g}_a B_a. \tag{22.18}$$

The mechanism can work in much the same way if $\partial_\mu \phi$ is replaced by any 4-vector field. Candidates that have been proposed are $\partial_\mu R$ where R is the spacetime curvature, and a gauge field. Also, the mechanism can give leptogenesis instead of (or in addition to) baryogenesis.

Exercises

22.1 Verify the values given after Eq. (22.14), for the elements of the 2×2 submatrix of M_{ij}.

22.2 Show that according to the Standard Model the effective number of particle species g_* before the electroweak phase transition is 106.75. Using the quark and lepton masses in Table 16.2, calculate g_* at $T = 200\,\mathrm{MeV}$, just before the quark–hadron phase transition.

References

[1] S. Davidson, M. Losada and A. Riotto. Baryogenesis at low reheating temperatures. *Phys. Rev. Lett.*, **84** (2000) 4284.

[2] S. Weinberg. *Cosmology* (Oxford: Oxford University Press, 2008).

[3] A. D. Sakharov. Violation of CP invariance, C asymmetry, and baryon asymmetry of the Universe. *Pisma Zh. Eksp. Teor. Fiz.*, **5** (1967) 32.

[4] J. Garcia-Bellido, D. Y. Grigoriev, A. Kusenko and M. E. Shaposhnikov. Non-equilibrium electroweak baryogenesis from preheating after inflation. *Phys. Rev. D*, **60** (1999) 123504.

[5] D. H. Lyth. Constraints on TeV-scale hybrid inflation and comments on non-hybrid alternatives. *Phys. Lett. B*, **466** (1999) 85.

[6] E. J. Copeland, D. Lyth, A. Rajantie and M. Trodden. Hybrid inflation and baryogenesis at the TeV scale. *Phys. Rev. D*, **64** (2001) 043506.

[7] I. Affleck and M. Dine. A new mechanism for baryogenesis. *Nucl. Phys. B*, **249** (1985) 361.

[8] M. Dine, L. Randall and S. D. Thomas. Baryogenesis from flat directions of the supersymmetric Standard Model. *Nucl. Phys. B*, **458** (1996) 291.

[9] K. Enqvist and A. Mazumdar. Cosmological consequences of MSSM flat directions. *Phys. Rept.*, **380** (2003) 99.

[10] A. de Gouvea, T. Moroi and H. Murayama. Cosmology of supersymmetric models with low-energy gauge mediation. *Phys. Rev. D*, **56** (1997) 1281.

[11] D. Jeong, K. Kadota, W.-I. Park and E. D. Stewart. Modular cosmology, thermal inflation, baryogenesis and predictions for particle accelerators. *JHEP*, **0411** (2004) 046.

[12] K. Enqvist and J. McDonald. B-ball baryogenesis and the baryon to dark matter ratio. *Nucl. Phys. B*, **538** (1999) 321.

[13] A. Kusenko and M. E. Shaposhnikov. Supersymmetric Q-balls as dark matter. *Phys. Lett. B*, **418** (1998) 46.

[14] A. G. Cohen and D. B. Kaplan. Thermodynamic generation of the baryon asymmetry. *Phys. Lett. B*, **199** (1987) 251.

[15] S. M. Carroll and J. Shu. Models of baryogenesis via spontaneous Lorentz violation. *Phys. Rev. D*, **73** (2006) 103515.

23

Cold dark matter and dark energy

In this chapter we look at possible mechanisms for creating the cold dark matter (CDM). As with baryogenesis, we pay particular attention to mechanisms that directly involve a scalar field, because they offer the best chance of a primordial isocurvature perturbation. We go on to look at some relics from the early Universe that might be forbidden by observation. We end with some current thinking about the dark energy.

23.1 Axion CDM

The only decay channel of the axion is into photons, and its lifetime is much longer than the age of the Universe making it a cold dark matter (CDM) candidate. In a significant region of parameter space, the interaction of the axion with the magnetic field allows a detection of axion CDM in the foreseeable future. Most of the issues mentioned in this section are reviewed in Refs. [1, 2, 3].

23.1.1 Switching on the potential

The potential (16.26) of the axion, generated by the QCD instanton effect, switches on gradually as the temperature falls towards the QCD scale $\Lambda_{\rm QCD} \simeq 200\,{\rm MeV}$. The gradual growth is given by

$$\frac{m_{\rm a}(T)}{m_{\rm a}} = 0.08\left(\frac{\Lambda_{\rm QCD}}{T}\right)^{3.7}. \tag{23.1}$$

The mass rises above the Hubble parameter at the epoch

$$\tilde{T} = 0.8\left[\frac{f_{\rm a}}{10^{12}\,{\rm GeV}}\right]^{-0.18}\,{\rm GeV}. \tag{23.2}$$

This is the epoch at which the axion field starts to oscillate, corresponding to the presence of axion particles. After the oscillation starts, axion number is conserved. Therefore, $\tilde{T}$ may be regarded as the epoch of axion creation.

23.1.2 Axionic strings

Since we are dealing with a global $U(1)$ symmetry, there can exist axionic strings. If Peccei–Quinn (PQ) symmetry is restored at early times, either by finite temperature effects or by the coupling of the PQ fields to a scalar field displaced from its vev, the strings will form when PQ symmetry is spontaneously broken.

Let us first suppose that the potential leading to PQ symmetry breaking is of the general form (14.32) (with $\lambda \sim 1$). Then, if the symmetry restoration is thermal, the condition for symmetry restoration is that the reheat temperature exceeds f_a. The symmetry breaks as soon as T falls below f_a, leading to the formation of axionic strings. Even if the PQ symmetry is never restored, axionic strings will form at the end of inflation if the vacuum fluctuation of $|\phi|$ during inflation is big enough to kick $|\phi|$ to the top of its potential. According to the discussion of Section 24.6, the condition for string formation in this way is roughly $H_*/f_a \sim 1$. An investigation using just one field shows that H_*/f_a is within a factor 3 or so of this value, for $\lambda \sim 1$ and $|\dot{H}|/H^2 \lesssim 0.1$ [4].

Now suppose instead that the potential is very flat, of the general form (21.17) with $m \sim 100\,\mathrm{GeV}$. Then, if the symmetry restoration is thermal, the symmetry breaks only when T falls below $100\,\mathrm{GeV}$, and there is some thermal inflation before the breaking.

Studies of the evolution of the string network show that an initial network of strings settles down to a 'scaling' solution that is more or less independent of the initial conditions. It has of order one string stretching across each horizon volume, and lots of loops of string within each horizon volume.

23.1.3 Axion CDM: no-string scenario

We suppose first that PQ symmetry is never restored, and that no strings are present. Before the axion mass switches on the axion field will have some time-independent value $a = \sqrt{2} f_a \theta_0$, where θ_0 is called the misalignment angle. When the axion mass switches on the axion field oscillates, corresponding to the presence of non-relativistic axion particles. Using Eqs. (23.1), (23.2) and (4.37) one finds[1]

$$\boxed{\frac{\Omega_a}{0.2} = 2\gamma\theta_0^2 \left(\frac{f_a}{10^{12}\,\mathrm{GeV}}\right)^{1.2}} \qquad \text{(no strings)}, \tag{23.3}$$

[1] The coefficient allows for the continuous switching on of the mass.

where γ is the possible entropy dilution factor.

This estimate assumes a quadratic potential, but it turns out to be quite accurate over most of the range $0 < \theta < \pi$. The regime where θ is very close to π is forbidden because the quantum fluctuation of θ would produce unacceptable bubbles of domain wall, as described in Section 21.4.1.

Taking $\theta \sim 1$ as a typical value within the allowed range $0 < \theta < \pi$, and setting $\gamma = 1$, this requires $f_a \sim 10^{12}$ GeV, Allowing maximal entropy production (matter domination from when the axion mass switches on at $T \sim 1$ GeV until reheating at $T \sim 1$ MeV) one can have $f_a \sim 10^{15}$ GeV, still with a typical value for θ. (See Ref. [5] for details on the effect of entropy production.)

That is not the end of the story though. Since θ is a light field, it will acquire a perturbation from the vacuum fluctuation during inflation. What we are here calling θ_0 is actually the average value of $\theta(\mathbf{x})$ within the observable Universe. This average will vary randomly as we alter the assumed location of the observable Universe within a larger volume. There will be particular locations for which θ_0 is very small, and if we live in one of them we need $f_a \gg 10^{12}$ GeV for the axion to be the CDM. Such a value might be expected if the axion is a string axion.

23.1.4 Axion CDM: string scenario

Now we estimate Ω_a if there are axionic strings. Axion strings radiate axions before the axion mass switches on, but the dominant contribution to Ω_a is generated after the axion mass switches on. Domain walls then form, which connect the strings. Some number N of domain walls meet at each string, that number depending on the potential of the fields responsible for the breaking of PQ symmetry. In the probably unrealistic case that a single field is responsible, N is the number of oscillations of the axion potential going around the rim of the Mexican hat potential, which in Eq. (16.26) we set equal to 1.

The case $N > 1$ is almost certainly unviable, because analytic and numerical estimates show that the network of strings and domain walls will survive to the present. The energy per unit area of the walls is of order ${f_a}^3$, and even one wall across the observable Universe would vastly outweigh everything else. The only hope of rescuing the situation would be to make the domain walls unstable by introducing PQ symmetry breaking at tree level, but as explained in Section 16.6 such breaking has to be very small and a calculation shows that it can hardly get rid of the walls [3].

The case $N = 1$ is perfectly viable. Because each string is the boundary of only one wall, it is pulled in the direction of that wall. As a result, each piece of domain wall quickly disappears and so does the string that bounds it. In this case, axions are created in three ways. First: away from the strings and domain walls, the axion field at each position oscillates to produce axions according to Eq. (23.3).[2] The appropriate value of θ_0^2 in Eq. (23.3) is now the average $\overline{\theta^2} = \pi^2/3$. Second: before the walls form (more precisely, before the wall tension becomes significant) long strings oscillate and string loops collapse. In both cases, axions are radiated. Most of the axions are produced in the few Hubble times before the domain walls stop the radiation. Third: the energy of each domain wall and its bounding string is released mostly in the form of axions.

Numerical calculations of the second effect are difficult, but they suggest that the number of axions from this source is roughly the same as from the first source. As for the domain walls, they release axions only when they annihilate and a straightforward estimate of the energy in a domain wall (bounded by its string) shows that the axion number created in this way is only a small fraction of the total. A reasonable estimate of the axion CDM density in the string scenario is therefore

$$\boxed{\frac{\Omega_{\rm a}}{0.2} \simeq 10 \left(\frac{f_{\rm a}}{10^{12}\,\mathrm{GeV}} \right)^{1.2}} \qquad \text{(with strings).} \tag{23.4}$$

In the string scenario, the density of axions is very inhomogeneous on the horizon scale at the time of their formation. In regions where the density contrast is of order $+1$, gravitational collapse of the axion CDM occurs, forming mini-clusters which might be relevant for the detection of axion CDM.

23.2 The LSP as a CDM candidate

The axion is a CDM candidate with or without supersymmetry. If there is supersymmetry, another candidate is the Lightest Supersymmetric Particle (LSP). We consider the LSP in this section, and then mention some other candidates. As described in Chapter 17, the LSP candidates are the lightest ordinary supersymmetric particle (LOSP), the gravitino and the axino.

23.2.1 Neutralino CDM

We begin by considering the LOSP, the lightest superpartner of a Standard Model particle. To be the CDM it has to be stable, which means that it has to be lighter than the gravitino. This requires $m_{3/2} \gtrsim 100\,\mathrm{GeV}$, which means that *the neutralino is not a CDM candidate if there is gauge-mediated supersymmetry breaking with* $m_{3/2} \ll 100\,\mathrm{GeV}$.

The LOSP has to be neutral to be the CDM, and it turns out that the sneutrino is not viable leaving only a neutralino (combination of gauginos and higgsinos).

[2] The oscillation is roughly a standing wave, because the axion field is roughly homogeneous on the scale of the horizon.

To accurately calculate the LSP abundance one should follow the gradual failure of thermal equilibrium using the Boltzmann equation (see for instance Refs. [6, 7]). Here we assume instant freeze-out which gives a reasonable approximation.

A neutralino is its own antiparticle which means that in thermal equilibrium it has the generalized blackbody distribution. When T falls below the mass m of the lightest neutralino, its number density n is given by Eq. (4.27) with $g_i = 2$.

The neutralino falls out of equilibrium at a 'freeze-out' temperature $T_{\rm F}$, given approximately by

$$n\langle\sigma(v)v\rangle \simeq \frac{T_{\rm F}^2}{M_{\rm Pl}}. \tag{23.5}$$

In this equation the left-hand side is the interaction rate per neutralino at freeze-out, and the right-hand side is an estimate of the Hubble parameter. On the left-hand side, σ is the annihilation cross-section, $v \sim T_{\rm F}/m$ is the relative speed of the colliding neutralinos and $\langle\rangle$ is the thermal average. The subscript F denotes the freeze-out epoch.

After freeze-out, n/s achieves a constant value, which through Eq. (4.37) determines the CDM density $\Omega_{\rm c}$. Taking freeze-out to occur instantly at a temperature given by Eq. (23.5), Eqs. (4.27) and (4.37) determine $\Omega_{\rm c}$ and $T_{\rm F}/m$ in terms of $\langle\sigma v\rangle$ and m. The dependence on the mass m of the neutralino is very weak, and for any m roughly of order 100 GeV one finds that the observed $\Omega_{\rm c}$ requires $T_{\rm F} \simeq m/30$ and

$$\langle\sigma v\rangle \sim \left(10^4\,{\rm GeV}\right)^{-2}. \tag{23.6}$$

This is a reasonable value, because the annihilation of a particle with mass 100 GeV by the exchange of particles with mass $M \sim 100$ GeV and couplings $\lambda \sim 10^{-2}$ can indeed give the right order of magnitude $\langle\sigma v\rangle \sim (\lambda/M)^2$.

What about precise calculations? The thermal average $\langle\sigma v\rangle$ depends on the soft supersymmetry breaking parameters of the MSSM, along with the value of the μ parameter and the masses and self-couplings of the Higgs fields. The dependence of $\langle\sigma v\rangle$ upon these parameters is complicated because the kinetic energy of the collision can be of the same order as other relevant dimensionful quantities (formed from masses and couplings). Exploration of the parameter space of the MSSM has identified a region in which it is possible for a neutralino to be the CDM, and in part of that region direct detection of the CDM will be possible.

All of this can work for any neutral particle with $\langle\sigma v\rangle \sim (10^4\,{\rm GeV})^{-2}$ and a mass not too many orders of magnitude away from 100 GeV. Such a particle is called a Weakly Interacting Massive Particle (WIMP). The neutralino is the best-motivated candidate but others have been suggested, not always relying on supersymmetry.

23.2.2 Gravitino CDM and unstable gravitinos

We saw in Section 17.5.3 that the gravitino mass might be anywhere in the range $10^{-3}\,\text{eV} \lesssim m_{3/2} \lesssim 10\,\text{TeV}$. The lowest few decades are disfavoured because they would require an implausibly low supersymmetry breaking scale, and the top three decades correspond to gravity- or anomaly-mediated supersymmetry breaking which might be regarded as less attractive than gauge-mediated breaking. A best guess for the gravitino mass might therefore be something like $1\,\text{eV} \lesssim m_{3/2} \lesssim 10\,\text{GeV}$.

The MSSM gives a gravitino decay rate of order $10^{-2} m_{3/2}^3/M_{\text{Pl}}^2$ in a wide range of parameter space, and the decay channels can also be calculated. If $m_{3/2} \lesssim 10\,\text{MeV}$, the gravitino lifetime is longer than the age of the Universe, which means that the gravitino is a CDM candidate. If the gravitino has a bigger mass, it is stable provided that it is the LSP (assuming as usual that R-parity is conserved), making it again a CDM candidate. To be the LSP, the gravitino has to be lighter than the LOSP whose mass is roughly of order $100\,\text{GeV}$, and lighter than the axino.

Gravitinos are created by collisions and decays of the MSSM particles in thermal equilibrium, with an abundance n/s that can be estimated rather firmly. Assuming that this is the only source of gravitinos and that there is no subsequent entropy production, the following constraints can be derived.

Mass $m_{3/2} \lesssim 10\,\text{MeV}$

In the regime $m_{3/2} \lesssim 1\,\text{keV}$ the gravitinos are in thermal equilibrium before the electroweak phase transition. The present number of gravitinos per photon is therefore of order $1/g_* \sim 10^{-2}$, where g_* is the effective number of species when the gravitinos decouple. In the regime $m_{3/2} \lesssim 10\,\text{eV}$ the gravitinos have no significant effect. The regime $10\,\text{eV} \lesssim m_{3/2} \lesssim 1\,\text{keV}$ is forbidden [8], because the gravitinos constitute hot dark matter which is incompatible with the observation of the Lyman-α forest mentioned in Section 9.6.1.

In the regime $1\,\text{keV} \lesssim m_{3/2} \lesssim 1\,\text{MeV}$ the gravitinos come from the decay or collisions of thermalized particles, without themselves coming into thermal equilibrium. This gives the following bounds [9]:[3]

$$T_{\text{R}} < 10^6\,\text{GeV}\,\frac{m_{3/2}}{10\,\text{MeV}} \qquad (100\,\text{keV} \lesssim m_{3/2} \lesssim 10\,\text{MeV}), \tag{23.7}$$

$$T_{\text{R}} < 10^2 \text{ to } 10^3\,\text{GeV} \qquad (1\,\text{keV} \lesssim m_{3/2} \lesssim 100\,\text{keV}). \tag{23.8}$$

In the first regime, the gravitino is CDM if the bound is saturated, provided that the production mechanism we are considering is the only one. In the second regime, saturation of this bound would make the gravitino warm dark matter.

[3] In the first regime collisions are the dominant mechanism, while in the second regime decays dominate.

Mass $m_{3/2} \gtrsim 10\,\mathrm{MeV}$

In this regime, the constraint on T_{R} depends on whether the gravitino is stable (hence a CDM candidate). Stability presumably requires that the gravitino is the LSP. In that case, the LOSP can decay into the gravitino. The LOSP might be charged, and if so its production and decay at a collider might be detectable [10].

Since the LOSP can hardly be much heavier than $100\,\mathrm{GeV}$, a stable gravitino must have mass $m \lesssim 100\,\mathrm{GeV}$. In this regime Eq. (23.7) still applies, requiring $T_{\mathrm{R}} \lesssim 10^{10}\,\mathrm{GeV}$.

Finally, consider an unstable gravitino with $m_{3/2} > 10\,\mathrm{MeV}$. For $10\,\mathrm{MeV} < m_{3/2} < 100\,\mathrm{GeV}$ it decays too late to affect BBN, but it can distort the CMB blackbody spectrum [11] or generate a diffuse γ ray (or neutrino) background [12]. For $100\,\mathrm{GeV} < m_{3/2} < 10\,\mathrm{TeV}$ it is constrained by BBN [13]. For $m_{3/2} > 10\,\mathrm{TeV}$ it decays well before BBN, but its decay can overproduce the LOSP [12]. Going from the smallest to the biggest mass, these considerations require $T_{\mathrm{R}} \lesssim 10^{6}$ to $10^{10}\,\mathrm{GeV}$.

Comments on the constraints

The constraints assume continuous radiation domination after reheating. If instead there is a second reheating, entropy production will dilute the pre-existing gravitinos. Then the bound on T_{R} may refer only to the second reheat temperature.

The constraints also assume that gravitino production comes only from scattering and decay processes involving MSSM particles in thermal equilibrium. There could instead be significant gravitino creation from the decay of particles that do not belong to the MSSM, such as the particle which causes the first or second reheating. In that case the bound on T_{R} is tightened. Also, the random motion of the gravitino may be bigger or smaller, which would narrow or widen the range of $m_{3/2}$ corresponding to CDM as opposed to warm/hot dark matter.

Finally, we note that if the gravitino is the LSP with a mass $\gtrsim 10\,\mathrm{GeV}$, then the gravitinos produced by the decay of the LOSP will have very roughly the required Ω_{c}, without any other source of gravitino production. This is because (i) there will be about one gravitino per LOSP and (ii) the LOSP has about the mass and abundance required for it to be a CDM candidate if it were stable.

23.2.3 Axino dark matter

The case of axino dark matter [14] is quite similar to that of the gravitino. The main difference is that its interactions are stronger, by roughly a factor $M_{\mathrm{Pl}}/f_{\mathrm{a}}$. If we are taking on board the axino we should think about constraints that may come from saxion production. The saxion is similar to a modulus, but its decay rate is bigger and its displacement from the vacuum is more likely to be f_{a} than M_{Pl}. In a significant region of parameter space, the saxion gives a second reheating and significant entropy generating, without actually causing a problem. For simplicity we ignore the saxion in the following brief account.

As noted in Section 17.6, the axino mass is typically of order the gravitino mass, though it can also be much lower. For the axino to be stable its mass must anyway not exceed that of the gravitino.

The parameter space for axino dark matter is explored in, for instance, Ref. [15]. They take the saxion and axino masses to be of order $m_{3/2}$, and assume that the axino comes just from scattering processes in thermalized radiation with no entropy dilution factor. To avoid the axion being the CDM, it is reasonable to consider a low value $f_a \sim 10^{10}\,\mathrm{GeV}$ for the PQ scale. With these assumptions, the gravitino bound on T_R prevents enough axino production if $m_{\tilde{a}} \lesssim 1\,\mathrm{MeV}$. For a higher axino mass, the axino is the CDM if $T_R \sim 10^3\,\mathrm{GeV}$ or so. Taking into account axino production from the decay of the LOSP, the reheat temperature can be lower [14]. Like the gravitino, the axino can hardly be the CDM if its mass is much bigger than $100\,\mathrm{GeV}$, since it would then be heavier than the LOSP.

These considerations assume a fairly standard cosmology. One can also consider the case of thermal inflation, caused by a flat PQ symmetry breaking potential. This has been explored on the assumption of gravity mediated supersymmetry breaking [16]. As explained in Section 17.6 there are now at least two axinos, the lightest of which is the CDM candidate. This scenario is quite similar to the standard one, except that the reheat temperature is necessarily fairly small.

Perhaps the most important conclusion to draw from all this is that *neither the gravitino nor the axino can easily account for the CDM if the gravitino mass is less than* $100\,\mathrm{keV}$ *or so.*

23.3 Supermassive CDM candidates

Apart from the LSP, the possibility most usually considered for the CDM is that it consists of particles with negligible interaction, so that they never thermalize. The axion is such a particle, but it is exceptionally light. Usually one goes to the other extreme to suppose that the mass is many orders of magnitude bigger than $100\,\mathrm{GeV}$.

Because of its large mass the supermassive particle is unlikely to thermalize because that would require a huge cross-section. Instead one supposes that it is produced in some other way. It may be produced by the decay of some particle. It could also be created from the vacuum fluctuation as described in Section 24.8. The huge entropy production coming from thermal inflation might allow the supermassive particle to be the inflaton [17].

We have been assuming that the mass of the CDM particle is the same in the early

Universe as it is now. Instead, it might be much bigger now, due to its coupling to some field with a large vev. This idea has been proposed in connection with both instant preheating [18] and thermal inflation [19]. Before thermal inflation the CDM particle is in thermal equilibrium, and afterwards it acquires a large mass to become the CDM.

23.4 Primordial black holes

As reviewed for instance in Ref. [20], primordial black holes may be created in the early Universe. Depending on their mass, they may be CDM candidates or else they may decay.

Black holes form when a comoving region enters the horizon, if the average curvature perturbation ζ within that region is of order $+1$. The mass of the black hole is of order the total energy within a horizon volume, which is related to the temperature at formation by

$$\frac{M}{10^4 M_\odot} \sim \left(\frac{1\,\text{MeV}}{T_\text{form}}\right)^2 \sim \frac{t}{1\,\text{s}}. \tag{23.9}$$

When a given scale k enters the horizon, the mean square curvature perturbation within the observable Universe is of order $\mathcal{P}_\zeta(k)$. If $\mathcal{P}_\zeta(k) \sim 1$ a large fraction of the energy density collapses to black holes. If instead $\mathcal{P}_\zeta(k) \ll 1$, black hole formation occurs at those rare positions where $\zeta \sim 1$. The abundance of black holes is then given by the Press–Schechter formula, and is completely negligible if ${\mathcal{P}_\zeta}^{1/2}(k)$ is much less than 10^{-1}.

To obtain observational constraints on the black hole abundance, we have to remember that a black hole with mass M evaporates by emitting radiation at temperature $T = M_\text{Pl}^2/M$. It evaporates completely at the time[4]

$$\begin{aligned} t_\text{evap} &\sim 1\left(\frac{M}{10^{-24}M_\odot}\right)^3 \text{s} \sim 10\left(\frac{M}{10^{-18}M_\odot}\right)^3 \text{Gyr} \\ &\sim \left(\frac{10^{11}\,\text{GeV}}{T_\text{form}}\right)^6 \text{s} \sim \left(\frac{10^{8}\,\text{GeV}}{T_\text{form}}\right)^6 \text{Gyr}. \end{aligned} \tag{23.10}$$

Black holes that have yet to evaporate are CDM candidates. For black holes that have already evaporated, there are constraints from BBN, the diffuse γ ray background, and the blackbody spectrum of the CMB. Because the black hole abundance is so sensitive to $\mathcal{P}_\zeta$, the upper bound on $\mathcal{P}_\zeta$ is around 10^{-2} to 10^{-3} over the whole range of scales [20].

Starting with $\mathcal{P}_\zeta(k_0) = (5 \times 10^{-5})^2$ at the pivot scale, what does it take to

[4] There are strong arguments that the evaporation is complete, as reviewed for instance in Ref. [21].

get $\mathcal{P}_\zeta(k) \sim 10^{-2}$ at the scale leaving the horizon at the end of inflation? The relation between the scales is $\ln(k_{\rm end}/k_0) = N$ with N typically around 50. With a constant spectral index we would need $n = 1.3$ which is ruled out. With instead a constant running n' we need $n' = 10^{-2}$, which is just about allowed by observation at the time of writing.

One can consider a more complicated behaviour of the spectrum, such as a break in the spectral index caused by a change in the mechanism of producing the curvature perturbation [22]. Of course, if the primordial curvature perturbation is created only after inflation ends, it will exist only on scales that are outside the horizon at the time of its creation. Then $\mathcal{P}_\zeta$ will have to increase faster with k than the estimates of the previous paragraph, in order to achieve a value of order 10^{-2} on the shortest scale for which ζ exists.

23.5 Dark energy

In this final section we present two ideas about the origin of the dark energy. Both of them assume Einstein gravity, ascribing the dark energy to the scalar field potential. For comprehensive reviews of dark energy at the time of writing, see Ref. [24, 25].

23.5.1 Cosmological constant

To obtain a strictly constant dark energy density, one has to suppose that every scalar field has by now successfully relaxed to its vev. Then the fields are at a minimum of the potential. The potential has some value ρ_Λ, which discounting the quantum fluctuation is the cosmological constant.[5] Accepting that this is the case, one would like to understand the small value of the cosmological constant. This has so far proved impossible within the context of quantum field theory. What is very surprising is that a zero value would have been just as hard to understand [23]. It is surprising because zero values in field theory can usually be explained by a symmetry.

In the absence of a preferred zero value, one might expect $\rho_\Lambda^{1/4}$ to be of order the cutoff for field theory, say $M_{\rm Pl}$. In particular, radiative corrections to ρ_Λ will be of this order, as in Eq. (15.39). Supersymmetry controls those corrections, but still one expects a value of order the typical mass splitting within supermultiplets, which is at least of order $100\,\mathrm{GeV}$.

In view of all this, anthropic considerations are widely invoked. It is supposed that there is a multiverse, containing universes with practically all possible values of the cosmological constant. (This supposition receives some support from string

[5] As described in Section 15.4, we absorb the vacuum fluctuation into the potential.

theory.) Then we can see why $\rho_\Lambda^{1/4}$ is far below $M_{\rm Pl}$ or even $100\,\mathrm{GeV}$: we cannot live in a universe with such a value because it contains no galaxies.

Going further, one might try to estimate the value of ρ_Λ, assuming that all values within the multiverse are found with equal probability within the small range allowed by the anthropic requirement. Then the probability $\mathbf{P}(\rho_\Lambda)\,d\rho_\Lambda$ of us observing ρ_Λ in a given range may be something like

$$\mathbf{P}(\rho_\Lambda)\,d\rho_\Lambda = \frac{N(\rho_\Lambda)\,d\rho_\Lambda}{\int_0^\infty N(\rho_\Lambda)\,d\rho_\Lambda}, \tag{23.11}$$

where $N(\rho_\Lambda)$ is the number of galaxies that ever form in a given comoving region. The function $\mathbf{P}$ defined by this expression can be estimated using the Press–Schechter approximation.

Sophisticated calculations along these general lines have been done, in some cases allowing other quantities to vary such as $\mathcal{P}_\zeta$ or the axion CDM density. As one would expect, the predicted value of ρ_Λ is never orders of magnitude away from the observed one, but the precise value is very dependent on the assumptions.

23.5.2 Dark energy from a rolling scalar field

Instead of assuming that the potential is non-vanishing at the minimum, one can assume that it vanishes but that the dark energy density is nevertheless non-zero because some field some ϕ has not yet arrived at the minimum. This scenario is poorly motivated because a zero value for the potential at its minimum is no easier to understand than a non-zero value. Nevertheless it has been widely considered because it leads to many interesting possibilities for both theory and observation.

The scenario was first suggested under the assumption that ϕ has the canonical kinetic term. The resulting dark energy was later called **quintessence**. Quintessence is like single-field slow-roll inflation, with the difference that the slope of the potential has to be extraordinarily small. More recently a generic kinetic term has been allowed in which case the dark energy is called K-essence, and the even more general inflation scenarios mentioned in Section 25.7 have been considered (not necessarily invoking Einstein gravity). It has also been suggested that ϕ is the field actually responsible for inflation in which case inflation is called quintessential inflation.

Dark energy from a rolling scalar field has at least two observable consequences. One of them is time dependence of the dark energy density, corresponding to equation of state $w_\Lambda \equiv P_\Lambda/\rho_\Lambda > -1$. The other is the ability of the dark energy to be perturbed, affecting thereby the evolution of the matter density perturbation. Neither effect has been observed at the time of writing. Assuming that w_Λ is more or less constant, observation currently requires $|1 + w_\Lambda| \lesssim 10^{-1}$.

Exercises

23.1 Verify Eq. (23.6) by using the following approximate method. First use Eq. (23.5) to estimate $n_{\rm F}/s_{\rm F}$, and put the result into Eq. (4.37) assuming $m \sim T_{\rm F}$.

23.2 Verify Eq. (23.9), using Appendix B and working back from the epoch of matter–radiation equality.

23.3 Verify Eq. (23.10), using the black hole temperature $M_{\rm Pl}^2/M$ and assuming blackbody radiation.

23.4 Taking 1 GeV as the energy unit, calculate upper bounds on the first and second derivatives of the potential needed for quintessence. Work out the relation between $w_\Lambda \equiv P_\Lambda/\rho_\Lambda$ and V'. Hence calculate V' in units of GeV^3 in the case $w_\Lambda = -0.9$.

References

[1] J. E. Kim. Light pseudoscalars, particle physics and cosmology. *Physics Reports*, **150** (1987) 1.

[2] E. W. Kolb and M. S. Turner. *The Early Universe* (California: Addison-Wesley, 1990).

[3] P. Sikivie. Axion cosmology. *Lect. Notes Phys.*, **741** (2008) 19.

[4] D. H. Lyth and E. D. Stewart. Axions and inflation: string formation during inflation. *Phys. Rev. D*, **46** (1992) 532.

[5] M. Kawasaki, T. Moroi and T. Yanagida. Can decaying particles raise the upper bound on the Peccei–Quinn scale? *Phys. Lett. B*, **383** (1996) 313.

[6] V. Mukhanov. *Physical Foundations of Cosmology* (Cambridge: Cambridge University Press, 2006).

[7] S. Weinberg. *The Quantum Theory of Fields, Volume III* (Cambridge: Cambridge University Press, 1996).

[8] M. Viel, J. Lesgourgues, M. G. Haehnelt, S. Matarrese and A. Riotto. Constraining warm dark matter candidates including sterile neutrinos and light gravitinos with WMAP and the Lyman-alpha forest. *Phys. Rev. D*, **71** (2005) 063534.

[9] T. Moroi. Effects of the gravitino on the inflationary universe. arXiv:hep-ph/9503210.

[10] A. De Roeck, J. R. Ellis, F. Gianotti, F. Moortgat, K. A. Olive and L. Pape. Supersymmetric benchmarks with non-universal scalar masses or gravitino dark matter. *Eur. Phys. J.*, C **49** (2007) 1041.

[11] M. Kawasaki, K. Kohri and T. Moroi. Big-bang nucleosynthesis and hadronic decay of long-lived massive particles. *Phys. Rev. D*, **71** (2005) 083502.

[12] K. Kohri, T. Moroi and A. Yotsuyanagi. Big-bang nucleosynthesis with unstable gravitino and upper bound on the reheating temperature. *Phys. Rev. D*, **73** (2006) 123511.

[13] T. Kanzaki, M. Kawasaki, K. Kohri and T. Moroi. Cosmological constraints on neutrino injection. *Phys. Rev. D*, **76** (2007) 105017.

[14] L. Covi, J. E. Kim and L. Roszkowski. Axinos as cold dark matter. *Phys. Rev. Lett.*, **82** (1999) 4180.

[15] M. Kawasaki, K. Nakayama and M. Senami. Cosmological implications of supersymmetric axion models. *JCAP*, **0803** (2008) 009.

[16] E. J. Chun, H. B. Kim, K. Kohri and D. H. Lyth. Flaxino dark matter and stau decay. *JHEP*, **0803** (2008) 061.

[17] A. R. Liddle, C. Pahud and L. A. Ureña-López. Triple unification of inflation, dark matter, and dark energy using a single field. *Phys. Rev. D*, **77** (2008) 121301(R).

[18] G. N. Felder, L. Kofman and A. D. Linde. Instant preheating. *Phys. Rev. D*, **59** (1999) 123523.

[19] L. Hui and E. D. Stewart. Superheavy dark matter from thermal inflation. *Phys. Rev. D*, **60** (1999) 023518.

[20] B. J. Carr. Primordial black holes: do they exist and are they useful? arXiv:astro-ph/0511743.

[21] G. T. Horowitz. Black holes, entropy, and information. arXiv:0708.3680 [gr-qc].

[22] K. Kohri, D. H. Lyth and A. Melchiorri. Black hole formation and slow-roll inflation. *JCAP*, **0804** (2008) 038.

[23] S. Weinberg. The cosmological constant problem. *Rev. Mod. Phys.*, **61** (1989) 1.

[24] E. J. Copeland, M. Sami and S. Tsujikawa. Dynamics of dark energy. *Int. J. Mod. Phys. D*, **15** (2006) 1753.

[25] P. J. E. Peebles and B. Ratra. The cosmological constant and dark energy. *Rev. Mod. Phys.*, **75** (2003) 559.

24

Generating field perturbations at horizon exit

In Part II of this book, we described the perturbations at the 'primordial' epoch $T \sim 1\,\mathrm{MeV}$, when they first become directly accessible to observation. At that stage the dominant perturbation is the curvature perturbation ζ. There may also be a tensor perturbation h_{ij}, and isocurvature perturbations S_i (with $i =$ c, b or ν).

Now we broaden the definition of 'primordial' and ask about perturbations at earlier times. In this chapter we see how the perturbations of light scalar fields are generated during inflation.

The idea is quite simple. Let us focus on some comoving wavenumber k. Well before horizon exit the curvature of spacetime is negligible and we are dealing with flat spacetime field theory where the particle concept should be valid. The particle number is assumed to be negligible, so that each field is in the vacuum state.

The crucial point now is that the vacuum fluctuation of a light field will 'freeze in' at horizon exit, to become a classical perturbation. The process was understood in the 1970s, before inflation was proposed as a physical reality, and has nothing to do with gravity. It occurs simply because the timescale a/k of the would-be vacuum fluctuation becomes bigger than the Hubble time H^{-1}. We will see in some detail how this intuitive picture can emerge from a proper calculation. We will also look at the vacuum fluctuation of the tensor perturbation, which freezes at horizon exit in precisely the same manner.

24.1 Quantum theory of a massless free scalar field during inflation

In this section we consider the first-order perturbation of a light scalar field during almost-exponential inflation. The metric perturbation is ignored so that the field perturbation lives in unperturbed spacetime. No theory of gravity is needed, just the time dependence of the scale factor.

24.1.1 Field equation

The classical field equation for a set of fields $\phi_n(\mathbf{x},t)$ living in unperturbed spacetime is

$$\ddot{\phi}_n + 3H\dot{\phi}_n - a^{-2}\nabla^2\phi_n + V_n = 0, \tag{24.1}$$

where $V_n \equiv \partial V/\partial\phi_n$. The first-order perturbation satisfies

$$\ddot{\delta\phi}_{\mathbf{k}n} + 3H\dot{\delta\phi}_{\mathbf{k}n} + \left(\frac{k}{a}\right)^2 \delta\phi_{\mathbf{k}n} + V_{nm}\delta\phi_{\mathbf{k}m} = 0, \tag{24.2}$$

where $V_{nm} \equiv \partial^2 V/\partial\phi_n\partial\phi_m$ and a summation over m is understood.

We are concerned with the few Hubble times either side of horizon exit. We are going to see that heavy fields do not acquire a significant perturbation, and therefore keep only light fields. We drop V_{nm} on the grounds that it is much less than H^2 because the field is light, and much less than $(k/a)^2$ because that quantity will not be very much less than H^2 until well after horizon exit.

We therefore deal with a massless free field, whose classical field equation is

$$\ddot{\delta\phi}_{\mathbf{k}} + 3H\dot{\delta\phi}_{\mathbf{k}} + \left(\frac{k}{a}\right)^2 \delta\phi_{\mathbf{k}} = 0. \tag{24.3}$$

As we are concerned only with the few Hubble times around horizon exit, we set H equal to its value H_k at horizon exit. Often the slight scale-dependence of H_k can be ignored, and we can set H_k equal to a constant that we denote by H_*.

We can convert this equation into that of a harmonic oscillator with a time-dependent frequency by going to conformal time, and working with $\varphi \equiv a\delta\phi$. With constant H, conformal time can be defined by $\eta = -1/aH$, so that $\eta = 0$ corresponds to the infinite future. We use this in the present case, with $H = H_k$. Then Eq. (24.3) becomes

$$\boxed{\frac{d^2\varphi_{\mathbf{k}}(\eta)}{d\eta^2} + \omega_k^2(\eta)\varphi_{\mathbf{k}}(\eta) = 0}, \tag{24.4}$$

with

$$\omega_k^2 = k^2 - \frac{2}{\eta^2} \equiv k^2 - 2(aH_k)^2. \tag{24.5}$$

24.1.2 Quantization

Well before horizon exit the harmonic oscillator has constant angular frequency k. Then we can work in a small spacetime region with size $\Delta\eta \sim \Delta x$ satisfying $k^{-1} \ll \Delta\eta \ll (aH_k)^{-1}$. Such a region is big enough to contain many oscillations and yet is small enough that spacetime curvature is negligible. We will assume that flat spacetime field theory applies in such a region, by virtue of the equivalence principle.

We made the same assumption at the beginning of Section 21.2. As in that case, we can set $a = 1$ over the short time interval $\Delta\eta$, so that k during this interval can be identified with the physical wavenumber. As in Eq. (15.36) we define the mode function φ_k in terms of the Fourier component $\varphi_{\mathbf{k}}$ by

$$\hat{\varphi}_{\mathbf{k}}(\eta) = \varphi_k(\eta)\hat{a}(\mathbf{k}) + \varphi_k^*(\eta)\hat{a}^\dagger(-\mathbf{k}). \tag{24.6}$$

It satisfies Eq. (24.4) and we need the solution with the initial condition

$$\varphi_k(\eta) = \frac{1}{\sqrt{2k}} e^{-ik\eta}. \tag{24.7}$$

It is

$$\varphi_k(\eta) = \frac{e^{-ik\eta}}{\sqrt{2k}} \frac{(k\eta - i)}{k\eta}. \tag{24.8}$$

Well after horizon exit it approaches

$$\varphi_k(\eta) = -\frac{i}{\sqrt{2k}} \frac{1}{k\eta}. \tag{24.9}$$

We assume that the state corresponds initially to the vacuum (no ϕ particles with momentum $\mathbf{k}$). As we saw in Section 21.2, this must be the case if a reasonable amount of inflation has occurred before the epoch of horizon exit. Let us suppose that the Fourier components $\varphi_{\mathbf{k}}$ are measured at some particular instant. Since the vacuum state corresponds to the ground state of an harmonic oscillator, the outcome of the measurement has a gaussian distribution for the real and imaginary part of each component, with no correlation except the reality condition. We therefore deal with a gaussian random field whose ensemble average may be identified with the vacuum expectation value. The mean $\langle\varphi_{\mathbf{k}}\rangle$ vanishes, and the spectrum is defined by

$$(2\pi)^3\langle\varphi_{\mathbf{k}}\varphi_{\mathbf{k}'}\rangle = \frac{2\pi^2}{k^3}\mathcal{P}_\varphi(k)\delta^3(\mathbf{k}+\mathbf{k}'). \tag{24.10}$$

Inserting Eq. (24.6) and the commutator (15.38), and remembering that the expectation value refers to the vacuum state, we find

$$\mathcal{P}_\varphi(k,\eta) = \frac{k^3}{2\pi^2}|\varphi_k(\eta)|^2. \tag{24.11}$$

Dividing by a^2 and evaluating it a few Hubble times after horizon exit we find a time-independent quantity

$$\boxed{\mathcal{P}_{\delta\phi}(k) = \left(\frac{H_k}{2\pi}\right)^2}. \tag{24.12}$$

This was first derived in Ref. [1].

24.1.3 Random walk

From the spectrum (24.12), let us derive the mean-square perturbation. We take H_k to be a constant H_*, and deal only with the classical perturbation, so that the spectrum is set equal to zero on scales $k \gtrsim aH_*$. Equivalently, we smooth ϕ on a fixed (not comoving) scale H_*^{-1}. We work in a comoving box with size aL. Before the box leaves the horizon, it is smaller than the smoothing scale and ϕ has no classical perturbation. Well after horizon exit, the mean-square field perturbation is

$$\langle \delta\phi^2(\mathbf{x},t)\rangle = \left(\frac{H_*}{2\pi}\right)^2 \int_{L^{-1}}^{aH_*} \frac{dk}{k} \tag{24.13}$$

$$= \left(\frac{H_*}{2\pi}\right)^2 \ln(LH_*a) = \left(\frac{H_*}{2\pi}\right)^2 N(t), \tag{24.14}$$

where $N(t)$ is the number of e-folds of inflation after the box leaves the horizon.

This expression has a very simple interpretation. At each position, the smoothed field is doing a random walk, with step length $H_*/2\pi$ and one step per Hubble time. The random walk occurs because comoving scales are continually leaving the horizon, generating random contributions to the classical field perturbation.

24.1.4 Perturbing the action

For future reference, we mention an alternative method of calculation. To derive Eqs. (24.6) and (24.8), we calculated the vacuum fluctuation well before the horizon exit, and then propagated the result forward in time using the perturbed classical field equation. The latter was obtained from the full field equation, which came ultimately from the full action (13.55). Schematically,

action $\to$ field equation $\to$ perturbed field equation.

Instead, we can perturb the full action:

action $\to$ perturbed action $\to$ perturbed field equation.

It is easy to check that this gives the same result. Up to a 4-divergence, the action for $\varphi \equiv a\delta\phi$ is [3]

$$S = \frac{1}{2}\int \left[(\partial_\eta\varphi)^2 - (\partial_i\varphi)^2 + 2\frac{\varphi^2}{\eta^2}\right] d\eta\, d^3x. \tag{24.15}$$

This is the same as Eqs. (13.22) and (15.55), with φ instead of ϕ and η instead of t, with $-4/\eta^2$ as the time-dependent mass-squared.

To arrive at the quantum theory corresponding to this action, we we redefine the Lagrangian, Lagrangian density and Hamiltonian operators, so that Eqs. (13.1), (13.21) and (13.30) become

$$S = \int L d\eta = \int \mathcal{L} d^3x d\eta, \tag{24.16}$$

$$\Pi \equiv \frac{\partial \mathcal{L}}{\partial \dot{\varphi}}, \tag{24.17}$$

$$H \equiv \int d^3x \Pi \dot{\varphi} - L, \tag{24.18}$$

with the dot standing for $\partial/\partial\eta$. Using a finite box we see that H is the sum of harmonic oscillator contributions at early times, and after imposing the appropriate commutation relation and going to the continuum limit we arrive again at the commutation relation (15.38). The spectrum $\mathcal{P}_\varphi$ then follows as before.

In this context, the direct derivation of the perturbed field equation is easier than the direct derivation of the perturbed action (which we did not consider). For higher correlators the reverse turns out to be true, and practically all calculations start with the perturbed action as described in Section 24.4 (see Ref. [2] for an exception).

24.2 Quantum to classical transition

Well before horizon exit, $\varphi_{\mathbf{k}}$ is the vacuum fluctuation of a free field in flat spacetime and is an essentially quantum object. If $\varphi_{\mathbf{k}}$ were somehow measured, a repetition of the measurement would give a quite different value unless it were made almost immediately, after a time $\Delta t \ll a/k$. In the Heisenberg picture that we are adopting, this is because the operator $\varphi_{\mathbf{k}}$ changes with time while the state vector after the measurement is time independent. The state vector is an eigenvector of $\varphi_{\mathbf{k}}$ just after the measurement, but after a short time $\varphi_{\mathbf{k}}$ becomes a completely different linear combination of the (time-independent) creation and annihilation operators. The state vector is then no longer an eigenvector of $\varphi_{\mathbf{k}}$.

The situation is quite different well after horizon exit, because the mode function (24.8) then becomes purely imaginary and Eq. (24.6) reads

$$(2\pi)^3 \hat{\varphi}_{\mathbf{k}}(t) = \varphi_k(t) \left[\hat{a}(\mathbf{k}) - \hat{a}^\dagger(-\mathbf{k}) \right]. \tag{24.19}$$

The time dependence of $\hat{\varphi}_{\mathbf{k}}(t)$ is now trivial, and the state continues to be an eigenvector. We conclude that if $\varphi_{\mathbf{k}}$ is measured at some instant well after horizon exit, it will continue to have a definite value. In this sense, $\varphi_{\mathbf{k}}$ can be regarded as a classical object [4].

There remains the usual interpretation problem, arising whenever we talk about measurement in quantum theory. We would like to know why eigenstates of the field operator have been selected by Nature, since after all they provide only one from an infinity of possible ways of expanding the state vector. And we would like to be assured that one of the eigenstates was indeed selected in the early Universe, since otherwise it would seem that the pattern of say the cosmic microwave anisotropy may not have existed before it was observed. (Remember that we are going to use one or more of the light field perturbations to generate the primordial curvature perturbation.) Following the attitude taken to the interpretation of quantum mechanics in the laboratory situation, the first of these problems might be addressed by considering the interaction of the field with its environment to invoke what is called decoherence [5]. The second problem though is more controversial, being a version of the Schrödinger Cat problem.

24.3 Linear corrections to the calculation

The calculation so far ignores the potential V, and it also ignores the perturbation of the spacetime metric. Let us see how to handle these effects to first order in the field perturbations.

24.3.1 Effect of the potential

The effect of the potential corresponds to the last term of Eq. (24.2). To first order in $\delta\phi$ it gives the linear equation (24.2). Let us focus on one light field $\phi \equiv \phi_n$, and suppose that the last term of Eq. (24.2) involves only $\delta\phi$:

$$\ddot{\delta\phi} + 3H\dot{\delta\phi} + \left(\frac{k}{a}\right)^2 \delta\phi + m^2(\phi)\delta\phi = 0. \tag{24.20}$$

Here $m^2(\phi) \equiv \partial^2 V/\partial\phi^2$ is the effective mass-squared of the perturbation $\delta\phi$. It will be equal to the mass-squared of ϕ if the latter is a practically free field corresponding to $V = \frac{1}{2}m^2\phi^2$, and more generally it may have a practically constant value m_k^2 during the few Hubble times at horizon exit. Let us work out that case.

Writing Eq. (24.20) in terms of $\varphi = a\delta\phi$ we have

$$\frac{d^2\varphi(\mathbf{k},\eta)}{d\eta^2} + \left[(am_k)^2 + k^2 - \frac{2}{\eta^2}\right]\varphi(\mathbf{k},\eta) = 0, \tag{24.21}$$

with $aH_k = -1/\eta$. The field operator can be written in the form (24.6), and the mode function φ_k satisfies Eq. (24.21). The solution with the initial condition (24.7) is

$$\varphi(k,\eta) = e^{i(\nu+\frac{1}{2})\pi/2}\sqrt{\frac{\pi}{4k}}\,\sqrt{k\eta}\,H_\nu^{(1)}(k\eta), \tag{24.22}$$

where $H_\nu^{(1)}$ is the Hankel function and

$$\nu = \sqrt{\frac{9}{4} - \frac{m_k^2}{H_k^2}} \simeq \frac{3}{2} - \frac{m_k^2}{3H_k^2}. \tag{24.23}$$

The quantum-to-classical transition occurs provided that ν is real, which corresponds to $m_k^2 < (9/4)H_k^2$.

Well after horizon exit,

$$\varphi(k,\eta) = e^{i(\nu-\frac{1}{2})\pi/2} \frac{2^\nu \Gamma(\nu)}{2^{3/2}\Gamma(\frac{3}{2})} \frac{1}{\sqrt{2k}} (k\eta)^{\frac{1}{2}-\nu}. \tag{24.24}$$

Calculating the spectrum from Eq. (24.11) gives

$$\boxed{\mathcal{P}_{\delta\phi}(k,\eta) \simeq \left(\frac{H_k}{2\pi}\right)^2 \left(\frac{k}{aH_k}\right)^{2m_k^2/3H_k^2}}. \tag{24.25}$$

This equation is valid for as long as $m^2(\phi)$ and H have negligible variation. More generally, the time dependence of $m^2(\phi)$ should be calculated by solving the unperturbed field equation (13.57) with a specified initial condition. Then the mode function can be determined as before, giving $\mathcal{P}_\phi$.

With the spectrum in place we can calculate the mean-square field $\langle \delta\phi^2(\mathbf{x}) \rangle$. As in the massless case (Eq. (24.13)), we ignore the time dependence of H, and smooth ϕ with a fixed smoothing scale H^{-1} so that all modes have the classical perturbation.[1] Also, we assume that ϕ is a free field so that m^2 has no time dependence. Then

$$\langle \delta\phi^2(\mathbf{x}) \rangle = \left(\frac{H}{2\pi}\right)^2 \int_{L^{-1}}^{aH} \frac{dk}{k} \left(\frac{k}{aH}\right)^{2m^2/3H^2} \tag{24.26}$$

$$\rightarrow \left(\frac{H}{2\pi}\right)^2 \int_0^{aH} \frac{dk}{k} \left(\frac{k}{aH}\right)^{2m^2/3H^2} \tag{24.27}$$

$$= \frac{3H^4}{8\pi^2 m^2}. \tag{24.28}$$

In contrast with the massless case, the mean square tends to a finite limit as the box size increases.

The unperturbed value of ϕ rolls down towards the vev $\phi = 0$. Given enough time it will be negligible, so that the mean-square field is $\overline{\phi^2} = \langle \delta\phi^2 \rangle$. We will see in Section 24.6 how to extend this calculation to the case of a general potential $V(\phi)$.

A similar treatment works if Eq. (24.2) involves two or more fields. If V_{nm} is

[1] To be precise, we smooth over a fixed scale a few times bigger than H^{-1} to achieve this.

time independent it can be diagonalized to recover the case of constant m^2. Otherwise, the system (13.58) of unperturbed equations has to be solved with specified initial conditions to give $V_{nm}(t)$, and then Eq. (24.2) has to be solved to give the mode functions, giving the spectra $\mathcal{P}_{\phi_n}(k, \eta)$.

24.3.2 *Effect of the metric perturbation*

The scalar field equation in a generic spacetime is Eq. (13.41), with $\Box$ given by Eq. (3.22). To define the metric perturbation and the field perturbation, we have to specify a slicing and threading of spacetime. Once this has been done, we can define the first-order perturbation $\delta\Box$. Then, to first order, Eq. (24.1) becomes

$$\delta\ddot{\phi}_n + 3H\delta\dot{\phi}_n + \left(\frac{k}{a}\right)^2 \delta\phi_n + V_{nm}\delta\phi_m = (\delta\Box)\phi_n(t). \quad (24.29)$$

The left-hand side is the one in Eq. (24.2). The right-hand side is the effect of the metric perturbation at first order. From Eq. (3.22), we see that $\delta\Box$ to first order is a function times $\partial/\partial t$. It therefore vanishes in the limit $\dot{\phi}_n \to 0$. As we are dealing with light fields, $\dot{\phi}_n$ will be small and we may hope that the effect of the metric perturbation is negligible.

From now on, we take field perturbations to be defined on the flat slicing, defined as the one with $D = 0$ where D is defined by Eq. (8.8). (This generalizes to all scales the super-horizon definition of 'flat slicing' given after Eq. (5.14).) To quantify the effect of the metric perturbation, we assume that the light fields are responsible for inflation with Einstein gravity. In other words, we assume that there is (single- or multi-field) slow-roll inflation. Then the metric perturbation is generated by the field perturbations themselves, and we are dealing with what is called back-reaction.

For single-component inflation the field equation becomes

$$\delta\ddot{\phi} + 3H\delta\dot{\phi} + \left(\frac{k}{a}\right)^2 \delta\phi + m^2(\phi)\delta\phi = M_{\text{Pl}}^{-2}\left[\frac{1}{a^3}\frac{d}{dt}\left(\frac{a^3\dot{\phi}^2}{H}\right)\right]\delta\phi. \quad (24.30)$$

Since $\delta\phi$ is a light field, the square bracket is much less than H^2, which means that back-reaction is small until well after horizon exit. The multi-field generalization of Eq. (24.2) is

$$\delta\ddot{\phi}_n + 3H\delta\dot{\phi}_n + \left(\frac{k}{a}\right)^2 \delta\phi_n + V_{nm}\delta\phi_m = M_{\text{Pl}}^{-2}\left[\frac{1}{a^3}\frac{d}{dt}\left(\frac{a^3\dot{\phi}_n\dot{\phi}_m}{H}\right)\right]\delta\phi_m. \quad (24.31)$$

Ignoring back-reaction, we find that each light field perturbation a few Hubble times after horizon exit is of the same order $H_k/2\pi$. We see again that back-reaction is small until well after horizon exit.

We will see in Section 25.6 how Eq. (24.30) can be used to calculate the curvature perturbation generated by $\delta\phi$. For that purpose it is usually written for $\varphi \equiv a\delta\phi$ using conformal time;

$$\boxed{\frac{\partial^2\varphi(\mathbf{k},\eta)}{\partial\eta^2} + \left(k^2 - \frac{1}{z}\frac{d^2z}{d\eta^2}\right)\varphi(\mathbf{k},\eta) = 0}\,, \tag{24.32}$$

where z is given in terms of the unperturbed field by $z \equiv a\dot{\phi}/H$. It is then called the Mukhanov–Sasaki equation [6, 7]. Using the field equation (13.57) one can show that

$$\frac{1}{z}\frac{d^2z}{d\eta^2} = 2a^2H^2\left(1 + \epsilon_{\rm H} - \frac{3}{2}\eta_{\rm H} + \frac{1}{2}\eta_{\rm H}^2 - \frac{1}{2}\epsilon_{\rm H}\eta_{\rm H} + \frac{1}{2H}\frac{d\epsilon_{\rm H}}{dt} - \frac{1}{2H}\frac{d\eta_{\rm H}}{dt}\right), \tag{24.33}$$

with $\epsilon_{\rm H}$ and $\eta_{\rm H}$ defined by Eqs. (18.24) and (18.25).

24.4 Non-gaussianity of the field perturbations

We so far included only linear corrections to the basic calculation described in Section 24.1. At this level the Fourier components with different $\mathbf{k}$ remain uncoupled, so that the field perturbations remain gaussian. Now we see how to calculate the non-gaussianity coming from non-linear corrections. To do this, one considers the perturbation in the action.

The calculation is done using the conformal time Hamiltonian, Eq. (24.18). In linear theory the action is given by Eq. (24.15), corresponding to a Hamiltonian $\hat{H}_0$ that we do not need. We are going to consider effects which give a small additional contribution $\hat{H}_I$ to the Hamiltonian. The strategy is to treat $\hat{H}_I$ as a small perturbation, analogous to an interaction term in ordinary quantum field theory. As in that case, the perturbation is handled in basically the same way as a perturbation in ordinary quantum mechanics, but the field theory case is much more complicated.

24.4.1 Interaction Picture

The calculation is done in what is called the Interaction Picture. This is defined by the requirement that $\hat{\varphi}$ has the same time dependence as in the absence of $\hat{H}_I$:

$$\dot{\hat{\varphi}} = i[\hat{H}_0, \hat{\varphi}], \tag{24.34}$$

where the dot denotes $\partial/\partial\eta$. Its mode expansion is therefore given by Eqs. (24.6) and (24.8). The time dependence of the corresponding Heisenberg Picture operator is given by

$$\hat{\varphi}_{\rm hp} = \hat{U}^{-1}\hat{\varphi}\hat{U}, \tag{24.35}$$

with

$$\dot{\hat{U}} = -i\hat{H}_I\hat{U}. \tag{24.36}$$

Well before horizon exit, $\hat{H}_I$ is negligible and the Interaction Picture coincides with the Heisenberg Picture. In the Heisenberg Picture the time-independent vacuum state is the one annihilated by the annihilation operators $\hat{a}_{\mathbf{k}}$. To first order in $\hat{H}_I$, the solution of Eq. (24.36) is

$$\hat{U}(\eta) = I - i\int_{-\infty}^{\eta} \hat{H}_I(\eta), \tag{24.37}$$

and Eq. (24.35) becomes

$$\boxed{\hat{\varphi}_{\rm hp}(\eta) = \hat{\varphi}(\eta) + i\int_{-\infty}^{\eta} d\eta'[\hat{H}_I(\eta'), \hat{\varphi}(\eta)]}\,. \tag{24.38}$$

We now see how this expression allows one to work out the correlators of φ in the Heisenberg Picture, with the vacuum initial condition imposed.

24.4.2 Non-gaussianity from the self-interaction of the field

Ignoring at first the metric perturbation, let us consider the effect of keeping the quadratic term in the perturbed field equation:

$$\ddot{\delta\phi} + 3H_*\dot{\delta\phi} - a^{-2}\nabla^2\delta\phi + \frac{1}{2}V_*'''(\delta\phi)^2 = 0, \tag{24.39}$$

where the dot denotes $\partial/\partial t$, and the star denotes the value during inflation which is taken to be constant. The corresponding term in the Hamiltonian given by Eq. (24.18) is

$$\hat{H}_I(\eta) = \frac{1}{2}V_*'''\int d^3x\hat{\varphi}^3(\mathbf{x},\eta). \tag{24.40}$$

Putting this into Eq. (24.38) we can calculate the bispectrum to first order in H_I. The result is [8]:

$$B_{\delta\phi} = \frac{H_*^2V_*'''}{4\prod_i k_i^3}I_3, \tag{24.41}$$

where

$$I_3 \equiv {\rm Re}\left(-ie^{-k_t\eta}\right)\left[\prod_k(1+ik_i\eta)\right]\int_{-\infty}^{\eta}\frac{d\eta'}{\eta'^4}e^{ik_t\eta'}\prod_i\left(1-ik_i\eta'\right), \tag{24.42}$$

where $k_t \equiv k_1 + k_2 + k_3$.

The integral diverges. This kind of divergence is often encountered in field theory, and to make sense of it one adops what is called the $i\epsilon$ prescription [9]. According to this prescription, η' is given a small positive imaginary part, which is taken to zero at the end of the calculation. For simplicity, we take all k_i to have roughly the same order of magnitude.[2] Then, a few Hubble times after horizon exit, we find $I_3 = \Delta N \sum k_i^3$, where $\Delta N(\eta)$ is the number of Hubble times. The contribution to the bispectrum of $\delta\phi$ from the self interaction is therefore

$$\boxed{B_{\delta\phi}^{\text{self}} = -\frac{H_*^2 V_*'''}{12 \prod_i k_i^3} \sum k_i^3 \Delta N(\eta)}\,. \tag{24.43}$$

In contrast with the two-point correlator specified by the spectrum, the three-point correlator is not slowly varying on the Hubble timescale.

The first-order calculation goes through in a similar way for terms in Eq. (24.20) of higher order in $\delta\phi$. A cubic term would give the four-point correlator and so on for higher powers.

24.4.3 Non-gaussianity from the gravitational interaction

Now we see how the calculation may be extended, to include in $\hat{H}_I$ the effect of the metric perturbation that is generated by the field perturbation (back-reaction). As in the previous section we work to first order in $\hat{H}_I$.

The crucial step is to identify the degrees of freedom. For this purpose one writes the metric perturbation in what is called the ADM form [10]:[3]

$$\boxed{ds^2 = -N^2 dt^2 + g_{ij}\left(dx^i + N^i dt\right)\left(dx^j + N^j dt\right)}\,. \tag{24.44}$$

When this metric is put into the Einstein–Hilbert action (13.60), the action principle gives four constraint equations obtained by varying N and N^i. Those quantities can therefore be eliminated as degrees of freedom, leaving only the spatial metric g_{ij}.

As in Eq. (5.6) the spatial metric will be of the form

$$\boxed{g_{ij} = e^{2\psi(\mathbf{x},t)}\left[I e^{2h(\mathbf{x},t)}\right]_{ij}}\,, \tag{24.45}$$

with $h_{ii} = 0$ (traceless). It can be shown [10] that it is possible to choose $\partial_i h_{ij} = 0$. Then the perturbation h_{ij} is the tensor perturbation (or to be more precise a non-linear generalization of the tensor perturbation as defined in linear theory).

[2] The opposite case corresponds to the 'squeezed' triangle considered in Section 25.4.3.
[3] This is the standard notation; N has nothing to do with a number of e-folds.

As the cosmic fluid is supposed to consist of a single scalar field with a more or less flat spectrum, the separate Universe assumption will apply after smoothing on a comoving scale which has left the horizon. In this regime, the metric components defined by Eq. (24.44) will have negligible spatial gradients. The threads of the ADM gauge will therefore become comoving in the super-horizon regime, so that Eq. (24.45) coincides with Eqs. (5.12) and (5.13).

Let us choose the flat slicing $\psi = 0$, so that the degrees of freedom are $\delta\phi$ and h_{ij}. As we are working to first order in H_I and are interested only in $\delta\phi$, we can consistently drop terms in H_I that involve the tensor perturbation. For single-field slow-roll inflation, the bispectrum of $\delta\phi$ on the flat slicing is found to be [11]

$$B_{\delta\phi}^{\rm grav} = -\frac{1}{8}H_*^4\left(\frac{V'}{V}\right)g(k_1,k_2,k_3), \quad (24.46)$$

$$g(k_1,k_2,k_3)\prod_i k_i^3 \equiv \frac{1}{2}\sum_{i\neq j} k_i k_j^2 + \frac{4}{k_{\rm t}}\sum_{i<j} k_i^2 k_j^2 - \frac{1}{2}\sum_i k_i^3. \quad (24.47)$$

We see that $B_{\delta\phi}^{\rm grav}$ vanishes in the limit $V'/V \to 0$, where H is constant. We will see that the physical effect of $B_{\delta\phi}^{\rm grav}$ is unlikely to be measurable.

Going to multi-field inflation [11] using the basis $\{\phi, \sigma_i\}$ defined in Section 26.2, the three-point correlator of the adiabatic perturbation $\delta\phi$ with itself is as in the single-field case. The other correlators vanish except for the correlator $\langle\delta\phi\delta\sigma_i\delta\sigma_i\rangle$ (no sum), which is the same as the correlator of $\delta\phi$ with itself.

The trispectrum of the field perturbations has also been calculated [12]. It has the form

$$T(\mathbf{k}_1,\mathbf{k}_2,\mathbf{k}_3,\mathbf{k}_4) = H_*^6 \sum_{\rm perms} \delta_{\alpha\beta}\delta_{\gamma\delta}\, g(\mathbf{k}_1,\mathbf{k}_2,\mathbf{k}_3,\mathbf{k}_4), \quad (24.48)$$

where the four subscripts label the four fields, and g is a rather complicated function of the momenta which is of order 1 for generic momenta. The physical effect of the trispectrum is almost certainly unmeasurable.

24.5 Higher orders of perturbation theory

It is easy to write down the generalization of Eq. (24.38) to all orders in the perturbation H_I;

$$\hat\varphi_{\rm hp}(\eta) = \hat\varphi(\eta) + \sum_{N=1}^{\infty}\int_{-\infty}^{\eta} d\eta_1 \int_{-\infty}^{\eta_1} d\eta_2 \ldots \int_{-\infty}^{\eta_{N-1}} d\eta_N \left\langle [\hat H_I(\eta_N), \ldots [\hat H_I(\eta_2), [\hat H_I(\eta_1), \hat\varphi(\eta)]]] \right\rangle . \quad (24.49)$$

Each term can be evaluated in principle, by inserting the interaction Hamiltonian given by Eq. (24.40), or by a similar expression including higher powers of φ.

The formalism that we have described applies quite generally, allowing one to calculate the expectation value of any observable in a state corresponding to a time-dependent background. If the state is $|\text{in}\rangle$ and the operator corresponding to the observable is $\hat{A}$, one is calculating $\langle\text{in}|\hat{A}|\text{in}\rangle$. The formalism is similar to that used to calculate transition amplitudes. There, one evaluates $\langle\text{in}|\hat{S}|\text{out}\rangle$, where the state vectors correspond to the initial and final states and S is the scattering operator (S-matrix). Again, the formalism applies quite generally, for example to transition amplitudes encountered in condensed matter physics as well as in particle physics.

The first formalism is often called the in–in formalism and one may call the second the in–out formalism. In both cases, the expressions obtained become complicated as one goes to higher orders of perturbation theory, being a sum of many terms. Each term is conveniently represented by a Feynman graph [13] which in general contains loops. Each loop corresponds to an integration over a 'momentum'. In the case of a transition amplitude it really is a momentum, but in the case of a correlator it is just a wave-vector.[4]

There are big differences between the use of the in–in formalism for cosmological perturbations, and the in–out formalism to describe particle physics transition amplitudes. In the particle physics case a great variety of accurate data has to be confronted, and the calculations have been carried through to high order involving Feynman diagrams with many loops. In the cosmology case the data are and will always be very limited, and are unlikely to ever require anything beyond a tree-level calculation. This last statement does however require a qualification as we now describe.

The qualification comes when we consider the divergences encountered when we evaluate integrals over loop momenta. They can occur in the limit of infinite momentum (ultra-violet) or in the limit of zero momentum (infra-red). In the particle physics cases they are well understood. The ultra-violet divergences are handled by renormalization. The infra-red divergences have to do with the fact that a finite momentum can be carried by indefinitely many particles as in photon bremsstrahlung, and are also well understood.

In the cosmology cases little work has been done, but there will presumably be divergences of the same type as in particle physics. In addition though, one expects additional infra-red divergences, coming from the fact that cosmological perturbations of indefinitely long wavelength may be significant. A concrete example is

[4] In the case at hand we are evaluating the expectation value of a product of field perturbations, so that the wave-vector actually can be identified with the momentum of the corresponding particles, but that is exceptional. In particular it is no longer the case if we replace the field perturbation by the curvature perturbation. We see in Section 25.5.2 that the in–in formalism applies equally to that case.

given in Section 25.5. To avoid these infra-red divergences, one should work in a box not too much bigger than the observable Universe.

24.6 Stochastic field evolution

The approach we have developed so far works only if the perturbation $\delta\phi(\mathbf{x},t)$ is small, so that it makes sense to perturb Eq. (24.1) as in Eqs. (24.20) and (24.39). Now we describe an approximate approach that works even for large perturbations after smoothing on a fixed scale somewhat bigger than H_*^{-1}.[5] It is generally known as the stochastic approach, and it applies to any light field. As the field need not be the inflaton we will call it σ instead of ϕ.

24.6.1 Fokker–Planck equations

A useful account of the stochastic approach may be found in Ref. [14]. It starts with the observation that Fourier modes of the light field will be continually crossing the smoothing scale to become classical perturbations. We saw in Section 24.1.3 that this causes the field at each point to perform a random walk, superimposed on its slow-roll evolution. To calculate this process we make the decomposition

$$\sigma(\mathbf{x},t) = \bar\sigma(\mathbf{x},t) + \frac{1}{(2\pi)^3}\int d^3k\,\theta(k-\epsilon a H_*)\left[\hat a_{\mathbf{k}}\sigma_k e^{i\mathbf{k}\cdot\mathbf{x}} + \text{h.c.}\right], \qquad (24.50)$$

$$\sigma_k = \frac{i}{\sqrt{2k}}\frac{H_*}{k}. \qquad (24.51)$$

(The super-horizon mode function σ_k is equal to φ_k/a, where φ_k is as given by Eq. (24.9).) Here θ is the step function, and ϵ is a constant somewhat less than 1. With this choice of ϵ, $\bar\sigma$ contains only super-horizon modes and can be regarded as a classical quantity. Note that $\bar\sigma$ is smoothed over a fixed physical scale, not over a comoving scale.

Differentiating Eq. (24.50) we find

$$\boxed{\dot{\bar\sigma}(\mathbf{x},t) = -\frac{1}{3H_*}V'(\bar\sigma) + f(\mathbf{x},t)}, \qquad (24.52)$$

with

$$f(\mathbf{x},t) = \frac{a(t)H_*^2}{(2\pi)^3}\int d^3k\,\delta(k-\epsilon a H_*)\left[\hat a_{\mathbf{k}}\sigma_k e^{i\mathbf{k}\cdot\mathbf{x}} + \text{h.c.}\right]. \qquad (24.53)$$

The first term of Eq. (24.52) accounts for the classical motion. The term f is a

[5] We take the inflationary Hubble parameter to be constant, though the approach goes through with slowly varying H.

stochastic noise term caused by the quantum fluctuation. Its two-point correlator is

$$\boxed{\langle f(\mathbf{x},t)f(\mathbf{x},t')\rangle = \frac{H_*^3}{4\pi^2}\delta(t-t')}\,, \tag{24.54}$$

so that Eq. (24.52) is a Langevin equation. (Note that we considering stochastic behaviour as a function of time at fixed position.)

To estimate the relative importance of the quantum and classical contributions in Eq. (24.52), we may proceed as follows. During a Hubble time, σ receives a quantum kick $\Delta\sigma_{\rm q} = H_*/2\pi$, and changes classically by an amount $\Delta\sigma_{\rm c} = \dot\sigma/H_*$. We assume that the random motion is small:

$$\frac{\Delta_{\rm q}}{\Delta_{\rm c}} = \frac{H_*}{2\pi}\frac{H_*}{|\dot\sigma|} = \frac{3H_*^3}{2\pi|V'|} \ll 1. \tag{24.55}$$

The function $F(\sigma)$, such that $Fd\sigma$ is the probability of finding σ in a given interval, satisfies the Fokker–Planck equation which according to standard methods is

$$\boxed{\frac{\partial F}{\partial t} = -\frac{V'}{3H_*}\frac{\partial F}{\partial\sigma} + \frac{H_*^3}{8\pi^2}\frac{\partial^2 F}{\partial\sigma^2}}\,. \tag{24.56}$$

24.6.2 Late-time probability distribution

Using the Fokker–Planck equation, one can calculate a probability distribution for σ, which will hold within a comoving region if there has been enough inflation and if the variation of H really is sufficiently slow. One finds

$$\boxed{F = A\exp\left(-\frac{8\pi^2}{3H_*^4}V(\sigma)\right)}\,, \tag{24.57}$$

where A is a normalization constant making $\int_0^\infty F d\sigma = 1$. For a quadratic potential, this reduces to Eq. (24.28). Otherwise, it represents a non-gaussian probability distribution.

To see what it implies, suppose first that the potential has a minimum at $\sigma = 0$, and increases monotonically away from the minimum. In that case the late-time probability distribution is centred on the origin, and σ is practically confined to a region around the origin given by[6]

$$V(\sigma) \lesssim \frac{3H_*^4}{8\pi^2}. \tag{24.58}$$

Now suppose instead that $V(\sigma)$ spontaneously breaks a discrete symmetry, given by Eq. (14.28) with σ instead of ϕ. If H_*^4 is much less than the height V_0 of the

[6] We set $V = 0$ at the minimum, because adding a constant to V has no effect on the probability distribution at fixed H_*.

potential, the probability distribution will peak at the vevs $\sigma = \pm f$, the width of each peak being given by Eq. (24.57). If instead $H_*^4 \gtrsim V_0$, then σ can climb over the barrier between the vacua. A similar analysis holds if σ is the radial part of a complex field that spontaneously breaks a $U(1)$ symmetry. As we saw in Section 23.1, that case is relevant for the potential that spontaneous breaks PQ symmetry.

Finally, suppose that σ is a pseudo-Nambu–Goldstone boson (PNGB) with the periodic potential (14.34). If $H_* \ll \Lambda$, then σ is again confined to one of the minima. If instead $H_* \gg \Lambda$, the late-time probability distribution is practically flat, and any value of the field is equally likely. In the latter case it is reasonable to expect the flat distribution to hold even if the variation of H isn't particularly slow. We invoke the flat distribution for the axion field in Sections 16.6 and 21.4.1.

All of this depends on the existence of one or more light fields that have nothing to do with the dynamics of inflation. It also depends on inflation lasting for exponentially many Hubble times before the observable Universe leaves the horizon. Given these things we have found something remarkable. We have found that the average values of the light fields within the observable Universe may depend strongly on its location. Different locations may correspond to different cosmologies, and if the parameters in the field-theory Lagrangian depend on some light fields (moduli) different locations may even correspond to different fundamental laws of physics.

In the context of the landscape, we are saying that the different universes within a multiverse need not correspond to isolated minima of the potential. Instead, the region of field space around a minimum may contain a continuous family of universes, in addition to the discontinuous members of the multiverse that can be reached only by going over a potential barrier.

24.7 Primordial tensor perturbation

The calculation that we described for a light scalar field can work for a bosonic field of any spin. All that we require, for the conversion of the quantum fluctuation to a classical perturbation, is that the mode function after horizon exit becomes a fixed combination of the creation and annihilation operators, up to an overall function.

The case of a vector field has been widely studied, as a way of generating a primordial magnetic field. Such a field might be required by observation, and might give an observable signal in the CMB anisotropy. More generally, the possible contribution of a vector field to the curvature perturbation has been considered. Such a contribution will usually exhibit statistical anisotropy, which may be detected in the future [15].

We shall not study primordial vector fields. Instead, we focus on the primordial tensor perturbation, whose existence at some level is an automatic consequence of inflation with Einstein gravity.

24.7.1 The spectrum

At linear order, we defined the tensor perturbation h_{ij} at the beginning of Section 12.5. Putting the perturbed metric (12.31) into the Einstein–Hilbert action (13.60), the action of h_{ij} to second order may be calculated. For each of the components $h_{+,\times}$ it is found to be the same as the action of a free scalar field given by

$$\phi_{+,\times} \equiv \sqrt{2} M_{\rm Pl}\, h_{+,\times}. \tag{24.59}$$

According to the calculation we presented earlier, the vacuum fluctuation of $\phi_{+,\times}$ becomes classical a few Hubble times after horizon exit, and has the spectrum $(H_k/2\pi)^2$. We saw after Eq. (5.7) that h_{ij} will remain constant until the approach of horizon entry, when it becomes accessible to observation on cosmological scales. The spectrum defined in Eq. (12.33) is given by

$$\mathcal{P}_h(k) = \frac{8}{M_{\rm Pl}^2} \left(\frac{H_k}{2\pi} \right)^2 . \tag{24.60}$$

Combined with the observed value of the spectrum of the curvature perturbation, this gives for the tensor fraction (Eq. (12.34))

$$\boxed{r = \left(\frac{\rho^{1/4}}{3.3 \times 10^{16}\,{\rm GeV}} \right)^4} . \tag{24.61}$$

The tensor spectral index is

$$n_{\rm T} \equiv \frac{d \ln \mathcal{P}_h(k)}{d \ln k} = -2\epsilon_{\rm H}, \tag{24.62}$$

where $\epsilon_{\rm H} \equiv -\dot{H}/H^2$.

24.7.2 Non-gaussianity of the tensor perturbation

What about the non-gaussianity of the primordial tensor perturbation? It corresponds to coupling between different Fourier modes, which means that we are no longer dealing with linearized gravity. One may therefore anticipate that the non-gaussianity will be completely unobservable, and this is confirmed by a calculation of the three-point correlator [16]. We will describe the main points.

Once the slicing has been chosen, Eq. (24.45) defines the non-linear tensor perturbation h_{ij}. In general h_{ij} depends on the choice of slicing. In particular, its value for the flat slicing is different from its value for the uniform-density slicing.

It is found though that after horizon exit during inflation h_{ij} becomes the same for these slicings, and time independent up to second order. The time independence concurs with our non-perturbative proof of the time independence in Section 5.4.1, which also shows that the time independence persists until the approach horizon entry.

Doing the calculation on the uniform-density slicing, it is found that

$$\langle h^{s_1}_{\mathbf{k}_1} h^{s_2}_{\mathbf{k}_2} h^{s_3}_{\mathbf{k}_3} \rangle = -\frac{4\pi^3}{\prod(k_i^3)} \delta_{\mathbf{k}_1+\mathbf{k}_2+\mathbf{k}_3} \frac{H_*^4}{M_{\rm Pl}^4} \epsilon^{s_1}_{ii'} \epsilon^{s_2}_{jj'} \epsilon^{s_3}_{kk'} t_{ijk} t_{i'j'k'} I, \quad (24.63)$$

$$I \equiv = -k_{\rm t} + \frac{\sum_{i>j} k_i k_j}{k_{\rm t}} + \frac{k_1 k_2 k_3}{k_{\rm t}^2} \quad (24.64)$$

$$t_{ijk} = k_2^i \delta_{jk} + k_3^j \delta_{ik} + k_1^k \delta_{ij}. \quad (24.65)$$

Following the discussion of Section 6.4, we see that the amount of non-gaussianity characterized by say the skewness is tiny, of order $\mathcal{P}_h^{1/2} = 5 \times 10^{-5} r^{1/2}$.

24.7.3 Cosmic gravitational wave background

The gravitational waves that affect the CMB anisotropy have enormous wavelengths, but within the inflationary cosmology the primordial tensor perturbation will create gravitational waves also on smaller scales, right down to the scale leaving the horizon at the end of inflation. On such scales, the wavelength may be within the range of direct detection.

A gravitational wave entering the horizon at temperature T will now have frequency

$$f(T) = \frac{T}{10^7\,\text{GeV}}. \quad (24.66)$$

Let $\Omega_h(k) d\ln k$ be the energy density of the gravitational waves with angular frequency k in a given interval. After horizon entry, the energy density of the wave falls like $1/a^4$, just like any other form of radiation. This gives (taking $\mathcal{P}_h$ to be flat)

$$\Omega_h(k) = \times 10^{-15} r, \quad (24.67)$$

where r is the tensor fraction. If instead horizon entry is before reheating Ω_h is reduced.

Although $\Omega_h(k)$ is very small, it might be observable in the future if r isn't too small. If so, observation might determine the reheat temperature $T_{\rm R}$.

That isn't the end of the story of the cosmic gravitational wave background. According to Eq. (12.32), any time-dependent anisotropic stress Σ^T_{ij} is a source of gravitational waves. This is a universally valid formula for linearized gravity, applying to the usual astrophysical (point-like) sources of gravitational waves, but also to the early Universe. Calculations show that a contribution to $\Omega_h(k)$ exceeding the primordial contribution can be generated during the phase transition and preheating events that we describe in Chapter 21. In contrast with the primordial contribution, these early-Universe contributions to $\Omega_h(k)$ will be broadly peaked around a characteristic frequency. Fortunately, it is found that in some cases proposed gravitational detectors will probe frequencies within the peak. Further study of these matters may begin with Refs. [17, 18].

24.8 Particle production from a perturbation created during inflation

A light field perturbation σ, created during inflation from the vacuum fluctuation, will oscillate after horizon entry corresponding to the presence of σ particles. The σ particles will be present with all momenta k, up to the scale $k_{\rm end}$ leaving the horizon at the end of inflation. This production of σ particles from the vacuum fluctuation is irrelevant if σ has interactions of ordinary strength (gauge or Yukawa), because the σ particles will then sooner or later come into thermal equilibrium making their initial number irrelevant. Even if the σ particles do not thermalize, their production from the vacuum fluctuation has to compete with their production by the decay and collision processes of other particles. Taking the σ particles to have interactions of only gravitational strength, a rough guess [19] suggests that their production from the vacuum fluctuation may be the dominant effect for a high inflation scale.

We now discuss particle creation from the vacuum fluctuation, closely following Ref. [19]. We set $m_k^2 = m_*^2$ and $H_k = H_*$ (independent of the epoch of horizon exit). For a light field, corresponding to $m_*^2 \ll H_*^2$, each mode has occupation number $|\beta_k|^2 \gg 1$, in accordance with the fact that the perturbation $\delta\phi_{\mathbf{k}}$ can be regarded as classical. For a heavy field, corresponding to $m_*^2 \gg H_*^2$, the perturbation never becomes classical and no particles are created. In the intermediate case where m_*^2/H_*^2 is fairly close to 1, we can therefore expect $|\beta_k|^2 \sim 1$. As we saw in Section 20.2, this case may be regarded as generic in the context of supergravity. Then the number density is

$$n = \frac{1}{2\pi^2}\frac{1}{a^3}\int_0^{a_{\rm end}H_*} k^2 dk \sim 10^{-2} H_*^3 \left(\frac{a_{\rm end}}{a}\right)^3 . \tag{24.68}$$

This estimate is very rough, because it ignores the displacement of the unperturbed field from the vacuum, which will be negligible only for a heavy field.

The case of a spin-$1/2$ species is more interesting. It can be created from the vacuum fluctuation in the same way, and there is no question of a classical field displacement. The occupation number is at most 1 in accordance with the Pauli exclusion principle. It vanishes in the limit $m_*/H_* \to \infty$ for the same reason as in the scalar case. It also vanishes in the limit $m_*/H_* \to 0$, because the action is then conformally invariant and we can make a conformal transformation to flat spacetime. In the intermediate case $m_* \sim H_*$, the abundance will again be given by Eq. (24.68). We conclude that Eq. (24.68) is a reasonable estimate for the abundance of a spin-$1/2$ particle created from the vacuum fluctuation, but that creation is only possible if the effective mass during inflation is of order H_*.

In contrast with the scalar case, the mass of order H_* for a spin-$1/2$ field is not generic in supergravity, but might be expected in some cases. (The situation is similar to that of spin-$1/2$ particles in the MSSM with gravity-mediated supersymmetry breaking.) Let us suppose anyway that for some reason $m_* \sim H_*$. If the created particle has sufficiently-weak interaction with other particles, the number density (24.68) will apply until it decays. If it is stable it becomes a CDM candidate. The true mass m of the particle is presumably less than the mass m_*, and using Eqs. (4.37) and (24.68) we find

$$\frac{\Omega_c}{0.2} \sim \left(\frac{H_*}{10^{14}\,\mathrm{GeV}}\right)^2 \left(\frac{m}{10^4\,\mathrm{GeV}}\right) \left(\frac{T_R}{10^9\,\mathrm{GeV}}\right). \tag{24.69}$$

The right-hand side should be close to 1 if the particle is to be the CDM. This is the possible mechanism for the production of supermassive particles that we mentioned in Section 23.3.[7]

In the context of supersymmetry, gravitinos are also created from the vacuum fluctuation. The predicted number density is very model dependent, and has to be considered along with the number density of gravitinos coming from the decay and interaction of other particles.

Exercises

24.1 Prove Eq. (24.36) by differentiating Eq. (24.35) and comparing the result with $\dot{\hat{\phi}}_{\rm hp} = i[\hat{H}_{\rm hp}, \hat{\phi}_{\rm hp}]$.

24.2 Work out the function defined by Eq. (24.47) in the squeezed configuration, and in the equilateral configuration.

24.3 For a potential with monotonically increasing V' starting from a minimum at $\phi = 0$, verify that the width $\phi_{\rm max}$ of the late-time probability distribution (given by Eq. (24.58)) lies well within the regime given by Eq. (24.55) provided that $\phi_{\rm max} < H$.

[7] With this mechanism in mind, it is sometimes stated that the mass of the supermassive particle is likely to be of order $m \simeq 10^{12}$ GeV. This estimate comes from setting $m = H_*$ in Eq. (24.69) and then assuming gravitational-strength decay for the inflaton to get T_R. The estimate is without foundation, because there is no reason to expect the true mass to satisfy $m \simeq H_*$, and because there are other mechanisms for creating supermassive particles.

24.4 Show that a gravitational wave entering the horizon at temperature T will now have frequency $(T/10^7\,\mathrm{GeV})$ Hz. (Use Appendix B, working out first the result if horizon entry is just before matter–radiation equality.)

References

[1] T. S. Bunch and P. C. W. Davies. Quantum field theory in de Sitter space: renormalization by point splitting. *Proc. R. Soc. Lond. A*, **360** (1978) 117.

[2] D. Seery, K. A. Malik and D. H. Lyth. Non-gaussianity of inflationary field perturbations from the field equation. *JCAP*, **0803** (2008) 014.

[3] V. Mukhanov. *Physical Foundations of Cosmology* (Cambridge: Cambridge University Press, 2006).

[4] A. A. Starobinsky. Dynamics of phase transition in the new inflationary universe scenario and generation of perturbations. *Phys. Lett. B*, **117** (1982) 175.

[5] D. Polarski and A. A. Starobinsky. Semiclassicality and decoherence of cosmological perturbations. *Class. Quant. Grav.*, **13** (1996) 377.

[6] V. F. Mukhanov. Gravitational instability of the universe filled with a scalar field. *JETP Lett.*, **41** (1985) 493.

[7] M. Sasaki. Large scale quantum fluctuations in the inflationary universe. *Prog. Theor. Phys.*, **76** (1986) 1036.

[8] T. Falk, R. Rangarajan and M. Srednicki. The angular dependence of the three point correlation function of the cosmic microwave background radiation as predicted by inflationary cosmologies. *Astrophys. J.*, **403** (1993) L1.

[9] S. Weinberg. *The Quantum Theory of Fields, Volume I* (Cambridge: Cambridge University Press, 1995).

[10] R. Arnowitt, S. Deser and C. W. Misner. The dynamics of general relativity. In *Gravitation: An Introduction To Current Research*, Louis Witten ed. (Wiley, 1962). Available at arXiv:gr-qc/0405109.

[11] D. Seery and J. E. Lidsey. Primordial non-gaussianities from multiple-field inflation. *JCAP*, **0509** (2005) 011.

[12] D. Seery, J. E. Lidsey and M. S. Sloth. The inflationary trispectrum. *JCAP*, **0701** (2007) 027.

[13] S. Weinberg. Quantum contributions to cosmological correlations. *Phys. Rev. D*, **72** (2005) 043514.

[14] A. A. Starobinsky and J. Yokoyama. Equilibrium state of a self-interacting scalar field in the De Sitter background. *Phys. Rev. D*, **50** (1994) 6357.

[15] K. Dimopoulos, D. H. Lyth and Y. Rodriguez. Statistical anisotropy of the curvature perturbation from vector field perturbations. arXiv:0809.1055 [astro-ph].

[16] J. M. Maldacena. Non-Gaussian features of primordial fluctuations in single field inflationary models. *JHEP*, **0305** (2003) 013.

[17] R. Easther and E. A. Lim. Stochastic gravitational wave production after inflation. *JCAP*, **0604** (2006) 010.

[18] R. Easther, J. T. Giblin, E. A. Lim, W. I. Park and E. D. Stewart. Thermal inflation and the gravitational wave background. *JCAP*, **0805** (2008) 013.

[19] D. H. Lyth, D. Roberts and M. Smith. Cosmological consequences of particle creation during inflation. *Phys. Rev. D*, **57** (1998) 7120.

25

Generating ζ at horizon exit

We have seen how the vacuum fluctuation of each light scalar field is converted to a classical perturbation at the time of horizon exit. One or more of these perturbations should in turn generate the curvature perturbation ζ, that is probed by observation when cosmological scales begin to enter the horizon.

In this chapter we are going to consider the simplest scenario, where the curvature perturbation already achieves its final value a few Hubble times after horizon exit. In other words, we are going to consider the scenario in which ζ is constant during the entire era when the smoothing scale is outside the horizon. According to Section 5.4.2, this means that the locally defined pressure is a unique function of the locally defined energy density throughout that era. This is achieved in a single-field inflation model, if at each position the field value a few Hubble times after horizon exit determines the subsequent pressure and energy density. During inflation, this is the same thing as saying that the trajectory $\phi(\mathbf{x}, t)$ is independent of $\mathbf{x}$, up to a shift in t. In other words, it is the same as saying that the inflationary trajectory is an attractor. After inflation though, it would be possible for some other light field to play a significant role. We shall assume in this chapter that such is *not* the case.

We begin with the case of slow-roll inflation, which we refer to as the standard paradigm. We then see how to calculate a correction to the slow-roll prediction, and how to handle some cases where we have a canonically normalized field that doesn't satisfy the slow-roll approximation. In all of these cases, the required attractor behaviour is guaranteed by the theorem of Section 18.6.2. Then we consider a couple of quite different single-field models, known as K-inflation and warm inflation, which however still have the attractor behaviour in a suitably-defined 'slow-roll' regime.

Before starting these calculations, we want to consider the extent to which anthropic considerations may be relevant for the curvature perturbation.

25.1 Anthropic constraints on the curvature perturbation

The single-field prediction, which we are going to obtain for the spectrum and non-gaussianity of ζ, will depend on parameters of the Lagrangian that determine the inflationary potential. In a deeper theory, those parameters may become functions of the unperturbed values of one or more light fields, such as the moduli of string theory. Within the more general setup of Chapter 26, the prediction for the spectrum and non-gaussianity of ζ may depend directly on the unperturbed values of one or more light fields (which, in contrast with the single-field case, are not determined by the number of e-folds to the end of inflation).

The unperturbed value of a light field may be identified with its average within the comoving box surrounding the observable Universe that is used to calculate the field correlators, and we noticed at the end of Section 24.6.2 that such averages may depend on the location of the observable Universe within a much larger inflated patch. With that in mind we may ask what anthropic constraints can be placed on ζ. Such constraints might, for instance, justify a fine-tuning of parameters or unperturbed field values.

Anthropic considerations for the spectrum $\mathcal{P}_\zeta$ of the curvature perturbation can certainly play an important role. Viable galaxy formation cannot take place if the spectrum is not within at least a few orders of magnitude of the observed value. Assuming a flat a priori probability distribution for $\mathcal{P}_\zeta$, one can calculate an anthropic probability distribution along the lines that we indicated for the cosmological constant Λ. Taking a given form for the potential, one may also be able to feed in a prior probability distribution for $\mathcal{P}_\zeta$. Calculations along these lines seem to give sensible results.

The situation is quite different for the spectral index, the tensor fraction, and non-gaussianity of the curvature perturbation. For them, anthropic considerations do not seem to narrow the range that would be permitted by theory. The community was originally quite happy to consider a large spectral tilt, as well as strong running, when those were allowed by observation. The observed smallness of these quantities doesn't seem to be required by any anthropic consideration. The same is true of non-gaussianity. Even a completely non-gaussian curvature perturbation (say the square of a gaussian) doesn't seem to be forbidden by anthropic considerations, and was certainly considered as a possibility before observation ruled it out.

The bottom line then is very simple. Fine-tuning of the parameters to obtain the correct magnitude for the spectrum of the curvature perturbation might be justified on anthropic grounds. But anthropic considerations cannot justify fine-tuning to give the spectral tilt its observed small negative value, or to get the small (perhaps negligible) running and non-gaussianity that is demanded by observation.

25.2 Prediction of the standard paradigm for the spectrum

In this section we calculate the curvature perturbation ζ, that is generated by the inflaton field perturbation $\delta\phi$ in a slow-roll model.

The perturbation $\delta\phi(\mathbf{x},t)$ is defined on the flat slicing, whereas ζ is defined on the uniform energy density slicing. Since $\phi(\mathbf{x},t)$ is independent of position up to a time shift, the value of $\phi(\mathbf{x},t)$ at a given instant determines the energy density (equivalently the Hubble parameter) through Eq. (18.9). Therefore, ϕ is uniform on a slice of uniform density.

Repeating the argument leading to Eq. (5.26), with ϕ instead of ρ, we find that ζ is given to first order by

$$\zeta = -H\frac{\delta\phi}{\dot\phi}, \tag{25.1}$$

where $\delta\phi$ is defined on the flat slicing. We need only evaluate this a few Hubble times after horizon exit, because ζ is time independent from that point onwards. A few Hubble times after horizon exit, $\delta\phi(\mathbf{k})$ has the spectrum $(H_k/2\pi)^2$ and

$$\mathcal{P}_\zeta(k) = \frac{1}{4\pi^2}\left(\frac{H^2}{\dot\phi}\right)^2\Bigg|_k . \tag{25.2}$$

The subscript k indicates the epoch of horizon exit, and indeed the quantities H and $\dot\phi$ might as well be evaluated then because they are slowly varying on the Hubble timescale.

Formulas equivalent to Eqs. (25.1) and (25.2) were given in [1]. After Guth's inflation paper came out equivalent formulas were given by several groups using different approaches [2, 3, 4, 5]. Our treatment, using the separate universe assumption to justify the conservation of ζ, is basically that of Ref. [5].

Using the slow-roll approximation $\dot\phi = -V'/3H$ we obtain [6]

$$\zeta(\mathbf{x}) = \frac{1}{M_{\mathrm{Pl}}^2}\frac{V}{V'}\delta\phi(\mathbf{x}), \tag{25.3}$$

and

$$\boxed{\mathcal{P}_\zeta(k) = \frac{1}{24\pi^2 M_{\mathrm{Pl}}^4}\frac{V}{\epsilon}\bigg|_k} . \tag{25.4}$$

The right-hand side of Eq. (25.4) is evaluated when $k = aH$. Using Eq. (18.17) with $d\ln(aH) \simeq H dt$, this gives the spectral tilt [7]:

$$\boxed{n(k) - 1 = -6\epsilon + 2\eta}, \tag{25.5}$$

with again the right-hand side evaluated at $k = aH$. Differentiating again using

Eq. (18.18) we find the running $n' \equiv dn/d\ln k$

$$\boxed{n' = 16\epsilon\eta - 24\epsilon^2 - 2\xi}\,. \tag{25.6}$$

Invoking the observational results (6.68)–(6.70), we obtain constraints on the potential and its derivative, at the epoch of horizon exit for the pivot scale:

$$\left(\frac{V}{\epsilon}\right)^{1/4} = 6.6 \times 10^{16}\,\text{GeV}, \tag{25.7}$$

$$2\eta - 6\epsilon \simeq -0.04 \pm 0.03, \tag{25.8}$$

$$-0.07 \lesssim 2\xi \lesssim 0.01. \tag{25.9}$$

In writing the last line we discounted the possibility of an extremely accurate cancellation in the previous line.

Let us estimate the error in the predictions. The slow-roll approximation for $\dot{\phi}$ incurs a fractional error $\mathcal{O}(\epsilon, \eta)$ (we take this notation to mean 'at most of order $\max\{\epsilon, \eta\}$'), and we will see in Section 25.4.1 that the first-order relation (25.1) incurs a fractional error of the same order. Finally, we see from Eq. (24.30) that the neglect of back-reaction incurs again a fractional error of the same order. The fractional error in the prediction for $\mathcal{P}_\zeta$ is therefore $\mathcal{O}(\epsilon, \eta)$ which is guaranteed to be small.

Differentiating the error in $\ln \mathcal{P}_\zeta$ we find that the error in $n-1$ has a contribution $\sim \xi$. It is small if $|\xi| \ll |\eta|$. Similarly, using again Eq. (18.18), the error in n' has a contribution $\sim \sigma$, which will be small if $|\sigma| \ll |\xi|$. These conditions are usually satisfied by inflationary potentials.

Finally, we note that Eqs. (25.4) and (24.60) give the tensor fraction as [7]

$$\boxed{r = 16\epsilon}\,. \tag{25.10}$$

Combining this with Eq. (24.62) for the tensor spectral index (using the slow-roll result $\epsilon_{\rm H} = \epsilon$) we find $n_{\rm T} = -r/8$. This 'consistency condition' for slow-roll inflation is a striking prediction, but $n_{\rm T}$ will be difficult to measure even if r turns out to be of order 10^{-1}.

25.3 Tensor fraction and constraints on small-field models

25.3.1 Tensor fraction and field variation

Using Eqs. (18.30) and (25.10), the tensor fraction r can be related to $\Delta\phi$, the change in ϕ after the pivot scale leaves the horizon [8]. Suppose first that $\ln V$ is concave-downward throughout observable inflation. Then $|V'/V|$ is continuously increasing, and $2\epsilon < N^{-2}(\Delta\phi/M_{\rm Pl})^2$. (As always N is the number of e-folds after the pivot scale leaves the horizon and ϵ is evaluated then.) This can be written

$$\boxed{16\epsilon = r < 0.003 \left(\frac{50}{N}\right)^2 \left(\frac{\Delta\phi}{M_{\rm Pl}}\right)^2}. \tag{25.11}$$

The assumption that V'/V continually increases is satisfied for practically all of the potentials that have been considered in the literature. But we can obtain a useful result without making that assumption, by considering just the change in ϕ while cosmological scales leave the horizon. Taking this to be ten e-folds, we denote the change by $\Delta\phi_{10}$. During this small number of e-folds it should be reasonable to ignore the variation in ϵ (given by Eq. (18.17)) which gives

$$16\epsilon = r = 0.08 \left(\frac{\Delta\phi_{10}}{M_{\rm Pl}}\right)^2 < 0.08 \left(\frac{\Delta\phi}{M_{\rm Pl}}\right)^2. \tag{25.12}$$

In Figure 28.2 (page 469), the r–n plane is divided by the sloping straight lines into three regions, according to whether V and $\ln V$ are concave-upward or concave-downward when the observable Universe leaves the horizon.

With these constraints in mind, it is useful to classify inflation models according to the variation $\Delta\phi = |\phi_0 - \phi_{\rm end}|$ of the inflaton field during observable inflation. We will call a model small-field if $\Delta\phi \ll M_{\rm Pl}$, medium-field if $\Delta\phi \sim M_{\rm Pl}$ and and large-field if $\Delta\phi \gg M_{\rm Pl}$.

According to Eq. (25.11), small-field and even medium-field models are likely to have $r \lesssim 0.003$, making *the tensor fraction probably too small to ever observe.* Also, Eq. (25.12) shows that for small-field models $6\epsilon \ll 0.04$, which means that the predicted tilt will be just $n - 1 = 2\eta$ if it agrees with observation. Even for medium-field models, Eq. (25.11) suggests that the same will be true. We conclude that *small-field models, and probably also medium-field models, must have* $\eta \simeq -0.02$ *while cosmological scales leave the horizon.* This suggests that inflation for small- and medium-field models will take place near a hilltop of the potential. We have more to say about hilltop inflation in Chapter 28.

According to a widely held view, the Lagrangian density will contain all terms consistent with the symmetries of the theory. That means that it will contain an infinite number of terms whose coefficients have negative energy dimension (non-renormalizable terms). It is also widely supposed that the coefficients of these terms will, in a generic theory, be roughly of order 1 in units of the ultra-violet cutoff. If that view is adopted, small-field models will be regarded as the norm, with medium- and large-field models requiring special justification to keep the non-renormalizable terms of the potential under control.

25.3.2 Constraints on small-field models

Let us see what it takes to make $V(\phi)$ flat enough for slow-roll inflation in a small-field model. The tree-level potential will have a power series expansion

$$V(\phi) = V_0 + \frac{1}{2}m^2\phi^2 + M\phi^3 + \frac{1}{4}\lambda\phi^4 + \sum_{d=5}^{\infty} \lambda_d M_{\rm Pl}^4 \left(\frac{\phi}{M_{\rm Pl}}\right)^d . \tag{25.13}$$

The coefficients can have either sign. The terms of low dimension Eq. (25.13), which do not involve $M_{\rm Pl}$, are renormalizable terms (i.e. they correspond to a renormalizable quantum field theory).

The field variation while cosmological scales leave the horizon cannot be too small, or the perturbation of ϕ would generate a curvature perturbation bigger than the one observed. Indeed, Eqs. (25.7) and (25.12) require

$$\Delta\phi_{10} = 1 \times 10^4 H_* . \tag{25.14}$$

The large variation makes it difficult to cancel any term of Eq. (25.13) with another term or with the loop correction. Let us assume that there is no such cancellation. The constant term then has to dominate, and each of the terms has to respect the flatness condition $|\eta| \ll 1$. The other condition $\epsilon \ll 1$ is then automatic by virtue of our small-field assumption $\phi \ll M_{\rm Pl}$.

We shall not consider the cubic term, which usually is forbidden by a symmetry. For the quadratic term, $|\eta| \ll 1$ just means $|\eta_0| \ll 1$ where $|\eta_0|$ is the value at the origin. For the quartic and higher terms, we have using Eq. (25.14)

$$\lambda \ll \left(\frac{H}{\phi}\right)^2 \lesssim 10^{-8}, \tag{25.15}$$

$$\lambda_d \ll \left(\frac{M_{\rm Pl}^2}{\phi^2}\right)^{(d-2)/2} \lesssim 10^{-8} \left(\frac{10^{16}\,{\rm GeV}}{V^{1/4}}\right)^{2(d-4)} . \tag{25.16}$$

In a generic field theory, radiative corrections tend to drive all scalar masses up to the ultra-violet cutoff, in gross violation of the flatness condition $|\eta_0| \ll 1$. Even if the bare mass is fine-tuned to cancel the radiative correction in the vacuum, the loop correction (equivalently the running of the mass) may spoil the fine tuning. To avoid the latter problem the coupling of ϕ to other fields has to be sufficiently small. Similarly, the self-coupling of ϕ represented by λ has to be very small, and the same goes for λ_d unless $V^{1/4}$ is well below 10^{16} GeV.

These very small couplings may be regarded as fine tuning. Taking that attitude, let us see how fine tuning may be avoided. If ϕ is a pseudo-Nambu–Goldstone boson (PNGB) then $V(\phi)$ is perfectly flat in the limit where the relevant global symmetry is unbroken. If the breaking is under control one can then arrange that V is sufficiently flat.[1] The most usual approach though is to invoke supersymmetry.

Supersymmetry can easily suppress the quartic coupling λ; as we saw in Sections 17.3 and 17.5.1, there will typically be flat directions in field space, where the quartic couplings are absent in the global supersymmetry limit, and with the inclusion of supergravity they are still very suppressed. The coefficients λ_d up to finite order can be suppressed by a discrete R symmetry, as mentioned in another context in Section 17.2.1. Finally, we can impose $|\eta_0| \ll 1$ using global supersymmetry, though as we see in Section 28.4 supergravity corrections typically raise the value to $|\eta_0| \sim 1$ implying mild fine-tuning (the η problem).

25.4 Prediction of the standard paradigm for non-gaussianity

25.4.1 δN formula

To investigate the non-gaussianity of ζ we need to go beyond the first-order expression (25.3). For this purpose we use the δN formula (5.15). According to that formula, $\zeta(\mathbf{x}, t)$ is the perturbation in the number of e-folds of local expansion, starting from any flat slice and ending on a slice of uniform density at time t. In this chapter we are assuming that N at a given location depends only on ϕ. It follows that

$$\boxed{\zeta(\mathbf{x}) = \delta N = N'\delta\phi(\mathbf{x}) + \frac{1}{2}N''(\delta\phi(\mathbf{x}))^2 + \ldots}, \tag{25.17}$$

where $\delta\phi(\mathbf{x})$ is evaluated on the initial slice, $N' \equiv dN/d\phi$ and $N'' \equiv d^2N/d\phi^2$. To make the spatial average of $\zeta(\mathbf{x})$ vanish, the second term should actually be $\frac{1}{2}N''[(\delta\phi(\mathbf{x}))^2 - \overline{(\delta\phi)^2}]$ (where the overline denotes the spatial average) and similarly for higher powers.

The derivatives N' and N'' are evaluated on the unperturbed trajectory. They are independent of the final epoch, making $\zeta(\mathbf{x})$ independent of that epoch. This is because a change in ϕ just represents a shift back and forth along the same trajectory. Notice that N is here defined as the number of e-folds *from* the initial slice where ϕ is specified, *to* a final slice of uniform energy density. If we took that slice to be just before the end of inflation, $N(\phi)$ would therefore be exactly the object we discussed in Section 18.4, whose value when the observable Universe leaves the horizon is typically about 50. For the present purpose the location of the final slice is irrelevant, since changing that slice just changes $N(\mathbf{x})$ by a constant.

What we need are the derivatives of $N(\phi)$, which in each case (slow-roll inflation, K-inflation, etc.) are given by the inflationary dynamics. To evaluate them

[1] In the non-hybrid case one has to be careful that the entire potential doesn't vanish, as seen in Section 28.11.

one uses

$$N' = -\frac{H(t)}{\dot{\phi}}. \tag{25.18}$$

For the slow-roll case this gives

$$N' = M_{\rm Pl}^{-2}\frac{V}{V'}. \tag{25.19}$$

Differentiating Eq. (25.19) we find

$$N'' = (2\epsilon - \eta)N'^2. \tag{25.20}$$

It follows that to second order

$$\boxed{\zeta = N'\delta\phi + \frac{1}{2}(2\epsilon - \eta)(N'\delta\phi)^2}. \tag{25.21}$$

We see that the relative contribution of the quadratic terms vanishes in the slow-roll limit.

25.4.2 Bispectrum of ζ

Let us calculate the three-point correlator of ζ, characterized by $f_{\rm NL}$. If $\delta\phi$ were gaussian we could use Eqs. (6.72) and (25.21) to find

$$\frac{3}{5}f_{\rm NL} = \frac{2\epsilon - \eta}{2}. \tag{25.22}$$

If instead we kept only the linear term of Eq. (25.21) but kept the non-gaussianity of $\delta\phi$ we would get

$$B_\zeta = N'^3\left(B_\phi^{\rm self} + B_\phi^{\rm grav}\right), \tag{25.23}$$

where the terms on the right-hand side are given by Eqs. (24.43) and (24.46). Taking both sources of non-gaussianity into account we find

$$\frac{6}{5}f_{\rm NL} = 3\epsilon - \eta + \zeta\Delta N + \epsilon f(k_1, k_2, k_3), \tag{25.24}$$

$$\left(f + \frac{3}{2}\right)\sum_k k_i^3 = \frac{1}{2}\sum_{i\neq j} k_i k_j^2 + \frac{4}{k_{\rm t}}\sum_{i<j} k_i^2 k_j^2. \tag{25.25}$$

It can be shown that $0 \leq f \leq 5/6$, the lower limit corresponding to the squeezed configuration $k_1 \ll k_2 \simeq k_3$ and the upper limit to the equilateral configuration $k_1 = k_2 = k_3$.

The third term changes linearly with ΔN, which seems contrary to the fact that ζ goes to a constant value after horizon exit. But according to Eq. (18.18), the change is cancelled by an identical change in η. Within the accuracy of the calculation we can therefore drop the third term and declare that η is evaluated just a few Hubble times after horizon exit. Either way, we have to assume that the change in η during a few Hubble times is negligible. According to Eq. (18.18) the assumption about η is equivalent to $|\xi| \ll |\eta|$, which in turn is equivalent to $|n'| \ll |n-1|$. Current observations are not strong enough to test this, and if it is not the case then the second-order calculation that we have described is invalid. Still, its conclusion that $|f_{\rm NL}|$ is well below 1 is unlikely to be changed.

In terms of the spectral index n and the tensor fraction r, the final result is

$$\boxed{\frac{3}{5} f_{\rm NL}(k_1, k_2, k_3) = -\frac{n-1}{4} + \frac{r}{32} f(k_1, k_2, k_3)}\,. \tag{25.26}$$

Hence $f_{\rm NL}$ is at most of order 10^{-2} which is probably too small to measure.

Following Refs. [9] and [10], we have derived Eq. (25.26) by first calculating the bispectrum of $\delta\phi$ using Eq. (24.45) on the flat slicing, going afterwards to ζ with the δN formula. The original calculation by Maldacena [11] used the $\delta\phi = 0$ slicing, where ζ is itself is the degree of freedom. He also calculated the correlators $\langle \zeta_{\mathbf{k}_1} \zeta_{\mathbf{k}_2} h^{+,\times}_{\mathbf{k}_3} \rangle$, and the three-point correlator of the tensor with itself given in Eq. (24.65), which are almost certainly too small to observe.

Using the δN method along with Eq. (24.48), the trispectrum of ζ has been calculated [12]. It is almost certainly too small to observe.

25.4.3 Non-gaussianity in the squeezed limit

Equation (25.26) depends in general on the momenta k_i. But in the 'squeezed' limit corresponding to (say) $k_3 \ll k_1 \simeq k_2$ the last term vanishes and we have simply $3f_{\rm NL}/5 = -(n-1)/4$. As we now see, this occurs because the perturbation on the large scale $1/k_3$ has become classical by the time that the shorter scales leave the horizon [11]. The result is of great interest because it applies also to the more general single-field inflation models considered in Section 25.7, where $f_{\rm NL}$ for generic k_i may be large. A similar result applies also to higher correlators of ζ, and it has been worked out for the trispectrum [12, 13]. Here we give the calculation for the bispectrum.

To do the calculation we choose an intermediate scale L:

$$\frac{1}{k_1} \simeq \frac{1}{k_2} \ll L \ll \frac{1}{k_3}, \tag{25.27}$$

and split ζ into a contribution ζ_+ containing inverse wavenumbers bigger than L, and a contribution ζ_- containing inverse wavenumbers smaller than L. Using translation invariance we can write

$$\langle \zeta_{\mathbf{k}} \zeta_{\mathbf{k}'} \rangle = \frac{1}{(2\pi)^3} \delta_{\mathbf{k}+\mathbf{k}'} \int e^{i\mathbf{k}\cdot\mathbf{x}} \langle \zeta(\mathbf{x}) \zeta(0) \rangle d^3x, \tag{25.28}$$

$$\langle \zeta_{\mathbf{k}_1} \zeta_{\mathbf{k}_2} \zeta_{\mathbf{k}_3} \rangle = \frac{1}{(2\pi)^3} \delta_{\mathbf{k}_1+\mathbf{k}_2+\mathbf{k}_3} \int e^{i(\mathbf{k}_1\cdot\mathbf{x}+\mathbf{k}_3\cdot\mathbf{z})} \langle \zeta_-(\mathbf{x}) \zeta_-(0) \zeta_+(\mathbf{z}) \rangle d^3x d^3z. \tag{25.29}$$

We can describe the generation of $\zeta_-(\mathbf{x})$ by using a box of size L. Centering the box at the origin, ζ_+ can be taken to have a homogeneous value $\zeta_+(0)$. Recalling the definition of ζ, we see that its average value within the box can be absorbed into the unperturbed scale factor, giving a new unperturbed scale factor $a_L \equiv [1 + \zeta_+(0)]a(t)$. A given location has displacement $a\mathbf{x} = a_L \mathbf{x}_L$, where $\mathbf{x} = [1 + \zeta_+(0)]\mathbf{x}_L$. In terms of these,

$$\zeta_-(\mathbf{x}) = \zeta_-(\mathbf{x}_L) + \zeta_+(0) \nabla_L \zeta(\mathbf{x}_L), \tag{25.30}$$

where ∇_L denotes the gradient with respect to $\mathbf{x}_L$. This gives

$$\langle \zeta_-(\mathbf{x}) \zeta_-(0) \zeta_+(\mathbf{z}) \rangle = \nabla_L \langle \zeta_-(\mathbf{x}_L) \zeta_-(0) \rangle \langle \zeta_+(0) \zeta_+(\mathbf{z}) \rangle. \tag{25.31}$$

To first order we can write $\mathbf{x}_L = \mathbf{x}$, and we drop the subscripts $\pm$. Differentiating Eq. (6.29) gives

$$\frac{d}{dx} \langle \zeta(\mathbf{x}) \zeta(0) \rangle = (n-1) \langle \zeta(\mathbf{x}) \zeta(0) \rangle. \tag{25.32}$$

Then, using Eq. (25.28) to get the Fourier transform in Eq. (25.29), we find the advertised result

$$\boxed{\frac{3}{5} f_{\rm NL} = -\frac{n-1}{4}}. \tag{25.33}$$

25.4.4 Transplanckian effect

The fundamental assumption, made in all of the calculations that we have described, is that flat spacetime field theory applies well before horizon exit. The assumption seems harmless because it is simply an application of the equivalence principle, but there is a difference between this application and ordinary applications of the principle.

The difference is that during inflation, the energy density and Hubble parameter might be just a few orders of magnitude below the Planck scale. Specifically, $H/M_{\rm Pl}$ may be as big as 10^{-5}. One might imagine that the equivalence principle fails at the level $H/M_{\rm Pl}$, giving non-gaussianity at the same level. If so it would invalidate the calculation of non-gaussianity that we presented. On the other hand, the failure might be only at the level $(H/M_{\rm Pl})^2$ or some higher power, making it negligible. This kind of failure of the equivalence principle has been termed the **transplanckian effect**.

25.5 Loop contributions to the correlators of ζ

25.5.1 Spectrum

To calculate the spectrum of ζ we dropped the second term of Eq. (25.17). We justified this approximation by noticing that the second term is non-gaussian, and that observation places strong limits on the non-gaussianity of ζ in the observable Universe.

In this subsection we calculate the effect of the second term, verifying that indeed it is small. The calculation is important because it raises issues of principle, and because an analogous calculation for the bispectrum could give an observable effect (Section 26.3.2).

Consider again Eq. (25.17). The full expression for the two-point correlator is

$$\langle \zeta_{\mathbf{k}} \zeta_{\mathbf{k}'} \rangle = N'^2 \langle \delta\phi_{\mathbf{k}} \delta\phi_{\mathbf{k}'} \rangle + \frac{N''^2}{4} \langle (\delta\phi)^2_{\mathbf{k}} (\delta\phi)^2_{\mathbf{k}'} \rangle \tag{25.34}$$

$$\equiv \langle \zeta_{\mathbf{k}} \zeta_{\mathbf{k}'} \rangle_{\text{tree}} + \langle \zeta_{\mathbf{k}} \zeta_{\mathbf{k}'} \rangle_{\text{one-loop}}. \tag{25.35}$$

The first term is the one we considered already, and the second is the one that we are going to calculate. We call them the tree-level and one-loop contributions for a reason that will become apparent.

Using the convolution theorem, the second term gives a contribution [14]

$$\mathcal{P}_\zeta^{\text{one-loop}}(k) = \frac{N''^2}{4} \frac{k^3}{2\pi} \int d^3p \frac{\mathcal{P}_{\delta\phi}(p)}{p^3} \frac{\mathcal{P}_{\delta\phi}(|\mathbf{k}-\mathbf{p}|)}{|\mathbf{k}-\mathbf{p}|^3}. \tag{25.36}$$

The result contains an integration because the convolution theorem generates two integrals, and only one of them is killed by the delta functions that come with the two-point correlators. If $\delta\phi$ is taken to be massless and we ignore the time dependence of H, the spectrum $\mathcal{P}_{\delta\phi}$ has the scale-independent value $(H/2\pi)^2$. Then the integral diverges at the points $\mathbf{p} = 0$ and $\mathbf{p} = \mathbf{k}$. The same thing happens if $\mathcal{P}_{\delta\phi}(k)$ increases as k decreases.

To handle the possible divergence, one should work in a box of finite size L, much bigger than the region of interest. Then no wavelength is bigger than L, and we can set the integrand equal to zero within a sphere of radius L^{-1} around each singular point. Denoting the use of the box by a subscript L^{-1}, we conclude that

$$\boxed{\mathcal{P}_\zeta^{\text{one-loop}}(k) = \frac{N''^2}{4} \frac{k^3}{2\pi} \int_{L^{-1}} d^3p \frac{\mathcal{P}_{\delta\phi}(p)}{p^3} \frac{\mathcal{P}_{\delta\phi}(|\mathbf{k}-\mathbf{p}|)}{|\mathbf{k}-\mathbf{p}|^3}}. \tag{25.37}$$

Physical wavenumbers have to satisfy $L^{-1} \ll k$. If $\mathcal{P}_{\delta\phi}$ is scale-independent, one finds in this regime

$$\mathcal{P}_\zeta{}^{\text{one-loop}}(k) = N''^2 \mathcal{P}_{\delta\phi}^2 \ln(kL). \tag{25.38}$$

This expression gives the correct variance $\langle \delta\phi^4 \rangle = 3\langle \delta\phi^2 \rangle^2$, confirming the consistency of the finite-box approach.

Setting $\mathcal{P}_\zeta^{\text{tree}} \simeq \mathcal{P}_\zeta$ we find

$$\mathcal{P}_\zeta{}^{\text{one-loop}} = \left(\frac{N''}{N'^2} \sqrt{\mathcal{P}_\zeta} \right)^2 \ln(kL) \mathcal{P}_\zeta^{\text{tree}}. \tag{25.39}$$

The term in the large bracket is a contribution $(3 f_{\text{NL}}/5)\mathcal{P}_\zeta^{1/2}$. As seen in Section 6.4 this is a measure of the non-gaussianity of ζ, and current observation makes it less than 10^{-3} or so.[2] We conclude that $\mathcal{P}_\zeta{}^{\text{one-loop}}$ is much less than $\mathcal{P}_\zeta{}^{\text{tree}}$ provided that $\ln(kL) \ll 10^{-6}$ or so.

Now comes an issue of principle. If inflation lasts for more than 10^6 e-folds before the observable Universe leaves the horizon, we could choose an extremely large box corresponding to $\ln(kL) \gtrsim 10^6$ and then $\mathcal{P}_\zeta{}^{\text{one-loop}}$ would be the dominant contribution. That would be very bad news, because it would mean that the second term of Eq. (25.17) is bigger than the first one causing one to suspect that the next term will be even bigger. The fundamental step of writing $N(\mathbf{x}) = N + \delta N$ would even be called into question. These and other problems mean that it would be difficult to do reliable calculations within such a huge box. Such a calculation would in any case of debatable use, since it refers to what is seen by a typical observer within that box, whose experience might be very different from our own.

Even if we just choose a moderately large box, say with $\ln(kL)$ of order 100, we introduce uncertainty in regard to the relevant physics many Hubble times before the observable Universe leaves the horizon, and we increase the chance of our location being untypical. The conclusion seems to be that one should stick to a box where $\ln(kL)$ is just of order a few. The dependence of the spectrum on the box size is not surprising, when we remember that the correlator in position space can be regarded as a spatial average within the box (Section 6.5).

25.5.2 Higher-order graphs and higher correlators

If we include more terms in Eq. (25.17) there will be more integrals. A similar thing occurs if we calculate three-point or higher correlators. Finally, if the non-gaussianity of the field perturbations is taken into account, the correlators of those perturbations will have to be taken into account.

[2] In arriving at this conclusion we didn't use the actual values of N' and N''. As a result it remains valid if $\delta\phi$ is replaced by a different light field perturbation, as described in Chapter 26.

All of this can be organized by Feynman graphs [15], which we might call δN graphs. The tree-level graphs involve no integration, the one-loop graphs involve one integration, and so on. The δN graphs are obviously unrelated to the usual Feynman graphs that organize the perturbative calculation of transition amplitudes. They are also different from the Feynman graphs that organize the perturbative calculation of the in–in formalism for the correlators of the inflaton field perturbation. To calculate the correlators of ζ, we need both the δN graphs and the in–in graphs, since the correlators of the inflaton field perturbation provides the input for the δN calculation.

The matter doesn't stop there. Instead of using the δN formula, we could have worked on the uniform-density slicing where $\delta\phi$ vanishes. Taking ζ itself as the degree of freedom, we could have directly calculated the spectrum and the higher correlators. (As we noted earlier, this is how the tree-level calculation of the bispectrum of ζ was first done.) The in–in Feynman graphs for the $\delta\phi = 0$ slicing are therefore equivalent to the in–in Feynman graphs for the flat slicing *plus* the δN Feynman graphs that arise from the use of that slicing. At the time of writing this issue has not been explored in detail but one remark is worth making.

One might object that there can be no connection between an integral involving a classical perturbation and an integral occurring in quantum field theory. But that is wrong, because each classical field perturbation is generated from the vacuum fluctuation and comes with a factor $\hbar$, that is absent in our formulas only because we chose units in which $\hbar = 1$. Therefore, each loop in a Feynman graph arising from the δN formalism comes with a factor $\hbar$, just like each loop of a Feynman graph of quantum field theory.

25.6 The standard paradigm beyond slow roll

Now we drop the assumption of slow roll, while still assuming canonical normalization and general relativity. This allows us to calculate corrections to the slow-roll approximation, and to handle situations where the slow-roll approximation is not valid at all. We stick to linear perturbation theory which remains an excellent approximation even without the slow-roll assumption. Within the linear approximation, ζ is gaussian and we need only calculate its spectrum.

Using Eqs. (24.11) and (25.1), the spectrum is given in terms of the mode function $\varphi_k(\eta)$ by

$$\boxed{\mathcal{P}_\zeta(k) = \frac{k^3}{2\pi^2}\left|\frac{\varphi_k}{z}\right|^2}, \tag{25.40}$$

where $z \equiv a\dot{\phi}/H$. The mode function is the solution of the Mukhanov–Sasaki equation (24.32), with the initial condition (24.7). It can be calculated once the unperturbed inflationary trajectory $\phi(\eta)$ is specified. By virtue of the attractor theorem, we expect that the unperturbed trajectory will in turn be determined more or less uniquely by the potential.[3]

Well after horizon exit, there is a solution $\varphi_k \propto z \equiv a\dot{\phi}/H$ of the Mukhanov–Sasaki equation, which makes ζ constant. The other solution is proportional to $z\int d\eta/z^2$ which decays and presumably becomes negligible after a few Hubble times. In this way, we establish the constancy of ζ during inflation without reference to the separate universe assumption.

We now give some examples of the use of Eq. (25.40), following Ref. [16]. The starting point is to assume that $\epsilon_{\rm H}$ and $\eta_{\rm H}$ have negligible variation. Then one finds

$$\mathcal{P}_\zeta{}^{1/2} = 2^{\nu-\frac{3}{2}}\frac{\Gamma(\nu)}{\Gamma(\frac{3}{2})}(1-\epsilon_{\rm H})^{\nu-\frac{1}{2}}\left.\frac{H^2}{2\pi|\dot{\phi}|}\right|_k, \tag{25.41}$$

$$\nu \equiv \frac{1+\epsilon_{\rm H}-\delta_H}{1-\epsilon_{\rm H}}+\frac{1}{2}. \tag{25.42}$$

One application of this formula is to hilltop inflation with a quadratic potential, $V = V_0(1+\eta_0\phi^2/2M_{\rm Pl}^2)$ and $\phi \ll M_{\rm Pl}$, with η_0 negative and $|\eta_0|$ not necessarily small. This kind of inflation has been called fast-roll inflation, though we are about to see that $\dot{\phi}$ is small just as in slow-roll inflation. To determine $\dot{\phi}$, we assume that H is practically constant and then check the regime of validity of that assumption. With H constant,

$$\phi = Ae^{FHt} + Be^{-(F+3)Ht}, \tag{25.43}$$

where

$$F = \frac{3}{2}\left(\sqrt{1-\frac{4}{3}\eta_0}-1\right). \tag{25.44}$$

The first term is the growing mode. A small departure from this mode, represented by the second term, will quickly become negligible in accordance with the attractor theorem. We keep only the growing mode. The regime where this solution is consistent is $\phi \ll M_{\rm Pl}/F$ (corresponding to $\epsilon_{\rm H} \ll 1$). In this regime $\eta_{\rm H} \simeq -F$ is practically constant making $\epsilon_{\rm H}$ practically zero. Using Eq. (25.41) one then finds

$$n-1 = -3\left(\sqrt{1-\frac{4}{3}\eta_0}-1\right). \tag{25.45}$$

In the regime $|\eta_0| \gtrsim 1$ we have $F \sim \sqrt{|\eta_0|}$. The cosmic microwave background normalization corresponding to Eq. (25.41) gives roughly $\phi_0 \sim 10^4 H/F$, and

[3] The attractor theorem refers to solutions that are sufficiently close to a given solution, but it doesn't give a criterion for sufficiency.

$\phi_{\text{end}} \sim M_{\text{Pl}}/F$. This gives

$$N \sim 2F^{-1} \ln \frac{10^{16}\,\text{GeV}}{V_0^{1/4}}. \tag{25.46}$$

Since nucleosynthesis requires $V_0^{1/4} > 1\,\text{MeV}$ we find $N \lesssim 40$ and it will typically be much less. Fast-roll inflation therefore can hardly replace slow-roll inflation, even if we generate the curvature perturbation from some other perturbation as in Chapter 26. We also see that $|\eta_0|$ cannot be extremely large, or there will not even be an e-fold of inflation.

Equations (25.43)–(25.45) are valid also for $0 < \eta_0 < 3/4$, corresponding to hybrid inflation.

Another application of Eq. (25.41) is to power-law inflation, $a \propto t^p$, that we encountered in Section 19.3. This gives $\epsilon_{\text{H}} = \eta_{\text{H}} = 1/p$ and spectral tilt $n - 1 = -1/(p-1)$. The slow-roll regime corresponds to $p \gg 1$ but the exact solution is valid for the entire inflationary regime $p > 1$.

Finally, Eq. (25.41) can be used to calculate a correction to the slow-roll spectrum, assuming that ϵ_{H} and η_{H} both have negligible variation. Using the fact that they are both much less than 1 for slow roll, one finds

$$\mathcal{P}_\zeta^{1/2}(k) = \left[1 - (2C+1)\,\epsilon_{\text{H}} + C\eta_{\text{H}}\right] \frac{H^2}{2\pi|\dot{\phi}|}, \tag{25.47}$$

where $C = -2 + \ln 2 + b \simeq -0.73$, with b the Euler–Mascheroni constant. As always, the right-hand side is evaluated at $k = aH$. Assuming also that $\dot{\eta_{\text{H}}}$ is practically constant, and expressing everything in terms of the potential and its derivatives, one finds

$$\frac{n-1}{2} = -3\epsilon + \eta - \frac{5+36C}{3}\epsilon^2 + (8C-1)\,\epsilon\eta + \frac{1}{3}\eta^2 - \frac{3C-1}{3}\xi^2. \tag{25.48}$$

25.7 K-inflation

25.7.1 General case

K-inflation [17] considers the most general form of the action consistent with general relativity, taking the inflaton to be the only field. It corresponds to

$$\mathcal{L} = P(X, \phi), \tag{25.49}$$

where $X \equiv -\partial_\mu\phi\partial^\mu\phi$ and P is an arbitrary function.

For the unperturbed Universe, $X = \dot{\phi}^2$. Working out the energy–momentum tensor from Eq. (13.32) one finds

$$P = P(X, \phi), \qquad \rho + P = 2X\frac{\partial P}{\partial X}. \tag{25.50}$$

As is implied by the notation, P is the pressure. The continuity equation $\dot{\rho} = -3H(\rho + P)$ still applies, and there is almost-exponential inflation if $|\dot{\rho}/H\rho| \ll 1$.

To calculate the curvature perturbation one can define

$$\varphi \equiv z\zeta, \qquad z \equiv \frac{a\sqrt{2\epsilon_{\rm H}}}{c_{\rm s}}, \qquad c_{\rm s}^2 \equiv \frac{P_X}{\rho_X}, \tag{25.51}$$

where the subscript X denotes $\partial/\partial X$. In the limit $c_{\rm s} = 1$ this reduces to the earlier definition (before Eq. (24.4)). For any $c_{\rm s}(\eta)$, it can be shown [18] that the Mukhanov–Sasaki equation is satisfied with $\varphi \to v \equiv z\zeta$ and $k \to c_{\rm s}k$. Well after horizon exit, when $z \gg 1$, the solution ζ_k tends to a constant. This shows that the K-inflation trajectories $\phi(\mathbf{x}, t)$ have the attractor behaviour, at least to first order.

The action for φ is Eq. (24.15), with ∂_i replaced by $c_{\rm s}\partial_i$ and φ replaced by v. To quantize the theory we assume that $c_{\rm s}$ is slowly varying on the Hubble timescale:

$$\boxed{|s| \ll 1, \qquad s \equiv \frac{\dot{c}_{\rm s}}{Hc_{\rm s}}}. \tag{25.52}$$

Until k leaves the 'sound horizon' corresponding to $c_{\rm s}k = aH$, the mode function oscillates with angular frequency $c_{\rm s}k$. We impose the initial condition Eq. (24.7) with $k \to c_{\rm s}k$. Well before sound horizon exit, the Fock space describes free 'particles' with momentum and energy related by $E = c_{\rm s}k$. As before, the state vector has to be the vacuum, corresponding to no 'particles', because 'particles' would give positive pressure which would spoil inflation. Solving the Mukhanov–Sasaki equation, the late-time spectrum of φ is given by Eq. (24.11) from which the spectrum $\mathcal{P}_\zeta(k)$ follows.

We obtain an almost scale-invariant spectrum in a slow-roll regime, defined by

$$\boxed{|\epsilon_{\rm H}| \ll 1}, \qquad \boxed{|\eta_{\rm H}| \ll 1}, \tag{25.53}$$

together with Eq. (25.52). Indeed, the mode function of v in this regime is given by the right-hand side of Eq. (24.8), leading to

$$\mathcal{P}_\zeta(k) = \frac{1}{2M_{\rm Pl}^2\epsilon_{\rm H}c_{\rm s}}\left(\frac{H}{2\pi}\right)^2, \tag{25.54}$$

with the right-hand side evaluated at horizon exit. The spectral tilt is

$$n - 1 = 2\eta_{\rm H} - 4\epsilon_{\rm H} - s. \tag{25.55}$$

In the limit $c_{\rm s} = 1$ we recover the standard slow-roll spectrum.

So far the result is a modest generalization of standard slow roll. Now comes the big difference. When the non-gaussianity parameter $f_{\rm NL}(k_1, k_2, k_3)$ is calculated, it is found to be of order $1/c_s^2$ for $c_s \ll 1$. The calculation follows the line we sketched for standard slow roll, but the result [19] is much more complicated.

K-inflation is the most general possibility for inflation with a single field if we stick to Einstein gravity. Allowing non-Einstein gravity, a rather general form for the action is given in Ref. [20]. It includes K-inflation as a special case. Using this action, it has been shown [21] that ζ is conserved during inflation, up to second order of perturbation theory. This implies that the inflationary trajectories have the attractor behaviour up to that order. Consistently with that fact, the calculated bispectrum for K-inflation in the squeezed limit has the behaviour found in Section 25.4.3 ([19], later version).

25.7.2 Colliding brane inflation

Motivation for considering K-inflation comes from a braneworld scenario, in which ϕ is the distance between two colliding branes. The action describing inflation is a special case of K-inflation:

$$\mathcal{L} = -f^{-1}(\phi)\sqrt{1 + f(\phi)\partial_\mu\phi\partial^\mu\phi} - V(\phi). \tag{25.56}$$

Inflation with this action is called colliding brane inflation, or DBI inflation.

If f is sufficiently small we recover canonical normalization. The crucial point though is that $\dot\phi$ cannot exceed a maximum 'speed limit', corresponding to $\dot\phi^2 < 1/f(\phi)$. As a result, almost-exponential inflation can occur even if V is very steep. The function $f(\phi)$ is under relatively good theoretical control, corresponding to what is called the warp factor of the extra dimension. The potential $V(\phi)$ is not under good control at present. Further study of colliding brane inflation may begin with Refs. [22, 23].

25.8 Warm inflation

Warm inflation [24, 25] differs in a radical way from all other paradigms, because it invokes a significant component of radiation during inflation. The radiation comes from the time dependence of the inflaton field, and it slows down the motion of the inflaton field.

Taking the inflaton field to be canonically normalized, the radiation is taken into account by writing

$$\ddot\phi + (3H + \Gamma)\dot\phi + V' = 0, \tag{25.57}$$

where Γ is a function of time which vanishes in the limit $\dot\phi \to 0$. Ordinary inflation assumes that Γ is negligible, but warm inflation assumes that it is significant, or even dominant ($\Gamma \gg H$).

It is clear that the time-dependent ϕ will produce radiation at some level, analogously with the situation during preheating. But the calculation of Γ within a given model is more difficult than for preheating, and there have been far fewer studies. On the basis of existing studies (for instance Ref. [26]) it seems that warm inflation is possible if the Lagrangian describing inflation is carefully chosen.

The curvature perturbation is given in terms of $\delta\phi$ by the usual formula (25.1). The vacuum fluctuation still contributes to $\delta\phi$, giving the usual gaussian contribution with spectrum $(H/2\pi)^2$, but now $\delta\phi$ receives a contribution from the thermal fluctuation. That fluctuation freezes out at horizon exit, and its contribution to the spectrum of the curvature perturbation is [27]

$$\mathcal{P}_\zeta(k) \simeq \frac{H^4}{\dot{\phi}^2}\frac{T}{H}\left(\frac{\Gamma}{H}\right)^{1/2}, \tag{25.58}$$

with the right-hand side evaluated at horizon exit. This contribution can dominate, and it can have the required small spectral tilt. One might argue that the Schrödinger's Cat problem associated with the vacuum fluctuation is absent in this case, since the thermal fluctuation $\delta T(\mathbf{x}, t)$ is something that does occur in reality once the gas is in place. Of course, it is still unclear how the particular $\delta T(\mathbf{x}, t)$ for our Universe is chosen, on the reasonable hypothesis that the state vector is translation invariant at some very early time.

Warm inflation can give large non-gaussianity [28]. The non-gaussianity becomes small in the squeezed limit, consistent with the fact that the slow-roll solution has the attractor behaviour. At the time of writing it is not quite clear whether the attractor behaviour holds to second order so that the squeezed limit of Section 25.4.3 applies.

Exercises

25.1 Starting from Eq. (25.4), verify the predictions Eqs. (25.5) and (25.6) for n and n'.

25.2 Derive the general form of the potential that, within the slow-roll approximation, gives scale-invariant density perturbations.

25.3 For the case $V \propto \phi^2$, work out the corrections to the slow-roll formulas for $\mathcal{P}_\zeta$ and n. For the same case, work out the maximum value of the non-gaussianity parameter $f_{\rm NL}$.

25.4 Verify that power-law inflation gives $\epsilon_{\rm H} = \eta_{\rm H} = 1/p$.

25.5 By requiring $1\,\text{MeV} < V^{1/4} < M_{\rm Pl}$, calculate the maximum number of e-folds of power-law inflation for a given p, and show that it is only of order 10 if the slow-roll condition $p \gg 1$ is badly violated. Say why this makes such inflation unviable. Explain why $p \sim 1$ also makes it difficult to generate a viable curvature perturbation, even using a field perturbation different from $\delta\phi$.

25.6 Calculate the next term of Eq. (25.21) in terms of ϵ, η and ξ.

25.7 Verify Eq. (25.38), giving a one-loop contribution to the spectrum of the curvature perturbation.

References

[1] V. F. Mukhanov and G. V. Chibisov. Quantum fluctuations and a nonsingular universe. *JETP Lett.*, **33** (1981) 532.

[2] A. A. Starobinsky. Dynamics of phase transition in the new inflationary universe scenario and generation of perturbations. *Phys. Lett. B*, **117** (1982) 175.

[3] S. W. Hawking. The development of irregularities in a single bubble inflationary universe. *Phys. Lett. B*, **115** (1982) 295.

[4] A. H. Guth and S. Y. Pi. Fluctuations in the new inflationary universe. *Phys. Rev. Lett.*, **49** (1982) 1110.

[5] J. M. Bardeen, P. J. Steinhardt and M. S. Turner. Spontaneous creation of almost scale-free density perturbations in an inflationary universe. *Phys. Rev. D*, **28** (1983) 679.

[6] D. H. Lyth. Large Scale Energy Density Perturbations And Inflation. *Phys. Rev. D* **31** (1985) 1792.

[7] A. R. Liddle and D. H. Lyth. COBE, gravitational waves, inflation and extended inflation. *Phys. Lett. B*, **291** (1992) 391.

[8] D. H. Lyth. What would we learn by detecting a gravitational wave signal in the cosmic microwave background anisotropy? *Phys. Rev. Lett.*, **78** (1997) 1861.

[9] D. Seery and J. E. Lidsey. Primordial non-gaussianities from multiple-field inflation. *JCAP*, **0509** (2005) 011.

[10] D. Seery, K. A. Malik and D. H. Lyth. Non-gaussianity of inflationary field perturbations from the field equation. *JCAP*, **0803** (2008) 014.

[11] J. M. Maldacena. Non-gaussian features of primordial fluctuations in single field inflationary models. *JHEP*, **0305** (2003) 013.

[12] D. Seery, J. E. Lidsey and M. S. Sloth. The inflationary trispectrum. *JCAP*, **0701** (2007) 027.

[13] M. x. Huang and G. Shiu. The inflationary trispectrum for models with large non-Gaussianities. *Phys. Rev. D*, **74** (2006) 121301.

[14] D. H. Lyth. Axions and inflation: vacuum fluctuation. *Phys. Rev. D*, **45** (1992) 3394.

[15] C. T. Byrnes, K. Koyama, M. Sasaki and D. Wands. Diagrammatic approach to non-Gaussianity from inflation. *JCAP*, **0711** (2007) 027.

[16] E. D. Stewart and D. H. Lyth. A More accurate analytic calculation of the spectrum of cosmological perturbations produced during inflation. *Phys. Lett. B*, **302** (1993) 171.

[17] C. Armendariz-Picon, T. Damour and V. F. Mukhanov. K-inflation. *Phys. Lett. B*, **458** (1999) 209.

[18] J. Garriga and V. F. Mukhanov. Perturbations in K-inflation. *Phys. Lett. B*, **458** (1999) 219.

[19] X. Chen, M. x. Huang, S. Kachru and G. Shiu. Observational signatures and non-Gaussianities of general single field inflation. *JCAP*, **0701** (2007) 002. [A later version is at arXiv:hep-th/0605045].

[20] C. Cheung, P. Creminelli, A. L. Fitzpatrick, J. Kaplan and L. Senatore. The effective field theory of inflation. *JHEP*, **0803** (2008) 014.

[21] C. Cheung, A. L. Fitzpatrick, J. Kaplan and L. Senatore. On the consistency relation of the 3-point function in single field inflation. *JCAP*, **0802** (2008) 021.

[22] E. Silverstein and D. Tong. Scalar speed limits and cosmology: acceleration from D-cceleration. *Phys. Rev. D*, **70** (2004) 103505.

[23] S. H. Henry Tye. Brane inflation: String theory viewed from the cosmos. *Lect. Notes Phys.*, **737** (2008) 949.

[24] I. G. Moss. Primordial inflation with spontaneous symmetry breaking. *Phys. Lett. B*, **154** (1985) 120.

[25] A. Berera and L. Z. Fang. Thermally induced density perturbations in the inflation era. *Phys. Rev. Lett.*, **74** (1995) 1912.

[26] I. G. Moss and C. Xiong. Dissipation coefficients for supersymmetric inflationary models. arXiv:hep-ph/0603266.

[27] L. M. H. Hall, I. G. Moss and A. Berera. Scalar perturbation spectra from warm inflation. *Phys. Rev. D*, **69** (2004) 083525.

[28] I. G. Moss and C. Xiong. Non-gaussianity in fluctuations from warm inflation. *JCAP*, **0704** (2007) 007.

26

Generating ζ after horizon exit

In this chapter we see how the curvature perturbation may be generated after inflation is over. We also see how the curvature perturbation may be generated during multi-field inflation. In the next chapter we will consider the related issue of how an isocurvature perturbation may be generated.

26.1 The generic δN formula for ζ

In Section 5.4.3, we saw that the curvature perturbation $\zeta(\mathbf{x}, t)$ is equal to δN, where $N(\mathbf{x}, t)$ is the number of e-folds of expansion starting from an initial flat slice and ending on a slice of uniform density. In Section 25.4.1 we used this result for single-field inflation. We took the initial epoch to be during inflation, and we assumed that the value $\phi(\mathbf{x})$ at that epoch determines N so that we have $N(\phi(\mathbf{x}), t)$. This allowed us to write ζ as a power series in the initial field perturbation $\delta\phi(\mathbf{x})$.

The same approach works without any need to assume slow-roll inflation, and without any need to consider just one field. We just need to assume that the values of some set of light fields determine the local evolution of N. Then

$$\zeta(\mathbf{x}, t) = \delta N(\phi_1(\mathbf{x}), \phi_2(\mathbf{x}), \ldots, t) \quad (26.1)$$

$$\equiv N(\phi_1(\mathbf{x}), \phi_2(\mathbf{x}), \ldots, t) - N(\phi_1, \phi_2 \ldots, t). \quad (26.2)$$

The formula is very general. It does not invoke a theory of gravity, or a particular model of inflation. It also makes no assumption about the dynamics of the fields (in particular, the kinetic terms need not be canonical).

As in the single-field case, the initial slice can be any flat slice during inflation, because the expansion going from one flat slice to another is uniform. Also as in that case, the final slice is one of uniform energy density. Differently from the single-field case though, ζ will in general depend on the final time t. An equivalent statement is that the local pressure will not in general be a unique function of the local energy density.

Another difference from the single-field case concerns the unperturbed values of the fields. As in the single-field case, we are invoking the separate universe assumption, so that the expansion $N(\mathbf{x}, t)$ refers to a family of unperturbed universes. Also as in that case, no knowledge of cosmological perturbation theory is required, only a knowledge of the evolution of the family of unperturbed local universes. The difference is that we have to put in by hand the unperturbed values of the light fields, which apply to *the* unperturbed Universe in which we live. In the single field case the unperturbed value ϕ_0 (evaluated when say the pivot scale leaves the horizon) is determined by the number N of e-folds of subsequent inflation, which in turn is determined by the evolution of the scale factor after inflation. There is no corresponding result in the general case.

26.2 Spectrum of ζ

To calculate the spectrum we expand δN to first order in the field perturbations;

$$\boxed{\zeta(\mathbf{x}, t) = \sum N_i(t)\delta\phi_i(\mathbf{x})} . \tag{26.3}$$

(In this chapter a subscript i denotes $\partial/\partial\phi_i$.) We displayed the time dependence for clarity, but from now on we suppose that the expression has been evaluated after ζ settles down to the final constant value that is to be compared with observation. We take the initial slice just a few Hubble times after horizon exit, so that each field has an almost gaussian perturbation with spectrum $(H_k/2\pi)^2$. This gives

$$\boxed{\mathcal{P}_\zeta(k) = \left(\frac{H_k}{2\pi}\right)^2 \sum N_i^2(k)} . \tag{26.4}$$

The k-dependence of N_i occurs because we chose the initial epoch to depend on k.

To obtain a formula for the spectral index [1, 2], we assume that the fields ϕ_i satisfy the slow-roll equations of Section 20.1, without assuming that they necessarily have a significant effect on the inflationary trajectory. It is convenient to define

$$\bar{\bar{\eta}} \equiv \frac{\sum_{ij} N_i N_j \eta_{ij}}{\sum_m N_m^2} , \tag{26.5}$$

where $\eta_{ij} \equiv V_{ij}/3H_k^2$ as in Section 20.1. In a basis where η_{ij} is diagonal, $\bar{\bar{\eta}}$ is a weighted mean of η_{ii}. The weighting is different from the one in Eq. (20.8), but $|\bar{\bar{\eta}}| \ll 1$ will be more or less equivalent to the flatness condition $|\eta_{ij}| \ll 1$ unless there are many fields. In terms of $\bar{\bar{\eta}}$, and the tensor fraction r given by Eqs. (12.34) and (24.60),

$$\boxed{n(k) - 1 = 2\bar{\bar{\eta}} - 2\epsilon_{\rm H} - r/4} . \tag{26.6}$$

This formula follows from the chain rule applied to $dN = -Hdt$, which gives, using Eq. (20.4) and $V \simeq 3M_{\rm Pl}^2 H^2$ which is equivalent to Eq. (20.3),

$$\frac{d}{d\ln k} = -\frac{M_{\rm Pl}^2}{V} V_i \frac{\partial}{\partial \phi_i}, \qquad N_i V_i = \frac{V}{M_{\rm Pl}^2}, \tag{26.7}$$

and the derivative of the second expression which gives

$$V_i N_{ij} + N_i V_{ij} = \frac{V_j}{M_{\rm Pl}^2}. \tag{26.8}$$

With just one field ϕ_1,

$$n(k) - 1 = 2\eta_{11} - 2\epsilon_{\rm H} - \frac{r}{4} \qquad \text{(single field)}. \tag{26.9}$$

If r is negligible, and a single field ϕ_1 dominates the spectrum,

$$\boxed{n(k) - 1 = 2\eta_{11} - 2\epsilon_{\rm H}}\,. \tag{26.10}$$

If η_{11} is positive we need $\epsilon_{\rm H} \gtrsim 0.02$ to get the observed tilt. As described in Chapter 28 that would require a large-field inflation model. As seen in Section 28.13, the best-motivated case of ϕ^2 inflation would give only $\epsilon_{\rm H} \simeq 0.01$, but $\epsilon_{\rm H}$ can be increased by modifying the potential or by going to multi-field inflation. (Note that the observational constraints discussed in Section 28.13 don't apply here, since they assume that ζ is generated by the inflaton perturbation.)

We may be dealing with single-component inflation where only one of the light fields has significant variation, or multi-component inflation where two or more vary. In any case, it is useful to choose at horizon exit a field basis $\phi_i = \{\phi, \sigma_i\}$, where ϕ points along the inflationary trajectory. The adiabatic perturbation $\delta\phi$ represents a shift along the inflationary trajectory, leading to a time-independent contribution $\zeta_{\rm ad}$ to the curvature perturbation. The spectrum $\mathcal{P}_\zeta^{\rm ad}$ of $\zeta_{\rm ad}$ is given by the usual formula Eq. (25.3) with $V' \equiv \partial V/\partial\phi$. In contrast, the orthogonal perturbations $\delta\sigma_i$ represent a shift onto a nearby trajectory. Such a shift initially has no effect on the curvature perturbation (i.e. the orthogonal perturbations initially are isocurvature perturbations). In the previous chapter we supposed that their effect on ζ remains negligible at all times. In this chapter we suppose instead that one or more of the orthogonal perturbations eventually gives a significant contribution to ζ.

Pulling out the adiabatic contribution

$$\zeta(\mathbf{k}) = \zeta_{\rm ad}(\mathbf{k}) + \sum N_i(k)\delta\sigma_i(\mathbf{k}), \tag{26.11}$$

$$\mathcal{P}_\zeta(k) = \mathcal{P}_\zeta^{\rm ad}(k) + \left(\frac{H_k}{2\pi}\right)^2 \sum N_i^2(k). \tag{26.12}$$

The adiabatic contribution $\mathcal{P}_\zeta^{\rm ad}$ is given by the standard paradigm (Eq. (25.4)). Additional contributions add to the spectrum, and the tensor fraction is correspondingly reduced;

$$\boxed{r = 16\epsilon\,\frac{\mathcal{P}_\zeta^{\ \rm ad}}{\mathcal{P}_\zeta}}\,. \tag{26.13}$$

The spectral index is given by Eq. (26.6). If the adiabatic perturbation dominates (the standard paradigm), Eq. (26.9) applies with $r = 16\epsilon$ and we recover Eq. (25.5) in a different notation. If a single isocurvature perturbation dominates, Eq. (26.10) applies.

26.3 Non-gaussianity

To calculate the non-gaussianity we expand δN to higher order in the field perturbations [3]:

$$\zeta(\mathbf{x},t) = \sum N_i(t)\delta\phi_i(\mathbf{x}) + \frac{1}{2}\sum N_{ij}(t)\delta\phi_i(\mathbf{x})\delta\phi_j(\mathbf{x}) + \ldots, \tag{26.14}$$

where a subscript i denotes $\partial/\partial\phi_i$, and it is usually enough to keep just the linear and quadratic terms.

Non-gaussianity will come from two sources; the non-gaussianity of the field perturbations, and the appearance of quadratic and higher-order terms in the expansion (26.14). We will assume that the perturbed fields have canonical kinetic terms and negligible interaction. As we saw in Section 24.4, the non-gaussianity of the field perturbations vanishes in the slow-roll limit. One expects that their effect will be very small, which has been verified in specific cases. We therefore consider only the non-gaussianity coming from quadratic and higher-order terms of Eq. (26.14).

26.3.1 Single field

Let us assume that only one field perturbation is significant. We already dealt with the case that this is the perturbation $\delta\phi$ of the inflaton in a single-field inflation model. We now suppose that it is a different field denoted by σ (or else that inflation is not driven by a scalar field so that there is no inflaton). We keep just the quadratic term so that

$$\zeta = N'\delta\sigma + \frac{1}{2}N''\delta\sigma^2 \tag{26.15}$$

$$= N'\delta\sigma + \frac{1}{2}\frac{N''}{N'^2}\left(N'\delta\sigma\right)^2, \tag{26.16}$$

where a prime denotes $\partial/\partial\sigma$.

This is exactly the expression that we considered in Section 25.4.1. The second term must be small because the non-gaussianity is small. To first order in that term,

$$\boxed{\frac{3}{5}f_{\rm NL} = \frac{1}{2}\frac{N''}{N'^2}, \qquad 2\mathcal{T}_\zeta = \left(\frac{6f_{\rm NL}}{5}\right)^2} \qquad \text{(single field).} \tag{26.17}$$

Now comes the big difference from the inflaton case. *For a generic field the non-gaussianity predicted by these formulas may easily be big enough to observe.* This is because, in contrast with the inflaton case, there is no reason for N'' to be suppressed relative to N'^2. We will see that in many models there is a definite prediction $|f_{\rm NL}| \gtrsim 1$, barring fine tuning of the parameters.

26.3.2 Uncorrelated non-gaussianity

In the language of Section 25.5, Eq. (26.17) is the tree-level approximation. Suppose now that two different fields are significant in the δN formula, but that only one of them contributes to the quadratic term:

$$\zeta = N_\phi\delta\phi(\mathbf{x}) + N_\sigma\delta\sigma(\mathbf{x}) + \frac{1}{2}N_{\sigma\sigma}(\delta\sigma(\mathbf{x}))^2. \tag{26.18}$$

In the extreme case $N_\sigma = 0$, the (practically) gaussian term $N_\phi\delta\phi$ is uncorrelated with the final non-gaussian term. Then the bispectrum of ζ vanishes at tree level so that the leading term is the one-loop contribution.

Repeating the calculation in Section 25.5 of the one-loop contribution to the spectrum, we find [4]

$$8B_\zeta^{\text{one-loop}} = (2\pi)^3 N_{\sigma\sigma}^3 \int_{L^{-1}} d^3p \frac{\mathcal{P}_{\delta\sigma}(p)}{p^3}\frac{\mathcal{P}_{\delta\sigma}(p_1)}{p_1^3}\frac{\mathcal{P}_{\delta\sigma}(p_2)}{p_2^3}, \tag{26.19}$$

where $p_1 \equiv |\mathbf{p} - \mathbf{k}_1|$ and $p_2 \equiv |\mathbf{k}_2 + \mathbf{p}|$.

As we found earlier, the spectrum is expected to have an almost scale-independent value $(H_*/2\pi)^2$. With absolute scale independence the integral diverges logarithmically when each of the denominators vanishes. The subscript L^{-1} indicates that a sphere of radius L^{-1} is cut out from the integral, around each of these singularities. The integral cannot be done analytically, but with all momenta comparable a reasonable estimate is [4]

$$8B_\zeta^{\text{one-loop}} \simeq (2\pi)^3 \times 4\pi N_{\sigma\sigma}^3 \frac{\ln(kL)}{k^6}\mathcal{P}_{\delta\sigma}^3. \tag{26.20}$$

If $N_\sigma = 0$, so that there is no tree-level contribution, this gives

$$\boxed{\frac{6}{5}f_{\rm NL}\mathcal{P}_\zeta^2 \simeq N_{\sigma\sigma}^3\mathcal{P}_{\delta\sigma}^3\ln(kL)}. \tag{26.21}$$

To obtain the prediction for the observable Universe, one should choose $\ln(H_0 L)$ not too much bigger than 1. The non-gaussian fraction for a given value of $f_{\rm NL}$ is bigger than the estimate (6.73). If the loop contribution dominates the fraction is of order 10% for any $f_{\rm NL} \sim 1$ to 100.

The coefficient $N_{\sigma\sigma}$ depends on the size and location of the box, because it should be evaluated with σ set equal to its unperturbed value and that value should be set equal to the spatial average within the box. Of course, the factor $\ln(kL)$ depends on the box size too. These dependences cease to be puzzling if we remember the ergodic theorem. As shown in Section 6.5, each correlator of ζ (as a function of the relative positions) represents a spatial average, which obviously will depend on the size and location of the box. If the box size is not too big, the dependence will be weak and we have a prediction that can usefully be compared with observation on the assumption that our location within the box is typical. From this viewpoint, the dependence of the prediction on the size and location of the box is just a kind of cosmic variance.

26.4 Curvaton paradigm

According to the curvaton paradigm, all or part of the curvature perturbation comes from the perturbation of a light oscillating 'curvaton' field, that is not the inflaton. We will focus on the cleanest case, that the curvaton accounts for practically all of the curvature perturbation. A review of the curvaton paradigm with references is given in Ref. [5].

During inflation the curvaton acquires the usual almost scale-independent and gaussian perturbation with spectrum $(H_k/2\pi)^2$. As each scale comes inside the horizon, the corresponding Fourier component of σ oscillates and redshifts away. Therefore, σ at a given epoch is nearly homogeneous on the horizon scale.

The curvaton oscillation starts during a radiation-dominated era as in Section 21.1.2. At this stage the curvaton still has a negligible effect, both on the energy density and on the curvature perturbation. But the curvaton energy density goes like a^{-3}, whereas the initially dominant radiation contribution goes like a^{-4}. As a result, the curvaton contribution to the energy density grows, and with it the curvaton contribution to the curvature perturbation.

Well before cosmological scales start to enter the horizon, the curvaton decays leaving behind the curvature perturbation which is assumed to undergo no further change. If the curvaton comes to dominate the energy density, then its decay corresponds to a second reheating.

The curvaton paradigm doesn't require Einstein gravity, and doesn't postulate any particular model of inflation. If there is single-field inflation though, which makes the inflaton contribution to ζ time-independent, the curvaton field has to be different from the inflaton field.

To calculate $\zeta = \delta N$, we need the number of e-folds N up to an epoch after which the curvature perturbation is constant. If the curvaton density dominates at some epoch before it decays, we can take that as the final epoch, because the pressure then vanishes making ζ constant. Otherwise, we will assume that the curvaton decays suddenly and take the final epoch to be just before curvaton decay. The sudden-decay approximation may be checked by following the evolution of the initial radiation density, and it turns out to be quite adequate.

26.4.1 General predictions

Using the sudden-decay approximation, the spectrum of the curvature perturbation can be calculated from the δN formula [3]. We denote the curvaton field by σ, and its value a few Hubble times after horizon exit for the scale k under consideration by σ_*. The initial amplitude of the curvaton oscillation will be some function $g(\sigma_*)$ of σ_*.

Since the curvature perturbation is supposed to be negligible before the oscillation starts, we can take the initial slice in the δN formula to be just after the curvaton oscillation begins, when the energy density ρ_1 is still dominated by the practically unperturbed radiation density. At that stage the curvaton energy density is $\frac{1}{2}m_\sigma^2 g^2$.

Just before decay, let us denote the total energy density by ρ_2, the radiation energy density by ρ_{rad} and the curvaton energy density by ρ_σ. Then

$$\rho_2(\mathbf{x}) = \rho_{\text{rad}}(\mathbf{x}) + \rho_\sigma(\mathbf{x}), \qquad \rho_\sigma(\mathbf{x}) = \frac{1}{2}m_\sigma^2 g^2(\sigma_*(\mathbf{x})) \left(\frac{\rho_{\text{rad}}(\mathbf{x})}{\rho_1}\right)^{3/4}. \tag{26.22}$$

Calculating $N(g(\sigma_*), \rho_1, \rho_2)$ and taking the partial derivative with respect to σ_* one finds

$$N_\sigma = \frac{2}{3}f\frac{g'}{g}, \qquad f \equiv \frac{3\rho_\sigma}{3\rho_\sigma + 4\rho_{\text{rad}}} \simeq \Omega_\sigma. \tag{26.23}$$

In the definition of f, the energy densities are unperturbed, and $\Omega_\sigma \equiv \rho_\sigma/\rho$ is evaluated just before the curvaton decay starts to be significant. The final equality becomes exact in the limit $\Omega_\sigma \to 1$, and $f \simeq \Omega_\sigma$ is always an adequate approximation because it generates an error no bigger than that of the sudden-decay approximation.

To first order in $\delta\sigma$,

$$\boxed{\zeta = \frac{2\Omega_\sigma g'}{3g}\delta\sigma(\mathbf{x})}, \tag{26.24}$$

and

$$\boxed{\mathcal{P}_\zeta = \frac{4}{9}\Omega_\sigma^2 \left(\frac{g'}{g}\right)^2 \left(\frac{H_*}{2\pi}\right)^2}. \tag{26.25}$$

The spectral index is given by Eq. (26.10). It depends only on $(\partial^2 V/\partial\sigma^2)/3H^2$ and $\dot{H}/H^2$, evaluated at horizon exit during inflation.

Calculating $N_{\sigma\sigma}$ and taking $\delta\sigma$ to be gaussian, we find the non-gaussianity parameter

$$\boxed{\frac{3}{5}f_{\rm NL} = \frac{1}{2}\frac{N_{\sigma\sigma}}{N_\sigma^2} = -1 - \frac{2}{3}\Omega_\sigma + \frac{3}{4}\Omega_\sigma^{-1}\left(1 + \frac{g''g}{g'^2}\right)}. \tag{26.26}$$

We see that $|f_{\rm NL}| \gtrsim 1$ unless there is accurate cancellation. If $|f_{\rm NL}| \gg 1$, Eq. (26.26) reduces to

$$\boxed{\frac{3}{5}f_{\rm NL} = \frac{3}{4}\Omega_\sigma^{-1}\left(1 + \frac{g''g}{g'^2}\right) \gg 1}. \tag{26.27}$$

Instead of using δN, Eq. (26.24) can be derived more simply [6] from Eq. (5.28). It gives

$$\zeta(\mathbf{x}) = \frac{1}{3}\Omega_\sigma \frac{\delta\rho_\sigma}{\rho_\sigma}. \tag{26.28}$$

To first order in δg this gives

$$\zeta = \frac{2}{3}\Omega_\sigma \frac{\delta g}{g}, \tag{26.29}$$

which is equivalent to Eq. (26.24). Going to second order in δg gives Eq. (26.27), which is valid if $|f_{\rm NL}| \gg 1$ [6]. These results treat $\delta\rho$ to first order. To derive the full expression (26.26) (needed for $|f_{\rm NL}| \lesssim 1$) one has to go to second order in $\delta\rho$ [7]. That means going to second-order cosmological perturbation theory, which is not described in this book, and is more complicated than using the δN formula.

26.4.2 Linear evolution of the curvaton field

In writing these formulas, we kept $g(\sigma_*)$ as an unspecified function. It may indeed be highly non-trivial. In particular if σ is a pseudo Nambu–Goldstone boson (PNGB) corresponding to the distance in field space around the bottom of a Mexican-hat potential, then strong evolution of the symmetry-breaking potential might occur. On the other hand, it could easily happen that $g(\sigma_*)$ is just σ_* (corresponding to negligible evolution of the field) or at least that $g(\sigma_*)$ is linear.

Taking $g = \sigma_*$, the spectrum is

$$\mathcal{P}_\zeta = \frac{4}{9}\Omega_\sigma^2 \left(\frac{H_*}{2\pi\sigma_*}\right)^2 . \tag{26.30}$$

Using Eq. (21.5) one finds that the observed value of $\mathcal{P}_\zeta$ cannot be reproduced unless $H_* \gtrsim 10^7$ GeV. Assuming only that $g(\sigma_*)$ is linear, the non-gaussianity parameter is $f_{\rm NL} = (5/4)\Omega_\sigma^{-1}$, and the observational bound $|f_{\rm NL}| \lesssim 10^2$ therefore requires $\Omega_\sigma \gtrsim 10^{-2}$; the curvaton must account for more than one percent of the energy density before it decays.

To estimate σ_*, we may invoke the stochastic estimate Eq. (24.58). For a quadratic potential this gives Eq. (24.28), leading to

$$\mathcal{P}_\zeta \sim \Omega_\sigma^2 \frac{m_\sigma^2}{H_*^2}. \tag{26.31}$$

The observed value of $\mathcal{P}_\zeta$ requires $\eta_{\sigma\sigma} \equiv m_\sigma^2/3H_*^2 \sim 10^{-9}\Omega_\sigma^{-2}$.

Instead we may suppose that σ is a PNGB with the sinusoidal potential (14.34). This would allow $\eta_{\sigma\sigma} < 0$, and one might hope that $\eta_{\sigma\sigma} \simeq -0.02$ can be achieved, giving the required negative tilt $n - 1 \simeq -0.04$ with negligible $\epsilon_{\rm H}$. The maximum tilt is obtained if σ_* corresponds to the maximum of the potential, giving

$$n - 1 \sim -\frac{\Lambda^2}{H_*^2}\frac{\Lambda^2}{f^2}. \tag{26.32}$$

We need $\Lambda^2 \ll H_*^2$ to have a reasonable probability of being at the top, and we expect $\Lambda^2 \ll f^2$ so that the self-coupling of σ is not too strong. A rather small tilt is therefore predicted, but it might be compatible with the observed value.

26.5 Inhomogeneous decay rate

In the curvaton model, inhomogeneous reheating occurs because the amplitude of the oscillating curvaton field is perturbed. Instead, reheating might be inhomogeneous because the lifetime of the decaying particle is perturbed by a dependence on a light field. Then the reheating may be the first one, corresponding to the decay of the inflaton field.

In this section we calculate the curvature perturbation generated by this mechanism, following Ref. [8] where references to earlier work are given. As with the curvaton paradigm, spacetime is supposed to be practically unperturbed until σ starts to decay. We again make the sudden-decay approximation, that the decay occurs on a slice of uniform H at say $H = \Gamma$. We consider the number of e-folds N, from an initial epoch before decay when spacetime is unperturbed, to a final epoch after decay when the energy density has an assigned value. Denoting these epochs by subscripts i and f, and the epoch of decay by 'dec',

$$e^N = \frac{a_\mathrm{f}}{a_\mathrm{i}} = \frac{a_\mathrm{f}}{a_\mathrm{dec}}\frac{a_\mathrm{dec}}{a_\mathrm{i}}. \tag{26.33}$$

For illustration we will assume that the Universe is matter dominated by the time of reheating. Then we can take the initial epoch to be during matter domination. During matter domination, $a \propto H^{-2/3}$, while after the decay we have radiation domination with $a \propto H^{-1/2}$. But the decay occurs when $H = \Gamma$, therefore

$$\frac{a_\mathrm{dec}}{a_\mathrm{i}} \propto \Gamma^{-2/3}, \qquad \frac{a_\mathrm{f}}{a_\mathrm{dec}} \propto \Gamma^{1/2}. \tag{26.34}$$

This gives $e^N \propto \Gamma^{-1/6}$, and

$$\frac{\partial N}{\partial \Gamma} = -\frac{1}{6}\frac{1}{\Gamma}, \qquad \frac{\partial^2 N}{\partial \Gamma^2} = \frac{1}{6}\frac{1}{\Gamma^2}, \tag{26.35}$$

and hence

$$\delta N = \zeta = -\frac{1}{6}\left[\frac{\delta\Gamma}{\Gamma} - \frac{1}{2}\left(\frac{\delta\Gamma}{\Gamma}\right)^2\right]. \tag{26.36}$$

To complete the calculation we need the function $\Gamma(\sigma_*)$, giving the decay rate as a function of the curvaton-like field at horizon exit. Expanding $\delta\Gamma$ to second order we find

$$\boxed{\mathcal{P}_\zeta^{1/2} = \frac{1}{6}\frac{\sigma_*\Gamma'}{\Gamma}\frac{H_*}{2\pi\sigma_*}}, \qquad \boxed{\frac{3}{5}f_\mathrm{NL} = 3\left[\frac{\Gamma''\Gamma}{\Gamma'^2} - 1\right]}. \tag{26.37}$$

Instead of (or as well as) the decay rate being inhomogeneous, it may be that the mass of the decaying particle is inhomogeneous. If the decay rate were homogeneous, this would generate a curvature perturbation in exactly the same way as in the curvaton scenario; in fact, if $m \propto \sigma_*^2$ the curvaton formulas would apply exactly. But the decay rate will itself depend on the mass, the dependence usually being $\Gamma \propto m^2$. This generates an inhomogeneity $\delta\Gamma/\Gamma = 2\delta m/m$, whose contribution to the curvature perturbation is given by the formulas that we just considered. The total curvature perturbation generated by the mass inhomogeneity is the sum of these two effects.

26.6 More ways of generating the curvature perturbation

In this section we mention some more mechanisms for generating a contribution to the curvature perturbation after horizon exit.

26.6.1 Multi-field inflaton

For multi-field slow-roll inflation there is a family of curved inflationary trajectories, and the perturbation of the fields that vary during inflation will select different members of the family at different spacetime points. As the potential dominates the energy density, the final spacetime slice defining $N(\mathbf{x}, t)$ during inflation can be taken to be an equipotential surface in field space. The curvature perturbation $\zeta = \delta N$ will change right up to the end of slow-roll inflation, unless the trajectories become straight as the end of slow-roll approaches. Whether or not that happens, it may be that the value of ζ at the end of slow-roll inflation is practically equal to its final value. One may call this the multi-field slow-roll paradigm.

26.6.2 End of inflation and preheating

If the end of inflation is abrupt, it may generate a contribution $\zeta_{\rm end}$ and this might be the dominant one. The generation will occur if the ending of inflation is inhomogeneous, in the sense that the energy density at the end has some perturbation $\delta\rho$. Remembering that $dN = Hdt$,

$$\zeta_{\rm end} = H\left(\frac{1}{\dot{\rho}_{\rm after}} - \frac{1}{\dot{\rho}_{\rm before}}\right)\delta\rho. \tag{26.38}$$

With Einstein gravity $\dot{\rho}$ will be smaller before the end of inflation than afterwards, and in a slow-roll model it will be much smaller so that the second term dominates. Keeping only the second term, the contribution from the end of inflation is taken care of by evaluating N up to the end of inflation (which is not a slice of uniform density). This scenario is realised in hybrid inflation, if the mass of the waterfall field during inflation depends on some light field.

Inhomogeneous preheating can also generate a curvature perturbation. For an example with ordinary preheating, see Ref. [10], and for one with instant preheating see Ref. [11].

Yet another mechanism for generating the curvature perturbation is provided by the production of fermions through the instant preheating mechanism, without the actual occurrence of preheating (see Section 21.2.3). For an example using the oscillation of an MSSM flat direction, see Ref. [12].

Exercises

26.1 Verify Eq. (26.19) giving the bispectrum for an uncorrelated non-gaussian contribution to ζ.

26.2 Verify Eq. (26.23) giving N_σ for the curvaton model.

26.3 Derive Eq. (26.27) for the non-gaussianity of the curvaton model, by the method mentioned after Eq. (26.29).

26.4 Suppose that slow-roll hybrid inflation takes place with potential $V(\phi)$, and ends inhomogeneously at a value $\phi_c(\mathbf{x})$. Use Eq. (26.38), keeping only the second term, to work out the curvature perturbation $\zeta_{\rm end}$ generated at the end of inflation to first order in $\delta\phi_c$. If ϕ_c is a function of a light field σ, use your formula to work out the spectrum of $\zeta_{\rm end}$ in terms of $d\phi_c/d\sigma$. (This calculation is done in Ref. [13].)

References

[1] M. Sasaki and E. D. Stewart. A general analytic formula for the spectral index of the density perturbations produced during inflation. *Prog. Theor. Phys.*, **95** (1996) 71.

[2] D. H. Lyth and A. Riotto. Particle physics models of inflation and the cosmological density perturbation. *Phys. Rept.*, **314** (1999) 1.

[3] D. H. Lyth and Y. Rodriguez. The inflationary prediction for primordial non-gaussianity. *Phys. Rev. Lett.*, **95** (2005) 121302.

[4] L. Boubekeur and D. H. Lyth. Detecting a small perturbation through its non-Gaussianity. *Phys. Rev. D*, **73** (2006) 021301.

[5] D. H. Lyth. Non-gaussianity and cosmic uncertainty in curvaton-type models. *JCAP*, **0606** (2006) 015.

[6] D. H. Lyth, C. Ungarelli and D. Wands. The primordial density perturbation in the curvaton scenario. *Phys. Rev. D*, **67** (2003) 023503.

[7] N. Bartolo, S. Matarrese and A. Riotto. On non-gaussianity in the curvaton scenario. *Phys. Rev. D*, **69** (2004) 043503.

[8] M. Zaldarriaga. Non-gaussianities in models with a varying inflaton decay rate. *Phys. Rev. D*, **69** (2004) 043508.

[9] A. A. Starobinsky. Multicomponent de Sitter (inflationary) stages and the generation of perturbations. *JETP Lett.*, **42** (1985) 152.

[10] A. Chambers and A. Rajantie. Non-Gaussianity from massless preheating. *JCAP*, **0808** (2008) 002.

[11] T. Matsuda. Generating the curvature perturbation with instant preheating. *JCAP*, **0703** (2007) 003.

[12] A. Riotto and F. Riva, Curvature perturbation from supersymmetric flat directions. *Phys. Lett. B*, **670** (2008) 169.

[13] D. H. Lyth. Generating the curvature perturbation at the end of inflation. *JCAP*, **0511** (2005) 006.

27

Generating primordial isocurvature perturbations

In Chapter 12 we defined the isocurvature perturbations that may exist at the 'initial' epoch just after the neutrino decoupling at $T \sim 1\,\text{MeV}$. These are the baryon, cold dark matter (CDM) and neutrino isocurvature perturbations S_B, S_c and S_ν. In this chapter we see how these perturbations might be generated from the perturbation in one or more light fields.

27.1 The δn_i formula

At the epoch $T \sim 1\,\text{MeV}$ the Universe is radiation dominated to high accuracy, and we may replace the definition (12.4) of S_i by

$$S_i(\mathbf{x}) = \frac{\delta n_i(\mathbf{x})}{n_i} \qquad \text{(uniform energy density slice).} \tag{27.1}$$

where the perturbation δn_i is defined on a slice of uniform energy density and n_i is the unperturbed quantity. This is an exact measure of the departure from the exact adiabatic condition Eq. (5.4), but in practice one treats S_i as a first-order cosmological perturbation. As we noted in Section 12.1, that will probably always be enough.

In Chapter 12 we used the definition (12.4) also after matter domination (on scales outside the horizon). That was convenient because it made S_i constant. But that is equivalent to defining $S_i = \delta n_i/n_i$ on slices where the combined number density of photons and neutrinos is uniform. Such a definition is not very useful in the early Universe, because we don't even want to assume that photons and neutrinos are present (still less that their number in a comoving volume is conserved). Instead, it will be far more useful to work on the slicing of uniform energy density, as in Eq. (27.1).

In order to define the CDM isocurvature perturbation S_c at early times we can use Eq. (27.1) as it stands. We can do the same for S_B, provided that n_B is the density of baryon number, as opposed to the number density of nucleons which makes sense only after the quark–hadron transition. For S_ν, we need Eq. (12.30) which relates it to the lepton number isocurvature perturbation $S_L \equiv \delta n_L/n_L$. That quantity makes sense in the early Universe, just like S_B.

As far as the origin of the isocurvature perturbations is concerned, then, we deal with S_c, S_B and S_L defined by Eq. (27.1). In each case, we will need to consider the number density n_i only after the epoch when the creation of the quantity is complete.

In terms of the locally defined scale factor, the number density after creation is given by

$$n_i(\mathbf{x}, t) = n_i(\mathbf{x}, t_{\rm cr})\frac{a^3(\mathbf{x}, t_{\rm cr})}{a^3(\mathbf{x}, t)}, \tag{27.2}$$

where ‘cr’ denotes the epoch of creation. This gives

$$S_i \equiv \frac{\delta n_i}{n_i} = \left.\frac{\delta n_i}{n_i}\right|_{\rm cr} + 3\left(\zeta_{\rm cr} - \zeta\right). \tag{27.3}$$

If ζ has attained its final value at the epoch of creation the last term vanishes.

Invoking the separate universe assumption, we suppose that $n_i(\mathbf{x}, t_{\rm cr})$ depends on the values $\phi_i(\mathbf{x})$ of one or more light fields. As with the δN formula, these values are to be evaluated on a flat slice during inflation after all scales of interest have left the horizon. Then we have

$$\boxed{\delta n_i(\mathbf{x}, t_{\rm cr}) = \sum_n n_{in}\delta\phi_n(\mathbf{x}) + \frac{1}{2}\sum_{n,m} n_{inm}\delta\phi_n\delta\phi_m + \ldots}, \tag{27.4}$$

where $n_{in} = \partial n_i/\partial\phi_n$ and $n_{inm} = \partial^2 n_i/\partial\phi_n\partial\phi_m$. In this formula, the number density $n_i(\phi_1, \phi_2, \ldots)$ is evaluated at the epoch of creation, and its partial derivatives are evaluated with ϕ_i at its unperturbed value.

This δn_i formula is quite analogous to the δN formula (26.14). Using both formulas, Eq. (27.3) gives the primordial isocurvature perturbations S_i, in terms of the perturbation in one or more light scalar fields. Those perturbations are practically gaussian with spectrum $(H_k/2\pi)^2$. If S_i is practically linear in the perturbations it too is practically gaussian, otherwise it can be strongly non-gaussian.

In most scenarios, S_i is created after ζ attains its final value. That presumably must be so if ζ is created during inflation. It must also be the case if n_i is created after the final reheating (or *the* reheating if there is only one), because that reheating is supposed to lead to practically complete radiation domination during which ζ cannot be created.

When S_i is created after ζ attains its final value, the second term of Eq. (27.3) vanishes. On the reasonable assumption that only one field σ is relevant, we then have

$$S_i(\mathbf{x}) = n_i'\delta\sigma(\mathbf{x}) + \frac{1}{2}n_i''(\delta\sigma(\mathbf{x}))^2 + \ldots, \tag{27.5}$$

where $n_i' = dn_i/d\sigma$ and $n_i'' = d^2n_i/d\sigma^2$.

In the special case that ζ is generated at horizon exit in a single-field inflation model, σ has to be different from the inflaton field. (This is because the local value of the inflaton field determines the local value of ρ. If it determined also the local value of n_i there would be no perturbation δn_i.) More generally, the perturbation $\delta\sigma$ that is responsible for S_i might also be wholely or partly responsible for ζ. That would require a special setup though, which is not usually considered.

Assuming, then, that $\delta\sigma$ has nothing to do with ζ, we conclude that *if the isocurvature perturbation S_i is created after the curvature perturbation ζ, the two will be uncorrelated.*

In principle, any mechanism for generating n_i might generate an isocurvature perturbation, because one or more of the parameters of the Lagrangian in the effective field theory might in reality depend on some field which is light (during inflation) and therefore perturbed. In particular a decay rate could depend on σ, as is supposed to be the case for the inhomogeneous decay rate mechanism for the creation of ζ. But the most straightforward way of generating an isocurvature perturbation is to have n_i depend directly on the value of σ, analogously with the curvaton mechanism for the generation of ζ. We now consider two examples of that.

27.2 Axion CDM isocurvature perturbation

As we saw in Section 23.1, axion CDM corresponds to the oscillation of the axion field. In the no-string scenario, the local value of this field at the onset of the oscillation is the same as it was during inflation (assuming that the form of the Peccei–Quinn (PQ) symmetry-breaking potential is the same). Since the axion is light, an axion CDM perturbation is then inevitable. We describe it now, drawing largely on Ref. [1].

Instead of the axion field a we use the misalignment angle $\theta = a/\sqrt{2}f_{\rm a}$. At the end of inflation $\theta(\mathbf{x})$ is perturbed on all scales outside the horizon:

$$\theta(\mathbf{x}) = \theta_* + \delta\theta(\mathbf{x}), \tag{27.6}$$

where θ_* is the unperturbed value. As each scale comes inside the horizon, θ oscillates corresponding to the presence of massless axions , and the oscillation dies as the axions redshift. At the epoch given by Eq. (23.2), the axion field starts to oscillate as a standing wave. Taking $\theta \ll \pi$, the axion potential will be quadratic so that the axion energy density $\rho_a(\mathbf{x})$ is proportional to $\theta^2(\mathbf{x})$, evaluated just before the oscillation begins.

As the axion field doesn't evolve on scales outside the horizon, the unperturbed value θ_* is the same as during inflation, and so is the spectrum of $\delta\theta$. The latter is given by

$$\mathcal{P}_{\delta\theta}^{1/2} = \frac{1}{\sqrt{2}f_a}\frac{H_*}{2\pi}. \tag{27.7}$$

27.2.1 Pure axion CDM

Let us first assume that the CDM consists entirely of axions. Then, since observation requires S_c to be very small, we need $|\delta\theta| \ll \theta_*$ leading to $S_c = 2\delta\theta/\theta_*$ as an excellent approximation, where θ_* is the unperturbed value during inflation. The spectrum of the isocurvature perturbation is given by

$$\mathcal{P}_{S_c}^{1/2} = \frac{2}{\theta_*}\mathcal{P}_{\delta\theta}^{1/2}. \tag{27.8}$$

Demanding that the axion density given by Eq. (23.3) has the observed value, this gives

$$\frac{H_*}{10^8\,\mathrm{GeV}} = \sqrt{\frac{\alpha_m}{\gamma}}\left(\frac{f_a}{10^{12}\,\mathrm{GeV}}\right)^{0.4}. \tag{27.9}$$

Here $\gamma \leq 1$ is the entropy dilution factor and $\alpha_m \equiv \mathcal{P}_{S_m}/\mathcal{P}_\zeta$ as in Section 12.1. (We dropped a factor Ω_c/Ω_m which is close to 1.)

These considerations apply in the no-string scenario. Assuming that PQ symmetry is broken by a single field with the Mexican-hat potential (14.32) and $\lambda \sim 1$, the no-string scenario will hold if the reheat temperature and H_* are both $\lesssim f_a$. Roughly similar conditions will be required if more than one field breaks PQ symmetry, unless the symmetry-breaking potential is very flat in which case the string scenario becomes more or less inevitable.

Assuming Einstein gravity so that $\rho = 3M_{Pl}^2H^2$, and taking for granted the condition on the reheat temperature, the excluded region for the energy scale during inflation implied by the bound $\alpha_m < 0.3$ is

$$\boxed{2\gamma^{-1/4}f_{12}^{0.2} \times 10^{13}\,\mathrm{GeV}\times \lesssim V_*^{1/4} \lesssim 10^{15}\,\mathrm{GeV} \times f_{12}^{0.5} \quad \text{(excluded)}}, \tag{27.10}$$

where $f_{12} \equiv f_a/10^{12}\,\mathrm{GeV}$. This is shown in Figure 27.1 assuming $\gamma = 1$ (no entropy dilution).

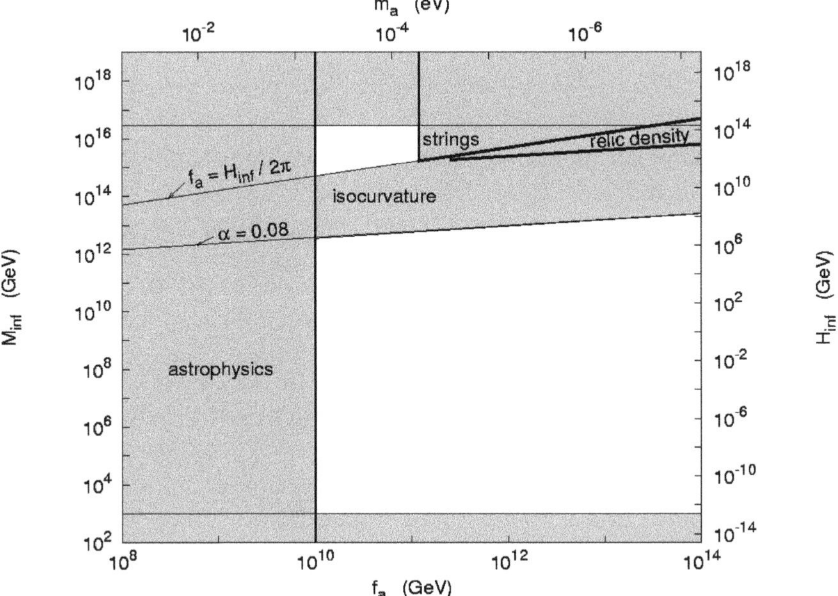

Fig. 27.1. The excluded region given by Eq. (27.10) with $\gamma = 1$ (no entropy dilution). Only $f_{\rm a} > 10^{11}$ GeV is shown because Eq. (23.3) for the axion density forbids smaller values.

27.2.2 A non-gaussian axion isocurvature perturbation

As PQ symmetry is rather well motivated, it makes sense to suppose that the CDM contains an axion component even if some other particle dominates. In that case direct detection of axion CDM might still be possible. As we will now see, a subdominant axion component of the CDM could even give a distinctive signal in the cosmic microwave background (CMB) anisotropy, namely non-gaussianity [1].

To see this, we keep the quadratic term of $\rho \propto \theta^2$ to get

$$S_{\rm c} = R_{\rm a}\left[2\frac{\delta\theta}{\theta} + \left(\frac{\delta\theta}{\theta}\right)^2\right] \tag{27.11}$$

$$= \left(2R_{\rm a}\frac{\delta\theta}{\theta}\right) + \frac{1}{4R_{\rm a}}\left(2R_{\rm a}\frac{\delta\theta}{\theta}\right)^2, \tag{27.12}$$

where $R_{\rm a} \equiv \Omega_{\rm a}/\Omega_{\rm c}$ is the axion fraction of the CDM. This is valid in the regime $\mathcal{P}_{\delta\theta/\theta_*} \ll 1$, where $\delta\theta/\theta_*$ is small almost everywhere.

The spectrum is now given by $\mathcal{P}_{S_{\rm c}}^{1/2} = 2R_{\rm a}\mathcal{P}_{\delta\theta}^{1/2}/\theta_*$, and H_* is a factor $R_{\rm a}^{-1/2}$ times the value given in Eq. (27.9). As seen in Chapter 6, the reduced bispectrum is $\mathcal{B}_{S_c} = 1/2R_a$. The non-gaussian fraction of S_c is

$$\frac{S_c^{\text{n.g.}}}{S_c^{\text{g.}}} \simeq \mathcal{P}_{S_c}^{1/2}\mathcal{B}_{S_c} = \frac{2}{\theta_*}\mathcal{P}_{\delta\theta}^{1/2}. \tag{27.13}$$

Fixing the spectrum $\mathcal{P}_{S_c}$, we can allow the non-gaussian fraction to to rise towards 1 by allowing R_a to fall.

Eventually though, we enter the regime $\mathcal{P}_{\delta\theta}^{1/2} \gg \theta_*$, where θ_* is negligible. Then

$$S_c = R_a \frac{(\delta\theta)^2}{\overline{\theta^2}}, \tag{27.14}$$

where the mean square $\overline{\theta^2}$ is of order $\mathcal{P}_{\delta\theta}$ as seen in Eq. (6.24) (assuming a minimal box size). The spectrum and reduced bispectrum, calculated as in Sections 25.5 and 26.3.2, are now given by

$$\mathcal{P}_{S_c}^{1/2} \simeq R_a, \qquad \mathcal{B}_{S_c} \simeq R_a^{-1}. \tag{27.15}$$

We see that the $\mathcal{P}_{S_c}$ can no longer be held fixed as R_a decreases. With fixed $\mathcal{P}_{S_c}$, the maximal non-gaussian fraction (of order 1) is attained when $\mathcal{P}_{\delta\sigma}^{1/2} \simeq \theta_*$. To achieve this, we need $R_a \simeq (\mathcal{P}_\zeta \alpha_m)^{1/2}$ and hence

$$\frac{H_*}{10^{10}\,\text{GeV}} \simeq \alpha_m^{1/4}\gamma^{-1/2}\left(\frac{f_a}{10^{12}\,\text{GeV}}\right). \tag{27.16}$$

What about detecting this maximal non-gaussianity? We see in Figures 12.1 (page 193) and 12.2 that the CMB and matter transfer functions, for the matter isocurvature mode, fall off quickly with increasing scale. The main signal will be in the CMB anisotropy on the Sachs–Wolfe plateau. A precise bound on isocurvature non-gaussianity has not been given at the time of writing, and the bound $\alpha_m^{1/2} \lesssim 0.3$ given in Section 12.1 was obtained on the assumption of gaussianity. If $\alpha_m^{1/2}$ is within a factor 10 or so of this bound (i.e. roughly of order 10%) then one can expect [2] the non-gaussianity to be detectable, since its effect on the CMB Sachs–Wolfe plateau will be similar to that of a 10% uncorrelated gaussian-squared contribution to ζ which, as saw in Section 26.3.2, ought to be detectable.

27.2.3 Bubbles of axion domain wall

We have been assuming $\theta \ll \pi$ so that the axion potential is quadratic. In the unperturbed case, allowing $\theta \simeq \pi$ just gives a modest enhancement of the axion density [1]. The perturbed case is quite different. If $\theta(\mathbf{x}) > \pi$ in some regions during inflation, the boundaries of those regions become domain walls when the axion mass switches on.

To avoid a conflict with observation, the regions with $\theta(\mathbf{x}) > \pi$ must be quite rare so that the domain walls are just small bubbles [3]. This corresponds to $\mathcal{P}_{\delta\sigma}^{1/2} \ll (\pi - \theta_*)$. That constraint will be relevant in part of the space of the parameters $(f_a, H_*, \gamma, \alpha_m)$, but it has not been explored at the time of writing.

27.3 Affleck–Dine isocurvature perturbation

In this section we see how an isocurvature perturbation could be generated by the Affleck–Dine mechanism. Let us recall its basic features.

The Affleck–Dine mechanism uses a field corresponding to a flat direction of the MSSM, which carries baryon number and/or lepton number. At some stage the phase of the field starts to change rapidly, corresponding to the presence of baryon and/or lepton number. In the simplest scenario the Affleck–Dine field oscillates and decays, releasing quarks and/or leptons which carry the baryon and/or lepton number that resided in the field. Alternatively the Affleck–Dine field may form Q-balls. The Q-balls may decay to give quarks and leptons as in the oscillation scenario. If the CDM is the neutralino and the decay of the Q-ball is after the freeze-out, the neutralinos from Q-ball decay can dominate those from freeze-out. Finally, the Q-balls might be stable, in which case they themselves could be the CDM.

In summary, the Affleck–Dine mechanism can create baryon number, lepton number, neutralino CDM or Q-ball CDM. In each case, there will be an isocurvature perturbation if the complex Affleck–Dine field ϕ is perturbed when its phase starts to change rapidly. Let us see if that can happen.

The expected form of $V(\phi)$ is given in Eq. (22.15). In a generic supergravity theory one expects $c_1 \sim \pm 1$. If that happens, we need the minus sign so that $|\phi|$ is well away from the origin when its phase starts to oscillate. Then $|\phi|$ will be fixed to minimize the potential and it will have no perturbation. The situation is different for c_2 though, because it would vanish if the relevant global $U(1)$ symmetry (corresponding to the conservation of baryon number or lepton number) were unbroken. The magnitude of c_2 is therefore quite model dependent, and we can easily have $|c_2| \ll 1$ leading to an isocurvature perturbation [4]. As with the axion case, the isocurvature perturbation can be highly non-gaussian.

27.4 Correlated CDM or baryon isocurvature perturbation

Now we consider two scenarios for getting an isocurvature perturbation which is fully correlated with the curvature perturbation.

27.4.1 Creation of n_i before creation of ζ

The simplest possibility is for the relevant number density, n_c or n_B, to be homogeneous when it is created, and for the Universe to have no curvature perturbation at that time. Then Eq. (27.3) gives $S_i = -3\zeta$. This is several times bigger than is allowed by observation, for both S_B and S_c. Allowing $\delta n_i \neq 0$ doesn't help, unless there is a cancellation. Instead *we need to create both CDM and baryon number when ζ has almost attained its final value* [5]. This is a strong restriction on scenarios of the early Universe that create the curvature perturbation after inflation is over.

If $\delta n_i = 0$ at the epoch of creation we have $S_i = -3(\zeta_{cr} - \zeta)$. In the usual case that a single perturbation $\delta\sigma$ is responsible for the curvature perturbation, the two terms of this expression are fully correlated which means that S_i is fully correlated with the observed curvature perturbation ζ.

What about a non-zero δn_i at the epoch of creation? Contrary to what one might think, some dependence of $n_i(\mathbf{x})$ on the value $\sigma(\mathbf{x})$ of the field responsible for creating ζ is quite natural. This is because σ can affect the relationship between energy density and other quantities, such as the temperature, upon which the generation mechanism might depend. Within the curvaton scenario, this dependence and the resulting δn_c is calculated in Ref. [6] for the neutralino, axion and supermassive CDM candidates.

If $n_i(\mathbf{x})$ at creation depends *only* on $\sigma(\mathbf{x})$ we again have a fully correlated isocurvature perturbation. That is a very reasonable supposition if the creation mechanism for n_i doesn't directly involve a light field, as with for example neutralino CDM or baryogenesis from out-of-equilibrium decay. Even if a light field is involved, as might be the case for say axion CDM or Affleck–Dine baryogenesis, the effect of the perturbation in that field could be negligible.

27.4.2 Generating CDM or baryon isocurvature from curvaton decay

If ζ is generated by the curvaton mechanism, the CDM or baryon number might be generated by the curvaton decay itself. Then we generate a CDM isocurvature perturbation at that time, provided that CDM or B is produced far out of thermal equilibrium, so that it is neither created nor destroyed after production. This perturbation is fully correlated with the curvature perturbation.

To calculate it, we just need the fact that the decay of a single curvaton particle will produce a fixed amount of CDM or B, averaged over many decays. (If there is only a single decay channel the number will be the same for every decay but we don't need to assume that.) The number density n_i ($i =$c or B) after the curvaton has completely decayed is therefore a constant times $n_\sigma(\mathbf{x})a_1^3/a_{cr}^3(\mathbf{x})$, where n_σ is the number density at an epoch when ζ is negligible and a_1 is evaluated at the same epoch [5]. Working to first order in δn_i and ζ (as is appropriate for first-order cosmological perturbation theory), Eq. (26.28) then gives

$$S_i = 3\,\frac{1-\Omega_\sigma}{\Omega_\sigma}\,\zeta. \tag{27.17}$$

To have $|S_i| \ll |\zeta|$, as required by obervation, Ω_σ has to be close to 1. Then, according to Eq. (26.27), ζ and S_i will be almost gaussian unless $gg'' \gg g'^2$.

27.5 Neutrino isocurvature perturbation

The neutrino isocurvature perturbation S_ν is given by Eq. (12.30). The corresponding value of $\alpha_\nu \equiv \mathcal{P}_{S_\nu}/\mathcal{P}_\zeta$ is given by

$$\alpha_\nu^{1/2} = 1 \times 10^4 \eta_L^2 \mathcal{P}_{S_L}^{1/2}, \tag{27.18}$$

where η_L is the lepton number per photon. As with S_B, there is a bound $\mathcal{P}_{S_L} \lesssim 1$ with the maximum corresponding to a completely non-gaussian S_L.

Observation requires roughly $\alpha_\nu \lesssim 10^{-2}$ (roughly the same as for $\alpha_{\rm m}$) and if the neutrino isocurvature perturbation is ever to be observable α_ν should not be many orders of magnitude below that value. On the other hand, successful Big Bang Nucleosynthesis (BBN) requires $|\eta_L| \lesssim 0.1$. To have an observable S_ν, we therefore need η_L to be not many orders of magnitude below its BBN bound, and hence many orders of magnitude *bigger* than $\eta_B = 6 \times 10^{-10}$.

As we saw in Chapter 22, thermal equilibrium before the electroweak transition is likely, fixing η_L/η_B at a ratio of order 1 given by Eq. (22.7). To avoid this conclusion one might try to generate L after the electroweak transition, but as we noticed in Chapter 22 it is difficult to come up with a suitable mechanism at such a late epoch.

There are however more possibilities for generating a large lepton number. One is to lock up the lepton number in Q-balls formed during Affleck–Dine baryogenesis, allowing it to leak out only after the electroweak transition [7]. Another is to have spontaneous leptogenesis, ending only after the electroweak transition has taken place [8]. In either case, the lepton number can easily be perturbed because a scalar field is directly involved.

As with S_B, the main signal for non-gaussianity of S_ν will be in the CMB anisotropy at low ℓ. Also as in that case, no upper bound has been published, but if there is extreme non-gaussianity corresponding to $\mathcal{P}_{S_L}$ then there should be a detectable signal if α_ν is not too far below the bound of order 10^{-2} implied by observation assuming gaussianity.

Finally, we note that a neutrino isocurvature perturbation is unlikely to be correlated with the curvature perturbation. This is because the most plausible mechanism for achieving correlation, described in the previous section, would give at most $|S_L| \sim |\zeta|$ and hence at most $\alpha_\nu^{1/2} \lesssim \eta_L^2$, which is unlikely to be big enough to observe.

Exercises

27.1 Verify Eq. (27.10), giving the excluded region for the inflation scale if the CDM consists of axions.

27.2 Verify Eq. (27.9), giving the inflationary scale needed to give axion CDM with a specified isocurvature fraction $\alpha_{\rm m}$.

27.3 Verify Eq. (27.16), giving the inflationary scale that is needed to maximse the non-gaussian signal in the CMB from an axion CDM isocurvature perturbation.

References

[1] D. H. Lyth. Axions and inflation: Vacuum fluctuations. *Phys. Rev. D*, **45** (1992) 3394.

[2] L. Boubekeur and D. H. Lyth. Detecting a small perturbation through its non-Gaussianity. *Phys. Rev. D*, **73** (2006) 021301.

[3] A. D. Linde and D. H. Lyth. Axionic domain wall production during inflation. *Phys. Lett. B*, **246** (1990) 353.

[4] K. Enqvist and J. McDonald. Observable isocurvature fluctuations from the Affleck–Dine condensate. *Phys. Rev. Lett.*, **83** (1999) 2510.

[5] D. H. Lyth, C. Ungarelli and D. Wands. The primordial density perturbation in the curvaton scenario. *Phys. Rev. D*, **67** (2003) 023503.

[6] D. H. Lyth and D. Wands. The CDM isocurvature perturbation in the curvaton scenario. *Phys. Rev. D*, **68** (2003) 103516.

[7] M. Kawasaki, F. Takahashi and M. Yamaguchi. Large lepton asymmetry from Q-balls. *Phys. Rev. D*, **66** (2002) 043516.

[8] F. Takahashi and M. Yamaguchi. Spontaneous baryogenesis in flat directions. *Phys. Rev. D*, **69** (2004) 083506.

28

Slow-roll inflation and observation

In Section 25.2 we considered the standard paradigm, according to which the curvature perturbation is generated by the vacuum fluctuation of the inflaton field during slow-roll inflation. We saw that observation gives considerable information about the potential when the pivot scale leaves the horizon. The normalization of the spectrum requires $(V/\epsilon)^{1/4} = 6.6 \times 10^{16}\,\text{GeV}$ and the spectral index determines $2\eta - 6\epsilon$. Also, the bound on the running n' constrains ξ, while the upper bound on the tensor fraction r constrains V.

If instead the curvature perturbation is generated after inflation, there is no particular reason to consider slow-roll inflation. If one does consider it, the only constraints are $\epsilon \ll 1$ and $(V/\epsilon)^{1/4} \ll 6.6 \times 10^{16}\,\text{GeV}$ (implying a small and probably negligible tensor fraction). In this chapter, we assume the standard paradigm and see how different models of slow-roll inflation are then constrained by observation.

By a model of inflation, we mean an effective field theory that is supposed to apply during inflation. In the end, it should be part of an effective field theory that takes us all the way from inflation to the present. A lot of work has been done in this direction, but so far no preferred model has emerged.

As we write, the Large Hadron Collider (LHC) is beginning operation, and may or may not find evidence for supersymmetry. If it doesn't, we will know that supersymmetry is too badly broken to be relevant for the Standard Model. It might still be relevant in the early Universe and in particular during inflation, but there is no doubt that increased emphasis will then be placed on non-supersymmetric inflation models. For the moment though, supersymmetry is generally assumed for inflation models, as it is in other proposals for field theory beyond the Standard Model.

28.1 Historical development

We begin with a brief account of the history. More detail with references is given in Refs. [1, 2, 3]. The first slow-roll inflation model was written down in 1982, and was called new inflation. It was a single-field non-hybrid model, making contact with particle physics because the inflaton was supposed to be a GUT Higgs field. As soon as the calculation of the curvature perturbation was understood, the model was seen to be ruled out because the predicted curvature perturbation was far too big.

Although the requirement of a small enough curvature perturbation proved to be rather demanding, viable models using GUTs and supersymmetry were developed, including what were later called hybrid inflation models. The models were rather complicated, in part because of a demand that the initial condition for observable inflation is set by an era of thermal equilibrium.

It was soon recognized that prior thermal equilibrium is not necessary. A second strand of model-building was initiated in 1983 with the proposal of chaotic inflation. This strand made little contact with particle physics, but included models with non-Einstein gravity theories. (The term 'particle physics' is here intended to include all proposed extensions of the Standard Model.)

In 1992, the detection of the cosmic microwave background (CMB) anisotropy by the COBE satellite gave the first observational constraints on the primordial perturbation, where previously only upper bounds had existed. Combined with galaxy surveys, the COBE data suggested that a primordial curvature perturbation existed and was the dominant effect. The spectral index, though not accurately measured, was seen to be no more that a few tenths away from the scale-independent value $n = 1$. From that time onwards, increasingly accurate measurements of n have continuously reduced the range of viable potentials.

With the formulation in 1994 of the simple hybrid inflation described in Section 18.7, attention went back to the connection with particle physics. Almost all proposals for field theory beyond the Standard Model were considered as arenas for inflation model-building, especially including GUTs, and almost always within the context of supersymmetry. Finally, at the time of writing, another phase of model-building is underway based directly on braneworld scenarios.

It should perhaps be emphasized that detailed inflation model-building has been strongly influenced by the standard paradigm for generating the curvature perturbation, even after several alternatives including the curvaton mechanism appeared starting around the year 2000. This has meant that strong selection has been exercised by model-builders, perhaps creating the impression that inflation inevitably predicts the state of affairs indicated by current observation. In fact inflation, even after adopting the standard paradigm, doesn't predict the small observed value of the spectrum, nor the small observed value of the spectral tilt. It also doesn't say much about the tensor fraction, and if we are prepared to extend it to include the more general single-field models like those in Section 25.7 it doesn't even predict small non-gaussianity. Turning this around, the situation is that the adoption of the standard paradigm (and more generally the single-field paradigm) allows observation to discriminate sharply between inflation models, giving information about the relevant fields and interactions at the very beginning of the history of the observable Universe.

28.2 Eternal inflation

We saw in Section 24.6 how classical slow-roll is modified by the quantum fluctuation. There we assumed that classical slow-roll dominates the quantum fluctuation. If instead the quantum fluctuation dominates one has what is called eternal inflation.

Classical slow-roll dominates if the ratio (24.55) is much less than 1. Then the ratio is equal to the inflaton contribution to $\mathcal{P}_\zeta(k)$, where $k = aH$ is the scale leaving the horizon. If instead the ratio is much bigger than 1, the vacuum fluctuation dominates. A detailed investigation [4] shows that the vacuum fluctuation dominates within any interval $\Delta\phi \gg H/2\pi$, such that

$$\frac{1}{2\pi^2 M_{\rm Pl}^6}\frac{V^3}{V'^2} > 1. \tag{28.1}$$

Within such a region, the random vacuum fluctuation overcomes the classical motion for an indefinitely long period, giving what is called **eternal inflation** [5, 6].

Once eternal inflation starts in a region, it creates an indefinitely large volume. It is therefore commonly supposed that we live in some such volume. In other words, it is typically thought that observable inflation is preceded by eternal inflation.

Eternal inflation provides a realization of the multiverse idea, according to which all possible universes consistent with fundamental theory (nowadays, string theory) will actually exist. This is because the indefinitely long duration of eternal inflation will allow time for tunneling to all local minima of the scalar field potential. Since eternal inflation creates an indefinitely large volume, it seems reasonable to suppose that we live in such a volume no matter how unlikely is the initial condition for eternal inflation.

Regarding the form of the potential during eternal inflation, two possibilities are commonly considered. These are eternal inflation near a maximum of the potential (hilltop eternal inflation, also called topological inflation) [5], and eternal inflation at large field values with a potential $\propto \phi^p$ (chaotic eternal inflation) [6].

Consider first hilltop eternal inflation. The eternal inflation scenario is derived on the assumption of classical slow roll. Taking that literally we need $|\eta_0| \ll 1$ where the subscript denotes the hilltop. Taking the potential to be quadratic so that

η is practically constant, the eternal inflation regime is

$$\phi < \frac{\sqrt{6}}{\eta_0}\frac{H}{2\pi}. \tag{28.2}$$

As it should be, the right-hand side is much bigger than the vacuum fluctuation $H/2\pi$. It is not known at present whether eternal inflation is possible for $|\eta_0| \gtrsim 1$. If so it would provide a reasonable initial condition for the 'fast-roll' inflation described in Section 25.6.

Now consider chaotic eternal inflation with $V \propto \phi^p$. Taking say $V = m^2\phi^2/2$ the eternal inflation regime is $(\phi/M_{\rm Pl})^2 > 4\pi M_{\rm Pl}/m$.

Since eternal inflation creates an indefinitely large volume, there is no need to demand that its occurrence is in some sense probable. Still, one may ask how eternal inflation may come about. As we saw in Section 18.3, inflation presumably begins with the emergence of four-dimensional spacetime. This initial era of inflation need have nothing to do with the observable inflation that takes place after the observable Universe leaves the horizon. There could be an intervening era of non-inflation corresponding to (say) a gaseous Universe. With this in mind, let us discuss in turn how chaotic eternal inflation and hilltop eternal inflation may come about.

In Ref. [7] it was suggested that four-dimensional spacetime emerges when the potential is at the Planck scale, with ϕ a random function of position and eternal inflation occurring in those presumably rare patches where ϕ is sufficiently homogeneous. This suggestion was made within the context of the chaotic inflation potential $V \propto \phi^p$, the term 'chaotic' referring to the random initial condition. If that potential does apply it will give eternal inflation until Eq. (28.2) is satisfied. It is clear though from Eq. (28.1) that the chaotic initial condition will give initial eternal inflation with a far steeper potential, which could correspond to a far smaller initial value of ϕ.

Coming to hilltop eternal inflation there are several possibilities. It could be that the field ϕ is quite inhomogeneous before eternal inflation begins. Then there will be regions where ϕ is at the hilltop and eternal inflation will begin there when the potential comes to dominate the energy density. Another possibility is for the field to arrive near the hilltop by quantum tunneling from a lower point on the potential, in particular from a minimum. Finally, if the hilltop is a fixed point of symmetries, the initial condition can be set by the restoration of symmetry, either thermally or through the action of another scalar field.

28.3 Field theory and inflation

In and of itself, field theory says almost nothing. It doesn't even predict the existence of particles, never mind their interactions. To use field theory, one has to write down a Lagrangian and work out the predictions. If there is sufficiently impressive agreement with observation, the Lagrangian is accepted by the community as correct. Factors counting towards a good impression include the variety of the data and the aesthetic appeal of model.

Taking this viewpoint, inflation models (i.e. Lagrangians) have been proposed and compared with observation. Of course one seeks in the end a single model, describing also the post-inflationary Universe and containing within it the Standard Model. In other words, one seeks a convincing extension of the Standard Model that describes all observations. In practice most investigations adopt a piecemeal approach, whereby a given range of phenomena are described by a model without much reference to the whole.

Where to begin in writing down an inflation model? One approach is to work with fields playing a direct role in string theory, which nowadays usually means working with a braneworld scenario. We will not have much to say about models based directly on the braneworld, because they are rapidly evolving as we write and because we have not provided the reader with the necessary theoretical background.

The other approach is to construct models that work within the framework generally used when seeking extensions of the Standard Model. This 'low-energy' approach was described in Chapters 15–17, and we are adopting it both for inflation and the subsequent evolution. An extensive review of models that were viable in 1999 was given in Ref. [2]. In recent years the range has been steadily eroded by improved observations. The situation was reviewed again in 2006 [8] and the erosion continues.

In this chapter we will give an updated and abbreviated version of the above reviews. Before doing so, we discuss a problem that arises in the context of supergravity.

28.4 The η problem

In this subsection, we see that a generic supergravity theory gives $|\eta| \gtrsim 1$, in violation of the flatness condition $|\eta| \ll 1$. This is called the η problem. We see also that there may be an analogous problem also after inflation.

28.4.1 The η problem during inflation

To see the η problem during inflation, recall the discussion of Section 17.4. The potential has a positive part V_+, and a negative part $-3M_{\rm Pl}^2 m_{3/2}^2(\phi_n)$. In the vacuum the two parts (practically) cancel for some unknown reason to give zero energy density. During inflation there is no reason for any cancellation and we expect $V \sim V_+$. The η problem is caused by the Kahler potential of the inflaton. It has the form (17.22), with the leading term determined by canonical normalization.[1] Without loss of generality we can take the inflaton field to be the canonically normalized real part of say ϕ_1, and choose $\phi_1 = 0$ when the pivot scale leaves the horizon. Then the factor $e^{K/M_{\rm Pl}^2}$ in Eq. (17.18) gives

$$V(\phi_1) = V(0)\left(1 + \frac{|\phi_1^2|}{M_{\rm Pl}^2} + \frac{1}{2}\frac{|\phi_1|^4}{M_{\rm Pl}^4} + \ldots\right). \tag{28.3}$$

The quartic term is very suppressed but the quadratic term gives a contribution $3H^2$ to the effective mass-squared V'', and a contribution $+1$ to η.

This is not the only contribution to η that is generated by supergravity. In particular, the contribution to the mass-squared of ϕ identified after Eq. (17.25) will typically give a contribution of order ± 1 to η. The mechanism producing this term is basically the same as the one that produces soft supersymmetry breaking terms in a globally supersymmetric theory. The analogues of the other soft supersymmetry breaking terms seen in Eq. (17.13) can also be produced. In particular, depending on the inflation model, one can have a term in the potential of the form $A\phi^3 +$ c.c., with typicallly $A \sim H$.

If we only want to achieve slow-roll inflation, a value of $|\eta|$ not far below 1 may be viable, and there is no η problem. The problem comes when we are generating ζ from the inflaton perturbation, so that the spectral index is $n = 1 + 2\eta - 6\epsilon$. To achieve the observed value of n we need $|\eta| \sim 10^{-2}$ which means that the mandatory contribution $\eta = +1$ must be cancelled to an accuracy of a few percent.[2]

It is a matter of taste whether this fairly mild cancellation truly represents a problem. If it is deemed a problem, it should be avoided by invoking a special form for the supergravity potential, such that it becomes completely flat in some limit. This corresponds to making the light field a pseudo-Nambu–Goldstone boson (PNGB) during inflation. That doesn't necessarily mean that it is a PNGB in the vacuum; the approximate shift symmetry present during inflation might be so badly broken in the vacuum as to be irrelevant. Models where that is the case typically start with the prescription suggested in Ref. [9], which begins by demanding $W = 0$. Models where the inflaton is a PNGB in the vacuum are considered in Section 28.11.

[1] To be precise, the leading term can always be brought into this form through a Kahler transformation.
[2] We are assuming that there is no severe cancellation between the η and ϵ contributions to n, which is the case for all of the potentials that we shall be considering.

We end this discussion of the inflationary η problem with the following important remark. As seen at the beginning of this subsection, there is no reason to expect that the strong cancellation that is needed to keep the cosmological constant small in the vacuum also holds during inflation. This is the cancellation between the positive supersymmetry breaking term V_+ of the potential, and the negative contribution $-3M_{\rm Pl}^2 m_{3/2}(\phi_1, \phi_2, \ldots)$. Suppose now that the inflation scale is rather low, corresponding to rather late inflation. Then, one might expect that the mechanism which breaks supersymmetry in the vacuum (i.e. at the present epoch) is the same one that breaks supersymmetry during inflation. In that case one expects V_+ during inflation will have its vacuum value, $M_{\rm S}^4 = 3M_{\rm Pl}^2 m_{3/2}^2$. To avoid the strong cancellation that operates in the vacuum, this requires $V \sim M_{\rm S}^4$. Thus, we expect V during inflation to be at least of order $M_{\rm S}^4$. Equivalently, *we expect H during inflation to be at least of order the gravitino mass.*

If supersymmetry breaking in the vacuum is gravity mediated, $m_{3/2} \sim 100\,\text{GeV}$ which means that we expect $V^{1/4} \gtrsim 10^{10}\,\text{GeV}$, a fairly strong limit. But with gauge mediated breaking, $m_{3/2}$ is far smaller which might give no significant constraint on $V^{1/4}$.

28.4.2 The η problem after inflation

Although light fields in general are not a concern for the present chapter, we should mention that the η problem exists for them too. Indeed, as we saw in Chapter 20.1 the flatness condition $|\eta_{nm}| \ll 1$ typically corresponds to $|\eta_{nn}| \ll 1$ after diagonalizing η_{nm}. Then the single-field argument can be repeated to show that $|\eta_{nn}| \sim 1$ is expected for generic supergravity.

Actually, the problem is somewhat more severe if we want to use the light field to generate a primordial perturbation after inflation. The field should remain light after inflation, or it will be driven to the minimum of the potential and its perturbation will disappear. Let us see why a value $|\eta_{nn}| \sim 1$ is expected for generic supergravity.

After inflation, the inflaton oscillates about its vev, and so does the waterfall field in a hybrid model. The potential still gives a contribution of order 1 to η_{nm}. One may expect that the oscillation also generates a contribution of the same order. Indeed, if χ is the oscillating field and ϕ is the light field the kinetic term of χ will include the terms

$$\mathcal{L}_{\rm kin} = \partial_\mu \chi \partial^\mu \chi \left(1 + \lambda \frac{\phi^2}{M_{\rm Pl}^2}\right), \tag{28.4}$$

and we expect $|\lambda| \sim \pm 1$ if $M_{\rm Pl}$ is the ultra-violet cutoff. In the presence of the oscillating field the light field therefore acquires an additional mass-squared contribution

$$m^2 \supset \frac{2\lambda\overline{\dot{\chi}^2}}{M_{\rm Pl}^2} \sim \pm H^2, \tag{28.5}$$

where the bar denotes an average over one oscillation. As described in Ref. [10], a similar contribution will be present if there is a gas of particles of the species χ, provided that we deal with matter as opposed to radiation.

28.5 Hilltop inflation

We argued in Section 25.3.1 that viable small-field inflation, and probably also medium-field inflation, is likely to take place near a hilltop of the potential. In this section we give a general analysis of hilltop inflation, assuming for simplicity that the potential remains concave-downward until near the end of inflation. We set $\phi = 0$ at the hilltop, even though it may not be a fixed point of symmetries. Also, we take $\phi > 0$ during inflation.

If we deal with a non-hybrid model, the potential has a minimum at $\phi = \langle\phi\rangle$, at which $V = 0$, as in Figure 28.1 (top left). In a hybrid model, inflation ends at some point $\phi_{\rm end}$, beyond which the potential has no meaning. Hilltop inflation is attractive because eternal inflation at the hilltop provides a reasonable initial condition, and because it automatically makes $n < 1$ in accordance with observation.

Near the hilltop there will be an expansion

$$V(\phi) = V_0\left(1 + \frac{1}{2}\eta_0\frac{\phi^2}{M_{\rm Pl}^2} + \dots\right) = V_0 - \frac{1}{2}m^2\phi^2 + \dots. \tag{28.6}$$

In this expression, η_0 is the value of η at $\phi = 0$, *not* its value when the pivot scale leaves the horizon. To achieve $\langle\phi\rangle \ll M_{\rm Pl}$, higher-order terms become significant well before the end of inflation. Still, the quadratic term might dominate while cosmological scales leave the horizon, provided that the potential steepens quickly after the quadratic term dominates. Then $n - 1 = -2|\eta_0|$, which can take on any value $0 \lesssim n < 1$. The tensor fraction is

$$r = 16\epsilon \simeq 2(1-n)^2 e^{-N(1-n)}\left(\frac{\phi_{\rm end}}{M_{\rm Pl}}\right) \simeq 10^{-2}\left(\frac{\phi_{\rm end}}{M_{\rm Pl}}\right), \tag{28.7}$$

which through Eq. (25.7) determines the inflation scale $V^{1/4}$. Since we are dealing with a small-field model r is too small to observed, though $V^{1/4}$ need not be extremely low. Similar remarks apply to the prediction of more general hilltop inflation potentials that we discuss now.

If the quadratic term is negligible during observable inflation, the potential might

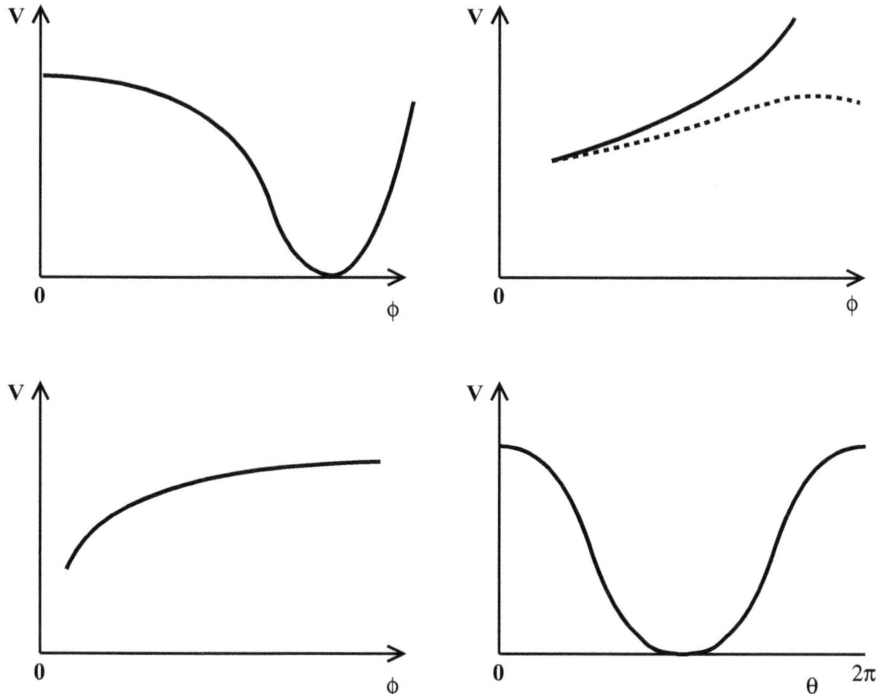

Fig. 28.1. The panels show hilltop inflation (top left), tree-level hybrid inflation (top right), inflation in a GUT/D-term/induced gravity/etc. model (bottom left), and natural inflation (bottom right). The dotted line in the top-right panel corresponds to running-mass inflation.

be approximated by

$$V \simeq V_0 \left[1 - \left(\frac{\phi}{\mu} \right)^p \right], \tag{28.8}$$

with $p > 2$. In a non-hybrid model, $\phi_{\rm end} \sim \mu$. Then, if $p \gtrsim 3$, the integral (18.30) is practically independent of $\phi_{\rm end}$. This gives spectral tilt

$$\boxed{n - 1 = -\frac{2}{N} \left(\frac{p-1}{p-2} \right)}. \tag{28.9}$$

With $N = 50$ and $p = 3, 4$ and ∞ this gives $n = 0.92, 0.94$ and 0.96. As discussed in Section 18.4, N can hardly be much bigger than 50, with certainly $N < 71$ if general relativity and standard field theory apply after inflation. At the time of writing, observation gives $n = 0.96 \pm 0.03$ at something like 2σ, which suggests that $p = 3$ is ruled out for a non-hybrid model satisfying Eq. (28.8).

In a hybrid model one can have $\phi_{\rm end} \simeq \phi_0$ (the value of ϕ when the pivot scale leaves the horizon). This reduces the tilt, and gives extra freedom to fit observation through the additional parameter $\phi_{\rm end}$.

Using Eq. (18.30) we find

$$\boxed{r = 8\left(\frac{\mu}{M_p}\right)^{2p/(p-2)}\left(\frac{p}{2N(p-2)}\right)^{2(p-1)/(p-2)}}. \tag{28.10}$$

In the exceptional case $p = 4$ it is more useful to write

$$V(\phi) = V_0 - \frac{1}{4}\lambda\phi^4. \tag{28.11}$$

Then Eq. (25.7) corresponds to $\lambda = 3 \times 10^{-13}$, independently of V_0.

Keeping both the quadratic term and a higher term gives

$$V(\phi) = V_0\left[1 + \frac{1}{2}\eta_0\frac{\phi^2}{M_{\rm Pl}^2} - \left(\frac{\phi}{\mu}\right)^p\right]. \tag{28.12}$$

The prediction for this case can still be worked out analytically [11]. The quadratic term can either increase or decrease the predicted value of n.

28.6 Ledge inflation

In the analysis of the previous section we assumed that the derivative of the potential vanishes at the chosen origin of ϕ, so that its power series expansion has no linear term. Now we take on board the possibility that the potential has a significant linear term. That is perfectly reasonable, provided that the origin of ϕ (near which inflation is supposed to take place) is well away from a fixed point of the symmetries.

We consider the leading three terms in a power series expansion:

$$V = V_0\left[1 - \lambda_1\frac{\phi}{M_{\rm Pl}} + \frac{1}{2}\eta_0\frac{\phi^2}{M_{\rm Pl}^2} - \lambda_3\phi^3\right]. \tag{28.13}$$

We take the cubic term negative. Also, we assume a non-hybrid model with ϕ rolling to the right. (Higher-order terms that we are not displaying will generate the minimum of V at which it vanishes.) Here, η_0 is the value of η at the chosen point, and with the same notation the values of the other flatness parameters are given by $\epsilon_0 = \lambda_1^2/2$ and $\xi_0 = \lambda_1\lambda_3$.

Dropping the linear term we recover the previous case with $p = 3$. Dropping the cubic term gives nothing new because we can redefine the origin of ϕ to eliminate the linear term. It might be equally reasonable though to drop the quadratic term, and then we get

$$V = V_0\left[1 - \lambda_1\frac{\phi}{M_{\rm Pl}} - \lambda_3\phi^3\right]. \tag{28.14}$$

Now there is a more or less localized feature around $\phi = 0$ superimposed on a steadily decreasing potential. Instead of calling this hilltop inflation it might be more appropriate to call it ledge inflation.

An analytic solution is available for this case [12]. For positive λ_1 it gives

$$n - 1 = -4\sqrt{\frac{\xi_0}{2}} \cot\left(N\sqrt{\frac{\xi_0}{2}}\right). \tag{28.15}$$

In the limit $\lambda_1 \to 0$ we recover the previous result $n - 1 = -4/N$, which seems to be too low.

Allowing positive λ_1 increases n, but the increase is negligible if we require eternal inflation as an initial condition. Indeed, imposing the CMB normalization one finds that eternal inflation is possible only if [12]

$$\xi_0 \lesssim \frac{\mathcal{P}_\zeta}{N} \sim 10^{-10}, \tag{28.16}$$

leading to an insignificant change in n. To have useful eternal inflation, it needs to be of the hilltop kind as before. This corresponds to negative λ_1, which decreases n

Without eternal inflation, one may be concerned that ϕ, starting from a generic negative value, will overshoot the origin so that inflation never takes place. That is not inevitable though, because Hubble drag can reduce $\dot{\phi}$ when it approaches the origin so that slow-roll inflation takes place. Whether that happens can only be decided in the context of a description of inflation that starts long before the observable Universe leaves the horizon.

Ledge inflation allows the inflaton to be a flat direction of the MSSM [13], and was first discovered in that context. Taking the origin of ϕ to be the fixed point of the gauge symmetries (the usual choice) the inflaton potential with ϕ a flat direction of the MSSM is expected to be well-approximated by

$$V = \frac{1}{2}m^2\phi^2 - A\frac{\lambda\phi^p}{pM_{\rm Pl}^{p-3}} + \lambda^2\frac{\phi^{2(p-1)}}{M_{\rm Pl}^{2(p-3)}}, \tag{28.17}$$

with p an integer bigger than 3. The last term comes from a non-renormalizable term in the superpotential, and the first two terms come from soft supersymmetry breaking. Assuming gravity-mediated supersymmetry breaking one expects $\lambda \sim 1$ and $m \sim A \sim 100\,\text{GeV}$. A parameter δ may be defined by

$$A^2 = 8(p-1)m^2(1+\delta). \tag{28.18}$$

If $\delta = 0$, both V' and V'' vanish at the point

$$\phi_0 = \left(\frac{mM_{\rm Pl}^{p-3}}{\lambda\sqrt{2p-2}}\right)^{1/(p-2)}. \tag{28.19}$$

Choosing now ϕ_0 as the origin and reversing the sign of ϕ, we arrive at the ledge inflation potential with $\lambda_1 = 0$. The observed normalization of the spectrum $\mathcal{P}_\zeta$ is obtained with say $p = 6$ and λ roughly of order 1.

More generally, $\lambda_1 = \delta m^2 \phi_0 M_{\rm Pl}/V_0$. To achieve ledge inflation with the observed $\mathcal{P}$ one needs $|\delta| \lesssim 10^{-20}$ [14]. This fine-tuning might be justified on anthropic grounds. The model does not explain why n is so close to 1, because by choice of δ it can give any value of n consistent with slow roll ($0 \lesssim n \lesssim 2$).

A ledge inflation potential has also been found [15] in an exploration of the parameter space for a version of the colliding-brane inflation mentioned in Section 25.7.2.

28.7 GUT inflation

The original new inflation model [16, 17] gave $V_0 \sim M_{\rm GUT}^4$, because the inflaton was a GUT Higgs field. As we see in Section 28.11, this doesn't work. One can however obtain the same result for V_0 by making the GUT Higgs field the waterfall field in hybrid inflation. This kind of GUT inflation has been developed over many years, to make contact with the MSSM and to describe the early Universe, though no single preferred model has yet emerged. A typical paper is Ref. [18], from which the reader can trace earlier and later works. Two kinds of potential are considered.

28.7.1 Potential with a loop contribution

One construction starts with

$$W = g\phi \left(\chi\bar{\chi} - M_{\rm GUT}^2\right) . \tag{28.20}$$

This is the superpotential (17.9) except that the square of the gauge singlet χ^2 is replaced by a gauge-invariant product $\chi\bar{\chi}$ of GUT Higgs fields.[3] This is analogous to the product $H_{\rm u}H_{\rm d}$ of MSSM Higgs fields, and as in that case we suppress indices that need to be summed over to get the gauge-invariant product.

Using global supersymmetry with a minimal Kahler potential, the tree-level potential is given by the analogue of Eq. (17.10). We will use that potential (i.e. replace $\chi\bar{\chi}$ by the square χ^2 of a single field) to get order of magnitude estimates.

We are dealing with hybrid inflation potential. From Eq. (17.12) the inflation regime is $|\phi^2| > |\phi_{\rm c}^2| \equiv 2M_{\rm GUT}^2$. Since $M_{\rm GUT} \simeq 10^{-2} M_{\rm Pl}$, we can have at best a marginally small-field model.

As it stands this model restores the GUT symmetry during inflation, which means that it may create unviable topological defects when inflation ends. It is

[3] To be in line with the literature we replaced $y/2$ by g.

not difficult to modify the model though, so that the waterfall fields are fixed at a symmetry-breaking value. Also, depending on the GUT symmetry group not all of the GUT fields need be waterfall fields.

At tree level $V(\phi)$ is constant, but including the loop contribution (17.17) gives a potential with the shape shown in Figure 28.1 (bottom left). Then the potential is[4]

$$V(\phi) = V_0 \left(1 + \frac{g^2}{8\pi^2} \ln \frac{\phi}{Q} \right), \tag{28.21}$$

with $V_0 = g^2 M_{\rm GUT}^2$. We then have

$$\eta = -\frac{g^2}{8\pi^2} \frac{M_{\rm Pl}^2}{\phi^2} = -\frac{16\pi^2}{g^2} \epsilon, \tag{28.22}$$

Because η increases during inflation, we can have $\eta \sim 1$ before $\phi = \phi_{\rm c}$. When that happens, ϕ oscillates until the oscillation amplitude falls below $\phi_{\rm c}$. This generates a few e-folds of inflation, which in a different context has been called locked inflation [19]. Locked inflation can last for only a limited number of e-folds. Indeed, the amplitude of the oscillation decreases like $a^{-3/2}$, which means that the number of e-folds is only $(3/2) \ln(\phi_{\rm end}/\phi_{\rm c})$. We shall ignore it, since we are just making rough estimates.

The integral Eq. (18.30) is dominated by the limit ϕ_0 giving

$$\frac{\phi_0}{M_{\rm Pl}} \simeq \sqrt{\frac{N g^2}{4\pi^2}} \simeq g. \tag{28.23}$$

We need $g^2 \sim 10^{-2}$, to get a small(ish) field value $\phi_0 \sim 10^{-1} M_{\rm Pl}$. The spectral tilt is $n = 1 - 1/N \simeq 0.98$, and the CMB normalization $M_{\rm GUT} \simeq 6 \times 10^{15}$ GeV (independently of g). The spectral index is more or less right and $M_{\rm GUT}$ is more or less at the GUT scale. Allowing g^2 to be smaller raises n and $M_{\rm GUT}$, making the agreement with observation less satisfactory.

So much for global supersymmetry. What we really should do is insert W into the supergravity potential. This will generically add a quadratic term so that V is of the form

$$V(\phi) = V_0 \left(1 + \frac{1}{2} \eta_0 \frac{\phi^2}{M_{\rm Pl}^2} + \frac{g^2}{8\pi^2} \ln \frac{\phi}{Q} \right), \tag{28.24}$$

with $|\eta_0| \sim 1$. To get the required spectral index we have to suppose that $|\eta_0|$ is actually much smaller (the η problem) but there is certainly no reason why it should be negligible.[5]

[4] This simple form of the loop contribution holds only for $g\phi \gg M_{\rm GUT}$, which is barely compatible with the small-field requirement, but we use it for illustration.

[5] Because W is linear in ϕ, it happens that the quadratic term coming from the e^K factor is cancelled by another

Choosing η_0 negative, we have a hilltop inflation model allowing eternal inflation to set the initial condition. Analytic expressions for the spectrum and spectral index are available [20], showing that the prediction for both n and $M_{\rm GUT}$ is reduced which seems to give better agreement with observation for n. Note that $|\eta_0|$ cannot be much bigger than 10^{-2} or ϕ could not roll towards the origin. As a result the model is predicting small tilt, in agreement with observation.

The main problem of the model is the large field value $\phi \sim 10^{-1}M_{\rm Pl}$, which tends to place higher-order terms out of control. Higher powers of ϕ in the superpotential up to ϕ^{N-1} can be forbidden by a Z_N R-symmetry, but a large value of N is needed if the coefficients are of order 1 in units of $M_{\rm Pl}$ as one might expect.[6] Higher powers of $|\phi|$ in K cannot be forbidden by a symmetry, nor can higher powers of ϕ coming from the e^K factor in the supergravity potential. As a result one can easily imagine that the potential Eq. (28.24) receives large corrections.

We should mention that a completely different calculation, having nothing to do with a GUT, can give the same superpotential Eq. (28.20) and the same kind of inflation model. This calculation invokes a suitable soft supersymmetry-breaking mechanism for the MSSM, such as the metastable vacuum mechanism of Ref. [21]. In this model the supersymmetry breaking is dynamical, which means that the waterfall fields are products of the fundamental fields. It predicts $V_0 \sim M_{\rm S}^4$, which is far below the value required to generate the curvature perturbation. The latter might instead be generated after inflation as described in Chapter 26. The scenario is described in Ref. [22], where earlier references can be found.

28.7.2 Mutated/smooth hybrid inflation

Instead of starting with the superpotential Eq. (28.20), one can start with

$$W = \Lambda_{\rm UV}^{-2}\phi\left[(\chi\bar{\chi})^2 - M_{\rm GUT}^4\right] . \tag{28.25}$$

The term $\chi\bar{\chi}$, that was present before, can be killed by a Z_2 R-symmetry which reverses $\chi\bar{\chi}$. This is a non-renormalizable potential, valid if $M_{\rm GUT}$ and all field values are much less than the cutoff $\Lambda_{\rm UV}$ of the effective theory. (The overall dimensionless factor multiplying W is expected to be order 1 and we omitted it for simplicity.)

As before we replace $\chi\bar{\chi}$ by the square of a single field, to get estimates. The potential drives the imaginary part of χ to zero, and going to the canonically nor-

contribution, but the quartic term of K still gives generically $|\eta_0| \sim 1$, and so does a modulus or other field with a vev of order $M_{\rm Pl}$.

[6] There is no need to worry about higher powers of $\chi\bar{\chi}$ since $\chi\bar{\chi}$ vanishes during inflation.

malized real parts of ϕ and χ we have

$$V = \Lambda^{-4}\left(\frac{1}{4}\chi^4 - M_{\rm GUT}^4\right)^2 + \Lambda^{-4}\phi^2\chi^6. \tag{28.26}$$

In contrast with ordinary hybrid inflation, the waterfall field for this potential has no mass term during inflation and is not fixed at the origin. Instead it adjusts to minimize the potential at each ϕ. Inflation with this feature is called mutated or smooth hybrid inflation.

Minimizing the potential at fixed ϕ we get to leading order in $M_{\rm GUT}/\phi$

$$V(\phi) = V_0\left(1 - \frac{1}{6}\frac{M_{\rm GUT}^4}{\phi^4}\right), \tag{28.27}$$

with $V_0^{1/4} = M_{\rm GUT}^2/\Lambda_{\rm UV}$. This potential has the shape shown in Figure 28.1 (bottom left). Inflation ends at $\phi \sim M_{\rm GUT}$, when $V(\phi)$ violates the flatness conditions. There is no phase transition at the end of inflation, which means that there is no production of monopoles or other topological defects.

The predictions are given by Eqs. (28.9) and (28.10) with $p = -4$. The spectral index $n = 1 - 5/3N$ is reasonable, but choosing $\Lambda_{\rm UV}$ to give the CMB normalization makes $M_{\rm GUT} > \Lambda_{\rm UV}$ which is not allowed.

As before, we should really be inserting W into the supergravity potential given by Eq. (17.18). We then expect a quadratic term so that the potential is

$$V(\phi) = V_0\left(1 + \frac{1}{2}\eta_0\frac{\phi^2}{M_{\rm Pl}^2} - \frac{2}{27}\frac{M_{\rm GUT}^4}{\phi^4}\right). \tag{28.28}$$

The analytic expression mentioned after Eq. (28.12) works for this case (with $p = -4$). Choosing $\eta_0 < 0$ to have hilltop inflation, we can get $M_{\rm GUT}$ and ϕ_0 marginally below $\Lambda_{\rm UV}$. Analogously with Eq. (28.24), higher-order terms may well be important, because $\phi \sim 10^{-1}M_{\rm Pl}$.

28.8 *D*-term inflation

There is an alternative way of arriving at the potential Eq. (28.21), again using global supersymmetry, in which the potential comes from the D-term and not the F-term [9, 23, 24]. The waterfall field is now one of the fields in the D-term given by Eq. (17.3). The inflaton is a gauge singlet, which couples to the waterfall field through an F-term.

The tree-level potential during inflation has the constant value $g^4\xi^4/2$, coming from the D-term [9]. The loop contribution gives Eq. (28.21). The CMB normalization gives [23, 24] $\xi = 6 \times 10^{15}$ GeV (as before with ξ instead of $M_{\rm GUT}$). The value of ξ is usually supposed to be determined by string theory and is perhaps of order the string scale, but its value depends heavily on the version of string theory that is supposed to apply. To allow the unification of couplings though, one would like the string scale to be no smaller than the GUT scale.

Several issues need to be considered in connection with D-term inflation, as reviewed for instance in Refs. [2, 25]. Here we just consider corrections to the potential, coming from higher-order terms in W and K and from supergravity.[7]

At first sight the corrections are under better control than in the F-term model, because V_0 dominates the potential and is given by the D-term. As a result, the effect of K is limited to its ability to give a non-canonical normalization for ϕ (which after going to the canonically normalized field will alter the shape of the potential). That leaves W, whose holomorphy allows the higher-order terms in Eq. (17.23) to be forbidden by an R-symmetry. As exact global symmetries are not supposed to exist, the R-symmetry should preferably be a discrete gauge symmetry, which means that it will forbid higher-order terms only up to a finite power of ϕ.

Now comes the bad news. In contrast with the GUT case, g is a gauge coupling. The general expectation from string theory is that $g^2/4\pi$ will be of order 10^{-1} like the GUT gauge coupling. Then $g^2 \sim 1$ giving $\phi \sim M_{\rm Pl}$. This makes it difficult to see how corrections to the model can be under control. Instead of the simple form (28.21), the potential is presumably

$$V(\phi) = V_0 \left(1 + f(\phi) + \frac{g^2}{8\pi^2} \ln \frac{\phi}{Q}\right), \tag{28.29}$$

This is like Eq. (28.24), except that it is difficult to see why $f(\phi)$ should be quadratic. But irrespective of the exact shape of $f(\phi)$, a decreasing function is favoured in that it allows hilltop inflation. This may be expected to reduce the spectral index, so that the situation is much like the one where f is quadratic. The spectral index can be lowered, and so can the height of the potential. The latter might be desirable to lower the tension of cosmic strings that may (but need not) be produced at the end of D-term inflation.

It has been suggested [26] that after all the series (17.23) may converge, because the expected values are really $|y_d| \sim 1/d!$ instead of $|y_d| \sim 1$. Then the discrete symmetry for W could work. If the coefficients in Eq. (17.22) are similarly suppressed, the correction from K might be small. Assuming that everything is under control, one can arrive at a well-motivated quadratic form for f [27].

[7] One should consider also the gauge kinetic function that multiplies the D-term. It is holomorphic like W and the same considerations apply to both functions.

28.9 Another potential

Another potential with the shape shown in Figure 28.1 (bottom left) is

$$V \simeq V_0 \left(1 - e^{-q\phi/M_{\rm Pl}}\right) . \tag{28.30}$$

We saw in Section 19.2 that it emerges from induced-gravity inflation, and from variable Planck mass inflation. In those cases, $q = \sqrt{2/3}$. The same potential is generated by a kinetic term passing through zero [9], with $q = \sqrt{2}$, and in a string-based model [28] with $q = 1/\sqrt{3}$. The prediction for the spectral tilt is given by Eq. (28.9) with $p = \infty$, leading to $n - 1 = -2/N$ which can agree with observation. The CMB normalization corresponds to

$$r = \frac{8}{q^2 N^2} = 0.003 \, \frac{2}{3q^2} \left(\frac{50}{N}\right)^2 . \tag{28.31}$$

This might eventually be observable.

As with GUT inflation and D-term inflation, one could reasonably suppose that modifications of this potential generate a maximum for the potential. In this way eternal inflation could provide the initial condition, without necessarily altering the predictions very much.

28.10 Running-mass inflation

Now we look at what is called running-mass inflation [29, 30]. This model has the unusual feature that it can give a running $n' \sim 10^{-2}$ of the spectral index, which we saw in Section 23.4 would be enough to generate primordial black holes at the end of inflation. Also, in contrast with the other models that we discuss, it suggests a rather low scale inflation scale.

The running-mass model is a hybrid inflation model, which invokes global supersymmetry in the inflaton/waterfall sector of the theory. The potential V_0 and the mass terms are generated by softly broken supersymmetry.

The starting point [31] is a tree-level hybrid inflation potential of the form

$$V(\phi, \chi) = V(\chi) + \frac{1}{2} m^2 \phi^2 + \frac{1}{2} \lambda' \chi^2 \phi^2, \tag{28.32}$$

$$V(\chi) = V_0 \, f(\chi / M_{\rm Pl}). \tag{28.33}$$

The value and low derivatives of the function f are equal to 1 at a generic point on the interval $0 < \phi < M_{\rm Pl}$, which means that the waterfall field χ is a modulus according to the definition of Section 21.7. For χ to serve as a waterfall field we need $\chi = 0$ to be a maximum of $V(\chi)$, which is a fixed point of symmetries so that $V'(\chi)$ vanishes. The vev $\langle\chi\rangle$ is of order M_{Pl}. The moduli problem described in Section 17.3.3 might be solved by thermal inflation.[8]

All of the terms except the coupling $\chi^2\phi^2$ are supposed to come from gravity-mediated supersymmetry breaking generated by V_0. For economy one assumes $V_0 \simeq M_S^4$ so that mechanism for the spontaneous breaking of supergravity is the same as in the vacuum.

Since they come from soft supersymmetry breaking, the masses m and m_χ are *roughly* of order H. But we need the inflaton mass to be significantly smaller to allow inflation. Also, we need the (tachyonic) waterfall mass to be significantly bigger so that the waterfall field descends quickly to the vacuum when inflation ends. The coupling $\phi^2\chi^2$ is supposed to be present in the limit of unbroken supersymmetry. It can be generated by a term $W = y\sigma\phi\chi$ in the superpotential, with additional terms setting $\sigma = 0$.

This tree-level setup gives $n > 1$, which is ruled out by observation. The running-mass model can solve this problem by invoking the loop contribution (17.15). The field denoted by χ in the derivation of Eq. (17.15) is supposed be a chiral or gauge supermultiplet, invoked in addition to the chiral supermultiplets of the inflaton and waterfall field. We use Eq. (17.15), but write $\tilde{m}^2$ instead of m_χ^2. In the regime $y\phi \gg |\tilde{m}|$ it gives

$$V(\phi) = V_0\left[1 + \frac{1}{2}\left(\eta_0 - c\ln\frac{\phi}{Q}\right)\frac{\phi^2}{M_{\text{Pl}}^2}\right], \tag{28.34}$$

where

$$\eta_0 \equiv \frac{M_{\text{Pl}}^2 m^2}{V_0}, \qquad c \equiv -\tilde{\eta}\frac{y^2}{16\pi^2}, \qquad \tilde{\eta} \equiv \frac{M_{\text{Pl}}^2\tilde{m}^2}{V_0}. \tag{28.35}$$

The formula for c holds if the loop contribution comes from a chiral supermultiplet, where $\tilde{m}^2$ can have either sign. For a gauge supermultiplet, y^2 is replaced by $-g^2$ and $\tilde{m}^2 > 0$ is the mass-squared of the gaugino. If $\tilde{m}^2$ refers to a scalar then a natural supergravity value would be $|\tilde{\eta}| \sim 1$. If instead it is a gaugino the supergravity value is more model dependent, but $|\tilde{\eta}| \sim 1$ can still be expected is some scenarios. With this estimate of $\tilde{\eta}$, we have roughly $c \sim y^2/16\pi^2$.

To permit initial eternal inflation, and to allow a spectral index below 1, we require $V(\phi)$ to have a maximum at some field value ϕ_*. This requires $c > 0$ for a viable model, which is automatic for a gauge supermultiplet and corresponds to $\tilde{m}^2 < 0$ for a chiral supermultiplet.

To include the whole range of ϕ during observable inflation one should probably be using the renormalization group improved potential $V = V_0 + \frac{1}{2}m^2(\phi)\phi^2$, with the running mass calculated from some specific set of renormalization group

[8] Alternatively it might be solved by making $\langle\chi\rangle$ a point of enhanced symmetry [31]. In the latter case the symmetry is presumably a Standard Model gauge symmetry, making χ a flat direction of the MSSM.

equations (RGEs). Around the maximum though one can use the loop correction with $Q \sim \phi_*$. Choosing $\ln(Q/\phi_*) = 1/2$ gives the convenient expression [32]

$$V(\phi) = V_0 \left[1 + \frac{1}{2} c \frac{\phi^2}{M_{\rm Pl}^2} \left(\ln \frac{\phi}{\phi_*} - \frac{1}{2} \right) \right], \tag{28.36}$$

with c now evaluated at the renormalization scale. Calculation of the RGEs in specific cases suggests that this will be a reasonable approximation while cosmological scales leave the horizon.

Differentiating Eq. (28.36) we have

$$M_{\rm Pl} \frac{V'}{V_0} = c \frac{\phi}{M_{\rm Pl}} \ln \frac{\phi_*}{\phi}. \tag{28.37}$$

Differentiating V again we find

$$\frac{n(k) - 1}{2} = s\, e^{c(N - N(k))} - c, \qquad s \equiv c \ln \frac{\phi_*}{\phi_0}, \tag{28.38}$$

where as usual ϕ_0 is the value when the pivot scale leaves the horizon. At the pivot scale,

$$\boxed{n - 1 = 2(s - c)} \qquad \boxed{n' = 2cs}. \tag{28.39}$$

To agree with observation we need $c = 0.02 + s$ and $cs \lesssim 10^{-2}$. For $s \ll 0.02$, this requires $c \simeq 0.02$ which gives negligible running n'. For $s \gtrsim 0.02$ we have $c \simeq s$. Then $n - 1$ passes through zero near the pivot scale, and there is large running $n' \simeq 2c^2$ which requires $c \lesssim 10^{-1}$ to agree with observation. The maximal allowed value $n' \sim 10^{-2}$ is about the value that would lead to black hole formation at the end of inflation, if n' remains constant. Depending on the RGE though, it is possible for n' to vary significantly in the running-mass model.

The value $c \sim 10^{-2}$ to 10^{-1} is consistent with the rough estimate $c = y^2/16\pi^2$, with the coupling y roughly of order 1. In other words, a coupling roughly of order 1 gives the small spectral tilt in accordance with observation, and can easily give also strong running $n' \sim 10^{-2}$.

What about the CMB normalization? We see from Eq. (28.37) that ϵ is equal to $\frac{1}{2} n'^2 \phi_0^2 / M_{\rm Pl}^2$. This can correspond to a scale $V_0^{1/4}$ many orders of magnitude below $M_{\rm Pl}$. Given specific interactions that determine the RGEs, the actual value of V_0 can be calculated by imposing the CMB normalization, in terms of the parameters c, $\phi_{\rm end}$ and ϕ_*. In the literature, a further constraint is imposed by requiring that $\eta = 1$ at some high value of $Q = M_{\rm Pl}$, where it is supposed that the tree-level supergravity potential will hold. It is found that a V_0 can be consistent with gravity mediation or gauge mediation (with a fairly high scale) in the MSSM. The main interest of the running-mass model though, lies in its ability to give strong running of the spectral index.

28.11 Small-field PNGB inflation

As an alternative to supersymmetry, or in addition to it, we can suppose that the inflaton is a pseudo-Nambu–Goldstone boson (PNGB). This can protect its mass against radiative corrections, and ensure that the quartic term is suppressed.

In this section we describe a non-hybrid and non-supersymmetric example [33]. Further (small-field) examples including hybrid inflation and/or supersymmetry are given in Refs. [33, 34, 35].

Since we are asking for small-field and non-hybrid inflation, the usual sinusoidal PNGB potential doesn't work. Indeed, it gives (with η_* the value of η at $\phi = 0$)

$$V = \frac{1}{2} V_0 \left[1 + \cos \left(\sqrt{2|\eta_*|} \frac{\phi}{M_{\rm Pl}} \right) \right] , \qquad (28.40)$$

leading to $\langle \phi \rangle = \pi M_{\rm Pl} / \sqrt{2|\eta_*|} \gg M_{\rm Pl}$.

We need the periodic PNGB potential to be exceptionally flat near the maximum, which can be achieved [33] using what is called the Little Higgs mechanism. Two complex fields ϕ and χ are used, and the potential near the origin is taken to be

$$V = f^2 \left[g_1^2 \left| \chi + i \frac{\phi^2}{f} \right|^2 - g_2^2 \left| \chi - i \frac{\phi^2}{f} \right|^2 \right] + \ldots . \qquad (28.41)$$

The first and second terms are invariant under the separate shift symmetries

$$\phi \to \phi + \epsilon, \qquad \chi \to \chi \pm 2i \frac{\epsilon \phi}{f}, \qquad (28.42)$$

with + for the second term and − for the first term. To break the shift symmetry $\phi \to \phi + \epsilon$ we need both g_1 *and* g_2 to be non-zero. Usually, a symmetry can be broken by making just one coupling non-zero.

This is the Little Higgs mechanism. Because two couplings are needed, the usual one-loop contribution to the mass-squared is absent and the radiative correction $m^2_{\rm rad}$ is well below $\Lambda^2_{\rm UV}$. Provided that the true mass is not too far below $\Lambda_{\rm UV}$, there is no need for a fine-tuned cancellation between the bare mass and the radiative correction. The mechanism was invoked for the Standard Model Higgs, to allow the ultra-violet cutoff to be of order 10 TeV which is the minimum compatible with observation. Here we are invoking it for the inflaton, which is taken to be a gauge singlet.

To achieve inflation one takes $g_2 \ll g_1$. At a given value of ϕ during inflation, the heavy field χ adjusts to a value $\chi(\phi)$ which minimizes the potential. One says that χ is integrated out (of the action). Taking ϕ real and making the usual replacement $\phi \to \phi/\sqrt{2}$, the effective potential is then

$$V(\phi) = V_0 - \frac{1}{2}m^2\phi^2 - \frac{1}{4}\lambda\phi^4 + \ldots, \tag{28.43}$$

$$m^2 \simeq \frac{g_1^2 g_2^2}{16\pi^2} f^2, \qquad \lambda \simeq \frac{4 g_1^2 g_2^2}{g_1^2 - g_2^2}, \qquad V_0 \simeq \lambda f^4. \tag{28.44}$$

The quadratic term is a two-loop contribution and the quartic term is the tree-level contribution. We see that after integrating out the heavy field, the origin has become the fixed point of a symmetry $\phi \to -\phi$.

This potential is valid for $\phi \ll f$, the full potential being periodic as expected for a PNGB. The period (hence $\langle\phi\rangle$) is of order f, which accordingly should be $\ll M_{\rm Pl}$. As advertised, the potential is flat unless both g_1 and g_2 are non-zero. The effect of the quadratic term cannot be too big when the pivot scale leaves the horizon, or the spectral index would disagree with observation. Taking it to be negligible corresponds to

$$g_1 \lesssim 10^7 \frac{f}{M_{\rm Pl}}, \qquad g_2 \sim 10^{-7}. \tag{28.45}$$

There is no need for g_1 to be very small. The required small coupling $\lambda \sim 10^{-13}$ corresponds to the square of the fundamental coupling g_2, so that the latter is less finely tuned than λ itself.

It is instructive to compare the Little Higgs model with the original new inflation model [16, 17]. That model is similar, in that it is a non-hybrid model with the origin the fixed point of symmetries. The inflaton was supposed to be a GUT Higgs field, but the mass and self-coupling of the Higgs were both supposed to be negligible. The potential $V(\phi)$ was generated from the loop contribution, corresponding to Eq. (15.70) with $m = \lambda = 0$. While cosmological scales leave the horizon, the potential is well approximated by the quartic potential Eq. (28.11), with $\lambda \sim g^2/32\pi^2$. This last relation is fatal because g is the GUT gauge coupling which the unification of couplings determines to be of order 1. The Little Higgs model doesn't have this problem because g is a Yukawa coupling. Also, it has much milder fine tuning coming from the fact that ϕ is a PNGB.

28.12 Modular inflation

Now we consider inflation using a modulus as defined in Section 21.7. This gives a medium-field model, corresponding to a variation $\Delta\phi \sim M_{\rm Pl}$.

During inflation two or more moduli will typically vary, but as a rough approximation the inflationary trajectory can usually be taken to be straight line in the space of canonically normalized moduli. Then we have a single-field model, which we assume from now on.[9]

Modular inflation is usually taken to be non-hybrid so that the potential $V(\phi)$ is of the form (21.20). To achieve inflation, the potential has to be flatter than would be typical, so as to make ϵ and $|\eta|$ much less than 1. The flattening might be achieved by a suitable symmetry [37], or by choosing a finely tuned location in the space of the moduli (see for instance Ref. [38]).

As this is a medium-field model, we expect the tensor fraction to be unobservable, and the spectral index to be just $1 - 2\eta$. Also, we expect inflation to take place near a hilltop (or ledge). Accordingly we expect the discussion of Section 28.5 to roughly apply, with two implications. First, since $\phi_{\rm end} \sim M_{\rm Pl}$, Eqs. (28.7) and (28.10) correspond to a fairly high CMB normalization, $V_0^{1/4} \sim 10^{15}\,\text{GeV}$. That corresponds to a modulus mass $m \sim 10^{12}\,\text{GeV}$, which presumably means that the potential of the modulus is not generated by supersymmetry breaking, in contrast with the case we discussed in Section 21.7. Second, the generic expectation is for a large spectral tilt, corresponding to $|\eta_0|$ not being particularly small. On the other hand, if $|\eta_0|$ is very small for some reason, then the expectation is for a non-negligible tilt roughly of order $1/N$, in accordance with observation. These estimates are in accordance with what is usually found in specific investigations.

28.13 Large-field models

28.13.1 Chaotic inflation

In this section we consider large-field models, which give an observable tensor fraction. We begin with the case $V \propto \phi^p$ (chaotic inflation [7]). The slow-roll parameters are

$$\epsilon = \frac{p^2}{2}\frac{M_{\rm Pl}^2}{\phi^2}, \qquad \eta = p(p-1)\frac{M_{\rm Pl}^2}{\phi^2}. \tag{28.46}$$

Inflation ends at $\phi_{\rm end} \simeq pM_{\rm Pl}$. When cosmological scales leave the horizon, we find from Eq. (18.30) that $\phi = \sqrt{2Np}\,M_{\rm Pl}$, giving $n - 1 = -(2+p)/2N$ and $r = 4p/N$. At the time of writing observation rules out $p \geq 4$ but allows $p = 2$. The prediction for $p = 2$ is shown in Figure 28.2 for $N = 50$ and 60. With the standard cosmology, corresponding to Eq. (18.8), this range of N corresponds to $10^6\,\text{GeV} < T_{\rm R} < V^{1/4}$.

As the potential $V = \frac{1}{2}m^2\phi^2$ is a large-field model it is legitimate to ask how higher-order terms in the potential might be controlled. A number of different ways have been suggested, which are mentioned in the following subsections.

[9] For a modular inflation model which is definitely not single-field see Ref. [36].

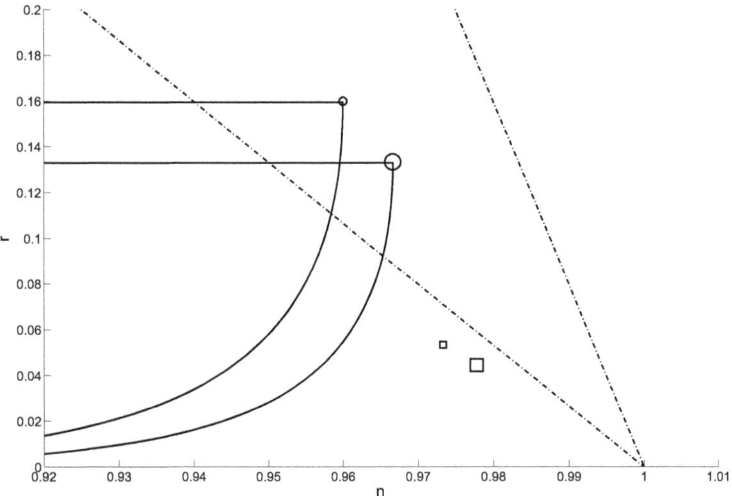

Fig. 28.2. The vertical axis is the tensor fraction r and the horizontal axis is the spectral index, measured at a given scale. The straight lines divide the plot into regions where (from left to right) V and $\ln V$ have the following behaviour: (i) both concave-down, (ii) $\ln V$ concave-down and V concave-up, (iii) both concave-up. The circles correspond to $V \propto \phi^2$ and the squares to $V \propto \phi^{2/3}$ (small/large symbol for $N = 60/50$). The curved lines are for natural inflation, and the horizontal lines are for $V = \frac{1}{2}\sum m_i^2 \phi_i^2$. Figure courtesy of L. Alabidi.

Finally, we mention that a non-integer power $p = 2/3$ has been derived from string theory [39]. It gives the prediction shown in Figure 28.2.

28.13.2 Natural inflation

Natural inflation [40] works with the sinusoidal potential (28.40), allowing $f \gg M_{\rm Pl}$. This potential is illustrated in Figure 28.1 (bottom right). The field value at the pivot scale is given by

$$\sin\left(\frac{\sqrt{|\eta_*|}}{2}\frac{\phi_0}{M_{\rm Pl}}\right) = \sqrt{\frac{1}{1+|\eta_*|}}\, e^{-N|\eta_*|}, \tag{28.47}$$

leading at the pivot scale to

$$\epsilon = \frac{1}{2N}\frac{2N|\eta_*|}{e^{2N|\eta_*|}-1}, \qquad \eta = \epsilon - |\eta_*|. \tag{28.48}$$

By varying the parameter $|\eta_*|$ we find a line in the r–n plane. This is shown in Figure 28.2 for $N = 50$ and 60. The observed value of n requires $r \gtrsim 10^{-2}$.

To justify the sinusoidal potential one might invoke string axions. The identification of ϕ with a single string axion doesn't seem to be viable, but it could correspond to a combination of two of them [41].

A different way of justifying the sinusoidal potential is called gauge inflation [35, 42]. Here, ϕ is the fifth component of a gauge field living in a five-dimensional theory. In the effective four-dimensional theory, ϕ becomes a fundamental scalar without a gauge coupling, with the sinusoidal potential. Supersymmetric realizations of gauge inflation have also been proposed [43].

28.13.3 N-flation

Another way of arriving at a large-field model exploits the fact that the canonically normalized field ϕ may lie in the space of a large number of canonically normalized fields, corresponding to say $\phi = \sum_{i=1}^{N} a_i\phi_i/\sqrt{\sum a_i^2}$. Then, with say all a_i equal, we can have $\phi \gg M_{\rm Pl}$ with each $\phi_i \ll M_{\rm Pl}$. This is known as assisted inflation [44].

At first sight one might think that the assisted inflation proposal lacks content, since a rotation of the field basis can always make ϕ one of the fields ϕ_i. The point though is that the field theory may select a particular basis as the one in which the power series (25.13) is expected to be relevant for each ϕ_i. It has been argued that this will be the case if each ϕ_i has a sinusoidal potential, leading to what is called N-flation [45].

In this scheme, more than one field might still be varying during observable inflation, which might correspond to multi-field inflation with the potential $\frac{1}{2}\sum_i m_i^2\phi_i^2$. Let us work out the prediction for that case.

For any slow-roll potential of the form $V = \sum V_i(\phi_i)$ the function N, whose perturbation gives the curvature perturbation just before the end of inflation, is given by [46]

$$N(\phi_i(t), \phi_2(t), \ldots) = M_{\rm Pl}^{-2} \sum_i \int_{\phi_i(t)}^{\phi_i} \frac{V_i}{V_i'} d\phi_i. \qquad (28.49)$$

Here $\phi_i(t)$ are the field values when the scale under consideration leaves the horizon, and ϕ_i are the values on a slice of uniform potential just before the end of inflation. With our potential, we expect each of the integrals to be dominated by the lower limit, giving $N(\phi_1, \phi_2, \ldots) = \frac{1}{4}\sum \phi_i^2(k)$. Then $\partial N/\partial\phi_i = \frac{1}{2}\phi_i(t)$. Using these results, one finds [8] that the prediction for $\mathcal{P}_\zeta$ is the same as in the single-field case that we discussed in Section 28.13.1. The spectral index on the other hand is lower. This is shown in Figure 28.2, as horizontal lines corresponding to $N = 50$ and $N = 55$. The tensor fraction r and non-gaussianity parameter $f_{\rm NL}$ turn out to be the same as in the single-field case.

Exercises

28.1 Verify Eqs. (28.9) and (28.10).

28.2 Recalculate Eq. (28.9) keeping $\phi_{\rm end}$ in Eq. (18.30), and verify that the change is small for $p \gtrsim 3$ and $\phi_{\rm end} \lesssim M_{\rm Pl}$.

28.3 (a) Taking $W = \phi$ times a constant and $K = |\phi|^2$, show that Eq. (17.20) gives zero mass for ϕ. (b) Show that an additional term $K = \phi^4/M_{\rm Pl}^2$ gives a mass of order $V/M_{\rm Pl}^2$. (c) Show that the same is true if there is a field χ with $\langle|\chi|\rangle = M_{\rm Pl}$ and $K = |\chi|^2$.

28.4 Show that the 'locked inflation' defined after Eq. (28.22) can last for no more than 7 e-folds.

28.5 Verify Eq. (28.27), the potential for smooth hybrid inflation.

28.6 Verify Eqs. (28.37) and (28.38), giving the prediction for running-mass inflation.

28.7 Considering in turn each of the models considered in this chapter, say to what extent the axion isocurvature bound shown in Figure 27.1 is relevant.

28.8 Calculate the slopes of the straight lines in Figure 28.2.

28.9 Assuming $V = \sum_i V_i(\phi_i)$ and the slow-roll approximation, show that

$$N = \frac{1}{3M_{\rm Pl}^2}\sum_i \int_{t_{\rm end}}^{t} \frac{V_i(\phi_i)}{H} dt$$

Then set $dt = d\phi_i/\dot{\phi}_i$ in each integral, to arrive at Eq. (28.49).

References

[1] K. A. Olive. Inflation. *Phys. Rept.*, **190** (1990) 307.

[2] D. H. Lyth and A. Riotto. Particle physics models of inflation and the cosmological density perturbation. *Phys. Rept.*, **314** (1999) 1.

[3] D. H. Lyth. Particle physics models of inflation. *Lect. Notes Phys.*, **738** (2008) 81.

[4] P. Creminelli, S. Dubovsky, A. Nicolis, L. Senatore and M. Zaldarriaga. The phase transition to slow-roll eternal inflation. *JHEP*, **0809** (2008) 036.

[5] A. Vilenkin. The birth of inflationary universes. *Phys. Rev. D*, **27** (1983) 2848.

[6] A. D. Linde. Eternally existing self-reproducing chaotic inflationary universe. *Phys. Lett. B*, **175** (1986) 395.

[7] A. D. Linde. Chaotic inflation. *Phys. Lett. B*, **129** (1983) 177.

[8] L. Alabidi and D. H. Lyth. Inflation models and observation. *JCAP*, **0605** (2006) 016.

[9] E. D. Stewart. Inflation, supergravity and superstrings. *Phys. Rev. D*, **51** (1995) 6847.

[10] D. H. Lyth and T. Moroi. The masses of weakly-coupled scalar fields in the early universe. *JHEP*, **0405** (2004) 004.

[11] G. German, G. G. Ross and S. Sarkar. Low-scale inflation. *Nucl. Phys. B*, **608**, 423 (2001).

[12] J. C. Bueno Sanchez, K. Dimopoulos and D. H. Lyth. A-term inflation and the MSSM. *JCAP*, **0701** (2007) 015.

[13] R. Allahverdi, K. Enqvist, J. Garcia-Bellido and A. Mazumdar. Gauge invariant MSSM inflaton. *Phys. Rev. Lett.*, **97** (2006) 191304.

[14] J. C. Bueno Sanchez, K. Dimopoulos and D. H. Lyth. A-term inflation and the MSSM. *JCAP* **0701** (2007) 015.

[15] D. Baumann, A. Dymarsky, I. R. Klebanov, L. McAllister and P. J. Steinhardt. A delicate universe. *Phys. Rev. Lett.*, **99** (2007) 141601.

[16] A. D. Linde. A new inflationary universe scenario: a possible solution of the horizon, flatness, homogeneity, isotropy and primordial monopole problems. *Phys. Lett. B*, **108** (1982) 389.

[17] A. Albrecht and P. J. Steinhardt. Cosmology for grand unified theories with radiatively induced symmetry breaking. *Phys. Rev. Lett.*, **48** (1982) 1220.

[18] R. Jeannerot, S. Khalil, G. Lazarides and Q. Shafi. Inflation and monopoles in supersymmetric SU(4)c x SU(2)L x SU(2)R. *JHEP*, **0010** (2000) 012.

[19] G. Dvali and S. Kachru. Large scale power and running spectral index in new old inflation. arXiv:hep-ph/0310244.

[20] L. Boubekeur and D. H. Lyth. Hilltop inflation. *JCAP*, **0507** (2005) 010.

[21] K. Intriligator, N. Seiberg and D. Shih. Dynamical SUSY breaking in metastable vacua. *JHEP*, **0604** (2006) 021.

[22] N. J. Craig. ISS-flation. *JHEP*, **0802** (2008) 059.

[23] P. Binetruy and G. R. Dvali. D-term inflation. *Phys. Lett. B*, **388** (1996) 241.

[24] E. Halyo. Hybrid inflation from supergravity D-terms. *Phys. Lett. B*, **387** (1996) 43.

[25] P. Binetruy, G. Dvali, R. Kallosh and A. Van Proeyen. Fayet–Iliopoulos terms in supergravity and cosmology. *Class. Quant. Grav.*, **21** (2004) 3137.

[26] C. F. Kolda and J. March-Russell. Supersymmetric D-term inflation, reheating and Affleck–Dine baryogenesis. *Phys. Rev. D*, **60** (1999) 023504.

[27] C. M. Lin and J. McDonald. Supergravity modification of D-term hybrid inflation: Solving the cosmic string and spectral index problems via a right-handed sneutrino. *Phys. Rev. D*, **74** (2006) 063510.

[28] M. Cicoli, C. P. Burgess and F. Quevedo. Fibre inflation: observable gravity waves from IIB string compactifications. arXiv:0808.0691 [hep-th].

[29] E. D. Stewart. Flattening the inflaton's potential with quantum corrections. *Phys. Lett. B*, **391** (1997) 34.

[30] E. D. Stewart. Flattening the inflaton's potential with quantum corrections. II. *Phys. Rev. D*, **56** (1997) 2019.

[31] L. Randall, M. Soljacic and A. H. Guth. Supernatural inflation: inflation from supersymmetry with no (very) small parameters. *Nucl. Phys. B*, **472** (1996) 377.

[32] L. Covi, D. H. Lyth, A. Melchiorri and C. J. Odman. The running-mass inflation model and WMAP. *Phys. Rev. D*, **70** (2004) 123521.

[33] N. Arkani-Hamed, H. C. Cheng, P. Creminelli and L. Randall. Pseudonatural inflation. *JCAP*, **0307** (2003) 003.

[34] E. D. Stewart and J. D. Cohn. Inflationary models with a flat potential enforced by non-abelian discrete gauge symmetries. *Phys. Rev. D*, **63** (2001) 083519.

[35] D. E. Kaplan and N. J. Weiner. Little inflatons and gauge inflation. *JCAP*, **0402** (2004) 005.

[36] K. Kadota and E. D. Stewart. Inflation on moduli space and cosmic perturbations. *JHEP*, **0312** (2003) 008.

[37] G. G. Ross and S. Sarkar. Successful supersymmetric inflation. *Nucl. Phys.*, B **461** (1996) 597.

[38] J. J. Blanco-Pillado *et al.* Inflating in a better racetrack. *JHEP*, **0609** (2006) 002.

[39] E. Silverstein and A. Westphal. Monodromy in the CMB: gravity waves and string inflation. *Phys. Rev. D*, **78** (2008) 106003.

[40] K. Freese, J. A. Frieman and A. V. Olinto. Natural inflation with pseudo-Nambu–Goldstone bosons. *Phys. Rev. Lett.*, **65** (1990) 3233.

[41] J. E. Kim, H. P. Nilles and M. Peloso. Completing natural inflation. *JCAP*, **0501** (2005) 005.

[42] N. Arkani-Hamed, H. C. Cheng, P. Creminelli and L. Randall. Extranatural inflation. *Phys. Rev. Lett.*, **90** (2003) 221302.

[43] R. Hofmann, F. Paccetti Correia, M. G. Schmidt and Z. Tavartkiladze. Supersymmetric models for gauge inflation. *Nucl. Phys. B*, **668** (2003) 151.

[44] A. R. Liddle, A. Mazumdar and F. E. Schunck. Assisted inflation. *Phys. Rev. D*, **58** (1998) 061301.

[45] S. Dimopoulos, S. Kachru, J. McGreevy and J. G. Wacker. N-flation. *JCAP*, **0808** (2008) 003.

[46] A. A. Starobinsky. Multicomponent de Sitter (inflationary) stages and the generation of perturbations. *JETP Lett.*, **42** (1985) 152.

29

Perspective

The recent excitement in cosmology has been due to the emergence of a precision description of the present Universe and its recent past, with the standard model of cosmology able to accurately reproduce a wide range of sensitive observations. The ambitions of this effort are to determine the material composition of the present Universe, to establish the basic properties of perturbations in the Universe, and to verify that standard physical laws can be applied on a Universal scale. All of these now appear to be well in hand, as we have described in the earlier parts of this book. In particular, the successful theoretical reproduction of the observed cosmic microwave anisotropies is a tour de force of general relativity, particle interactions, and detailed modelling of the Universe's composition, that leaves little room for doubt that the fundamentals of cosmology are in place and secure.

Our book has also followed a broader ambition — to use that knowledge to tell us about the very early Universe and about the nature of fundamental physical laws. According to standard belief, almost every quantity measured in the present Universe has its origin in the Universe's earliest stages, when the primordial perturbations were created, perhaps by inflation, and when various as-yet uncertain particle physics processes established the densities of baryons, of dark matter, and perhaps also dark energy. The challenge here is to use the rather limited information available observationally (essentially a handful of numbers) to meaningfully constrain a very wide range of conjectured physical mechanisms. The extent to which that can ultimately be achieved remains unclear. For instance, observation has certainly shown its ability to eliminate many of the inflationary models that existed a decade ago, but at the same time has inspired many new models to be proposed.

Theoretical cosmology currently sits at somewhat of a crossroads, its future direction being uncertain in at least two respects. Take first the astronomical observations. One possibility is that we have *already* identified all of the processes at work in our Universe that are amenable to observation. We might never detect non-gaussianity of the curvature perturbation, a tensor perturbation, isocurvature density perturbations and topological defects, and the dark energy might have no dynamics to be uncovered. In that event, future observations will be able to improve the precision on already-measured quantities, but in general such improvement is unlikely to bring new fundamental understanding. The extremely important exception to that statement is the spectral index of the primordial curvature perturbation, where a measurement of its running would allow strong discrimination between models.

Of course, astronomical observation may, for all we know, be on the verge of discovering new physical processes at work in our Universe, either amongst the set listed in the previous paragraph or something entirely different and perhaps even, best yet, completely unexpected. Only now is it becoming possible to properly probe the parameter space of such processes, particularly cosmic non-gaussianity, and one should remain optimistic that one or more will be revealed by precision experiments.

The second direction of great uncertainty concerns the outcome of observations at particle colliders, in particular the Large Hadron Collider (LHC), and at underground particle detectors. The identification of the cold dark matter particle would be a major triumph, and so would the discovery of supersymmetry at the LHC. Whatever their outcome, collider and detector observations are likely to have a major impact on current thinking about the early Universe, and may offer discrimination between different scenarios.

Any broadening of the cosmological model, in terms of the number of necessary parameters, enhances our ability to constrain models of the early Universe, and at least from that perspective complications of the model are to be welcomed rather than resented. Let's see what awaits!

Appendix A: Spherical functions

The Legendre polynomials are defined by

$$P_0(x) = 1, \quad P_1(x) = x, \quad P_2(x) = \frac{1}{2}(3x^2 - 1), \tag{A.1}$$

$$(\ell + 1)P_{\ell+1}(\mu) = (2\ell + 1)\mu P_\ell(\mu) - lP_{\ell-1}(\mu). \tag{A.2}$$

They are a complete orthogonal set on the interval $-1 < x < 1$;

$$\int_{-1}^{1} P_\ell(x) P_{\ell'}(x) dx = \frac{2}{2\ell + 1}\delta_{\ell\ell'}. \tag{A.3}$$

They satisfy

$$P_\ell(-x) = (-1)^\ell P_\ell(x) \qquad P_\ell(1) = 1. \tag{A.4}$$

The spherical harmonics are defined by

$$Y_{\ell m} = \left[\frac{2\ell + 1}{4\pi}\frac{(l - m)!}{(l + m)!}\right]^{1/2} P_\ell^m(\cos\theta)e^{im\phi}, \tag{A.5}$$

where the associated Legendre function is defined by[1]

$$P_\ell^m(x) = (-1)^\ell \frac{(1 - x^2)^{m/2}}{2^\ell \ell!}\frac{d^{\ell+m}}{dx^{\ell+m}}(1 - x^2)^\ell. \tag{A.6}$$

Their recurrence relation is

$$\mu P_\ell^2(\mu) = \frac{1}{2\ell + 1}\left[(\ell + 2)P_{\ell-1}^2 + (\ell - 2)P_{\ell+1}^2\right]. \tag{A.7}$$

The spherical harmonics are a complete orthonormal set, satisfying

$$\int Y_{\ell m}(\mathbf{e})Y_{\ell' m'}^*(\mathbf{e})\, d\Omega = \delta_{\ell\ell'}\delta_{mm'}, \tag{A.8}$$

[1] This is the widely used convention of Ref. [1]. Some authors differ by a factor $(-1)^\ell$, and/or by a factor $(-1)^m$ for negative m.

where $d\Omega \equiv d\cos\theta\, d\phi$, as well as

$$\sum_{\ell m} Y_{\ell m}(\mathbf{e}) Y^*_{\ell m}(\mathbf{e}') = \delta(\phi - \phi')\delta(\cos\theta - \cos\theta') \equiv \delta^2(\mathbf{e} - \mathbf{e}'), \tag{A.9}$$

where $\mathbf{e}$ is the unit vector in the direction (θ, ϕ).

The spherical harmonics satisfy $Y^*_{\ell m} = (-1)^m Y_{\ell,-m}$. Under a reversal of the direction (θ, ϕ), corresponding to a parity transformation,

$$Y_{\ell m} \to Y_{\ell m}(\pi - \theta, -\phi) = (-1)^\ell Y_{\ell,-m}(\theta, \phi). \tag{A.10}$$

For $m = 0$,

$$Y_{\ell 0}(\theta) = \sqrt{\frac{2\ell+1}{4\pi}} P_\ell(\cos\theta). \tag{A.11}$$

A rotation of the polar coordinate system leads to a unitary transformation for the $Y_{\ell m}$, with no mixing between different ℓ;

$$Y_{\ell m}(\theta, \phi) \to \sum_{m'} U_{mm'}(\ell) Y_{\ell m'}(\theta', \phi'). \tag{A.12}$$

Using a prime to denote the new system this gives

$$\sum_m Y^*_{\ell m}(\theta_1, \phi_1) Y_{\ell m}(\theta_2, \phi_2) = \sum_m Y^*_{\ell m}(\theta'_1, \phi'_1) Y_{\ell m}(\theta'_2, \phi'_2). \tag{A.13}$$

Using this expression, we can go to a coordinate system whose pole points along the direction (θ_1, ϕ_1) of the old coordinate system. Denoting the new coordinates simply by (θ, ϕ), and using $Y_{\ell 0}(0, 0) = \sqrt{(2\ell+1)/4\pi}$, we find

$$\sum_m Y^*_{\ell m}(\theta_1, \phi_1) Y_{\ell m}(\theta_2, \phi_2) = \sqrt{\frac{2\ell+1}{4\pi}} Y_{\ell 0}(\theta, \phi) = \frac{2\ell+1}{4\pi} P_\ell(\cos\theta_{12}), \tag{A.14}$$

where θ_{12} is the angle between the directions 1 and 2.

The spherical Bessel functions of integral order are defined by

$$j_0(x) = \frac{\sin x}{x}, \qquad j_1(x) = \frac{\sin x}{x}^2 - \frac{\cos x}{x}, \tag{A.15}$$

$$(2\ell+1)\frac{j_\ell(x)}{x} = j_{\ell-1}(x) + j_{\ell+1}(x). \tag{A.16}$$

They satisfy

$$(2\ell+1) j'_\ell = \ell j_{\ell-1} - (\ell+1) j_{\ell+1}. \tag{A.17}$$

Each function $j_\ell(kr)$ provides a complete orthonormal set for the interval $0 < r < \infty$, labelled by k:

$$\int_0^\infty \left[\sqrt{\frac{2}{\pi}}\, k j_\ell(kx)\right] \left[\sqrt{\frac{2}{\pi}}\, k' j_\ell(k'x)\right] x^2\, dx = \delta(k - k'). \tag{A.18}$$

A plane wave of unit amplitude has the expansion

$$e^{i\mathbf{k}\cdot\mathbf{x}} = 4\pi \sum_{\ell,m} i^\ell j_\ell(kx) Y_{\ell m}(\hat{\mathbf{x}}) Y^*_{\ell m}(\hat{\mathbf{k}}) \tag{A.19}$$

$$= \sum_\ell (2\ell+1) i^\ell j_\ell(kx) P_\ell(\hat{\mathbf{x}}\cdot\hat{\mathbf{k}}). \tag{A.20}$$

The **spin-weighted harmonics** may be given in terms of rotation matrices as

$${}_sY_{\ell m}(\theta,\phi) = \left(\frac{2\ell+1}{4\pi}\right)^{1/2} \mathcal{D}^\ell_{-s,m}(\phi,\theta,0). \tag{A.21}$$

An explicit expression is

$$\begin{aligned} {}_sY_{\ell m}(\theta,\phi) &= e^{im\phi}\left[\frac{(\ell+m)!)(\ell-m)!}{(\ell+s)!(\ell-s)!}\frac{2\ell+1}{4\pi}\right]^{1/2} \sin^{2\ell}(\theta/2) \\ &\times \sum_r \binom{\ell-s}{r}\binom{\ell+s}{r+s-m}(-1)^{\ell-r-s}\left[\cot(\theta/2)\right]^{2r+s-m}. \end{aligned} \tag{A.22}$$

The conjugation condition is ${}_sY^*_{\ell m} = (-1)^{m+s}{}_{-s}Y_{\ell m}$, and under a reversal of direction (parity transformation) ${}_sY_{\ell m}$ becomes $(-1)^\ell{}_{-s}Y_{\ell m}$.

The ordinary spherical harmonics have weight zero so that $Y_{\ell m} = {}_0Y_{\ell m}$. Each spin-weighted spherical harmonic transforms under rotations in the same way as $Y_{\ell m}$, except for a factor $e^{-is\psi}$, where ψ at a given point on the sphere is the rotation of the coordinate lines caused by the transformation. The transformation is unitary, and Eq. (A.13) generalizes to

$$\sum_m {}_rY^*_{\ell m}(\theta_1,\phi_1){}_sY_{\ell m}(\theta_2,\phi_2) = e^{i(r\psi_1 - s\psi_2)}\sum_m {}_rY^*_{\ell m}(\theta_1',\phi_1'){}_sY_{\ell m}(\theta_2',\phi_2'), \tag{A.23}$$

where the phase factors account for the rotation of the coordinate lines at the points 1 and 2 on the sphere. Taking the polar axis of the new coordinate system to be the point 1, and denoting the new coordinates simply by θ,ϕ, we find the analogue of Eq. (A.14);

$$\sum_m {}_rY^*_{\ell m}(\theta_1,\phi_1){}_sY_{\ell m}(\theta_2,\phi_2) = \sqrt{\frac{2\ell+1}{4\pi}}\,{}_rY_{\ell,-s}(\theta,\phi)e^{i(r\psi_1 - s\psi_2)}. \tag{A.24}$$

In the text we use this identity in conjunction with Table A.1.

We need the identity

$$\begin{aligned} \sqrt{\frac{4\pi}{3}} Y_{10}\,{}_sY_{\ell m} &= \frac{{}_s\kappa^m_\ell}{\sqrt{(2\ell+1)(2\ell-1)}}\,{}_sY_{\ell-1,m} - \frac{ms}{\ell(\ell+1)}\,{}_sY_{\ell m} \\ &+ \frac{{}_s\kappa^m_{\ell+1}}{\sqrt{(2\ell+1)(2\ell+3)}}\,{}_sY_{\ell+1,m}, \end{aligned} \tag{A.25}$$

Table A.1. *Quadrupole ($\ell = 2$) harmonics for spin-0 and spin-2*

m	Y_{2m}	${}_2Y_{2m}$
2	$\frac{1}{4}\sqrt{\frac{15}{2\pi}}\sin^2\theta\, e^{2i\phi}$	$\frac{1}{8}\sqrt{\frac{5}{\pi}}(1-\cos\theta)^2\, e^{2i\phi}$
1	$\sqrt{\frac{15}{8\pi}}\sin\theta\,\cos\theta\, e^{i\phi}$	$\frac{1}{4}\sqrt{\frac{5}{\pi}}\sin\theta\,(1-\cos\theta)\, e^{i\phi}$
0	$\frac{1}{2}\sqrt{\frac{5}{4\pi}}(3\cos^2\theta - 1)$	$\frac{3}{4}\sqrt{\frac{5}{6\pi}}\sin^2\theta$
−1	$-\sqrt{\frac{15}{8\pi}}\sin\theta\,\cos\theta\, e^{-i\phi}$	$\frac{1}{4}\sqrt{\frac{5}{\pi}}\sin\theta\,(1+\cos\theta)\, e^{-i\phi}$
−2	$\frac{1}{4}\sqrt{\frac{15}{2\pi}}\sin^2\theta\, e^{-2i\phi}$	$\frac{1}{8}\sqrt{\frac{5}{\pi}}(1+\cos\theta)^2\, e^{-2i\phi}$

where the Clebsch–Gordon coefficients are given by

$$ {}_s\kappa_\ell^m \equiv \sqrt{(\ell^2 - m^2)(\ell^2 - s^2)/\ell^2}. \tag{A.26}$$

Equations (A.2) and (A.7) are special cases.

Following Ref. [2] we define functions $j_\ell^{\ell' m}(kr)$, $\epsilon_\ell^m(kr)$ and $\beta_\ell^m(kr)$, such that

$$ e^{i\mathbf{k}\cdot\mathbf{x}} Y_{\ell' m}(\mathbf{n}) = \sum_\ell \sqrt{(2\ell+1)(2\ell'+1)}(-i)^{\ell+\ell'} j_\ell^{\ell' m} Y_{\ell m}(\mathbf{n}) \tag{A.27}$$

$$ e^{i\mathbf{k}\cdot\mathbf{x}} Y_{\ell' m}^{\pm}(\mathbf{n}) = \sum_\ell \sqrt{(2\ell+1)(2\ell'+1)}(-i)^{\ell+\ell'} \left(\epsilon_\ell^m \pm i\beta_\ell^m\right) Y_{\ell m}^{\pm}(\mathbf{n}), \tag{A.28}$$

where $\mathbf{n} = -\mathbf{e}$ and the polar axis $\theta = 0$ for the spherical harmonics is along the $\mathbf{k}$ direction. In the first relation we need only $\ell' = 0$, 1 and 2 corresponding to monopole, dipole and quadrupole sources, and in the second relation $\ell' = 2$. The relevant functions are

$$ j_\ell^{00} = j_\ell, \qquad j_\ell^{10} = j_\ell', \qquad j_\ell^{20} = \frac{1}{2}\left(3j_\ell'' + j_\ell\right), \tag{A.29}$$

$$ j_\ell^{11}(x) = \sqrt{\frac{\ell(\ell+1)}{2}}\frac{j_\ell(x)}{x} \qquad j_\ell^{21}(x) = \sqrt{\frac{3\ell(\ell+1)}{2}}\left(\frac{j_\ell(x)}{x}\right)', \tag{A.30}$$

$$ j_\ell^{22}(x) = \sqrt{\frac{3}{8}\frac{(\ell+2)!}{(\ell-2)!}}\frac{j_\ell(x)}{x^2}, \tag{A.31}$$

and

$$\epsilon_\ell^0 = \sqrt{\frac{3}{8}\frac{(\ell+2)!}{(\ell-2)!}}\frac{j_\ell(x)}{x^2}, \qquad \beta_\ell^0(x) = 0, \tag{A.32}$$

$$\epsilon_\ell^1(x) = \frac{1}{2}\sqrt{(\ell-1)(\ell+2)}\left[\frac{j_\ell(x)}{x^2} + \frac{j'_\ell(x)}{x}\right], \tag{A.33}$$

$$\beta_\ell^1(x) = \frac{1}{2}\sqrt{(\ell-1)(\ell+2)}\frac{j_\ell(x)}{x}, \tag{A.34}$$

$$4\epsilon_\ell^2(x) = -j_\ell(x) + j''_\ell(x) + 2\frac{j_\ell(x)}{x^2} + 4\frac{j'_\ell(x)}{x}, \tag{A.35}$$

$$2\beta_\ell^2(x) = j'_\ell(x) + 2\frac{j_\ell(x)}{x}. \tag{A.36}$$

References

[1] A. R. Edmonds. *Angular Momentum in Quantum Mechanics*. (Columbia and Princeton: University Presses of California 1996).

[2] W. Hu and M. White. CMB Anisotropies: Total Angular Momentum Method. *Phys. Rev. D,* **56** (1997) 596.

Appendix B: Constants and parameters

Conversion from natural units

$$
\begin{aligned}
1\,\mathrm{cm} &= 5.068 \times 10^{13}\,\mathrm{GeV}^{-1}\hbar \\
1\,\mathrm{s} &= 1.519 \times 10^{24}\,\mathrm{GeV}^{-1}\hbar/c \\
1\,\mathrm{g} &= 5.608 \times 10^{23}\,\mathrm{GeV}/c^2 \\
1\,\mathrm{erg} &= 6.242 \times 10^{2}\,\mathrm{GeV} \\
1\,\mathrm{K} &= 8.618 \times 10^{-14}\,\mathrm{GeV}/k_\mathrm{B}
\end{aligned}
$$

Constants

Reduced Planck constant	$\hbar$	$= 1.055 \times 10^{-27}\,\mathrm{cm}^2 \cdot \mathrm{g} \cdot \mathrm{s}^{-1}$
Speed of light	c	$= 2.998 \times 10^{10}\,\mathrm{cm} \cdot \mathrm{s}^{-1}$
Newton's constant	G	$= 6.672 \times 10^{-8}\,\mathrm{cm}^3\,\mathrm{g}^{-1}\,\mathrm{s}^{-2}$
Reduced Planck mass	M_Pl	$= 4.342 \times 10^{-6}\,\mathrm{g}$
		$= 2.436 \times 10^{18}\,\mathrm{GeV}/c^2$
(Planck mass	m_Pl	$= \sqrt{8\pi}\,M_\mathrm{Pl} = 2.177 \times 10^{-5}\,\mathrm{g}$)
Reduced Planck length	L_Pl	$= 8.101 \times 10^{-33}\,\mathrm{cm}$
Reduced Planck time	T_Pl	$= 2.702 \times 10^{-43}\,\mathrm{s}$
Boltzmann constant	k_B	$= 1.381 \times 10^{-16}\,\mathrm{erg}\,\mathrm{K}^{-1}$
		$= 8.618 \times 10^{-5}\,\mathrm{eV}\,\mathrm{K}^{-1}$
Thomson cross section	σ_T	$= 6.652 \times 10^{-25}\,\mathrm{cm}^2$
Electron mass	m_e	$= 0.511\ \mathrm{MeV}/c^2$
Neutron mass	m_n	$= 939.6\ \mathrm{MeV}/c^2$
Proton mass	m_p	$= 938.3\ \mathrm{MeV}/c^2$
Solar mass	$M_\odot$	$= 1.99 \times 10^{33}\,\mathrm{g}$
Megaparsec	$1\,\mathrm{Mpc}$	$= 3.086 \times 10^{24}\,\mathrm{cm}$
Year	$1\,\mathrm{yr}$	$= 3.156 \times 10^{7}\,\mathrm{s}$

Parameters

Hubble constant	H_0	$=$	$100\,h\,\mathrm{km\cdot s^{-1}\,Mpc^{-1}}$
	h	$\simeq$	0.70
Present Hubble distance	cH_0^{-1}	$=$	$3.00\,h^{-1}\,\mathrm{Gpc}$
Present Hubble time	H_0^{-1}	$=$	$9.78\,h^{-1}\,\mathrm{Gyr}$
Age of Universe	t_0	$=$	$13.7\,\mathrm{Gyr}$
Present critical density	$\rho_{\mathrm{c},0}$	$=$	$1.88\,h^2\times10^{-29}\,\mathrm{g\cdot cm^{-3}}$
		$=$	$2.78\,h^{-1}\times10^{11}\,\frac{M_\odot}{(h^{-1}\,\mathrm{Mpc})^3}$
		$=$	$\left(3.00\times10^{-3}\ \mathrm{eV}/c^2\right)^4\,h^2$
Present photon temperature	T_0	$=$	$2.73\,\mathrm{K}=2.36\times10^{-4}\,\mathrm{eV}$
Present photon density	Ω_γ	$=$	$2.48\times10^{-5}h^{-2}$
Present relativistic density[1]	Ω_R	$=$	$4.15\times10^{-5}h^{-2}$
Present baryon density	Ω_B	$\simeq$	0.046
Baryon-to-photon ratio	n_B/n_γ	$\equiv$	$\eta=2.68\times10^{-8}\,\Omega_B h^2$
		$\simeq$	5.9×10^{-10}
Present matter density	Ω_c	$\simeq$	0.23
Matter–radiation equality[1]	$1+z_\mathrm{eq}$	$=$	$24\,100\,\Omega_\mathrm{m}h^2\simeq3500$
Hubble length at equality[1]	$(a_\mathrm{eq}H_\mathrm{eq})^{-1}$	$=$	$14\,\Omega_\mathrm{m}^{-1}h^{-2}\,\mathrm{Mpc}\simeq95\,\mathrm{Mpc}$
Top-hat window function (units of $10^{12}\,M_\odot$)	$M(R)$	$=$	$1.16\,h^{-1}(R/1\,h^{-1}\,\mathrm{Mpc})^3$
Gaussian window function (units of $10^{12}\,M_\odot$)	$M(R)$	$=$	$4.37\,h^{-1}(R/1\,h^{-1}\,\mathrm{Mpc})^3$

[1] Includes three neutrino species, regarded for this purpose as massless.

Index

For EU product safety concerns, contact us at Calle de José Abascal, 56–1°,
28003 Madrid, Spain or eugpsr@cambridge.org.

www.ingramcontent.com/pod-product-compliance
Ingram Content Group UK Ltd.
Pitfield, Milton Keynes, MK11 3LW, UK
UKHW051147130726
473066UK00031BC/196
* 9 7 8 0 5 2 1 8 2 8 4 9 9 *